Phase-Modulated Optical Communication Systems

PHASE-MODULATED OPTICAL COMMUNICATION SYSTEMS

KEANG-PO HO
Institute of Communication Engineering and Department of Electrical Engineering
National Taiwan University, Taipei 106, Taiwan

Springer

Keang-Po Ho
Institute of Communication Engineering and Department of Electrical Engineering
National Taiwan University, Taipei 106, Taiwan

Phase-Modulated Optical Communication Systems

Library of Congress Cataloging-in-Publication Data

A C.I.P. Catalogue record for this book is available
from the Library of Congress.

ISBN 0-387-24392-5 e-ISBN 0-387-25555-9 Printed on acid-free paper.
ISBN 978-0387-24392-4

Printed in the United States of America.

9 8 7 6 5 4 3 2 1 SPIN 11055495

springeronline.com

This book is dedicated
to my wife Kate and
my daugther Caroline.

Contents

Preface

Currently, virtually all commercially available optical communication systems use on-off keying to carry information by the presence or absence of light. Neither the phase nor frequency of an optical signal is used to carry information. Phase-modulated optical communications, or coherent optical communications, have been studied for a long time since the early date of optical communications. However, early works focused on improving receiver sensitivity that have became less relevant after the widely deployment of optical amplifiers.

In the *Optical Fiber Communication Conference 2002 (OFC '02)*, Gnauck et al. (2002) and Griffin et al. (2002) revived phase-modulated optical communication systems based on direct-detection of return-to-zero differential phase-shift keying (RZ-DPSK) and differential quadrature phase-shift keying (DQPSK) signal. With 3 dB better receiver sensitivity and improved tolerance to fiber nonlinearities, RZ-DPSK signal becomes the emerging transmission format for long-haul and ultra-long-haul lightwave transmissions. DQPSK signal also improves the spectral efficiency of the lightwave systems.

Because the usage of optical amplifiers to maintain a high optical power along the fiber link, current optical communication systems are fundamentally limited by the balancing of both optical amplifier noises and fiber nonlinearities. Our initial studies focus on the impact of nonlinear phase noise to DPSK signal. The topics of nonlinear phase noise came to us by accident. Also in *OFC '02*, we published a paper about the capacity of constant-intensity modulation in lightwave systems (Ho and Kahn, 2002), mainly to clarify our comments of Kahn and Ho (2001) on the important paper of Mitra and Stark (2001). In Mitra and Stark (2001), the capacity of multichannel wavelength-division multiplexed (WDM) systems is limited by cross-phase modulation. Ideally, constant-intensity modulation, including phase and frequency modulation, gives a constant phase shift to other channels through cross-phase modulation. Constant-intensity modulation is more likely to be limited by four-wave mixing than cross-phase modulation. Because of Gnauck et al. (2002), many people mentioned that nonlinear phase noise of Gordon and Mollenauer (1990) is the major limitation to constant-intensity modulation. Without the knowledge of both Liu et al. (2002b) and Xu and Liu (2002) to compensate nonlinear phase noise using the received intensity, most people believed that nonlinear phase noise was the primary limitation with no practical solution.

Back to San Jose, within days, we realized that nonlinear phase noise is correlated with the received intensity and may be compensated by the received intensity using electronic circuits, functionally the same as both Liu et al. (2002b) and Xu and Liu (2002). After many revisions, the paper was published as Ho and Kahn (2004a), even after some of our contributions to other topics related to nonlinear phase noise had been published. Ho and Kahn (2004a) began our works on the research on phase-modulated signal for lightwave communications.

This book is originated from the notes for a seminar style class on coherent optical communications. The students in the class provided a great help on improving the manuscript and selecting the materials. For system in linear regime, the performance of phase-modulated signal is mainly studied for system dominated by amplifier noises. Nonlinear phase noise is the unquestionable limitations when the signal pulse maintains its shape along the fiber link. When the optical pulse is broadened by fiber chromatic dispersion, pulse overlap and the subsequent pulse-to-pulse interaction also degrades a DPSK signal. However, pulse-to-pulse interaction usually has less effect then nonlinear phase noise.

The materials of this book are suitable for researchers in the field of lightwave communications and graduate students in the class of advanced optical communication systems.

KEANG-PO HO

Acknowledgments

In my career in optical communications, Prof. Joseph Kahn is my mentor in UC Berkeley, colleague in StrataLight, and good friend outside business. He taught me the method and aesthetic to approach and solve problems. Earlier in my career, I would like to thank both Chinlon Lin and Paul Shumate as my manager in Bellcore (currently Telcordia Technologies), Frank Tong as my supervisor in IBM and colleague in the Chinese University of Hong Kong, and Lian Chen, Tony Lee, Robert Li, Kok Cheung, Wing Wong, and Raymond Yeung as my colleagues in the Chinese University of Hong Kong. The co-founding with Joseph Kahn and Terry Smith of StrataLight Communications was also a very exciting and rewarding experience. I would like to also thank Ted Schmidt, Gary Wang, and Anhui Liang as my colleagues in StrataLight. Gary was the one who made a copy of Gordon and Mollenauer (1990) from Stanford Library to me.

Prof. Jingshown Wu and Hen-Wai Tsao brought me to the field of optical communications in early 90's, especially introduced me to the field of coherent optical communications. Both Prof. Wu and Prof. Tsao are currently my senior colleagues in National Taiwan University. Prof. Tsao especially gave me with great encouragement during the preparation of this book. I also would like to thank my students in National Taiwan University, including to Hsi-Cheng Wang, Jen-An Huang, Alien Chen, Terry Yuan, Kevin Chen, Po-Yu Chen and others.

Many of my researches have been the results of the collaboration with Prof. Shien-Kuei Liaw from Bellcore to National Taiwan University. I would also like to thank collaboration with Prof. Min-Chen Ho, Kai-Ming Feng, and Ioannis Roudas.

Figures 6.1, 9.19, 11.1 are adapted from some initial drawings of Prof. Joseph Kahn. Ms. Hsiu-Huei Yen also provided lots of clerical supports in this book project.

Chapter 1

INTRODUCTION

Fiber-optic communication systems have been deployed worldwide and certainly revolutionized the current and future telecommunication infrastructures. Currently, virtually all telephone conversations, cellular phone calls, and Internet packets must pass through some pieces of optical fiber from source to destination. While initial deployment of optical fiber was mainly for long-haul or submarine transmission, lightwave systems are currently in virtually all metro networks. The future deployment of optical fiber is moving toward the home for broadband access. Fiber-to-the-premise (FTTP) and fiber-to-the-home (FTTH) are being considered seriously in most parts of the World right now (Abrams et al., 2005).

Since Kao and Hockham (1966) first proposed the usage of optical fiber to guide light for information transmission, the fiber loss has been reduced from the early date of 20 dB/km (Kapron et al., 1970) to about 0.15 dB/km (Kaiser and Keck, 1988, Kanamori et al., 1986, Murata and Inagaki, 1981, Nagayama et al., 2002). Most of the commercially available optical fiber has a loss of about 0.2 dB/km at the low-loss window around 1.55 μm[1]. Optical signal can be transmitted for a long distance without regeneration owning to the low-loss characteristic of optical fiber.

With great physical properties, Erbium-doped fiber amplifiers (EDFA) also provide gain at the low-loss window of 1.55 μm (Becker et al., 1999, Desurvire, 1994). Optical amplifiers are used to periodically amplify an optical signal to compensate for fiber loss. The low-loss window of optical

[1]Product information of optical fiber are available from `http://www.corning.com/opticalfiber/`, `http://www.ofsoptics.com`, and `http://www.alcatel.com/products`.

fiber can also partition to many channels for dense wavelength-division-multiplexed (WDM) systems. Only adding noise to the signal, EDFA amplifies many WDM channels together without crosstalk and distortion. Before the usage of phase-modulated signals, very high throughput fiber link has been constructed without electronic regeneration for trans-oceanic distances (Bakhshi et al., 2004, Bergano and Davidson, 1996, Cai et al., 2002, 2003b, Golovchenko et al., 2000, Suzuki and Edagawa, 2003). With the usage of phase-modulated signals, the system performance is further improved (Becouarn et al., 2003, Cai et al., 2004, Charlet et al., 2004b, Rasmussen et al., 2003, 2004). In all commercially available lightwave transmission systems, only the intensity of optical signal is used to carry information. Neither the phase nor frequency of an optical carrier is used. In order to transmit more information in a single optical carrier or a single WDM channel, the phase of an optical carrier must be explored. In this chapter, we will briefly explain the basic architecture of intensity modulated optical communication systems and the reason the phase should be used to converse information in an optical carrier.

1. Intensity-Modulated/Direct-Detection Systems

Currently, virtually all deployed fiber systems use the simple intensity modulation system in which the information is carried in the light intensity and recovered using a photodiode, so called on-off keying or intensity-modulated/direct-detection (IMDD) systems. Most textbooks of optical communications focus mainly on IMDD systems (Agrawal, 2002, Einarsson, 1996, Iannone et al., 1998, Kazovsky et al., 1996, Keiser, 1999, Kolimbiris, 2004, Mynbaev and Scheiner, 2001, Senior, 1992). Both transmitter and receiver for on-off keying systems are very simple, may be the simplest among all possibilities.

Figure 1.1 shows a typical long-haul IMDD system. The transmitted data are modulated into the optical carrier using an external intensity modulator that is basically a very fast switch to turn-on and -off the light path to carry either "0" or "1". After the modulator, the optical signal passes through an EDFA to boost up the power and then is launched into the optical fiber. In Fig. 1.1, EDFA is used periodically to compensate for fiber loss span after span.

After many spans of optical fiber, the optical signal is further amplified using a low-noise EDFA preamplifier. The optical signal is converted to electrical signal using a fast photodiode. Ideally, a photodiode converts a photon to an electron, i.e., the optical intensity to electrical current. Information in the phase or frequency of the optical carrier losses in the photodiode. The photodiode is followed by an electrical amplifier,

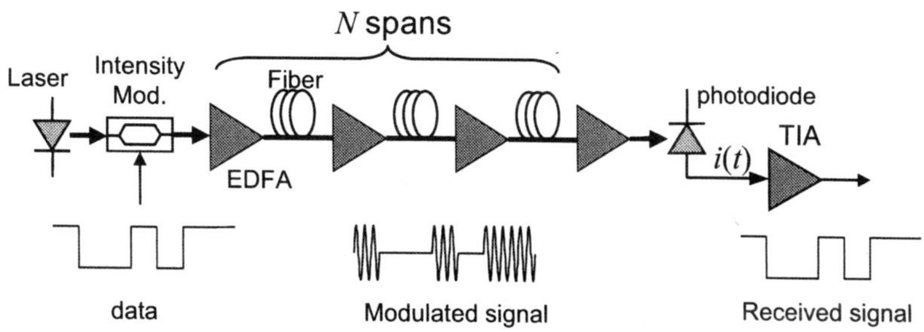

Figure 1.1. Typical configuration of an intensity-modulated/direct-detection (IMDD) system.

usually a trans-impedance amplifier (TIA) to convert photocurrent to voltage. The received signal after the TIA is the same as the transmitted signal but with noise mostly from optical amplification and the receiver circuitry.

The IMDD system illustrated schematically in Fig. 1.1 is very simple. The receiver decides whether the transmitted bits are either "0" or "1" based on the presence or absence of light. This class of system can use either non-return-to-zero (NRZ) or return-to-zero (RZ) pulses for digital transmission. Subcarrier multiplexing (Way, 1998) is also IMDD system to transmit either digital or analog modulated frequency-division multiplexed (FDM) channels in the intensity of the optical carrier. Subcarrier multiplexing is mostly for video distribution, but may also be used for high-speed digital data (Hui et al., 2002). Hybrid WDM systems can transmit some on-off keying and some subcarrier multiplexed channels (Ho et al., 1998b, Lee et al., 2002a, Way et al., 1990).

Some short distance IMDD systems do not need the usage of optical amplifiers. Typical semiconductor laser is also a very simple device in which light is generated with current injection. With inferior signal quality, direct-modulated semiconductor laser is a simple form of low cost transmitter. Light-emitting diode can also be used for low-speed applications, mostly for multimode fiber with a large core to alleviate the alignment requirement.

2. Phase-Modulated Optical Communications

The low-loss window of optical fiber transmits optical signal with a carrier frequency of about 190 THz at the wavelength around 1.55 μm. As an oscillator, laser for communication purpose is highly coherent with very pure spectrum. In digital communications (Proakis, 2000),

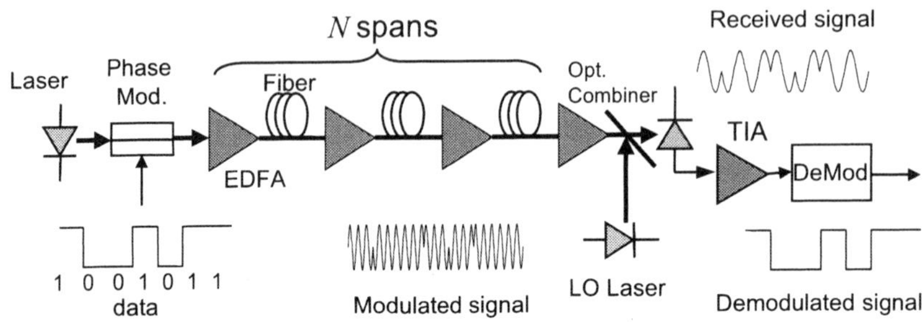

Figure 1.2. Typical schematic of a phase-modulated optical communication system.

the phase of the carrier is generally used in most wireline or wireless communication systems. While the coherence of the laser does not seriously affect the performance of an IMDD system, when the phase or frequency of an optical carrier is used to transmit information, the phase or frequency noise of the transmitter laser adds directly to the phase and frequency of the optical carrier. Both phase or frequency noise of a laser must be reduced for phase-modulated or coherent optical communications.

Figure 1.2 shows a typical schematic of a phase-modulated optical communication system based on phase-shift keying (PSK). In the transmitter, digital data are generally modulated to the amplitude, phase, or frequency of an optical carrier. More complicated system can modulate data into the combination of amplitude, phase, and/or frequency. In the receiver, the optical signal is first mixed with the light of a local oscillator (LO) laser to downconvert the signal from the optical carrier frequency to microwave carrier frequency in the range of GHz or tens of GHz. When the received signal is mixed with LO laser, an optical beating signal is generated at the photodiode, giving a beating signal having a frequency equal to an intermediate frequency (IF) that is the frequency difference between the optical carrier of the transmitter and the LO laser.

If the optical frequency of the signal is the same as that of the LO laser, the system is called a homodyne system. If the optical frequency of the signal differs with that of the LO laser, the system is called a heterodyne system with an IF of

$$\omega_{\mathrm{IF}} = \omega_c - \omega_{\mathrm{LO}}, \tag{1.1}$$

where ω_c and ω_{LO} are the optical frequency of the transmitted signal and LO laser, respectively. In homodyne systems, $\omega_{\mathrm{IF}} = 0$. Typical heterodyne systems have an IF larger than the data-rate. If the IF is less than the data rate, the system is called an intradyne system (Derr, 1992).

Compared the IMDD system of Fig. 1.1 with the PSK system of Fig. 1.2, the optical signal in the optical fiber for an IMDD system looks the same as that of the transmitted or received signal but that of a PSK system looks significantly different with the transmitted signal. In PSK systems, after the mixing of received signal with LO laser at the photodiode, the transmitted data are recovered using a demodulator. The demodulator generally converts an amplitude, phase or frequency modulated signal back to digital data.

In the 80's to early 90's, there were active researches on coherent optical communications to carry information in either the phase or frequency of the optical carrier. Some of those works were summarized in the books by Betti et al. (1995), Cvijetic (1996), Hooijmans (1994), Okoshi and Kikuchi (1988), and Ryu (1995), and reviewed by the papers of Brain et al. (1990), Kazovsky (1989), Linke and Gnauck (1988), Okoshi (1982, 1984, 1987), and Saito et al. (1991, 1993), and the collections by Henry and Personick (1990) and Shimada (1995). In that time, the goal was to achieve better receiver sensitivity and longer unregenerated distance using a coherent receiver, even for an intensity-modulated or amplitude-shift keying (ASK) signal. Beating with the LO laser to enhance the signal, the receiver sensitivity can be improved by up to 20 dB compared with simple direct detection. To certain extend, the mixing with LO laser serves as a signal amplifier. With the advances of optical amplifiers, especially EDFA (Desurvire et al., 1987, Mears et al., 1987, Nakazawa et al., 1989), longer unregenerated distance can be achieved by periodically amplifying the optical signal and better sensitivity can be achieved by optically preamplifing the received signal. Although coherent optical communication techniques may allow more efficient usage of optical bandwidth, fiber based coherent communications had lost its favor and relevant by the advances of optical amplifiers.

In digital communications (Proakis, 2000), a coherent demodulated system requires carrier recovery. In a homodyne system, carrier recovery requires phase-locking the LO laser to the received signal. In a heterodyne system, carrier recovery is conducted in the microwave signal at the IF of ω_{IF}. A heterodyne optical system is functionally the same as the superheterodyne receiver invented by Armstrong around 1920 (Brittain, 2004, Douglas, 1990) that is the dominant receiver in radio frequency (RF) communications for years. Homodyne RF receiver has become more popular for its low-power consumption (Abidi, 1995, Razavi, 1997).

Coherent optical communication systems use different terminology than that in digital communications. Conventionally, an optical communication system is called "coherent" as long as there is optical signal mixing even without carrier recovery. Even if the demodulator of Fig. 1.2 does not use carrier recovery but noncoherent or envelope detection, the system is called coherent optical communication systems. For example, differential phase-shift keying (DPSK) system is a noncoherent digital communication system (Proakis, 2000) but a coherent optical communication system (Betti et al., 1995, Henry and Personick, 1990, Okoshi and Kikuchi, 1988, Ryu, 1995). Following the traditional terminology of coherent optical communications, a coherent optical receiver with and without phase tracking is called synchronous and asynchronous receiver, respectively. Asynchronous receiver is usually based on power or envelope detection.

The mixing or heterodyning of two lasers for communication purpose was considered in the earliest date of optical communications (Goodwin, 1967, Oliver, 1961). Early systems operated in free space and used high power long-wavelength laser sources (DeLange, 1972, Goodwin, 1967, Nussmeier et al., 1974, Peyton et al., 1972). Even until today, coherent space communication still has its advantage as compared with on-off keying, especially for inter-satellite communications (Chan, 1987, 2000, 2003, Rochat et al., 2001). Coherent optical communication is also used for ultra dense radio-on-fiber signal (Kikuchi and Katoh, 2002a,b, Kuri and Kitayama, 2002, 2003).

Coherent optical communications in optical fiber were first proposed in early 80's (Favre et al., 1981, Favre and LeGuen, 1980, Kikuchi et al., 1981, Okoshi and Kikchi, 1980, Saito et al., 1980, 1981, Yamamoto, 1980, Yamamoto and Kimura, 1981). Although coherent optical communications were virtually disappear after the successful introduction of EDFA in early 90's, direct-detection DPSK has received renewed interested recently since the pioneer works of Gnauck et al. (2002) and Griffin et al. (2002). While early works focused on improving the receiver sensitivity, phase-modulated or coherent systems may be a candidate for advanced modulation scheme to improve the spectral efficiency. Direct-detection DPSK just provides a sensitivity gain of about 3 dB to on-off keying, significantly lower than the early claims of 10 to 20 dB in the 80's before the available of optical amplifiers (Okoshi and Kikuchi, 1988).

In term of receiver sensitivity, phase-modulated coherent optical communication systems provide the best performance among all types of modulation scheme. Here both PSK and DPSK systems are discussed further in details for their excellent sensitivity.

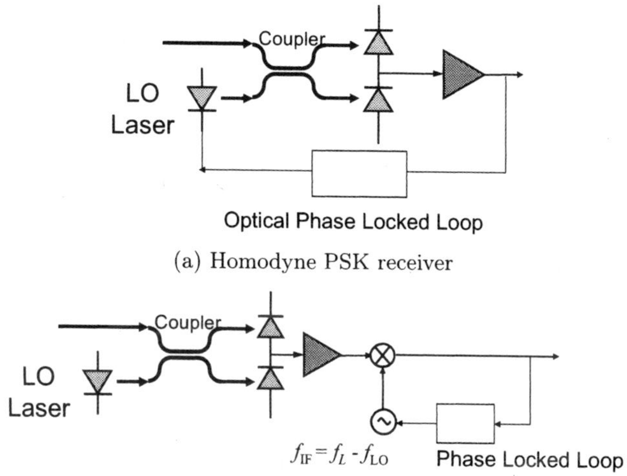

(a) Homodyne PSK receiver

(b) Heterodyne PSK receiver

Figure 1.3. Schematic diagram of (a) homodyne and (b) heterodyne PSK receiver.

2.1 PSK Systems

Optical PSK systems carry data in the phase of an optical carrier. Figure 1.2 shows a typical PSK transmitter consisting of a phase modulator following a semiconductor laser or other types of light source. An electrical driver amplifier is used to apply the data to the phase modulator, ideally providing a phase shift of either 0° or 180°. Figures 1.3 show the schematic diagram of both homodyne and heterodyne PSK receivers.

A homodyne PSK receiver uses an optical phase-locked loop (PLL) to lock the phase of the LO laser to that of the transmitter laser. In homodyne receiver, the optical frequency of LO laser should be the same as that of the transmitter laser, for example, by frequency tracking.

All receivers in Fig. 1.3 use a balanced receiver to sum the two beating optical signals after a 3-dB coupler. Among many advantages, a balanced receiver can suppress LO noise and provide larger signal power than single-branch receiver (Abbas et al., 1985). Similar to the combiner of Fig. 1.2, the coupler before the balanced receiver is also an 180° optical hybrid. The optical outputs of the coupler have phase difference of 180°.

A heterodyne receiver beats the optical signal of LO laser with the received signal to generate an IF signal. Frequency locking is necessary to provide a fixed IF. An electrical PLL, operating on the angular frequency of ω_{IF}, is used to recover the transmitted phase.

In the receivers of Fig. 1.3, the polarization of the LO laser must be the same as that of the transmitter laser or otherwise using a polarization-diversity receiver (Kazovsky, 1989). In heterodyne receiver, a 90° optical hybrid can be used for an image-rejection heterodyne receiver (Chikama et al., 1990a, Darcie and Glance, 1986, Glance, 1986b). Optically amplified heterodyne receiver has more or less the same performance as homodyne receiver (Jørgensen et al., 1992, Walker et al., 1990). A heterodyne receiver can also give quadrature components using a quadrature electrical mixer.

Due to the requirement of an optical PLL, homodyne PSK receiver is difficult to implement. With some successful demonstrations (Kahn et al., 1990, Kazovsky and Atlas, 1990, Norimatsu et al., 1990), homodyne receiver is not an active research area right now. To receive both in-phase and quadrature components, homodyne and heterodyne quadrature receivers in various configurations had also been demonstrated (Derr, 1990, Kahn et al., 1992, Norimatsu et al., 1992).

2.2 DPSK Systems

Optical DPSK signal carries data in the phase difference of an optical carrier between two consecutive symbols. Figures 1.4 show the schematic diagram of a DPSK transmitter and receiver. A DPSK transmitter is almost identical to the PSK transmitter in Fig. 1.2 other than the requirement of a precoder. In an RZ-DPSK transmitter, the laser is replaced by a pulse source that emits optical RZ pulses synchronized with the data.

When the differential phase is used to carry data, mathematically, the precoder should be the accumulative phase shift of the data stream. Because the phase of a signal is always confined to $[-\pi, \pi)$ and a phase difference of an integer multiple of 2π represents the same phase. The drive signal for the phase modulator can be the cumulative parity of the data and calculates by an exclusive-OR gate with a symbol time T of feedback as shown in Fig. 1.4(a). In Boolean variable, if $b_k \in \{0, 1\}$ is the binary data and $d_k \in \{0, 1\}$ is the drive signal, their relationship is $d_k = d_{k-1} \oplus b_k$, where $\oplus$ denotes exclusive-OR logic operation and the index k is for the data at the kth time interval. With the precoder, a DPSK receiver does not require special decoding circuitry.

Figures 1.4 show two types of DPSK receiver. A heterodyne receiver uses an electrical delay-and-multiplier circuit to find the differential phase. While frequency locking may be necessary, phase locking

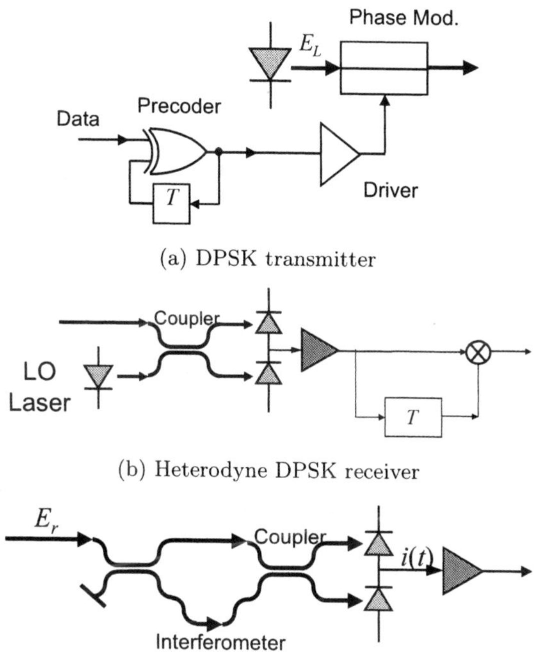

(a) DPSK transmitter

(b) Heterodyne DPSK receiver

(c) Direct-Detection DPSK receiver

Figure 1.4. Transmitter and receivers for optical DPSK signal.

is not required for DPSK signal. The delay-and-multiplier based receiver is functionally the same as DPSK receiver for RF communications.

The direct-detection DPSK receiver of Fig. 1.4(c) is the same as a heterodyne receiver in principle. An asymmetric Mach-Zehnder interferometer splits the signal to two paths and recombines these two signals after a path difference corresponding to the symbol time of T. A balanced receiver follows the interferometer as a multiplier to replace the electrical mixer. With optical amplifier to boost the signal before the receiver, the performance of heterodyne and direct-detection DPSK receiver is approximately the same (Tonguz and Wagner, 1991).

Heterodyne DPSK receivers were demonstrated in various configurations (Chikama et al., 1990a, Creaner et al., 1988, Gnauck et al., 1990). Recently, there is renewed interested of direct-detected DPSK signaling (Cai et al., 2004, Gnauck et al., 2002, 2003c, Rasmussen et al., 2003, Zhu et al., 2003) for long-haul transmission systems, mostly DPSK signal with RZ pulses. Differential quadrature phase-shift keying (DQPSK)

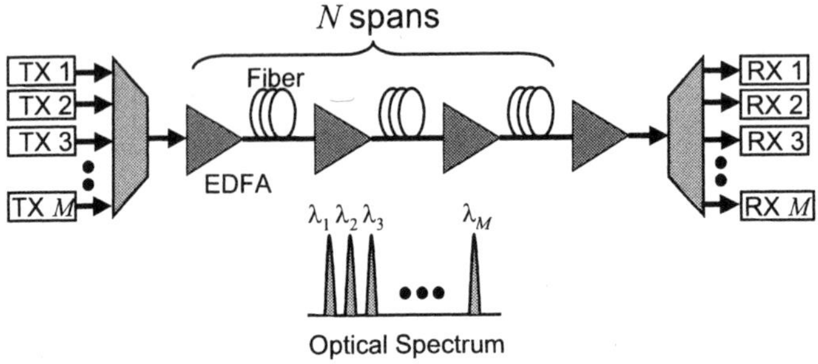

Figure 1.5. Many channels are multiplexed in a single fiber in a WDM system.

signal also improves the spectral efficiency of the systems (Cho et al., 2003, 2004a,b, Griffin et al., 2002, Kim and Essiambre, 2003, Tokle et al., 2004, Wree et al., 2003b, Yoshikane and Morita, 2004a,b, 2005). Using an interferometer to detect phase-modulated signal, direct-detection DPSK receiver is more complicated than the receiver for IMDD systems but far simpler than coherent receiver with a LO laser. Direct-detection DPSK signal is a very active research area right now. Later chapters of this book give more attention to DPSK signal than other modulation formats.

3. WDM Systems

The EDFA of Figs. 1.1 and 1.2 have a very wide gain bandwidth. Instead of just for a single channel, many channels can be amplified together. Figure 1.5 shows a schematic of a WDM system in which many WDM channels are multiplexed in a single fiber and amplified together using the same EDFA chain. Transparent to signal format, the EDFA chain can amplify either IMDD or phase-modulated signal. In principle, a mixed WDM system can be implemented in which some of the channels can be IMDD and others can be phase modulated.

EDFA operates more effectively in the conventional band or C-band from about 1.53 to 1.56 μm. International Telecommunication Union (ITU) standardizes WDM channel grid in either fraction or multiple of 100 GHz that corresponds to a wavelength separation of about 0.8 nm[2]. With an overall bandwidth of about 4 THz, the C-band can support fraction or multiple of 40 WDM channels. As shown in the spectrum

[2]ITU G.692 (1998): Optical interfaces for multichannel systems with optical amplifiers, and ITU G.694.1 (2002): Spectral grids for WDM applications: DWDM frequency grid.

of Fig. 1.5, a WDM system generally has uniform frequency spacing between channels. The wavelength spacing of each channel is also the same.

In addition to the C-band from 1.53 to 1.56 μm, WDM channels can extend to longer wavelength using L-band EDFA with a wavelength up to about 1.62 μm (Massicott et al., 1990, 1992, Ono et al., 1997, Sun et al., 1997). EDFA also has gain at wavelength shorter than 1.53 μm to support S-band transmission (Arbore et al., 2003, Ono et al., 2003, Yeh et al., 2004). Raman amplifiers can also be used alone or together with EDFA to further the usable bandwidth of an optical fiber (Bromage, 2004, Islam, 2002).

In the WDM systems of Fig. 1.5, a WDM multiplexer is used to combine all WDM channels. While it is preference to have the WDM multiplexer to reject some of the crosstalk from adjacent channels, the WDM multiplexer can be implemented as a passive combiner with a loss of at least $10 \cdot \log_{10} M$ in decibel, where M is the number of channels. To certain extend, a WDM multiplexer instead of passive combiner is used to multiplex many WDM channels mainly to reduce loss.

An IMDD system depends solely on the WDM demultiplexer to separate all WDM channels in Fig. 1.5. The WDM demultiplexer in IMDD system has the contradictory requirement to reject the crosstalk from adjacent channels without distort the signal channel. Therefore, WDM demultiplexer for IMDD WDM systems must have a good response in the pass band but high roll-off at the rejection band. Depending solely on the demultiplexer to reject crosstalk, IMDD WDM systems have a very restricted requirement on the WDM demultiplexer.

In the coherent systems of Fig. 1.2, the LO laser selects the WDM channel to be demodulated. Only the channel having an optical frequency close to the LO laser frequency gives a beating signal within the bandwidth of the receiver. In principle, using coherent receiver, the WDM demultiplexer can be a passive splitter. However, a WDM demultiplexer should be used to reduce loss. Too many coherent WDM channels may also over-load the photodiode. Because channel selection is mainly facilitated using the LO laser, high crosstalk rejection is preference but not essential for the WDM demultiplexer in homodyne and heterodyne systems.

The homodyne receiver of Fig. 1.3(a) requires the smallest receiver bandwidth and has the best channel selectivity for a WDM system. The heterodyne receivers of Figs. 1.3(b) and 1.4(b) require image-rejection to achieve the same channel selectivity of a homodyne receiver. The direct-detection DPSK receiver of Fig. 1.4(c) matches the frequency of a WDM channel to the path length to select a channel. Because adjacent channels may also have a good frequency match, direct-detection DPSK receiver does not guarantee good channel selection. For conservative system design to anticipate for the worst case, direct-detection DPSK receiver of Fig. 1.4(c) may require similar crosstalk rejection as IMDD signal.

Currently, in research laboratory, IMDD WDM systems can achieve an overall data-rate more than 10 Tb/s the distance of a couple hundreds of kilometers (Bigo, 2004, Bigo et al., 2001, Frignac et al., 2002, Fukuchi, 2002, Fukuchi et al., 2001). RZ-DPSK signals with a data rate of about 6 Tb/s can be transmitted over several thousands of kilometers (Charlet et al., 2004b, 2005, Zhu et al., 2003, 2004a). Commercial WDM systems can have an overall data rate over 1 Tb/s.

4. Comparison of Phase- and Intensity-Modulated Signals

In early day, coherent optical communications were investigated for better receiver sensitivity and channel selectivity. Before the advances of optical amplifiers in general and EDFA in particular, the mixing with LO laser served a function similar to a noiseless optical amplifier. A sensitivity gain of up to 20 dB was usually quoted in early literatures (Okoshi and Kikuchi, 1988). With optical amplifiers, coherent system has limited sensitivity gain. As shown later, PSK signal has around 3.5-dB sensitivity gain compared with on-off keying. DPSK signal has about 3.0-dB gain compared with on-off keying. The 3-dB gain may not worth the additional complexity of either a homodyne or heterodyne receiver.

For WDM systems, homodyne or heterodyne system with image-rejection provides good channel selectivity regardless of the quality of the WDM demultiplexer. Coherent optical receivers allow two WDM channels located very close to each other. To reduce the loss at channel demultiplexing, a WDM demultiplexer is desirable though not essential. For coherent WDM systems, crosstalk rejection is not a critical issue as compared with that in IMDD WDM systems.

With optical amplifiers, the biggest advantage of coherent system is to improve spectral efficiency using multilevel modulation. Figure 1.6 shows the signal constellation of binary and quaternary on-off keying signal (4-OOK), and quarter- and 64-ary quadrature-amplitude modulated (QAM) signal. The constellation or signal space is commonly used to study digital modulations (Proakis, 2000, Wozencraft and Jacobs, 1965). Although the optical carrier has both in- and quadrature-phase that was represented as a two-dimensional constellation in Fig. 1.6, on-off keying uses only the positive axis of a single dimension to carry information. In

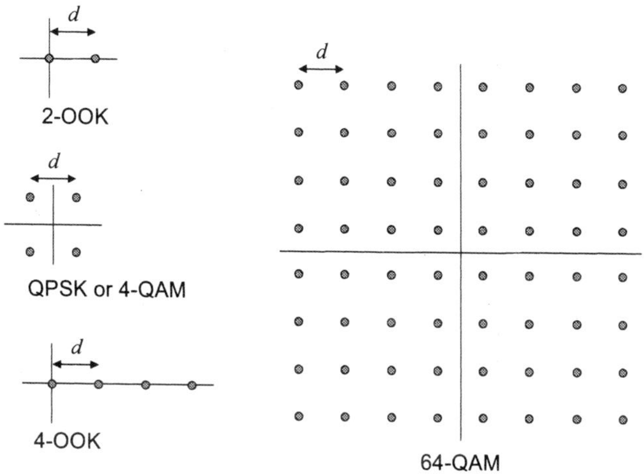

Figure 1.6. The signal space representation of 2- and 4-ary on-off keying (OOK), and 4- and 64-ary quadrature-amplitude modulation (QAM).

QAM scheme, positive and negative sides of both dimensions are used to carry information.

Table 1.1 summarizes the average energy per symbol, bits per symbol, and energy per bit of the constellations of Fig. 1.6. From Table 1.1, 4-OOK and 64-QAM has the same energy per bit of $1.75d^2$, where d is the Euclidean distance between two closest constellation points. The error probability of a signal is mainly determined by the minimum Euclidean distance of d (Proakis, 2000). From Table 1.1, coherent optical modulation using both in- and quadrature-phase can provide better spectral efficiency. Required the same energy per bit as 4-OOK, 64-QAM can transmit three times the data rate. Having 3-dB better energy per bit as binary on-off keying, 4-QAM or quadrature phase-shift keying (QPSK) can double the data rate.

In an alternative interpretation of Table 1.1, if the same data rate is transmitted, 64-QAM requires three times less bandwidth than 4-OOK but the same power. QPSK requires half the bandwidth of binary on-off keying and also half the power. The spectral efficiency of phase-modulated optical communications is rarely discussed in previous literatures for coherent optical communications. Although the spectral efficiency is also related to the better receiver sensitivity, the superior spectral efficiency may be the future driving force for coherent optical communications (Kahn and Ho, 2004).

Table 1.1. Comparison of the Signal in Fig. 1.6.

Modulation	Average Energy	bits per symbol	Energy per bit	Gain (dB)
2-OOK	$0.5d^2$	1	$0.5d^2$	0.00
4-OOK	$3.5d^2$	2	$1.75d^2$	-5.44
4-QAM (QPSK)	$0.5d^2$	2	$0.25d^2$	3.00
64-QAM	$10.5d^2$	6	$1.75d^2$	-5.44

5. Recent Advances in Direct-Detection DPSK Systems

Recently, direct-detection DPSK signaling has received great attention for long-haul transmission or high spectral efficiency systems. Table 1.2 summarizes recent experimental demonstrations of DPSK transmission with an overall capacity approaching 41 Peta-bits/s·km (Cai et al., 2003b). For DQPSK experiments in Table 1.3, the focus is to improve the spectral efficiency to 1.6 b/s/Hz or higher (Cho et al., 2004a,b). Other than Gnauck et al. (2003b), there are also activities to increase the data rate per channel up to 640 Gb/s (Beling et al., 2003, Kieckbusch et al., 2005, Marembert et al., 2004, Milivojevic et al., 2005, Möller et al., 2004).

Direct-detection DPSK signals had been around for years by directly modulated a semiconductor laser (Shirasaki et al., 1988, Vodhanel, 1989, Vodhanel et al., 1990). The main propose of those early works was to generate low-chirp optical signal to overcome fiber chromatic dispersion. Most recent DPSK systems use RZ pulses for better nonlinearity tolerance, adapted for long-haul transmission with optical amplifiers to boost the optical power. Before the wide usage of optical amplifiers, fiber nonlinearities usually did not have a major system impact. When optical signal is periodically amplified by a chain of optical amplifiers, a high power optical signal is maintained for high signal-to-noise ratio (SNR) before the signal is limited by fiber nonlinearities.

In phase-modulated optical systems, due to fiber Kerr effect, amplitude noise is converted to phase noise, generating nonlinear phase noise. The beating of the signal itself with amplifier noise gives the Gordon-Mollenauer effect (Gordon and Mollenauer, 1990), or more precisely, self-phase modulation (SPM) induced nonlinear phase noise. The beating of another WDM channel with amplifier noise gives cross-phase modulation (XPM) induced nonlinear phase noise. Added directly to

Table 1.2. Selected Recent DPSK Experimental Demonstrations.

Year	Data Rate (Gb/s)	Channel Number	Total Rate (Tb/s)	Distance (km)	Capacity (Tb/s·km)	Channel Space (GHz)	Reference	Comments
2002	42.7	64	2.5	4,000	10,000	100	Gnauck et al. (2002)	All-Raman
2002	42.7	80	3.2	5,200	16,600	100	Zhu et al. (2002)	
2002	43.0	40	1.6	400	640	100	Miyamoto et al. (2002)	S-band
2003	40.0	40	1.6	300	480	100	Bissessur et al. (2003)	
2003	40.0	25	1.0	1,200	1,200	50	Gnauck et al. (2003c)	co-pol.
2003	10.0	100	1.0	10,000	10,000	45	Ishida et al. (2003)	
2003	42.8	40	1.6	10,000	16,000	100	Rasmussen et al. (2003, 2004)	All-Raman
2003	42.7	160	6.4	3,200	20,800	50	Zhu et al. (2003, 2004a)	
2003	10.7	185	1.9	8370	15,485	25	Vareille et al. (2003)	22 dB span
2003	10.0	373	3.7	11,000	41,300	25	Cai et al. (2003b)	
2003	42.7	40	1.6	8,700	13,920	70	Tsuritani et al. (2003, 2004)	
2003	42.7	64	2.6	8,200	20,992	70	Morita and Edagawa (2003)	
2003	10.7	301	3.0	10,270	30,912	15.4	Becouarn et al. (2003)	
2003	43.0	100	4.0	6,240	24,960	62.5/100	Charlet et al. (2003)	
2003	170.6	6	1.0	2,000	2,000	200	Gnauck et al. (2003b)	
2004	10.0	96	0.96	13,000	12,480	33	Cai et al. (2004, 2005)	Field trial
2004	42.7	6	0.24	1,600	384	100	Gnauck et al. (2004a)	200-km span, EDFA only
2004	42.7	149	6.0	6,120	36,720	50	Charlet et al. (2004a, 2005)	
2004	42.7	42	1.68	4,820	8,098	100	Becouarn et al. (2004)	EDFA only
2004	10	64	0.64	6,000	3,840	50	Vaa et al. (2004)	No dispersion management
2004	42.7	40	1.60	9,180	14,688	100	Charlet et al. (2004b)	Alternating Polarization

Note: Total date-rate and capacity are calculated by discounting the redundancy due to forward-error-correction code.

Table 1.3. Selected Recent DQPSK Experimental Demonstrations.

Year	Data Rate (Gb/s)	Channel Number	Total Rate (Gb/s)	Distance (km)	Capacity (Tb/s·km)	Channel Space (GHz)	Reference	Comments
2002	20.0	1	20	200	2	nil	Griffin et al. (2002)	
2003	25.0	9	180	1,000	180	25	Cho et al. (2003)	0.8 b/s/Hz
2003	20.0	8	160	310	50	25	Kim and Essiambre (2003)	0.8 b/s/Hz
2003	40.0	8	320	300	96	25	Wree et al. (2003a)	1.6 b/s/Hz
2004	40.0	8	320	320	96	25	Zhu et al. (2004b)	1.6 b/s/Hz
2004	40.0	14	560	400	224	20	Cho et al. (2004b)	2 b/s/Hz
2004	85.4	50	4,000	300	1,200	70	Yoshikane and Morita (2004a, 2005)	1.14 b/s/Hz
2004	12.5	64	800	6,500	5,200	66	Tokle et al. (2004)	
2004	42.7	25	1,000	2,800	2,800	50	Gnauck et al. (2004b)	0.8 b/s/Hz
2004	85.4	64	5,120	320	1,638	50	Yoshikane and Morita (2004b)	1.6 b/s/Hz

Note: Total date-rate and capacity are calculated by discounting the redundancy due to forward-error-correction code.

the phase of a signal, nonlinear phase noise becomes a major limitation for phase-modulated optical communications (Ho, 2003b,e,g, 2004c, Ho and Kahn, 2004a, Kim, 2003, Kim and Gnauck, 2003, Mecozzi, 1994a,b, Ryu, 1992, Saito et al., 1993).

As a constant pulse train, DPSK signal has larger tolerance to interchannel nonlinearities induced mostly by XPM (Leibrich et al., 2002) or similar effects (Cho et al., 2004a, Lu et al., 2004). The periodic intensity of DPSK signal gives the same XPM distortion to adjacent pulses that does not degrade the differential phase.

In dispersive fiber, the optical pulse is broadened by chromatic dispersion with traveling distance. When adjacent pulses overlap with each other, their interaction with Kerr effect also induces phase noise to the pulse itself or other optical pulses. With a phase modulation into a pulse train, DPSK signal also has higher tolerance of intrachannel pulse-to-pulse interaction than on-off keying signal.

Recently, DPSK signal is also used to build a high dynamic range burst receiver (Nizhizawa et al., 1999, Su et al., 2004). For packet switching data, DPSK signal can also be used to label an on-off keying packet (Chi et al., 2003, Hung et al., 2004) or using an ASK signal as a label for DPSK packet (Liu et al., 2004a). To certain extend, the phase and amplitude are used to carry independent data in those applications.

6. Overview

Many books in coherent optical communications have been published, mostly in the 90's (Betti et al., 1995, Cvijetic, 1996, Hooijmans, 1994, Okoshi and Kikuchi, 1988, Ryu, 1995). Standard textbooks in optical communications also have a chapter in coherent optical communications (Agrawal, 2002, Kazovsky et al., 1996, Keiser, 1999, Liu, 1996, Senior, 1992). However, most of those works focused on coherent optical communications limited by shot noise when optical amplifiers were not yet widely deployed. Phase-modulated or coherent optical communication deserves a re-visit for the following reasons:

- For binary signals, sensitivity improvement for typical systems with optical amplification is limited to about 3 dB instead of much higher improvement quoted in early works.

- For system limited by LO-spontaneous beat noise, homodyne and heterodyne system has the same performance instead of 3-dB difference in shot-noise limited systems.

- Binary frequency-shift keying (FSK) system provides no performance improvement compared with on-off keying signal.

- While laser phase noise was the major impairment of earlier low-speed phase-modulated signals, contemporary 10- and 40-Gb/s signals are not likely to be limited by laser phase noise.
- Direct-detection DPSK and DQPSK signaling is the mainstream for its simple receiver design instead of FSK or continuous-phase FSK (CPFSK) signaling.
- Coherent WDM systems usually do not give closer channel spacing than IMDD WDM systems with a well-design WDM demultiplexer.
- With optical amplifiers to maintain high optical power along a fiber link, phase-modulated optical communication systems are more likely to be limited by various types of fiber nonlinearities.
- With optical amplifiers, the system is more likely to be limited by nonlinear phase noise induced by the interaction of fiber Kerr effect and amplifier noise, in contrast to the limitation by laser phase noise in early systems.
- In dispersive transmission, the interaction of overlapped pulses with fiber Kerr effect also limits the performance of the systems but its impact is typically smaller than nonlinear phase noise.
- With multi-level signaling, coherent WDM systems may mainly provide superior spectral efficiency without large receiver penalty as compared with on-off keying.
- Direct-detection DQPSK signals double the spectral efficiency and provide 1.3-dB sensitivity improvement over on-off keying.

Nonlinear phase noise affects mainly phase-modulated signals, including PSK and DPSK signals discussed in earlier parts of this chapter. Conventionally, phase-modulated signal is mostly analyzed for additive Gaussian noise channel with laser phase noise (Henry, 1982, Kazovsky, 1986a, 1989, Nicholson, 1984). With optical amplifiers to launch high optical power into the fiber, current phase-modulated signals are limited mostly by nonlinear phase noise or other types of fiber nonlinearity effects.

Later chapters of this book are organized as following: Chapter 2 reviews various methods to generate amplitude- and phase-modulated optical signals; Chapter 3 presents various types of coherent optical receiver and analyses the performance of those receiver as limited by optical amplifier noises; Chapter 4 mainly discusses the performance of an optical signal with other impairments in linear regime, mainly laser phase noise; Chapters 5 and 6 are both about nonlinear phase noise and method for its compensation; Chapter 7 investigates the performance of single-channel DPSK signal with pulse-to-pulse interaction in highly dispersive transmission systems; Chapter 8 studies WDM DPSK signals with homodyne crosstalk and XPM related fiber nonlinearities; Chapters 9 and 10 are about multilevel signaling and soliton DPSK transmission; Chapter 11 finishes the book with the discussion of the channel capacity of lightwave communication systems, mostly narrowband systems based on ultra-dense WDM technologies.

Chapter 2

DIGITAL MODULATION OF OPTICAL CARRIER

In phase-modulated optical communication systems, like other means of communication, an essential function is to modulate the transmitted signal with information. Digital modulation usually carries information using the amplitude, phase, or frequency of the carrier (Haykin, 1988, Proakis, 2000). Digital modulation for coherent optical communications is fundamentally the same as that for wireless or wireline communications. Optical carrier is used in optical communications, electromagnetic wave is used in radio frequency (RF) communications, and electrical signal is used in copper-wire based wireline communications. In this chapter, we will first review the basic modulation formats for digital communications and then the methods to modulate them to the optical carrier.

Phase-modulated or coherent optical communication systems, like intensity-modulated/direct-detection (IMDD) systems, mostly use semiconductor diode laser as the light source. The basic structure, modulation response, and noise properties of semiconductor laser are briefly reviewed. With low quality signal, a semiconductor laser can be directly modulated to generate on-off keying signal.

Both amplitude- and phase-modulated signals are usually generated by amplitude and phase modulator, respectively. The principle and basic structures of those modulators are presented here. An amplitude modulator can be used to generate on-off keying and binary phase-modulated signals. Various types of return-to-zero (RZ) differential phase-shift keying (RZ-DPSK) transmitter are designed with RZ duty cycle of 1/3, 1/2, and 2/3.

The frequency modulation characteristic of semiconductor laser is also discussed later in this chapter.

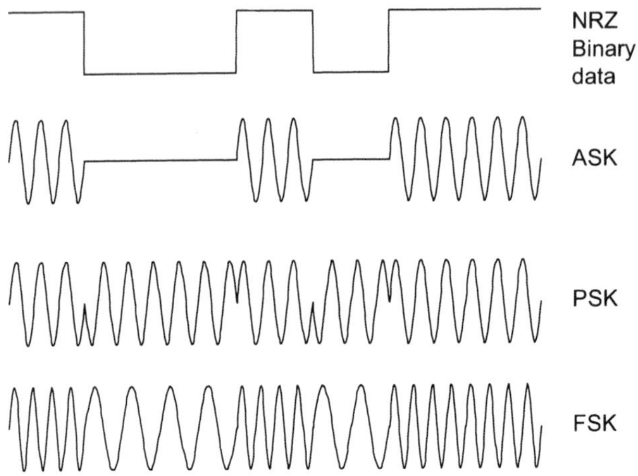

Figure 2.1. Basic modulation formats for optical communications.

1. Basic Modulation Formats

In addition to the amplitude or intensity, coherent optical communication systems utilize the phase or frequency of the optical carrier to carry information. Figure 2.1 shows basic modulation formats for coherent optical communications. The binary amplitude-shift keying (ASK) signal is the same as on-off keying. On and off states are represented by the presence or absence of light, the same as nonzero and zero amplitude, respectively. While the term of on-off keying is used for IMDD systems, ASK is the term commonly used for coherent optical communications. In binary ASK systems, the two transmitted signals to represent "0" and "1" states are

$$s_1(t) = A \cos \omega_c t, \qquad 0 < t \leq T \tag{2.1}$$

$$s_2(t) = 0, \qquad 0 < t \leq T \tag{2.2}$$

where A is a constant amplitude, ω_c is the angular frequency of the optical carrier, and T is the signal duration that is also equal to the bit interval for the binary signal.

Phase-shift keying (PSK) in Fig. 2.1 uses the phase of the carrier to carry information. In binary PSK systems, the two binary signals are

$$s_1(t) = A \cos \omega_c t, \qquad 0 < t \leq T \tag{2.3}$$

$$s_2(t) = -A \cos \omega_c t, \qquad 0 < t \leq T. \tag{2.4}$$

Ideally, the phase difference between the two antipodal signals is 180° for optimal performance. For the same signal-to-noise ratio (SNR) and assumed ideal signal demodulation, binary PSK signal can achieve the lowest error probability.

Frequency-shift keying (FSK) in Fig. 2.1 uses the frequency of the carrier to encode information. In binary FSK systems, the two FSK signals are

$$s_1(t) = A\cos\omega_1 t, \qquad 0 < t \leq T \tag{2.5}$$

$$s_2(t) = A\cos\omega_2 t, \qquad 0 < t \leq T \tag{2.6}$$

where ω_1 and ω_2 are two angular frequencies to represent "0" and "1" state, respectively. Ideally, the two signals of $s_1(t)$ and $s_2(t)$ should be orthogonal with each other.

In additional to the basic modulation formats of binary ASK, PSK, and FSK in Fig. 2.1, more complicated formats can be constructed by adding more levels into the basic signals or combining two components of, for example, amplitude and phase. More general discussion of digital modulation formats can be found in Proakis (2000) and Haykin (1988). The baseband signal can also be return-to-zero (RZ) instead of non-return-to-zero (NRZ) linecode of Fig. 2.1.

A single mode fiber can support two orthogonal polarizations, another alternative to the formats in Fig. 2.1 is polarization-shift keying (PolSK) (Benedetto and Poggiolini, 1992, Calvani et al., 1988, Imai et al., 1990b). Polarization-division multiplexing (PDM) utilizes the two orthogonal polarizations to transmit two independent data streams (Bigo et al., 2001, Frignac et al., 2002, Hill et al., 1992, Noé et al., 2001). In general, PDM gives better spectral efficiency than PolSK. PDM is similar to polarization multiplexing in microwave systems to utilize the two polarizations of electromagnetic wave as two, ideally, independent channels.

There are many methods to generate the basic modulation formats of Fig. 2.1 in optical domain. This chapter will study some popular methods to generate the digital modulated signal.

2. Semiconductor Diode Lasers

In all optical communication systems, a light source must be used to originate an optical signal. Because of its small size, low power consumption, reliability, and compatible with electronic circuits, semiconductor diode lasers are the most widely used light source for communication applications. Virtually all optical communication systems use semiconductor laser as light source. Erbium-doped fiber amplifiers (EDFA) are also commonly pumped by high-power semiconductor lasers (Becker et al., 1999, Desurvire, 1994, Nakazawa et al., 1989). Other than lightwave communications, semiconductor laser also find its applications for opti-

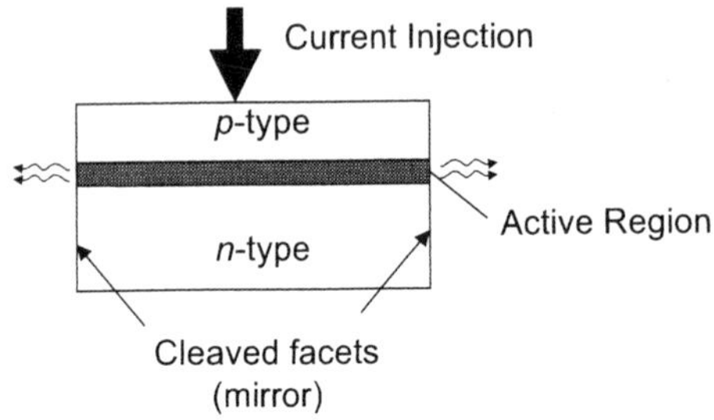

Figure 2.2. Structure of a semiconductor laser with a Fabry-Perot cavity.

cal data storage, like compact disk (CD) or digital versatile disk (DVD) machines.

2.1 Basic Structures

Figure 2.2 shows the basic structure of a semiconductor laser with a Fabry-Perot cavity. The active region is the media to provide optical gain sandwiched between the p- and n-type semiconductor materials. The p-n junction is preference to be a double-heterostructure junction in which the active region using a material having a band-gap smaller than both p- and n-type materials. The usage of heterostructure structure for semiconductor laser has two major advantages. Because of the band-gap difference, the active region can effectively confine electrons and holes into the active layer. Electron and hole recombination generates light or amplifies the passing through light through spontaneous and stimulated emission, respectively. The active region based on heterostructure also has larger refractive index. Similar to the principle of light guiding in an optical fiber, the active region acts as a dielectric slab waveguide to confine the light.

The laser cavity of the semiconductor laser of Fig. 2.2 is a Fabry-Perot resonator with two reflective mirrors at both sides. When light reaches one of the facets, part of the light reflects back to the active region as an optical feedback signal. In the most basic structure, the mirrors of semiconductor laser are formed simply by cleaving with a reflectivity of

$$r_m = \left(\frac{n_r - 1}{n_r + 1}\right)^2, \tag{2.7}$$

where n_r is the refractive index of the gain medium. Typical for most semiconductor lasers, $n_r = 3.5$ and $r_m = 30\%$. For a semiconductor laser to generate light, the round-trip optical gain must be equal to the

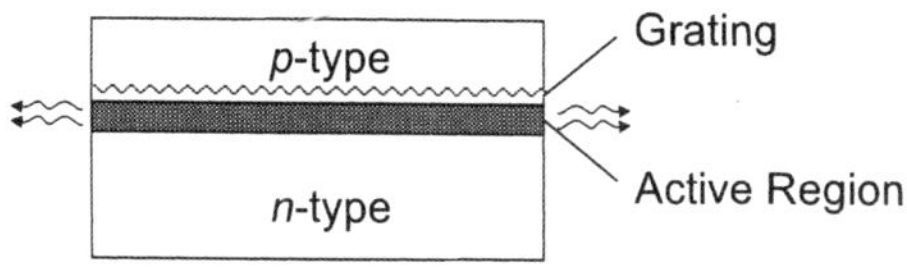

Figure 2.3. Structure of a distributed-feedback (DFB) semiconductor laser.

overall cavity loss. When current is injected into the p-n junction, the semiconductor laser reaches its lasing threshold when the optical gain is just equal to the cavity loss. In semiconductor laser, the optical gain is very large such that high facet loss can be tolerated.

For a cavity length of L_l, the modes of the Fabry-Perot cavity have a spacing of $c/(2n_r L_l)$, where c is the speed of light in free space. Note that the refractive index of n_r may change to n_{eff} to include the effect of the slab waveguide. For a typical length of $L_l = 200 - 400$ μm, the mode spacing of the laser is about 100 to 200 GHz, or about 1 to 2 nm at the wavelength around 1.55 μm.

The gain medium of semiconductor laser has a gain bandwidth on the order of 100 nm (about 12 THz). At least tens of modes can be supported within the gain bandwidth. In the semiconductor gain medium, due to heterogeneous broadening, the gain is reduced for the lasing wavelength with strong optical power. When the gain at the strong mode is reduced, the other modes can reach the same gain as the strong mode and start lasing. With tens of modes within the gain bandwidth, a semiconductor laser may have tens of lasing wavelengths. To limit the mode number in semiconductor laser, frequency selective loss can be introduced into the laser cavity. In additional to the Fabry-Perot resonator to limit the lasing wavelength to tones that are $c/2n_r L_l$ apart, frequency selective structure can be introduced close to the active region of Fig. 2.2.

Figure 2.3 shows a distributed-feedback (DFB) laser with a grating close to the active region. The grating gives periodic variation of the waveguide refractive index. Instead of two facets, optical feedback is provided in the whole cavity in a DFB laser. The waves propagating in the forward and backward directions are coupled with each other. With the grating, coupling occurs only for wavelength λ_B with the Bragg condition of

$$\Lambda_g = m\frac{\lambda_B}{2n_{\text{eff}}}, \tag{2.8}$$

where Λ_g is the grating period, n_{eff} is the effective refractive index of the active region of the DFB laser, and m is the order of Bragg diffraction. The coupling between the forward and backward waves is the strongest for the first-order Bragg diffraction of $m = 1$. In order to operate a laser at $\lambda_B = 1.55$ μm, Λ_g is about 235 nm for $m = 1$ for $n_{\text{eff}} = 3.3$. Such grating can be fabricated by the interferometric pattern of two short wavelength laser beams or electron beam writing.

First operated in the liquid nitrogen temperature of 77 K° (Hall et al., 1962) and then the room temperature (Hayashi et al., 1970), various types of semiconductor laser are described in more details in the books of Agrawal and Dutta (1986), Casey and Panish (1978), and Coldren and Corzine (1995). The DFB structure was first proposed by Kogelnik and Shank (1971). In additional to the DFB structure, there is also a distributed Bragg reflection (DBR) structure with the grating at one end of the waveguide of Fig. 2.3 (Suematsu et al., 1983). As a variation of DFB laser, $\lambda/4$-shifted DFB laser has a $\lambda/4$ shift in the middle of the grating to provide a $\pi/2$ phase shift (Akiba et al., 1987). Other advanced semiconductor laser structures were reviewed in Suematsu et al. (1992), including the low-cost surface-emitting semiconductor lasers (Iga et al., 1988). DFB laser for WDM applications was reviewed in Funabashi et al. (2004).

In additional to single wavelength DFB laser, there are also monolithically integrated widely tunable multiple wavelength semiconductor lasers (Coldren, 2000, Coldren et al., 2004), external cavity diode lasers tuned by micro-electro-mechanical systems (MEMS) (Anthon et al., 2002), or an array of DFB lasers selected by MEMS (Pezeshki et al., 2002). Those widely tunable lasers can cover the whole C- or L-band for EDFA.

The output intensity of a semiconductor laser follows its injected current. On-off keying signal may be generated by the presence or absence of injection current to turn-on and -off the semiconductor laser. Later in this section presents the rate equations to govern the laser dynamic.

Even when a semiconductor laser has a constant injection current, the output intensity of the semiconductor laser is noisy due to the spontaneous emission in the lasing material. External modulated system that uses a diode laser with constant injection current is still affected by the semiconductor laser noise. The laser phase noise directly affects a phase-modulated signal and laser intensity noise directly affects an on-off keying or amplitude-modulated signal. As a second-order effect, laser phase noise may convert to amplitude noise due to fiber chromatic dispersion.

2.2 Rate Equations and Laser Dynamic

The dynamic and noise characteristics of a semiconductor laser can be described by the laser rate equations. In its simplest way, a semiconductor laser converts electrons to photons. The laser electric field

E_L and the carrier density n_c as a function of time are governed by the following rate equations

$$\frac{dE_L}{dt} = -j\Delta\omega(n_c)E_L + \frac{1}{2}\left[G(n_c) - \frac{1}{\tau_p}\right]E_L + F_n(t), \tag{2.9}$$

$$\frac{dn_c}{dt} = I_c - \frac{n_c}{\tau_c} - G(n_c)|E_L|^2, \tag{2.10}$$

where $\Delta\omega(n_c)$ is the deviation of oscillation angular frequency from the natural frequency of the laser cavity as a function of carrier density of n_c, $G(n_c)$ is the power gain as a function of carrier density n_c, τ_p is the lifetime of the photon particle, $F_n(t)$ is the Langevin random force caused by spontaneous emission, and I_c is the rate of carrier injection, and τ_c is the carrier lifetime. In the equation of Eq. (2.9), the electric field of E_L should have a unit such that the photon density in the laser cavity is $p_c = |E_L|^2$.

The physical meaning of the above rate equations is obvious. The first term of the right hand side of Eq. (2.9) is the frequency detuning due to carrier injection. In the stationary condition, $\Delta\omega(n_c) = 0$. The second term of the right hand side of Eq. (2.9) is the gain of electric field provided by the lasing medium. Because the laser light losses due to material absorption, emitted out from the laser cavity, the electric field (or photon) has a photon lifetime of τ_p depending on the laser cavity. If the overall loss of the cavity is r_c due to mirror and internal material loss, the photon lifetime is equal to $\tau_p^{-1} = v_g r_c$ where v_g is the group velocity of light. The last term of the right hand side of Eq. (2.9) is due to spontaneous emission. The Langevin force of $F_n(t)$ is usually modeled as a complex value white Gaussian noise. The carrier rate equation of Eq. (2.10) also has simple physical meaning. I_c is the rate of carrier increase due to current injection, τ_c is the carrier lifetime, and the last term of Eq. (2.10) is the carrier converting to photons. For a current density of I_d, the rate of carrier injection is $I_c = I_d/e$, where e is the electron charge. The detuning of $\Delta\omega(n_c)$ just affects the phase of the electric field E_L and thus the term of $-j\Delta\omega(n_c)E_L$ with the multiplication of "j".

The refractive index of semiconductor material is equal to $n_r(n_c) = n_0 + \Delta n_r(n_c) + j\Delta n_i(n_c)$, where n_0 is the refractive index without carrier injection, $n_r(n_c)$ and $n_i(n_c)$ are the real and imaginary parts of the refractive index as a function of carrier density n_c. The gain of $G(n_c)$ is proportional to the imaginary refractive index of $-\Delta n_i(n_c)$. The lasing frequency should be aligned with the mode spacing of $n_r(n_c)/L_l c$. The term of $\Delta n_r(n_c)$ induces fluctuation in resonance frequency of the cavity and gives $\Delta\omega(n_c)$, generating frequency modulation by carrier injection.

When the semiconductor laser is biased well above the lasing threshold with a small signal injection current of

$$I_c = I_{c0} + \Re\{v_m e^{j\omega_m t}\}, \tag{2.11}$$

where v_m and ω_m are the modulation amplitude and frequency, respectively. The electric field and the carrier density have stationary values of $E_0 = \sqrt{P_{c0}}e^{j\phi_0}$ and N_0. The rate equations of Eqs. (2.9) and (2.10) can be linearized around E_0 and N_0 for $E_L(t) = \sqrt{P_{c0} + \Delta p(t)}e^{j\phi_0 + j\phi_n(t)}$ and $n_c(t) = N_0 + \Delta n(t)$. First of all, both frequency detuning $\Delta\omega(n_c)$ and medium gain $G(n_c)$ are linearized by

$$\Delta\omega(n_c) = -\frac{\partial\omega}{\partial n_c}\Delta n, \tag{2.12}$$

$$G(n_c) = G(N_0) + \frac{\partial G}{\partial n_c}\Delta n. \tag{2.13}$$

Ignores the Langevin noise of F_n, the rate equations become

$$\frac{d\phi_n}{dt} = \frac{\partial\omega}{\partial n_c}\Delta n, \tag{2.14}$$

$$\frac{d\Delta p}{dt} = P_{c0}\frac{\partial G}{\partial n_c}\Delta n, \tag{2.15}$$

$$\frac{d\Delta n}{dt} = \Delta i_c - \left(\frac{1}{\tau_c} + P_{c0}\frac{\partial G}{\partial n_c}\right)\Delta n - G(N_0)\Delta p, \tag{2.16}$$

where $\Delta i_c = \Re\{v_m e^{j\omega_m t}\}$. The photon density is

$$\Delta p(\omega_m) = \frac{P_{c0}\frac{\partial G}{\partial n_c}v_m}{\omega_R^2 - \omega_m^2 + j\gamma_e\omega_m}, \tag{2.17}$$

where

$$\omega_R^2 = P_{c0}G(N_0)\frac{\partial G}{\partial n_c}, \tag{2.18}$$

$$\gamma_e = \frac{1}{\tau_c} + P_{c0}\frac{\partial G}{\partial n_c} \tag{2.19}$$

as the resonance frequency and damping rate of relaxation oscillation. In semiconductor laser, the differential gain of $\partial G/\partial n_c$ is almost a constant above lasing threshold. The relaxation frequency of ω_R of Eq. (2.18) increases with both the photon density of P_{c0} and the gain of the laser $G(N_0)$. In general, ω_R increases almost linearly with P_{c0}.

The normalized modulation response of a semiconductor laser is the transfer function of

$$H_p(\omega_m) = \frac{\Delta p(\omega_m)}{\Delta p(0)} = \frac{\omega_R^2}{\omega_R^2 - \omega_m^2 + j\gamma_e\omega_m}. \tag{2.20}$$

The instantaneous frequency of a laser is $\frac{1}{2\pi}d\phi_n(t)/dt$. The response of the laser frequency can be written as

$$\dot{\phi}_n(\omega_m) = \frac{j\omega_m}{2P_{c0}} \frac{\frac{\partial\omega}{\partial n_c}}{\frac{1}{2}\frac{\partial G}{\partial n_c}} \Delta p(\omega_m). \tag{2.21}$$

The factor of

$$\alpha = \frac{\dfrac{\partial\omega_m}{\partial n_c}}{\dfrac{1}{2}\dfrac{\partial G}{\partial n_c}} \tag{2.22}$$

is the linewidth enhancement factor. The linewidth enhancement factor of α can also be rewritten as

$$\alpha = \frac{\dfrac{\partial n_r}{\partial n_c}}{\dfrac{\partial n_i}{\partial n_c}}, \tag{2.23}$$

where n_r and n_i represent the real and imaginary parts of the effective refractive index. The linewidth enhancement factor of commonly used DFB laser is about 3 to 4.

The response of the laser phase is

$$\dot{\phi}_n(\omega_m) = \frac{\alpha}{2P_{c0}} j\omega_m \Delta p(\omega_m). \tag{2.24}$$

The normalized transfer function of the laser frequency is

$$H_{\dot{\phi}}(\omega_m) = \frac{j\omega_m\omega_R^2}{\omega_R^2 - \omega_m^2 + j\gamma_e\omega_m}. \tag{2.25}$$

The transfer function of Eq. (2.25) will be revisited later in this chapter about direct frequency modulation of a semiconductor laser.

Figure 2.4 shows the photon response of direct-modulation of a high-speed semiconductor laser used for 40-Gb/s applications (Sato, 2002, Sato et al., 2002). The laser response has a second-order response of Eq. (2.20). Similar to Eq. (2.18), the relaxation oscillation frequency increases with the bias current or the output power.

The method to analyze laser dynamic here follows the books of Petermann (1991) and Okoshi and Kikuchi (1988). Laser dynamic can also be studied based on the rate equations of the carrier density and photon density (Lau and Yariv, 1985, Tucker, 1985, Tucker and Kaminow, 1984). Further increase of current injection generates more photons, but the optical gain is reduced to $G(n_c)(1 - \epsilon_{\mathrm{NL}}p_c)$, where ϵ_{NL} is a

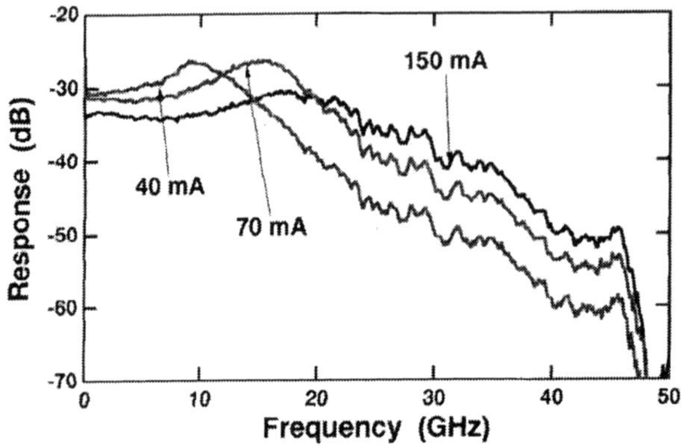

Figure 2.4. The measured small-signal amplitude response of a high-speed semiconductor laser. [From Sato (2002), © 2005 IEEE]

nonlinear-gain parameter due to saturation effect. In semiconductor material, due to heterogeneous broadening, the factor of ϵ_{NL} depends on wavelength. Nonlinear gain saturation is a major factor that affects the laser dynamic (Lau and Yariv, 1985, Olshansky et al., 1987, Ralston et al., 1993, Tucker, 1985). The relaxation peaks of the amplitude response of Fig. 2.4 is reduced due to the contribution of ϵ_{NL} (Ralston et al., 1993) or laser parasitic (Tucker and Kaminow, 1984). High-speed semiconductor laser with a speed of 20 to 40 GHz has been developed (Kjebon et al., 1997, Matsui et al., 1997, Morton et al., 1992, Ralston et al., 1993, Weisser et al., 1996). Semiconductor laser can be directly modulated for 40-Gb/s data rate (Sato, 2002, Sato et al., 2002). The parameters for the rate equations of Eqs. (2.9) and (2.10) can be found using parameter extraction methods (Cartledge and Srinivasan, 1997, Lee et al., 2002b, Salgado et al., 1997).

2.3 Laser Noises

When the semiconductor laser is biased well above the lasing threshold with a fixed current with constant carrier injection of I_{c0} without any modulation, with the Langevin noise, the rate equations of Eqs. (2.14) to (2.16) can be linearized to

$$\frac{d\phi_n}{dt} = \frac{\partial\omega}{\partial n_c}\Delta n + \frac{F_i}{\sqrt{P_{c0}}}, \tag{2.26}$$

$$\frac{d\Delta p}{dt} = P_{c0}\frac{\partial G}{\partial n_c}\Delta n + 2\sqrt{P_{c0}}F_r, \tag{2.27}$$

$$\frac{d\Delta n}{dt} = -\left(\frac{1}{\tau_c} + P_{c0}\frac{\partial G}{\partial n_c}\right)\Delta n - G(N_0)\Delta p, \tag{2.28}$$

where F_r and F_i are the real and imaginary parts of the Langevin force of $F_n(t)$.

In very low frequency, the time derivative of Eqs. (2.26) to (2.28) is equal to zero, this is similar to find the direct-current (d.c.) operating point of the linearized equations except that both the phase and amplitude equations of Eqs. (2.26) and (2.27) have noisy driving force of F_i and F_r, respectively.

From the equation of Eq. (2.27) with $d\Delta p/dt = 0$, we obtain

$$\Delta n = -\frac{F_r}{\dfrac{\sqrt{P_{c0}}}{2}\dfrac{\partial G}{\partial n_c}}, \tag{2.29}$$

that the number of carriers in the laser is fluctuated due to the spontaneous emission of F_r. As spontaneous emission depletes the injected carriers, the number of carriers becomes random. From the rate equation of Eq. (2.28), together with Eq. (2.29), we obtain

$$\Delta p = \frac{\dfrac{1}{\tau_c} + P_{c0}\dfrac{\partial G}{\partial n_c}}{G(N_0)\dfrac{\sqrt{P_{c0}}}{2}\dfrac{\partial G}{\partial n_c}}F_r \tag{2.30}$$

for the effect of spontaneous emission to the number of photons or the amplitude of the emitted electric field. From Eq. (2.30), the amplitude of the emitted electric field increases with spontaneous emission. The carrier fluctuation of Eq. (2.29) and the photon density fluctuation of Eq. (2.30) have opposite sign. As photon number increase, the carrier has to be consumed for photon creation.

The fluctuation of the angular frequency of Eq. (2.14) is

$$\frac{d\phi_n}{dt} = -\frac{\alpha F_r}{\sqrt{P_{c0}}} + \frac{F_i}{\sqrt{P_{c0}}}. \tag{2.31}$$

From Eq. (2.31), the linewidth enhancement factor of α gives more phase noise to the laser. The reason of linewidth enhancement of a factor of α

is due the fluctuation of amplitude of $|E_L|$ changing the carrier intensity because of the rate equations of Eq. (2.9) and the linearized version of Eq. (2.27). However, the refractive index of the lasing material also changes with carrier density through $\Delta\omega(n_c)$. The frequency or phase of the semiconductor laser changes due to the detuning effect from the variation of the refractive index of the lasing material.

From the phase equation of Eq. (2.31), the instantaneous frequency is a Gaussian process and the phase is

$$\phi_n(t) = \int_0^t n_\phi(t)\mathrm{d}t, \tag{2.32}$$

where

$$\sigma_{n_\phi}^2 = \frac{(1+\alpha^2)\sigma_F^2}{P_{c0}}, \tag{2.33}$$

where σ_F^2 is the spontaneous diffusion coefficient of $\sigma_F^2 = E\{F_r^2\} = E\{F_i^2\}$.

The phase noise of Eq. (2.32) is a Wiener process with autocorrelation of

$$E\{\phi_n(t_1)\phi_n(t_2)\} = \sigma_{n_\phi}^2 \min(t_1, t_2). \tag{2.34}$$

For a laser with only phase fluctuation, in normalized form, the electric field is $v(t) = e^{j\phi_n(t)}$ with autocorrelation of

$$R(\tau) = E\left\{v(t)v^*(t-\tau)\right\} = E\left\{e^{j\phi_n(t)-j\phi_n(t-\tau)}\right\}. \tag{2.35}$$

The phase difference of $\phi_n(t) - \phi_n(t-\tau)$ is zero-mean Gaussian distributed with a variance of $\sigma_{n_\phi}^2|\tau|$. The autocorrelation function of the electric field is

$$\begin{aligned} R(\tau) &= \frac{1}{\sqrt{2\pi|\tau|}\sigma_{n_\phi}} \int_{-\infty}^{+\infty} \exp\left(j\phi + \frac{\phi^2}{2\sigma_{n_\phi}^2|\tau|}\right) \mathrm{d}\tau \\ &= \exp\left(-\frac{1}{2}\sigma_{n_\phi}^2|\tau|\right). \end{aligned} \tag{2.36}$$

The laser linewidth is

$$\begin{aligned} S_{\phi_n}(f) &= \int_{-\infty}^{+\infty} R(\tau)e^{j2\pi f\tau}d\tau \\ &= \frac{\sigma_{n_\phi}^2}{\frac{1}{4}\sigma_{n_\phi}^4 + (2\pi f)^2} \end{aligned} \tag{2.37}$$

as the Lorentzian line-shape.

The full-width-half-maximum (FWHM) linewidth of the semiconductor laser is equal to

$$\Delta f_L = \frac{\sigma^2_{n_\phi}}{2\pi}. \tag{2.38}$$

Because the noise diffusion of $\sigma^2_{n_\phi}$ increases by a factor of $1+\alpha^2$. The Lorentzian line-shape of semiconductor laser increases by the same factor of $1+\alpha^2$. When the laser phase noise is white Gaussian noise with spectral density of $\sigma^2_{n_\phi}$, the frequency noise has a spectral density of

$$S_{\dot{\phi}}(f) = \frac{\Delta f_L}{2\pi f^2}. \tag{2.39}$$

Without the coupling from carrier density to phase change due to the real part of the refractive of n_r, i.e., $\alpha = 0$, we obtain $d\phi_n/dt = F_i/A_0$. The laser linewidth is given by the well-known Schawlow-Townes formula (Sargent et al., 1974, Yariv, 1997),

$$\Delta f_{L_0} = \frac{2\pi\hbar\omega_L(\Delta\nu_{1/2})^2 n_{\mathrm{sp}}}{P_0}, \tag{2.40}$$

and

$$\Delta f_L = (1+\alpha^2)\Delta f_{L_0}, \tag{2.41}$$

where ω_L is the angular frequency of the laser, P_0 is the laser output power, $\Delta\nu_{1/2}$ is the linewidth of the passive laser resonator, and n_{sp} is the spontaneous emission factor representing the population inversion. For a Fabry-Perot resonator, the linewidth of the laser cavity is

$$\Delta\nu_{1/2} = \frac{c}{2n_r L_l}\frac{1-r_c}{\sqrt{r_c}}, \tag{2.42}$$

where $c/2n_r L_l$ is the free-spectral range with c as the speed of light and L_l as the cavity length, r_c is the round-trip loss of the Fabry-Perot resonator, and $\pi\sqrt{r_c}/(1-r_c)$ is the finesse of the Fabry-Perot resonator.

The Schawlow-Townes formula is only for the case that the angular frequency of the laser of ω_L matches to that of the laser cavity. The laser linewidth is broadened due to frequency detuning (Lax, 1967a,b). In semiconductor laser, the refractive index of semiconductor material changes due to carrier injection from the factor of $\Delta\omega(n_c)$. The fluctuation of the laser cavity causes further linewidth broadening. In early 1980s, Mooradian and his coworkers measured the spectral characteristic of semiconductor laser in detail (Fleming and Mooradian, 1981a,b, Welford and Mooradian, 1982). The linewidth broadening by the factor of $1+\alpha^2$ is exactly due to carrier injection effect (Henry, 1982, 1986,

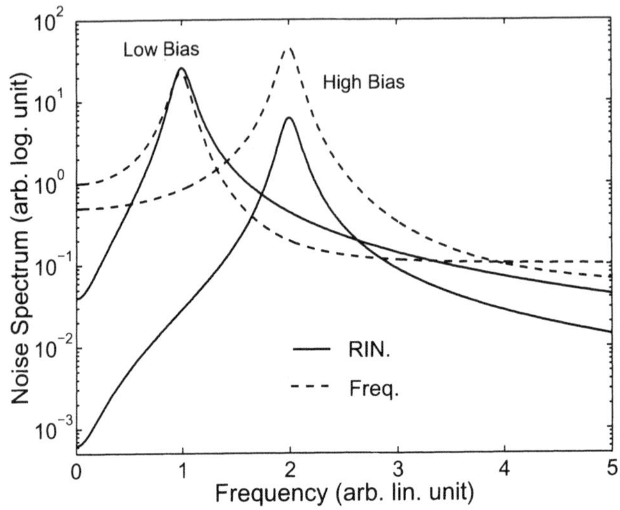

Figure 2.5. An illustration to show the change of RIN and frequency noise with bias current.

Henry et al., 1981). The linewidth increase by the factor of $1 + \alpha^2$ is confirmed by measurement.

The linearized equations of Eqs. (2.14) to (2.16) can be solved exactly without the low-frequency approximation. When d/dt is replaced by $j\omega_m$, with constant injection current, the noise spectrum becomes

$$S_{\phi_n}(\omega_m) = \Delta f_{L_0}\left[1 + \frac{\alpha^2\omega_R^4}{(\omega_R^2 - \omega_m^2)^2 + \gamma_e^2\omega_m^2}\right], \tag{2.43}$$

$$S_p(\omega_m) = 8\pi P_{c0}^2 \Delta f_{L_0}\frac{\omega_m^2 + \gamma_e^2}{(\omega_R^2 - \omega_m^2)^2 + \gamma_e^2\omega_m^2}. \tag{2.44}$$

The laser linewidth is equal to $\Delta f_L = (1 + \alpha^2)\Delta f_{L_0} = S_{\phi_n}(0)$. The amplitude noise of $S_p(\omega_m)$ is called the relative-intensity noise (RIN) of the laser.

Figure 2.5 shows the change of RIN and frequency noise with laser bias current or output power. The relaxation peaks of both amplitude and frequency noise are located at about the same frequency. Amplitude noise decreases rapidly with output optical power, frequency noises also decreases slightly with output power (or bias current). While amplitude noise decreases by a ratio of -40 dB per decade at very high frequency, the frequency noise approaches a constant of Δf_{L_0} at high frequency. At low frequency, the frequency noise is $(1+\alpha^2)\Delta f_{L_0}$, a factor of $1+\alpha^2$ larger

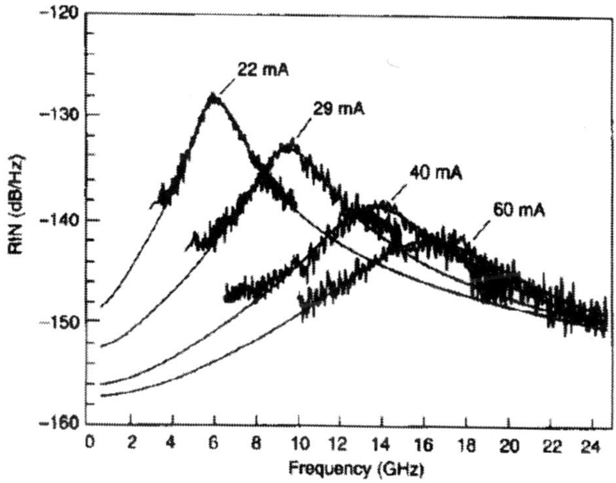

Figure 2.6. Comparison of measured and theoretical RIN from a high-speed semiconductor laser. [From Ralston et al. (1993), © 2005 IEEE]

than the high frequency limit. In Fig. 2.5, frequency noise decreases with bias current because Δf_{L_0} is inverse proportional to the laser output power.

Laser linewidth broadening in semiconductor laser was first observed by the group of Mooradian (Fleming and Mooradian, 1981a,b, Welford and Mooradian, 1982) and was first analyzed by Henry (1982). The impact of relaxation oscillation to the spectrum of semiconductor laser was first observed by Vahala et al. (1983) and analyzed using semi-classical methods by Henry (1983), Kikuchi and Okoshi (1985), Spano et al. (1983), and Vahala and Yariv (1983a,b). Yamamoto (1984a,b) analyzed both amplitude and frequency noise of semiconductor laser based on the quantum mechanical Langevin equations. The theory of laser linewidth was reviewed by the paper of Henry (1986) and the book of Petermann (1991). This section follows the books of Petermann (1991) and Okoshi and Kikuchi (1988).

Figure 2.6 shows a comparison of the measured and theoretical RIN from a high-speed semiconductor laser. The measurement results match well to the theoretical curves. For continuous-wave operation, DFB lasers for WDM applications typically have an output power more than 100 mW as laser chip, a RIN in the range lower than -150 to -160 dB/Hz, and a laser linewidth of few MHz (Funabashi et al., 2004). As RIN is inversely proportional to output power, high-speed laser of Ralston et al. (1993) has larger RIN than WDM laser due to an output power of only

10 to 20 mW. After coupling to an optical fiber, typical WDM laser has a power more than 20 mW.

Laser RIN is usually not an issue for common IMDD and phase-modulated high-speed optical communication systems. Laser phase noise adds directly to the signal phase and degrades a phase- or frequency-modulated signal. Laser phase noise is not an issue for high-speed 10- and 40-Gb/s phase-modulated systems, but for low-speed systems.

Laser RIN increases with optical feedback (Ho et al., 1993, Shikada et al., 1988, Tkach and Chraplyvy, 1986), an optical isolator must be used to eliminated optical feedback into the laser cavity for low-noise laser.

3. External Modulators

External modulators provide the best signal quality for both phase and amplitude modulated signals. An electro-optical crystal with proper orientation provides phase modulation with a voltage applied in the right direction. Lithium Niobate ($LiNbO_3$) is the most commonly used electro-optical crystal to fabricate external modulator. In this section, external modulator using $LiNbO_3$ is assumed by default.

3.1 Phase Modulator

The generation of a phase-modulated signal requires an external modulator capable of changing the optical phase in response to an applied voltage. In $LiNbO_3$, if an electric field is applied along the z-axis of the crystal, the refractive index of the material is changed by

$$\Delta n = \frac{1}{2} n_r^3 r_{33} E_z, \tag{2.45}$$

where r_{33} is the electro-optic coefficient for a change of refractive index n_r for light propagating along the z-direction, and E_z is the electric field along the z direction. In the structure of Fig. 2.7, when a voltage is applied to the electro-optical material, the electric field of E_z is approximately equal to V/d, where d is the distance between the two electrodes. The total phase shift over an interaction length of L_i is

$$\Delta \phi_0 = \frac{2\pi \Delta n L_i}{\lambda_c} = \pi n_r^3 r_{33} \frac{V L_i}{d \lambda_c}, \tag{2.46}$$

where λ_c is the wavelength of the optical signal in vacuum.

The voltage necessary to provide a phase shift of 180° is equal to

$$V_\pi = \frac{d \lambda_c}{n_r^3 r_{33} L_i}. \tag{2.47}$$

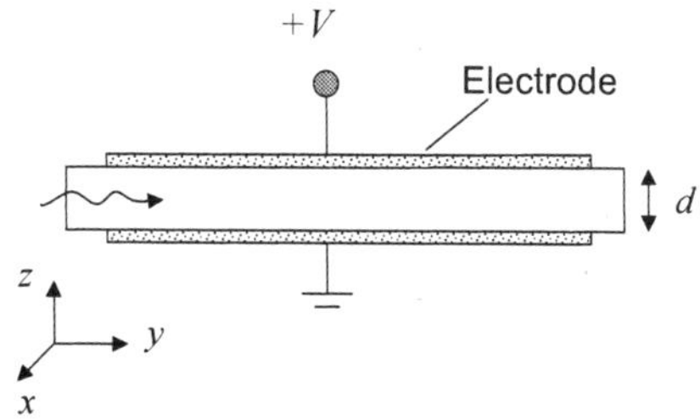

Figure 2.7. Structure to illustrate the operation of a phase modulator.

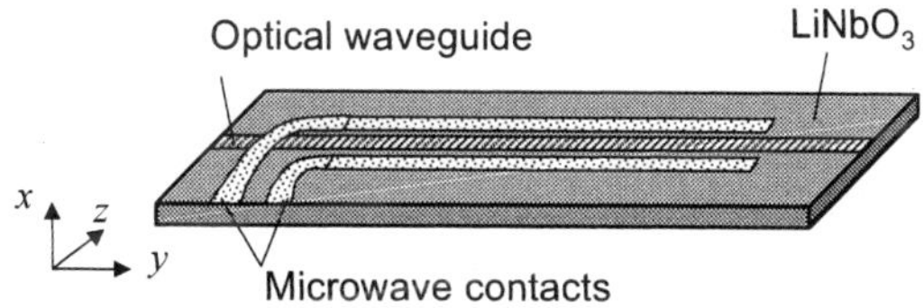

Figure 2.8. Illustration of a waveguide based traveling-wave phase modulator.

In the design of a phase modulator, one of the main objective is to reduce the voltage of V_π. The electro-optic coefficient of r_{33} in $LiNbO_3$ is the largest among all coefficients for various orientations (Turner, 1966). The parameter of $n_r^3 r_{33}$ of $LiNbO_3$ is equal to $328 \times 10^{-6} \mu m/V$(Alferness, 1982).

In the bulk modulator structure of Fig. 2.7, the modulator V_π can be reduced by increasing the interaction length of L_i or reducing the electrode distance of d. However, the capacitance of the capacitor formed between the two electrodes is proportional to L_i/d. The reduction of V_π increases the capacitance and slows down the operation of the modulator.

Figure 2.8 shows a traveling-wave phase modulator. In principle, if the light in the optical waveguide and modulated electrical signal in the microwave electrodes are traveling in the same speed, the traveling-wave phase modulator of Fig. 2.8 has infinite bandwidth. $LiNbO_3$ has a refractive index of about $n_r = 2.1$ to 2.2. The microwave signal must slow down to close to the speed of c/n_r of the optical signal. The design of the microwave waveguide/electrode must match to the traveling speed of optical signal in $LiNbO_3$ for high-speed operation.

In the waveguide phase modulator of Fig. 2.8, there are two methods to couple electric field into the optical waveguide. Figure 2.9 shows two waveguide structures using either x- or z-cut $LiNbO_3$ crystal. The electric field lines are along the z-axis in both cases. Using the x-cut

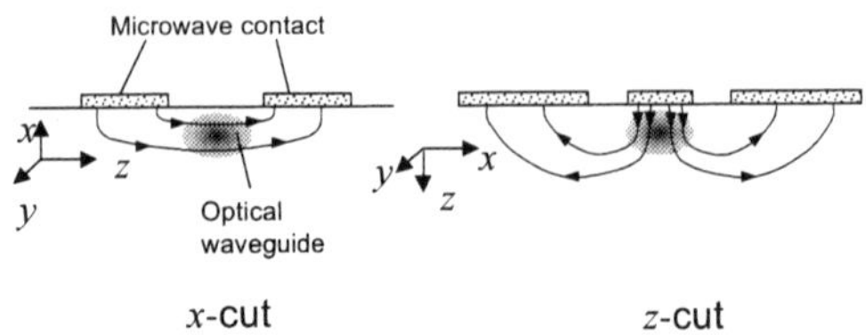

Figure 2.9. Waveguide based traveling-wave phase modulator using x- or z-cut $LiNbO_3$ materials.

material, the microwave electrodes for the transmission line are located in either side of the optical waveguide. Using the z-cut material, the electrode for applied drive signal is located exactly on the top of the optical waveguide. Usually, the z-cut phase modulator has better coupling efficiency between the electric field and the optical waveguide. The V_π of the modulator is still the same as that of Eq. (2.47) with the distance of d taking into account the overlapping between electric and optical fields, i.e., an effective separation of d instead of the physical separation.

Effects of Velocity Mismatch

If the microwave electrode impedance is matched to the connecting cable and signal source, the microwave drive signal along the transmission line is

$$V(z,t) = V_0 \Re\left\{ e^{j\beta_m z - j\omega_m t} \right\}, \tag{2.48}$$

where ω_m is the angular frequency of the microwave signal, and $\beta_m = \omega_m n_m / c + j\alpha_m$ is the propagation constant of the microwave signal with n_m as the effective refractive index of the microwave waveguide, z is the distance from the beginning of the microwave and optical waveguide, and α_m is the loss coefficient of the microwave signal. Traveling with a speed of c/n_r, the photon entering the optical waveguide at any time t_0 experiences a voltage of

$$\begin{aligned} V(z,t_0) &= V_0 \Re\left\{ \exp\left[j\beta_m z - j\omega_m \left(t_0 + \frac{z n_r}{c} \right) \right] \right\} \\ &= V_0 e^{-\alpha_m z} \Re\left\{ e^{j\omega_m d_{12} z - j\omega_m t_0} \right\} \end{aligned} \tag{2.49}$$

at a distance of z from the input of the waveguide, where $d_{12} = (n_m - n_r)/c$ is the walk-off parameter between microwave and optical fields.

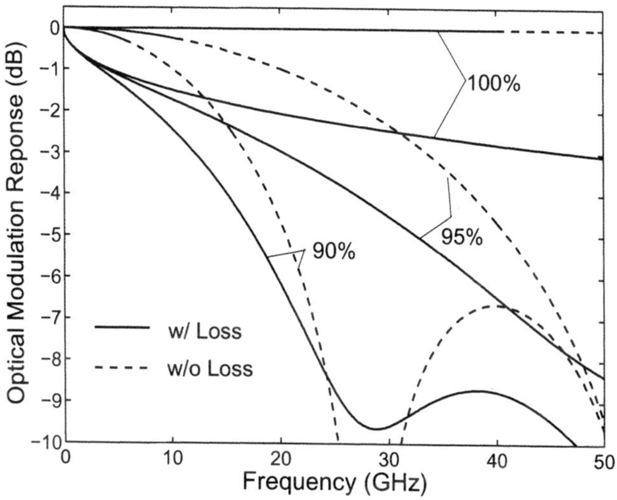

Figure 2.10. Effects of velocity mismatch of a phase modulation.

For a waveguide length of L, the overall phase shift is

$$\Delta\phi(\omega_m) = \frac{\Delta\phi_0}{L_i} \frac{1 - e^{-\alpha_m L + j d_{12}\omega_m L}}{\alpha_m - j d_{12}\omega_m} \tag{2.50}$$

as a function of the modulation frequency of ω_m, where $\Delta\phi_0$ is the phase shift from Eq. (2.46) and $L_i = (1 - e^{-\alpha_m L})/\alpha_m$ is the effective interaction length. For a d.c. voltage with $\omega_m = 0$, the phase shift is $\Delta\phi(0) = \Delta\phi_0$. Without waveguide loss of $\alpha_m = 0$, the frequency response is proportional to the sinc function of $\sin(x)/x$ with $x = d_{12}\omega_m L/2$.

Figure 2.10 shows the frequency response of a phase modulator with velocity mismatch. The microwave loss is assumed to be either without loss of $\alpha_m = 0$ or $\alpha_m = \alpha_{m0}\sqrt{f_m}$ due to skin effect, where $f_m = \omega_m/2\pi$ is the modulation frequency. The loss coefficient is $\alpha_{m0} = 0.2$ dB/cm/$\sqrt{\text{GHz}}$ (Kondo et al., 2002, Sugiyama et al., 2002). The length of the modulator is assumed to be $L = 5$ cm (Sugiyama et al., 2002). With microwave loss, Figure 2.10 shows that the velocity matching of only 95% reduces the bandwidth from 50 GHz to less than 22 GHz. In the design of external modulator, in additional to impedance match and loss reduction, the design of the microwave electrode for velocity matching is an important issue (Gopalakrishnan et al., 1994, Kawano et al., 1991, Koshiba et al., 1999).

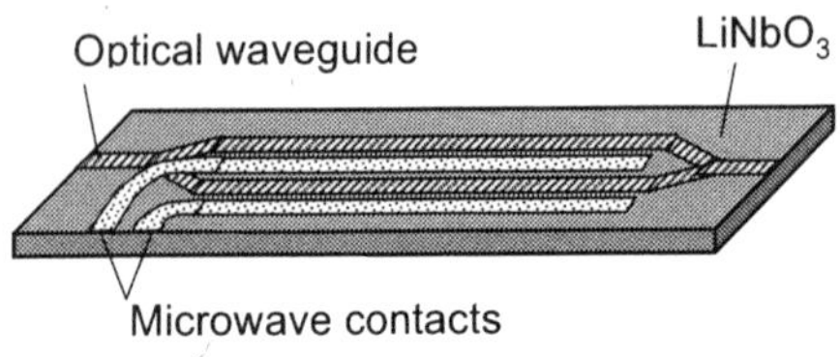

Figure 2.11. Basic schematic of an amplitude modulator based on Mach-Zehnder interferometer.

3.2 Amplitude Modulator

An amplitude-shift keying (ASK) signal of Fig. 2.1 can be generated using an amplitude or intensity modulator to turn-on and -off the light. Amplitude modulator functions as a very fast switch. While semiconductor laser can be directly modulated, direct modulation of a semiconductor laser comes with frequency chirp and limited the transmission distance (Corvini and Koch, 1987). An external amplitude modulator usually provides very a high quality signal.

Figure 2.11 shows an amplitude modulator based on the simple equal path-length Mach-Zehnder interferometer. The input optical signal is splitted into two paths via a **Y** junction. For illustration purpose, one of the optical paths is phase-modulated and another path remains unmodulated. If the **Y** junction splits the input signal of E_i into two equal electric fields of $E_i/\sqrt{2}$ each, ignored the path delay, some constant phase shifts, and waveguide loss, the combined signal at another end of the **Y** junction is

$$\begin{aligned} E_o &= \frac{E_i}{2}\left[1 + \exp\left(j\pi\frac{V(t)}{V_\pi}\right)\right] \\ &= E_i \cos\left[\frac{\pi}{2}\frac{V(t)}{V_\pi}\right] \exp\left[j\frac{\pi}{2}\frac{V(t)}{V_\pi}\right], \end{aligned} \tag{2.51}$$

where V_π is the parameter of the phase modulator at the lower path of Fig. 2.11. From Eq. (2.51), in term of optical intensity, the input and output relationship of a Mach-Zehnder intensity modulator is

$$\frac{|E_o|^2}{|E_i|^2} = \cos^2\left[\frac{\pi}{2}\frac{V(t)}{V_\pi}\right], \tag{2.52}$$

as a nonlinear sinusoidal transfer function shown in Fig. 2.12. The voltage to turn the modulator from minimum to maximum transmission points is V_π. Because of some amount of constant phase shift, $V(t) = 0$ in practical modulator is not necessary corresponding to the maximum transmission point as shown in Figure 2.12.

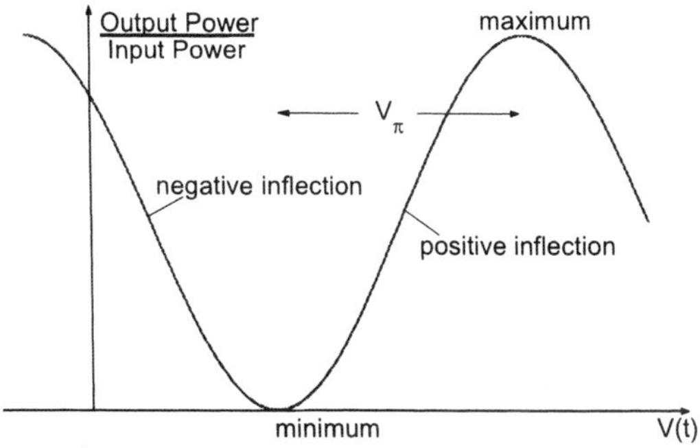

Figure 2.12. Input-output transfer characteristic of a Mach-Zehnder modulator.

However, in additional to the change of optical intensity, the input and output relationship of Eq. (2.51) also accompanies with phase modulation of $\exp[j\phi(t)]$ with $\phi(t) = \pi V(t)/2V_\pi$. The ratio of phase to amplitude modulation is called chirp coefficient (Koyama and Oga, 1988). The Mach-Zehnder modulator in the schematic of Fig. 2.11 has a chirp coefficient of ± 1 with the plus or minus sign depending on whether the amplitude modulator is operated in the positive or negative inflection points of Fig. 2.12, respectively. In additional to chirp, the unmodulated path of Fig. 2.11 reduces the modulation efficiency of the modulator. When the microwave electrodes are properly designed, both paths of the Mach-Zehnder modulator can be modulated to improve the modulator efficiency.

Figure 2.13 shows three different waveguide structures for a Mach-Zehnder modulator using x- and z-cut $LiNbO_3$ crystal. Using the x-cut crystal, the two paths of the Mach-Zehnder modulator are phase-modulated with opposite phase shifts in a push-pull structure. For the input-output relationship of Fig. 2.12, the V_π is reduced by half with respect to the single electrode schematic of Fig. 2.11. Because the two paths are modulated with opposite phases of $\pm\pi V(t)/2V_\pi$ with a parameter of V_π representing the combined V_π of the two phase modulators, the output field corresponding to Eq. (2.51) is

$$\begin{aligned} E_o &= \frac{E_i}{2}\left[\exp\left(-j\pi\frac{V(t)}{2V_\pi}\right) + \exp\left(j\pi\frac{V(t)}{2V_\pi}\right)\right] \\ &= E_i \cos\left[\frac{\pi}{2}\frac{V(t)}{V_\pi}\right]. \end{aligned} \tag{2.53}$$

In the single-drive x-cut push-pull modulator of Fig. 2.13, no phase modulation accompanies the amplitude modulation from Eq. (2.53) and

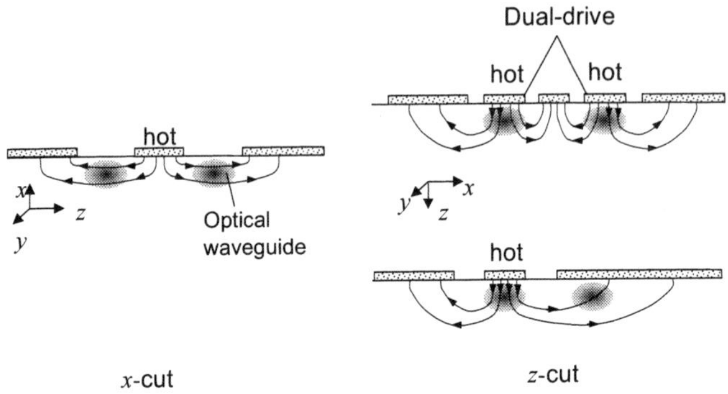

Figure 2.13. Three different structures for Mach-Zehnder modulator using x- or z-cut $LiNbO_3$.

the modulator has zero-chirp. In term of optical intensity, the transfer characteristic follows Eq. (2.52) or Fig. 2.12. The input and output transfer characteristic of Eq. (2.53) may also provide 0 and π phase modulation to an optical single. For $V(t)$ from 0 to V_π, E_o and E_i have the same phase. For $V(t)$ from V_π to $2V_\pi$, E_o and E_i has opposite phases. Phase modulation provided by zero-chirp modulator has the advantage that amplitude jitter does not give phase jitter, i.e., variation of drive voltage does not transfer to variation of output phase. However, the phase modulation is limited to 0 and π, equivalent to ± 1. For phase-modulation operation to minimize loss and obtain optimal amplitude jitter compression, the modulator should operate between two maximum transmission points for a peak-to-peak voltage swing of $2V_\pi$.

Figure 2.13 shows two waveguide configurations using z-cut $LiNbO_3$ crystal. Assume that the two paths have identical structure, the chirp coefficient is equal to (Koyama and Oga, 1988)

$$\alpha = \frac{V_1(t) + V_2(t)}{V_1(t) - V_2(t)}, \tag{2.54}$$

where $V_1(t)$ and $V_2(t)$ is the drive voltage of the two paths, respectively. When the two paths are driven by complementary signal with $V_1(t) = -V_2(t)$, the modulated signal has zero-chirp. In the dual-drive structure, the modulator chirp is adjustable (Gnauck et al., 1991).

The single-drive configuration of Fig. 2.13 for z-cut modulator usually has nonzero chirp coefficient because the electric field lines pass through the two optical waveguides are usually not the same. In normal operation, the chirp coefficient for single-drive z-cut modulator is about ± 0.75 (Schiess and Carlden, 1994). With proper waveguide arrangement, the single-drive z-cut modulator can be designed as zero-chirp modulator with reduced efficiency. If the dual-drive z-cut modulator is operated as a single-drive modulator with one hot electrode connected to ground, the chirp coefficient is also about the same as a single-drive z-cut modulator of about ± 0.75.

In on-off keying for IMDD system, the modulator chirp can be used to compensate for fiber dispersion (Gnauck et al., 1991). In ASK for coherent optical communications, additional phase modulation is not desirable and should be reduced. Currently, most commercially available amplitude modulator is $LiNbO_3$ Mach-Zehnder modulator of Fig. 2.13 (Wooten et al., 2000). Earlier amplitude modulator has various structures besides the Mach-Zehnder interferometer (Alferness, 1981, 1982, 1990, Korotky and Alferness, 1988, Kubota et al., 1980). Instead of just a phase modulator like Fig. 2.8 (Kawano et al., 1989, Tench et al., 1987), after 90's, high-speed modulators are designed mostly for amplitude modulation (Burns et al., 1999, Dolfi et al., 1988, Fujiwara et al., 1990, Howerton et al., 2000, Noguchi et al., 1998, 1995, Wooten et al., 2000). Currently, dual-drive 40 Gb/s modulator with a V_π of less than 2 V has been demonstrated experimentally (Sugiyama et al., 2002). Even experimental x-cut modulator has a V_π less than 3 V (Aoki et al., 2004, Kondo et al., 2002). Commercially available $LiNbO_3$ amplitude modulator has a V_π from 4 to 6 V. In additional to $LiNbO_3$, semiconductor materials also have electro-optical effect and can be used to fabricate external modulator (Alferness, 1981, Cites and Ashley, 1994, Leonberger and Donnelly, 1990, Tsuzuki et al., 2004, Walker, 1987, 1991, Yu et al., 1996). Even silicon has electro-optical effect that can be used for modulator (Dainesi et al., 2000, Jackson et al., 1998, Liu et al., 2004e). External modulator can also use the electro-optical effect in polymer (Chen et al., 1997b, Dalton et al., 1999, Lee et al., 2000, Oh et al., 2001, Shi et al., 1996, 2000). With a refractive index smaller than that of $LiNbO_3$, the speed of electrical and optical signals may be matched in polymer modulator without major engineering efforts. The electro-optic coefficient of other materials are listed in Liu (1996), Saleh and Teich (1991), and Yariv (1997).

The **Y** junction of the Mach-Zehnder modulator of Fig. 2.11 should split the optical signal into two equal parts. If the power splitting ratio is not equal to unity but a factor of, for example, γ_s, the ratio of the power at the maximum to the minimum transmission point of Fig. 2.12 is equal to

$$r_{\text{ex}} = \left(\frac{1+\gamma_s}{1-\gamma_s}\right)^2. \tag{2.55}$$

The ratio of r_{ex} is the extinction ratio. Most commercially available $LiNbO_3$ Mach-Zehnder modulator has an extinction ratio larger than 20 dB at d.c. However, the extinction ratio for the eye-diagram is usually limited to 10 to 12 dB because of waveform ripples of the drive signal. Finite d.c. extinction ratio also induces chirp to the optical signal (Kim and Gnauck, 2002, Walklin and Conradi, 1997).

Semiconductor materials can be used to make electroabsorption modulator (EAM) based on Frank-Keldysh effect for bulk semiconductor and quantum-confined Stark effect for quantum well (Bennett and Soref, 1987, Miller et al., 1986, Wood, 1988). Together with amplitude modulation, electroabsorption modulator always gives a chirped output because of the accompany frequency modulation. Electroabsorption modulator can be integrated with many other components (Akulova et al., 2002, Frateschi et al., 2004, Johansson et al., 2004, Mason et al., 2002) or operated in very high-speed (Akage et al., 2001, Choi et al., 2002, Kawanishi et al., 2001, Miyazaki et al., 2003).

3.3 Operation of Amplitude Modulator

When a Mach-Zehnder modulator is biased at the middle positive inflection point and the drive signal has a peak-to-peak voltage of V_π, the baseband representation of the electric field at the output of the modulator is

$$\begin{aligned} E_o(t) &= \frac{E_i}{2}\left\{\exp\left[\frac{j(1+\alpha)\pi}{2}\frac{V(t)}{V_\pi}\right]\right. \\ &\qquad \left. +j\exp\left[-\frac{j(1-\alpha)\pi}{2}\frac{V(t)}{V_\pi}\right]\right\}, \end{aligned} \tag{2.56}$$

where α is the chirp coefficient, and $V(t)$ is the binary drive signal of

$$V(t) = \sum_{k=-\infty}^{+\infty} \frac{b_k V_\pi}{2} p(t - kT), \tag{2.57}$$

where $b_k = \pm 1$ is the transmitted random data stream, $p(t)$ is the pulse shape of the drive signal, and T is the bit interval of the data. The two terms in Eq. (2.56) correspond to the two phase-modulated paths of the Mach-Zehnder modulator. The differential phase shift between the two phase modulators is $\pi V(t)/V_\pi$. The intensity of the modulator output of

$$|E_o(t)|^2 = |E_i(t)|^2 \sin\left[\frac{\pi}{4} + \frac{\pi}{2}\frac{V(t)}{V_\pi}\right]^2 \tag{2.58}$$

is independent of the chirp coefficient. When $V(t) = -V_\pi/2$, the output of $|E_o(t)| = 0$ locates at the minimum transmission point. When

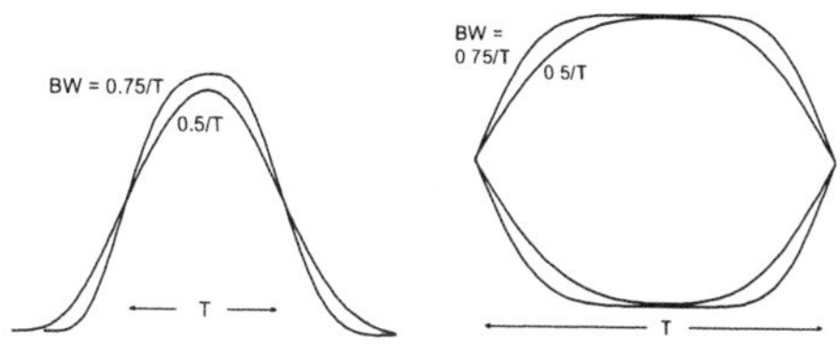

(a) $p(t)$ (b) $|E_o(t)|^2$

Figure 2.14. (a) Bessel-filtered pulse shape and (b) the corresponding eye-diagram of the optical intensity.

$V(t) = +V_\pi/2$, the output of $|E_o(t)| = |E_i(t)|^2$ is at the maximum transmission point. An external modulator is normally biased at the negative inflection point to give negative chirp for dispersion compensation (Gnauck et al., 1991). The bias at positive inflection point is used here for illustration propose.

The relation of Eq. (2.56) can be used to model all types of Mach-Zehnder modulators having different values of chirp coefficient α. For example, a dual-drive modulator has adjustable chirp, a single-drive z-cut modulator has a chirp coefficient of $\alpha = \pm 0.75$, and an x-cut push-pull Mach-Zehnder modulator has zero chirp. Because most external modulators have very large extinction ratio, the expression Eq. (2.56) assumes an infinite extinction ratio.

Ideally, the Mach-Zehnder modulator should be driven by a rectangular pulse that has been filtered by a Bessel low-pass filter. Figures 2.14 show both the pulse shape and the corresponding eye diagram when a fifth-order Bessel filter having a bandwidth of either $0.75/T$ or $0.5/T$ is used[1], where T is the bit interval for the binary signal. Owing to the nonlinear transfer characteristic of the Mach-Zehnder modulator, a bandwidth of $0.5/T$ is sufficient to provide an good eye-diagram in optical intensity, but then the receiver must have a very wide bandwidth to preserve the eye opening to the decision circuits.

While the eye-diagram of Fig. 2.14(b) does not change with the chirp coefficient, the phase of the output electric field of $E_o(t)$ of Eq. (2.56) varies with chirp coefficient. Figure 2.15 shows the electric field locus changing from the off state of $(0, 0)$ to the on state at the unit circle

[1] The fifth-order Bessel filter has a response of

$$H(s) = \frac{945}{945 + 945s + 420s^2 + 105s^3 + 15s^4 + s^5}$$

where $s = jf/f_0$ with f_0 adjust for different bandwidth.

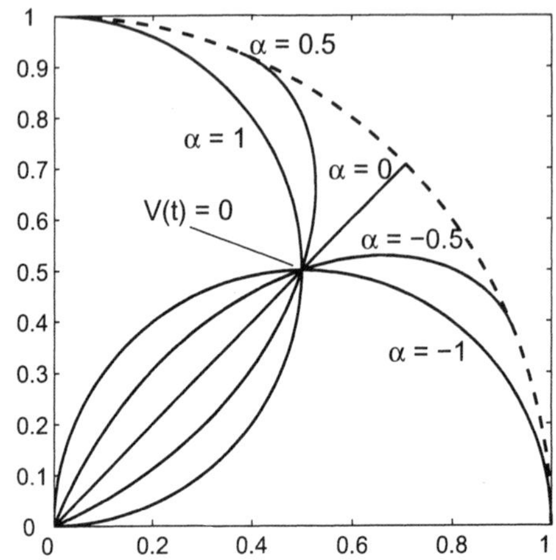

Figure 2.15. The phase change from turn-on to -off of an amplitude modulator biased at the positive middle inflection point. The dashed circle is the on state and the origin is the off state. The middle intersection corresponds the bias of $V(t) = 0$.

shown as dash-line. The intersection point corresponds to the bias point of $V(t) = 0$ located at $(1/2, 1/2)$. Using a zero-chirp modulator with $\alpha = 0$, the electric field has a constant angle of $\pi/4$. However, with a chirp modulator, the phase changes with the instantaneous drive voltage. With a chirp coefficient of $\alpha = \pm 1$, the output electric field changes along a circle ending at $(0, 1)$ and $(1, 0)$, respectively. The locus of Fig. 2.15 depends only on the chirp coefficient but independent to the waveform of the drive signal.

Figures 2.16 show the single-sided optical spectrum of the optical signal modulated by a random pulse stream with a Bessel-filtered pulse given by Fig. 2.14(a). Fifth-order Bessel filters have bandwidths of $0.75/T$ and $0.5/T$ for Figs. 2.16(a) and (b), respectively. Figs. 2.16 also show the optical spectra for different values of the chirp coefficient of $\alpha = 0, \pm 0.5, \pm 1$. For comparison, Figs. 2.16 also plot the electrical spectrum of the drive signal $V(t)$ for comparison.

Modulator chirp broadens the optical spectrum of the signal. Figures 2.16 show that the approximation of the optical spectrum using the electrical spectrum of the drive signal underestimates the spectral spreading. The differences between the power spectra of $E_o(t)$ and $V(t)$ are more significant when the low-pass filter having the small bandwidth of $0.5/T$ is used. Figures 2.16 show that chirp broadens the optical spectrum, causing more broadening when a Bessel filter having smaller bandwidth, corresponding to longer rise and fall times, is used.

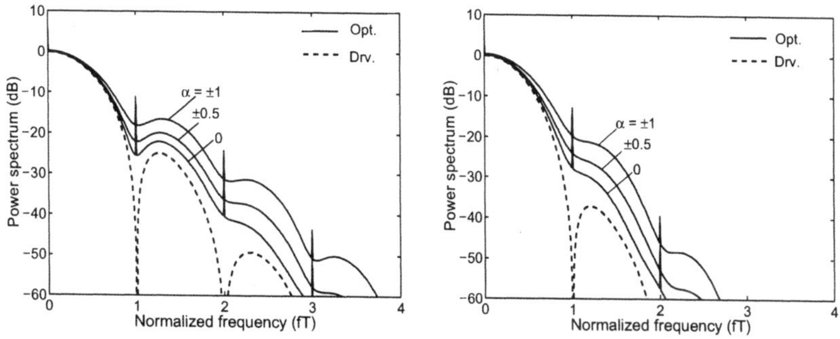

(a) BW $= 0.75/T$ (b) BW $= 0.5/T$

Figure 2.16. The power spectral density of an on-off keying optical signal when a Bessel-filtered pulse is used as the drive signal. The Bessel filter has a bandwidth of (a) $0.75/T$ and (b) $0.5/T$. The dashed lines are the electrical spectrum of the drive signal. [Adapted from Ho (2004e)]

In Figs. 2.16, there are major differences at the integer normalized frequencies of $fT = \pm 1, \pm 2, \ldots$. While the electrical spectrum of the drive signal of $V(t)$ has notches at those normalized frequencies, the optical spectrum has discrete tones at those same frequencies, and also has a tone at $f = 0$ due to the d.c. value of the electric field. The differences between the power spectra of $E_o(t)$ and that of $V(t)$ are more significant when the Bessel filtered pulse has smaller bandwidth. When the pulse bandwidth is $0.75/T$, the difference is about 3 dB at the second lobe of $fT = 1.5$ for $\alpha = 0$. When the pulse bandwidth is $0.5/T$, the second lobe is at about $fT = 1.25$, the difference increases to about 5 to 12 dB even for a chirp coefficient of $\alpha = 0$.

The effect of the chirp coefficient also depends on the pulse bandwidth. With a bandwidth of $0.75/T$, the chirp coefficient does not change the optical power spectral density as much as when the bandwidth is $0.5/T$. By comparing Figs. 2.16(a) and (b), we may also conclude that the modulator chirp has a greater effect for a drive signal having longer rise or fall times.

The operation of amplitude modulator is well-known in most optical communication textbooks (Agrawal, 2002, Keiser, 1999, Liu, 1996). The power spectrum of signal with modulator chirp was first analyzed by Ho and Kahn (2004c) using the method of Greenstein (1977), Ho (1999b), and Ho et al. (2000c). Of course, the optical spectrum is routinely measured or simulated for phase-modulated or IMDD optical communication systems.

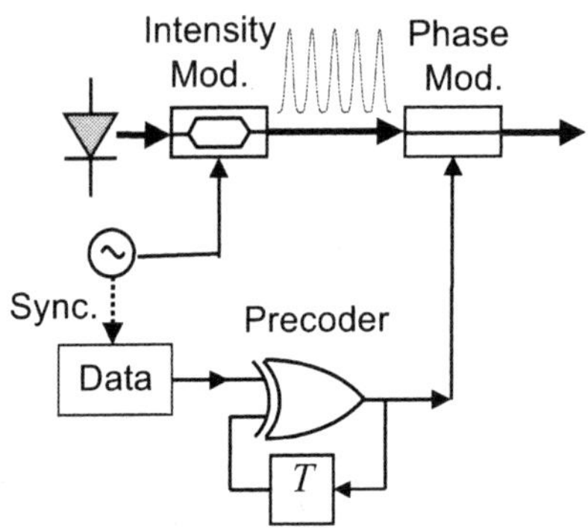

Figure 2.17. Schematic of transmitter to generate RZ-DPSK signals.

For the long-term operation of an amplitude modulator based on $LiNbO_3$, the bias voltage must be actively controlled to compensate for the d.c. drift (Korotky and Veselka, 1996, Nagata, 2000, Nagata et al., 2004). Low-speed controller is usually used for bias control to maintain the optimal biasing point for the modulator.

3.4 Generation of RZ-DPSK Signals

Among all digital modulation formats for phase-modulated optical communications, from Table 1.2, DPSK signal may be the most popular scheme using the phase difference of optical carrier to carry information. DPSK signals typically use return-to-zero (RZ) linecode.

Figure 2.17 is a transmitter to generate RZ-DPSK signal using the cascade of an amplitude and phase modulator. The amplitude or intensity modulator is preference to have zero-chirp and generate an RZ pulse train. The phase modulator modulates the phase of the RZ pulse train to generate the RZ-DPSK signal. Similar to Fig. 1.4(a), the drive signal of the phase modulator should be preceded by a precoder to calculate the cumulative parity of the input sequence. The precoder is required such that the phase difference can be either demodulated using the delay-and-multiplier circuits of Fig. 1.4(b) or the direct-detection interferometer of Fig. 1.4(c). In Fig. 2.17, the phases of all pulses given by the amplitude modulator are assumed to be identical or alternating.

The pulse generator of Fig. 2.17 is an amplitude modulator driven by a sinusoidal signal synchronized with the data source. The drive signal of the phase modulator must be arrived to the modulator at the same time as the pulse for optimal performance.

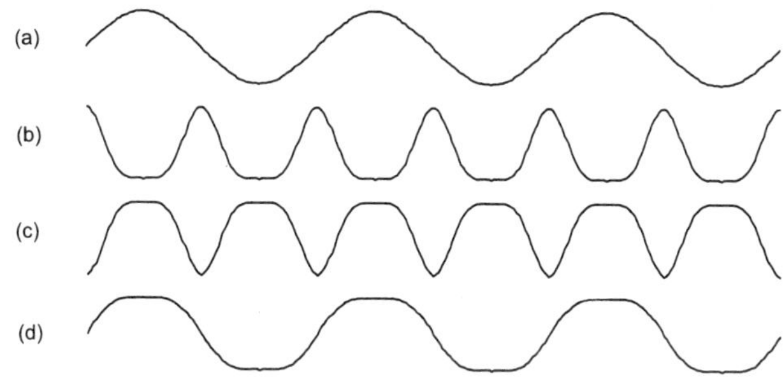

Figure 2.18. Pulse generation for RZ-DPSK modulation: (a) is the driving signal, (b), (c), and (c) are the output intensity. The output pulse trains of (b) and (c) use a driving swing of $2V_\pi$ and bias at maximum and minimum transmission points, respectively. The output pulse trains of (d) use a driving swing of V_π and biases at either the positive or negative inflection point.

Figure 2.18 shows three different methods to generate the pulse train for RZ-DPSK transmitter of Fig. 2.17. Figure 2.18(a) is the sinusoidal drive signal. Figures 2.18(b) and (c) are the output pulse train in optical intensity when the amplitude modulator is biased at the maximum and minimum transmission point of Fig. 2.12, respectively. From Figs. 2.18(b) and (c), when biased at either minimum or maximum transmission points, the pulse rate is twice the frequency of the sinusoidal drive signal of Fig. 2.18(a). In additional to pulse train generation, the frequency doubling property had also been used for electro-optical mixer (Ho et al., 1997, Sun et al., 1996).

The sinusoidal drive signal of Fig. 2.18(a) has a peak-to-peak voltage of $2V_\pi$ for the output pulse trains of Figs. 2.18(b) and (c). The duty cycles of the pulse train of Figs. 2.18(b) and (c) are 1/3 and 2/3, respectively. The pulse train of Fig. 2.18(b) biased at the maximum transmission point has the same optical phase at all pulses. However, the pulse train of Fig. 2.18(c) biased at the minimum transmission point has the opposite optical phase of 0 and π at adjacent pulses. The pulse train of Fig. 2.18(c) is commonly called carrier-suppressed RZ (CSRZ) pulse. Even with opposite phases at adjacent pulses, the precoder of Fig. 2.17 is still the same but operated with an inverted transmitted data, or the received signal can be inverted after the decision circuits.

The pulse train of Fig. 2.18(d) is generated by the sinusoidal drive signal of Fig. 2.18(a) with V_π of peak-to-peak voltage swing. The amplitude modulator is based at either the positive or negative inflection point. The rate of the pulse train of Fig. 2.18(d) is the same as the frequency of the sinusoidal drive signal of Fig. 2.18(a). The phase of all pulses in the pulse train of Fig. 2.18(d) is identical. The duty cycle of the pulse train is 1/2.

To obtain DPSK signal, the phase modulator in Fig. 2.17 may be a simple phase modulator or a zero-chirp amplitude modulator. From the transfer function of Eq. (2.53) for zero-chirp modulator, amplitude modulator can provide phase modulation. In steady state, if $V(t) = \{0, 2V_\pi\}$, the output electric field is $E_0 = \pm E_i$ with a phase difference of 180°. Note that amplitude ripples deviated from 0 and $2V_\pi$ do not give a phase other than 0 and 180°. The nonlinear transfer characteristic of Eq. (2.52) also gives intensity compression to the output signal. The amplitude ripples are compressed in the output intensity.

In additional to the dual-modulator configuration of Fig. 2.18 to generate RZ-DPSK signal, a single-drive zero-chirp modulator with push-pull drive signal can be used to generate NRZ-DPSK signal. The drive signal is a NRZ signal with a peak-to-peak voltage swing of $2V_\pi$. The modulator must be biased at the minimum transmission point. The drive voltage swings between two maximum transmission points of the transfer characteristic of Fig. 2.12 with phase difference of 180°. The operation of the modulator with and without the phase difference of 180° generates NRZ-DPSK signal. Normally, the modulator is operated with a bias at V_π with a signal swing between $\pm V_\pi$.

The transmitter of RZ-DPSK signal similar to Fig. 2.17 was first used by Gnauck et al. (2002) and explained in more detail in both Xu et al. (2004) and Gnauck and Winzer (2005). The locations of RZ and DPSK modulators of Fig. 2.17 can be interchanged without changing the function of the transmitter. Wen et al. (2004) proposed method to generate RZ-DPSK signal using only a single dual-drive modulator.

4. Direct Frequency Modulation of a Semiconductor Laser

The laser frequency is determined by the mode spacing of $c/2L_l n_r$, where the refractive index of $n_r = n_0 + \Delta n_r(n_c)$ is a function of carrier density of n_c. If the lasing mode remains the same, the frequency changes by a factor of $\Delta n_r(n_c)/n_0$ due the change of the laser refractive index. If the phase modulator discussed in Sec. 2.3.1 is placed inside the laser cavity, frequency modulated laser output can also be generated.

From the analysis of Sec. 2.2.2, a semiconductor laser can be directly frequency modulated with a frequency response given by Eq. (2.25). The frequency response of Eq. (2.25) was derived based on the rate equations of Eqs. (2.9) and (2.10) for the dynamic characteristic of a semiconductor laser. However, the thermal effect of carrier heating and cooling due to current injection is ignored in the rate equations of Eqs. (2.9) and

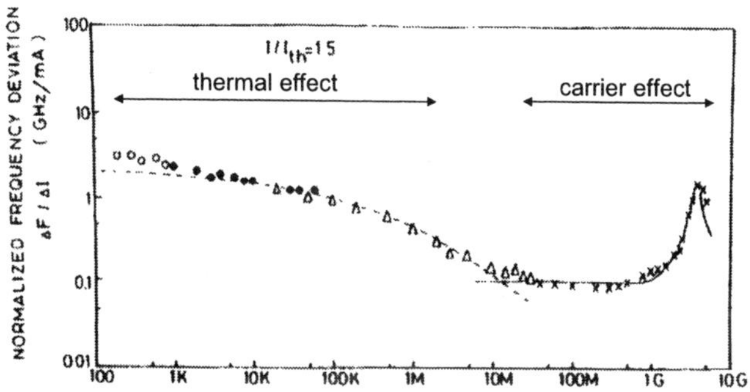

Figure 2.19. The measured small-signal frequency response of semiconductor lasers. [From Kobayashi et al. (1982), © 2005 IEEE]

(2.10). In Petermann (1991), taking into account the thermal effect, the response of Eq. (2.25) becomes

$$H_{\dot{\phi}}(\omega_m) = \frac{j\omega_m \omega_R^2}{\omega_R^2 - \omega_m^2 + j\gamma_e \omega_m} + k_{\rm th} \frac{1}{j\omega_m + \omega_{\rm th}}, \tag{2.59}$$

where $k_{\rm th}$ is the relative strength of the thermal effects and $\omega_{\rm th}$ is the first-order cut-off frequency due to thermal effect.

Figure 2.19 shows the measured frequency response of a semiconductor laser including both the thermal and carrier response, conformed closely to the transfer characteristic of Eq. (2.59). The thermal effect is similar to a first-order response from the second term of Eq. (2.59). The carrier effect is mainly the same as the first term of Eq. (2.59) or (2.25) but the low frequency part is constant instead of falling down to zero due to some large signal effect (Kobayashi et al., 1982).

In order to use the frequency response like that of Fig. 2.19 for frequency modulation, equalization circuits can be used to flatten the response (Alexander et al., 1989, Gimlett et al., 1987, Iwashita et al., 1986). Another method is to use coding to match the spectral density of the data with the response of frequency modulation (Emura et al., 1984, Hooijmans et al., 1990, Noé et al., 1989, Vodhanel and Enning, 1988, Vodhanel et al., 1988).

Some laser structures provide a flat frequency-modulation response. When the reflector section of a DBR laser is modulated by injection current, the frequency response is very flat. A multi-electrode DFB laser can inject complementary currents to different electrodes to eliminate the thermal effect and provide flat frequency-modulation (Kobayashi et al., 1991).

Direct frequency modulation of a semiconductor laser gives continuous-phase FSK (CPFSK) signal. Due to the availability of low-cost semiconductor laser as the signal source, CPFSK was one of the most popular signal formats in early studies of coherent optical communications (Emura et al., 1990a, Imai et al., 1991, Iwashita and Matsumoto, 1987, Iwashita and Takachio, 1988, Ryu et al., 1991b, Saito et al., 1992, Takachio et al., 1989). Recently, there are suggestions to use CPFSK signal for signal transmission (Idler et al., 2004). In the system of Idler et al. (2004), the CPFSK receiver and the optical fiber are both linear optical system. When the optical filter for the CPFSK signal followed the transmitter, a high quality on-off keying signal can be generated.

In additional to direct frequency modulation, a semiconductor laser can also be directly phase-modulated by injection locking (Kobayashi and Kimura, 1982, Kobayashi et al., 1991). However, the requirement of injection locking does not give a simple transmitter implementation.

5. Summary

This chapter briefly introduces methods to generate an optical signal with modulation in amplitude, phase, or frequency to carry digital information. Semiconductor lasers are widely used as the light source for typical lightwave communication systems that also enable directly either amplitude or frequency modulation.

To provide better signal quality, external modulator is used to apply amplitude or phase modulation to an optical signal. External modulator based on electro-optical effect is also discussed in this chapter. Amplitude modulator is mostly used for on-off keying signal to block or unblock an optical signal. Amplitude modulator is also able to generate either PSK or DPSK signals for a phase difference of either 0 or π.

The popular RZ-DPSK signals are generated by the cascade of a RZ and equivalently a phase modulator. The RZ duty cycle is typically 1/3, 1/2, and 2/3.

Chapter 3

COHERENT OPTICAL RECEIVERS AND IDEAL PERFORMANCE

Coherent detection of optical signal is first used for its superior receiver sensitivity compared to on-off keying. Equivalently speaking, the mixing of received signal with the local oscillator (LO) laser functions as an optical amplifier without noise enhancement. Even with the advances of Erbium-doped fiber amplifiers (EDFA), coherent detection can still provide better receiver sensitivity than amplified on-off keying. In this chapter, various structures and architectures of optical receiver are studied for phase-modulated or coherent optical communications. This chapter focuses on performance analysis to validate the receiver sensitivity improvement of phase-modulated optical communications, mainly for binary signals. However, for binary signals limited by amplifier noises, the improvement is only up to about 3 dB.

In a coherent receiver, phase-locked loop (PLL) may be required to track the phase of the received signal. In coherent optical communications, receiver with phase tracking is called synchronous receiver. Equivalent to the matched filter receiver in digital communications (Proakis, 2000), receivers with phase tracking always provide the optimal performance, at least at the linear regime.

Without phase tracking, the received signal has a random phase and can be detected based on the power or the envelope of the signal (Proakis, 2000). This type of noncoherent receiver is called asynchronous receiver in coherent optical communications. While the performance is typically inferior to synchronous receiver, asynchronous receiver has simple structure and provides low-cost implementation.

Coherent optical signal can also directly be detected without mixing with LO signal. While on-off keying or, equivalently, amplitude-shift keying (ASK) signal is directly detected by a photodiode, both phase-

or frequency-modulated signals have direct-detection receiver based on interferometer or optical filter. Direct-detection receiver may be the simplest receiver with low-cost implementation. However, direct-detection receiver for both phase- or frequency-modulated signal is still more complicated than a single photodiode to detect the presence or absence of light for on-off keying signal.

Another types of noncoherent receiver use phase-diversity techniques that combines two signals with a phase difference of 90°. Phase-diversity receiver is asynchronous receiver without the requirement of phase tracking. Equivalently speaking, phase-diversity receiver implements the envelope or power detection by combining optical and electrical techniques.

Other than polarization-diversity receivers, the mixing of two optical signals requires the alignment of their polarizations. In general, polarization alignment is provided by polarization controller. With optimal combining of the signal from both polarizations, polarization-diversity receivers have the same performance as receiver with polarization tracking.

Various types of coherent optical receiver will be discussed in this chapter.

1. Basic Coherent Receiver Structures

The basic structures of phase-shift keying (PSK) and differential phase-shift keying (DPSK) receivers have been shown in Figs. 1.3 and 1.4, respectively. This section studies each basic type of coherent optical receivers that mix the received signal with the LO laser. The signal-to-noise ratio (SNR) of each receiver type is derived, especially for systems limited by amplifier noises. The SNR is equal to the ratio of optical signal to the amplifier noise per polarization over an optical bandwidth equal to the data rate. For system without optical amplifiers, the SNR is equal to the number of photons per bit for heterodyne receiver.

1.1 Single-Branch Receiver

Figure 3.1 shows a typical structure of a single-branch coherent optical receiver. To enable optical signal mixing, the polarization of the received signal must be aligned to that of the LO laser. In Fig. 3.1, automatic polarization control (APC) is used to align the polarization of the received signal to that of the LO laser for optimal signal mixing. In general, the LO laser is phase or frequency locked to the received signal. Phase locking is used for homodyne receiver and frequency locking is required for heterodyne receiver for a fixed intermediate frequency (IF). Phase locking is facilitated by an optical PLL and frequency locking

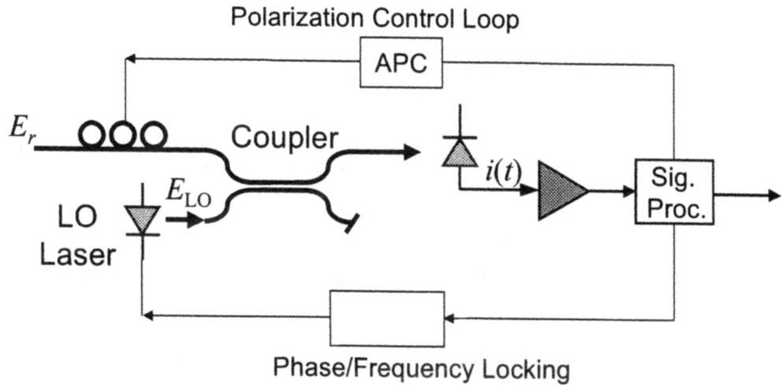

Figure 3.1. A single-branch coherent optical receiver.

is provided by an automatic frequency control (AFC) loop. The single-branch receiver is the simplest receiver structure. Most early heterodyne receivers used single-branch receiver for its simplicity (Goodwin, 1967, Nussmeier et al., 1974, Oliver, 1961, Peyton et al., 1972, Saito et al., 1981). Instead of 3-dB coupler, those early works used a power beam splitter or a power combiner, functioning as an 180° optical hybrid, to mix the received signal with the LO laser.

A 3-dB coupler is used in Fig. 3.1 to mix the received signal with the LO laser. The input and output relationship of a 3-dB coupler is[1]

$$S = \frac{1}{\sqrt{2}} \begin{bmatrix} 1 & 1 \\ 1 & -1 \end{bmatrix}. \tag{3.1}$$

The received signal is assumed to be

$$E_r(t) = [A_s(t)e^{j\phi_s(t)} + n_x(t)]e^{j\omega_c t}\mathbf{x} + n_y(t)e^{j\omega_c t}\mathbf{y}, \tag{3.2}$$

where $A_s(t)$ and $\phi_s(t)$ are the modulated amplitude and phase of the transmitted signal, respectively, ω_c is carrier frequency of the signal, $\mathbf{x}$ is the polarization of the signal, and $\mathbf{y}$ is the polarization orthogonal to $\mathbf{x}$, $n_x(t)$ and $n_y(t)$ are the amplifier noise in the polarizations of $\mathbf{x}$ and $\mathbf{y}$, respectively. The complex signal of $A_s(t)e^{j\phi_s(t)}$ is the low-pass representation of the signal. $A_s(t)e^{j\phi_s(t)} = \pm A$ for both PSK and DPSK signals

[1]The input and output relationship can also be

$$S = \frac{1}{\sqrt{2}} \begin{bmatrix} 1 & j \\ 1 & -j \end{bmatrix}.$$

without changing the phase relationship. Different input and output relationship for an 180° optical hybrid may be used in this book for convenience.

and $A_s(t)e^{j\phi_s(t)} = \{0, A\}$ for ASK signals, where A is the positive signal amplitude. Both $n_x(t)$ and $n_y(t)$ are the low-pass representation of the amplifier noise. Here, in this chapter, we assume that both the transmitted and LO lasers are a pure coherent source without phase noise. In next chapter, the impact of laser phase noise to phase-modulated signal is analyzed for further detail. The received signal may have a random phase of θ_0 of $A_s(t)e^{j\phi_s(t)+j\theta_0}$ due to propagation delay. For system with phase tracking using PLL, we assume that $\theta_0 = 0$ for simplicity when the phase of LO laser tracks out the phase of θ_0. For system without phase tracking, the phase of θ_0 is included when necessary.

The optical SNR of the received signal is defined as

$$\mathrm{SNR}_{o,s} = \frac{E\{|A_s(t)|^2\}}{E\{|n_x(t)|^2\} + E\{|n_y(t)|^2\}} = \frac{P_r}{2S_{n_s}\Delta f_{o,s}}, \tag{3.3}$$

where P_r is the received power, S_{n_s} is the spectral density of the received spontaneous emission in each polarization, and $\Delta f_{o,s}$ is the optical bandwidth of the received optical filter. In the complex representation of $n_x(t) = n_{x1}(t) + jn_{x2}(t)$, the spectral densities of $n_{x1}(t)$ and $n_{x2}(t)$ are the same and equal to $S_{n_s}/2$.

The LO signal is

$$E_{\mathrm{LO}}(t) = [A_L + n_L(t)]e^{j\omega_{\mathrm{LO}}t}\mathbf{x}, \tag{3.4}$$

where A_L is the continuous-wave amplitude of LO laser, $n_L(t)$ is the noise of LO laser in the same polarization as the signal, and ω_{LO} is the angular frequency of LO laser. The polarization of the received signal of Eq. (3.2) and the LO laser of Eq. (3.4) is assumed to be the same using APC. The noise of $n_L(t)$ may originate from the optical amplifier used to boost up the LO power or from the relative intensity noise (RIN) of the LO laser. In the LO electric field of Eq. (3.4), the noise at the polarization orthogonal to the signal is ignored here for simplicity. In practice, the noise from polarization orthogonal to the signal can be filtered by a polarizer, especially at a receiver having a well-controlled LO laser.

In the single-branch receiver of Fig. 3.1, the electric field at the input of the photodiode is $[E_r(t) + E_{\mathrm{LO}}(t)]/\sqrt{2}$, the photocurrent is

$$i(t) = \frac{R}{2}\left|E_r(t) + E_{\mathrm{LO}}(t)\right|^2 + i_{\mathrm{sh}} + i_{\mathrm{th}}, \tag{3.5}$$

where R is the responsivity of the photodiode, i_{sh} is the photocurrent shot-noise, and i_{th} is the thermal noise of the receiver. In later part of this chapter, other than specific, we ignored thermal noise for simplicity.

As an additive noise, the effect of thermal noise can be added to the signal afterward. In coherent optical communication systems with LO laser, thermal noise has less impact than both shot and amplifier noises.

The photodiode responsivity is equal to

$$R = \eta \frac{e}{\hbar \omega_c}, \tag{3.6}$$

where $e = 1.6 \times 10^{-19}$ C is the charge per electron, $\hbar\omega_c$ is the energy per photon, where $\hbar = h/(2\pi)$ with $h = 6.63 \times 10^{-34}$ J s as the Planck constant, η is the quantum efficiency of photodiode that is the average number of electrons generated per a photon by the photodiode.

The photocurrent of Eq. (3.5) is equal to

$$\begin{aligned} i(t) &= \frac{R}{2}\left\{|A_L + n_L(t)|^2 + |A_s(t)e^{j\phi_s(t)} + n_x(t)|^2 + |n_y(t)|^2\right\} \\ &\quad + RA_LA_s(t)\cos[\omega_{\mathrm{IF}}t + \phi_s(t)] \\ &\quad\quad + R\Re\left\{\left[A_L n_x(t) + A_s(t)e^{j\phi_s(t)} n_L(t)\right] e^{j\omega_{\mathrm{IF}}t}\right\} + i_{\mathrm{sh}}. \end{aligned} \tag{3.7}$$

In the photocurrent of Eq. (3.7), the intermediate frequency (IF) is $\omega_{\mathrm{IF}} = \omega_c - \omega_{\mathrm{LO}}$. Homodyne system has $\omega_{\mathrm{IF}} = 0$ but heterodyne system has $\omega_{\mathrm{IF}} \neq 0$. The photocurrent of Eq. (3.7) includes the intensity of LO laser with noise of $|A_L + n_L(t)|^2$, the intensity of received signal with noise of $|A_s(t)e^{j\phi_s(t)} + n_x(t)|^2$, the intensity of spontaneous emission from orthogonal polarization of $|n_y(t)|^2$, the beating of received signal with LO laser of $RA_LA_s(t)\cos[\omega_{\mathrm{IF}}t + \phi_s(t)]$, LO and spontaneous beating of $A_L n_x(t)$, signal and LO-spontaneous beating of $A_s(t)n_L(t)$, and together with shot noise i_{sh}. In the photocurrent of Eq. (3.7), the small effect of the beating of spontaneous noise with shot noise is ignored. If necessary, thermal noise can add to the photocurrent of Eq. (3.7).

In the photocurrent of Eq. (3.7), the signal component is

$$s(t) = RA_LA_s(t)\cos[\omega_{\mathrm{IF}}t + \phi_s(t)]. \tag{3.8}$$

For heterodyne system with $\omega_{\mathrm{IF}} = \omega_c - \omega_{\mathrm{LO}} \neq 0$, the signal power is

$$P_s = \frac{R^2}{2} P_{\mathrm{LO}} P_r, \tag{3.9}$$

where $P_{\mathrm{LO}} = A_L^2$ is the LO power, $P_r = E\{|A_s(t)|^2\}$ is the received power. In PSK homodyne system with $\omega_c = \omega_{\mathrm{LO}}$, we obtain

$$P_s = R^2 P_{\mathrm{LO}} P_r. \tag{3.10}$$

Comparing the signal power of Eq. (3.9) for heterodyne system with the signal power of Eq. (3.10) for homodyne system, homodyne system has twice the signal power of heterodyne system. Conventionally, it was generally believed that homodyne system is 3-dB better than the corresponding heterodyne system (Betti et al., 1995, Hooijmans, 1994, Okoshi and Kikuchi, 1988, Ryu, 1995). However, for system dominated by optical amplifier noise, heterodyne system generally has the same performance as similar homodyne system.

The dominant noise source for the single-branch receiver depends on the system configuration. In the usual case that the LO power is significantly larger then the received power, i.e., $P_{\mathrm{LO}} \gg P_r$, the dominant noise is usually LO-spontaneous beating noise, given by $R\Re\{A_L n_x(t) e^{j\omega_{\mathrm{IF}} t}\}$ in the photocurrent of Eq. (3.7).

If the spontaneous emissions of LO and received signal are not negligible with $n_L(t) \neq 0$ and $n_x(t) \neq 0$, the LO-spontaneous beating including two components of $R\Re\{A_L n_L(t)\}$ from $\frac{1}{2}R|A_L + n_L(t)|^2$ and $R\Re\left\{A_L n_x(t) e^{j\omega_{\mathrm{IF}} t}\right\}$. If the optical bandwidth of $n_L(t)$ is $\Delta f_{o,L}$ centered around ω_L, the bandwidth of $R\Re\{A_L n_L(t)\}$ is about $\Delta f_{o,L}/2$. For heterodyne systems with $f_{\mathrm{IF}} = \omega_{\mathrm{IF}}/(2\pi) \neq 0$ and $f_{\mathrm{IF}} > \frac{1}{2}\Delta f_{o,L} + B_d$, an electric filter can be designed such that the beating of A_L with $n_L(t)$ does not affect the performance of the system, where B_d is the symbol rate of the data channel. In homodyne system with $\omega_{\mathrm{IF}} = f_{\mathrm{IF}} = 0$, the beating of A_L with $n_L(t)$ always contributes to the noise of the system.

In the LO and spontaneous noise $n_x(t)$ beating of $R\Re\left\{A_L n_x(t) e^{j\omega_{\mathrm{IF}} t}\right\}$, if the optical bandwidth of $n_x(t)$ is $\Delta f_{o,s}$ centered around ω_c, the upper and lower frequency of LO-spontaneous beating noise is $f_{\mathrm{IF}} \pm \frac{1}{2}\Delta f_{o,s}$. Figure 3.2 illustrates the effect of the optical filter bandwidth of $\Delta f_{o,s}$ on the beating of LO laser and spontaneous emission of $n_x(t)$. The optical filter should have a center frequency align with the optical signal.

Without loss of generality and assume that $f_{\mathrm{IF}} > 0$ as in Fig. 3.2, the lower frequency of $f_{\mathrm{IF}} - \frac{1}{2}\Delta f_{o,s}$ may be a negative frequency for large $\Delta f_{o,s}$. The negative frequency noise affects the system if $f_{\mathrm{IF}} - \frac{1}{2}\Delta f_{o,s}$ falls into the data bandwidth of $-f_{\mathrm{IF}} + B_d$.

The upper two traces of Fig. 3.2 show the case when the optical bandwidth of $\Delta f_{o,s}$ is small and comparable to twice the data bandwidth of B_d. The beating noise is band-pass noise centered at f_{IF}. The lower two traces of Fig. 3.2 show the case when the optical bandwidth of $\Delta f_{o,s}$ is significantly larger than the data bandwidth of B_d. With $f_{\mathrm{IF}} - \frac{1}{2}\Delta f_{o,s} < 0$, the beating noise may extend to $-f_{\mathrm{IF}} + B_d$. In the worst case of having a wide-bandwidth receiver optical filter, the beating noise centered at f_{IF} with a bandwidth of $2B_d$ is doubled compared with case of narrow-bandwidth optical filter. In the diagram of Fig. 3.2, the data bandwidth is assumed to be $2B_d$. In practical system, depending on linecode or spectral filtering, the data bandwidth may vary.

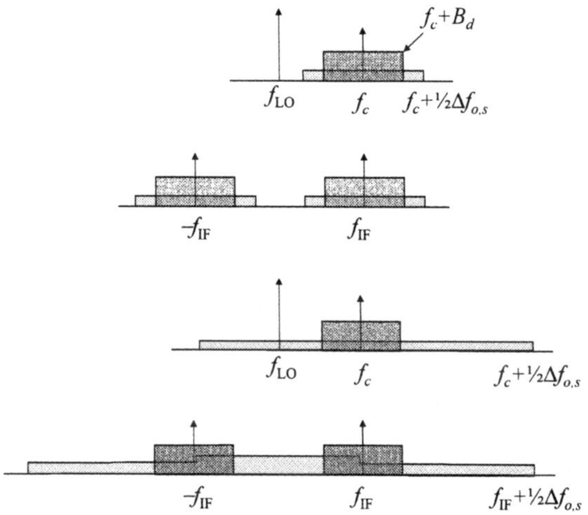

Figure 3.2. An illustration of the effects of receiver filter bandwidth of $\Delta f_{o,s}$ on LO-spontaneous beat noise. The optical bandwidth of $\Delta f_{o,s}$ is comparable (upper two traces) and larger than (lower two traces) than twice the data bandwidth.

The beating noise is not doubled if

$$-f_{\mathrm{IF}} + \frac{1}{2}\Delta f_{o,s} < f_{\mathrm{IF}} - B_d \tag{3.11}$$

or

$$f_{\mathrm{IF}} > \frac{1}{4}\left(\Delta f_{o,s} + 2B_d\right). \tag{3.12}$$

In homodyne system, the LO and spontaneous noise $n_x(t)$ beating is within the frequencies of $\pm\Delta f_{o,s}/2$ and contributes to the positive and negative frequency noise twice. Because the power of homodyne receiver is twice that of heterodyne receiver by comparing Eq. (3.9) and (3.10), with the condition of Eq. (3.12), the output SNR of a single-branch homodyne and heterodyne receivers is the same. However, without the condition of Eq. (3.12), the noise of homodyne and heterodyne system is the same, homodyne system is 3 dB better than heterodyne system due to its higher power.

Under the condition of Eq. (3.12), heterodyne and homodyne systems have the same performance. With proper design, a coherent system with optical amplifiers is generally limited by the beating of LO-spontaneous emission beating between A_L and spontaneous emission of $n_x(t)$. When the optical filter has a bandwidth of $\Delta f_{o,s} = 2B_d$, the IF must be larger than $f_{\mathrm{IF}} > B_d$. In the limit, the narrowest optical bandwidth is actually

$\Delta f_{o,s} = B_d$ for an optical match filter. However, for return-to-zero (RZ) signal with small duty cycle, the optical bandwidth of an optical match filter is larger than the data-rate of B_d.

In general, the condition of Eq. (3.12) is a good approximation for most practical cases. While the bandwidth of $f_{\mathrm{IF}} \pm B_d$ is a good approximation, ideal sinc pulse has the bandwidth of $f_{\mathrm{IF}} \pm B_d/2$ but RZ pulse has a bandwidth wider than $2B_d$.

Assuming the condition of Eq. (3.12), the LO-spontaneous beating noise is

$$RA_L n_{x1}(t)\cos(\omega_{\mathrm{IF}}t) - RA_L n_{x2}(t)\sin(\omega_{\mathrm{IF}}t), \tag{3.13}$$

where $n_s(t) = n_{x1}(t) + jn_{x2}(t)$ with $E\{n_{x1}^2(t)\} = E\{n_{x2}^2(t)\} = \frac{1}{2}S_{n_s}B_d$. The LO laser source of A_L beating with $n_x(t)$ induces an electrical noise with spectrum density of

$$N_{A_L-n_s} = \frac{1}{2}R^2 S_{n_s} P_{\mathrm{LO}}. \tag{3.14}$$

The LO laser source of A_L beating with $n_L(t)$ induces electrical noise with spectrum density of

$$N_{A_L-n_L} = \frac{1}{2}R^2 S_{n_L} P_{\mathrm{LO}}, \tag{3.15}$$

where S_{n_L} is the spectrum density of the spontaneous of $n_L(t)$ at optical domain.

The optical SNR of LO laser source is

$$\mathrm{SNR}_{o,\mathrm{LO}} = \frac{P_{\mathrm{LO}}}{2S_{n_L}\Delta f_{o,L}}, \tag{3.16}$$

where the amplifier noise is in the same polarization as the signal and the orthogonal polarization, in contrast to the LO electric field of Eq. (3.4).

Because the receiver by itself should not induce noise into the system, the system must be designed for $N_{A_L-n_s} > 10 \times N_{A_L-n_L}$, or $S_{n_s} > 10 \times S_{n_L}$, where the factor of 10 is for the condition that the additional penalty due to $n_L(t)$ is less than 0.5 dB. For optical SNR defined for the same bandwidth of $\Delta f_{o,s} = \Delta f_{o,L}$, the system requirement is

$$\frac{\mathrm{SNR}_{o,\mathrm{LO}}}{\mathrm{SNR}_{o,s}} > 10 \times \frac{P_{\mathrm{LO}}}{P_r}, \tag{3.17}$$

with the condition of $P_{\mathrm{LO}} \gg P_r$, the required optical SNR of the LO source must be far larger than that of the received signal.

If the optical SNR of LO source is not substantially larger than that of the signal source, the system is dominated by noise from the LO source.

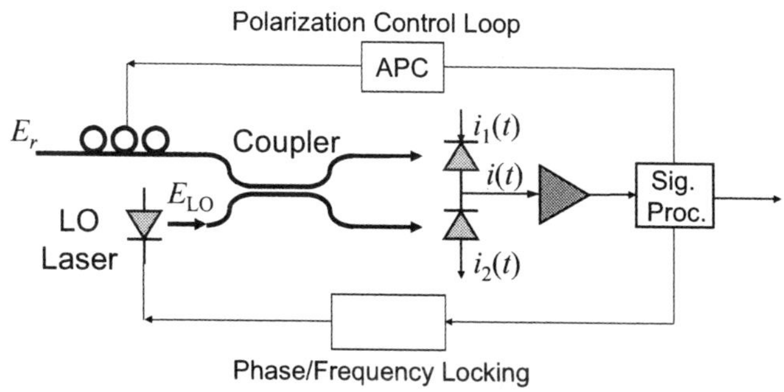

Figure 3.3. A dual-photodiode balanced receiver for phase-modulated optical communications.

The noise of the LO source may be induced by an optical amplifier that boosts up the power of the LO laser or the RIN of the LO laser. In order to eliminate the effects of the LO laser noise, a balanced receiver can be used instead. Because balanced receiver has better performance than single-branch receiver, the analysis of this book usually assumes a balanced receiver.

In the receiver of Fig. 3.1, many methods can control the polarization though APC (Aarts and Khoe, 1989, Martinelli and Chipman, 2003, Noé et al., 1988a, 1991, Okoshi, 1985, Walker and Walker, 1990). The polarization control algorithm should be able to provide endless polarization control without reset. While early homodyne or heterodyne coherent optical communication systems used single-branch receiver, balanced receiver is more popular for its superior performance.

1.2 Balanced Receiver

Figure 3.3 shows a dual-photodiode balanced receiver that increases the signal power and eliminates the noise from the LO laser source. Similar to single-branch receiver of Fig. 3.1, balanced receiver of Fig. 3.3 requires both polarization alignment using APC and phase or frequency locking. The electric field at the input of the upper photodiode is $[E_r(t)+E_{\mathrm{LO}}(t)]/\sqrt{2}$, the photocurrent is the same as Eq. (3.5) and is equal to

$$i_1(t) = \frac{R}{2}\left|E_r(t) + E_{\mathrm{LO}}(t)\right|^2 + i_{\mathrm{sh1}}, \tag{3.18}$$

where i_{sh1} is the shot noise. The electric field at the input of the lower photodiode is $[E_r(t) - E_{\mathrm{LO}}(t)]/\sqrt{2}$, the photocurrent is

$$i_2(t) = \frac{R}{2}\left|E_r(t) - E_{\mathrm{LO}}(t)\right|^2 + i_{\mathrm{sh2}}, \tag{3.19}$$

where $i_{\mathrm{sh}2}$ is the shot noise.

In both photocurrents of Eqs. (3.18) and (3.19), we assume that the two photodiodes are identical with the same responsivity and the 3-dB coupler or 180° optical hybrid is balanced 3-dB without excesses loss. The overall photocurrent is

$$i(t) = i_1(t) - i_2(t) = 2R\Re\{E_r(t) \cdot E_{\mathrm{LO}}^*(t)\} + i_{\mathrm{sh}} \tag{3.20}$$

or

$$\begin{aligned} i(t) = {} & 2RA_L A_s(t)\cos[\omega_{\mathrm{IF}}t + \phi_s(t)] \\ & + 2R\Re\left\{\left[A_L n_x(t) + A_s(t)e^{j\phi_s(t)}n_L(t)\right]e^{j\omega_{\mathrm{IF}}t}\right\} + i_{\mathrm{sh}}, \end{aligned} \tag{3.21}$$

where $i_{\mathrm{sh}} = i_{\mathrm{sh}1} - i_{\mathrm{sh}2}$ is the overall shot noise.

In the photocurrent of Eq. (3.21), the noise from the LO laser of $n_L(t)$ contributes to the system noise because of the beating term of $2R\Re\{A_s(t)n_L(t)e^{j\omega_{\mathrm{IF}}t+j\phi_s(t)}\}$ with a spectral density of

$$N_{A_s-n_L} = 2R^2 S_{n_L} P_r. \tag{3.22}$$

Comparing with the LO-spontaneous beating noise of

$$N_{A_L-n_s} = 2R^2 S_{n_s} P_{\mathrm{LO}}, \tag{3.23}$$

as long as the optical SNR has the relationship of $\mathrm{SNR}_{o,\mathrm{LO}} > 10 \times \mathrm{SNR}_{o,s}$ or $N_{A_L-n_s} > 10 \times N_{A_s-n_L}$, the signal and LO noise beating does not provide a penalty more than 0.5 dB. Even when a booster amplifier is used, the LO laser required low-gain optical amplifier but the signal usually passes through a chain of many optical amplifiers. The optical SNR of the LO signal is usually much larger than that of the received signal. The signal and spontaneous emission beating of $N_{A_s-n_L}$ is usually much smaller than the LO-spontaneous emission beating of $N_{A_L-n_s}$.

The signal component of the photocurrent of Eq. (3.21) is

$$s(t) = 2RA_L A_s(t)\cos[\omega_{\mathrm{IF}}t + \phi_s(t)] \tag{3.24}$$

with a signal amplitude twice of that in single-branch receiver of Eq. (3.8), giving four times larger received power. Since the noise is also increased by the same factor, balanced receiver does not increase the SNR for receiver limited by the beating of LO-spontaneous emission.

When the LO-spontaneous beating of $2R\Re\{A_L n_x(t)e^{j\omega_{\mathrm{IF}}t}\}$ is the dominant noise, similar to the illustration of Fig. 3.2 for single-branch receiver, a heterodyne balanced receiver also requires an optical filter with the condition of Eq. (3.12) without doubling the LO-spontaneous beating noise. For systems limited by optical amplifier noise, the performance of heterodyne and homodyne receiver is the same with the condition of Eq. (3.12).

Here, the SNR of a heterodyne system is evaluated with amplifier noise and the condition of Eq. (3.12) for a balanced receiver, the signal power is

$$P_s = 2R^2 P_{\mathrm{LO}} P_r. \tag{3.25}$$

The noise variance at a balanced receiver includes LO-spontaneous beating noise of

$$\sigma^2_{\mathrm{LO-sp}} = 2R^2 P_{\mathrm{LO}} S_{n_s} B_d, \tag{3.26}$$

and the signal-LO spontaneous beating noise variance of

$$\sigma^2_{\mathrm{sig.-LOsp}} = 2R^2 P_r S_{n_L} B_d. \tag{3.27}$$

The shot noise has a variance of

$$\sigma^2_{\mathrm{sh}} = 2eR(P_{\mathrm{LO}} + P_r + 2S_{n_s}\Delta f_{o,s} + 2S_{n_L}\Delta f_{o,L})B_d \tag{3.28}$$

for LO, signal, and spontaneous-emission induced shot noise.

For both heterodyne or homodyne systems, the SNR is

$$\rho_s = \frac{P_r}{S_{n_s}B_d + \frac{P_r}{P_{\mathrm{LO}}}S_{n_L}B_d + \frac{\hbar\omega_c}{\eta}\left(1 + \frac{P_r}{P_{\mathrm{LO}}}\right)B_d}, \tag{3.29}$$

when the shot noise of spontaneous emission and the beating of spontaneous emission with shot noise are ignored.

The LO spontaneous emission noise usually comes from a single amplifier with spectral density of

$$S_{n_L} = (G_L - 1)n_{\mathrm{sp}L}\hbar\omega_{\mathrm{LO}}, \tag{3.30}$$

where G_L and $n_{\mathrm{sp}L}$ are the gain and spontaneous emission factor of the LO amplifier.

The spontaneous emission noise of S_{n_s} together with the received signal is induced by a chain of optical amplifiers in the fiber link. Assume N_A identical fiber spans with loss of $1/G_s$, equal to the gain of the N_A identical optical amplifiers, the first optical amplifier has an amplified spontaneous emission noise of

$$(G_s - 1)n_{\mathrm{sp}s}\hbar\omega_c, \tag{3.31}$$

where $n_{\mathrm{sp}s}$ is the spontaneous emission factor of the optical amplifiers at each fiber span. The above spontaneous emission losses by the factor of $1/G_s$ at the fiber span, and is amplified by G_s by the second optical amplifier. The second optical amplifier also adds the same spontaneous emission as Eq. (3.31) to the optical signal. After N_A fiber spans, we obtain the overall amplifier noise spectral density of

$$S_{n_s} = N_A(G_s - 1)n_{\mathrm{sp}s}\hbar\omega_c. \tag{3.32}$$

The noise figure of optical amplifier is approximately equal to $2n_{\mathrm{sp}s}$. The spontaneous emission of Eq. (3.32) can be expressed using the noise figure of the optical amplifier as $S_{n_s} \approx \frac{1}{2}G_s F_n \hbar\omega_c$, where F_n is the noise figure of each optical amplifier.

Shot-noise limited systems

A heterodyne system without optical amplifiers is limited by the shot noise with $n_s(t) = 0$ and $n_L(t) = 0$, the signal power is that of Eq. (3.25) and the noise variance is that of Eq. (3.28). If the LO laser has significantly larger power than the receive power of $P_{\mathrm{LO}} \gg P_r$, the SNR of a heterodyne system becomes

$$\rho_s = \frac{RP_r}{eB_d}. \tag{3.33}$$

For binary signal, if the average number of photons per bit is N_s, $P_r = N_s\hbar\omega_c B_d$. For multilevel signal, N_s is the average number of photons per symbol. With the photodiode responsivity defined by Eq. (3.6), we obtain

$$\rho_s = \eta N_s. \tag{3.34}$$

In homodyne system, the power is twice that of Eq. (3.25) but the shot noise remains the same as that of Eq. (3.28), the SNR is

$$\rho_s = 2\eta N_s. \tag{3.35}$$

In the quantum limit with $\eta = 1$, the quantum-limited SNR is the photon number per bit of N_s and twice the photon number per bit of $2N_s$ for heterodyne and homodyne systems, respectively. The SNR of Eqs. (3.34) and (3.35) is usually used in the analysis of traditional coherent optical communication systems limited by shot noise (Betti et al., 1995, Okoshi and Kikuchi, 1988).

Amplifier-noise limited systems

If the amplifier noise is the dominant noise source, with the condition of Eq. (3.12), the SNR after the balanced receiver for heterodyne system with $\omega_{\text{IF}} \neq 0$ is

$$\rho_s = \frac{P_r}{S_{n_s} B_d}. \tag{3.36}$$

The SNR of Eq. (3.36) has a very simple and clear physical meaning of the received signal of P_r over the optical noise within a bandwidth of B_d in a single polarization. As the spontaneous emission has a spectral density of S_{n_s}, the noise power per polarization is $S_{n_s} B_d$. In a RF system, with proper filtering, mixing by upconversion or downconversion does not change the SNR. From Eq. (3.36), the SNR in optical domain, before the downconversion by balanced receiver, is also the SNR in the receiver. The definition of the SNR of Eq. (3.36) ignores the amplifier noise from orthogonal polarization that does not beat with the signal. In the photocurrent of Eq. (3.21), the optical amplifier noise from orthogonal polarization does not affect the system.

For a received power of $P_r = G_s N_s \hbar\omega_c B_d$, we obtain

$$\rho_s = \frac{N_s}{N_A n_{\text{sps}}} \frac{G_s}{G_s - 1} \approx \frac{N_s}{N_A n_{\text{sps}}}. \tag{3.37}$$

Comparing with Eq. (3.34), the SNR is degraded by the factor of $N_A n_{\text{sps}}$. The SNR of Eq. (3.37) is valid for both homodyne and heterodyne systems. For heterodyne systems, the condition of Eq. (3.12) must be satisfied such that the LO-spontaneous emission does not double.

The equivalent spontaneous emission factor can be defined as

$$n_{\text{eq}} = N_A n_{\text{sps}} (G_s - 1)/G_s \approx \frac{1}{2} N_A F_n \tag{3.38}$$

for N_A identical optical amplifiers where F_n is the noise figure of each optical amplifier. With the equivalent n_{eq} of Eq. (3.38), the spectrum density of Eq. (3.32) becomes

$$S_{n_s} = (G_s - 1) n_{\text{eq}} \hbar\omega_c, \tag{3.39}$$

and the SNR is

$$\rho_s = \frac{N_s}{n_{\text{eq}}}. \tag{3.40}$$

If the N_A optical amplifiers are not the same with different gain and noise figure, the equivalent spontaneous emission factor is

$$n_{\text{eq}} = \sum_{k=1}^{N_A} \frac{P_{\text{in}N}}{P_{\text{in}k}} \frac{G_k - 1}{G_k} n_{\text{sps}_k}, \tag{3.41}$$

where G_k, $k = 1, \ldots, N_A$, are the gain of each optical amplifier, $P_{\text{in}k}$, $k = 1, \ldots, N_A$, are the input power of each optical amplifier, and n_{sps_k} are the spontaneous emission factor of each optical amplifier. If all amplifiers are identical with $G_k = G_s$ and $n_{\mathrm{sps}_k} = n_{\mathrm{sps}}$, we obtain $n_{\mathrm{eq}} = N_A n_{\mathrm{sps}}(G_s - 1)/G_s$ of Eq. (3.38) again.

The SNR of Eq. (3.40) is the same as the optical SNR defined over a bandwidth of B_d for a single polarization of

$$\frac{P_r}{S_{n_s} B_d} = \frac{N_s}{n_{\mathrm{eq}}}. \tag{3.42}$$

In the quantum limit of a single optical amplifier of $N_A = 1$ and $F_n = 2$ (or 3 dB), $n_{\mathrm{eq}} = 1$, the quantum limited SNR is equal to the average number of photons per bit of N_s, the same as that of shot-noise limited heterodyne system.

In practice, optical SNR of $\mathrm{SNR}_{o,s}$ is measured over an optical bandwidth of $\Delta f_{o,s}$ using an optical spectrum analyzer. The SNR of Eq. (3.40) is equal to

$$\rho_s = \mathrm{SNR}o, s\frac{2\Delta f_{o,s}}{B_d}, \tag{3.43}$$

where the factor of 2 is for two polarizations in most optical SNR measurement. The optical bandwidth in typical measurements is $\Delta f_{o,s} =$ 12.5 GHz, corresponding to 0.1 nm or 1 Å in the wavelength around 1.55 μm. The difference between ρ_s and $\mathrm{SNR}_{o,s}$ is about 4.0 and -2.0 dB for 10 and 40-Gb/s signals, respectively.

If no optical filter is used or the optical filter has a wide bandwidth that does not conform to the condition of Eq. (3.12), the amplifier-noise limited SNR for heterodyne receiver is

$$\rho_s = \frac{N_s}{2n_{\mathrm{eq}}}. \tag{3.44}$$

In later section, we always assume a SNR of Eq. (3.40) for heterodyne receiver. With a balanced receiver, the performance of a system depends solely on the SNR. Both heterodyne and homodyne systems have the same performance if the SNR of the system is the same.

Balanced receiver for coherent optical communications was first analyzed in detail by Abbas et al. (1985), Yuen and Chan (1983), and Alexander (1987) to suppress the LO noise and to obtain signal power gain. Without LO noise and ignored thermal noise, the performance of balanced receiver should be the same as that of a single-branch receiver.

1.3 Quadrature Receiver

Figure 3.4 shows a quadrature receiver to recover both the in- and quadrature-phase components of the optical signal. The quadrature re-

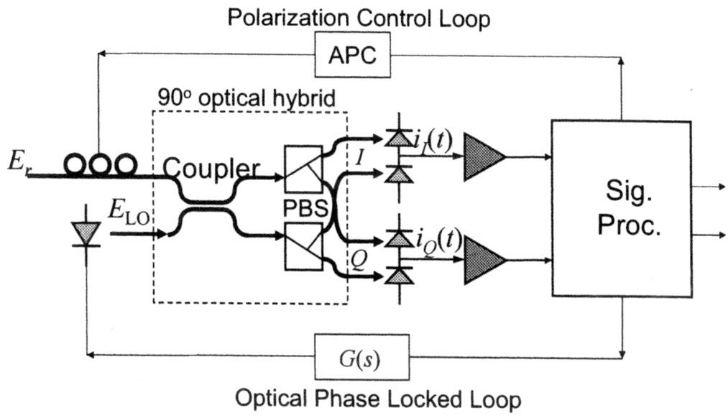

Figure 3.4. A quadrature homodyne receiver.

ceiver based on a 90° optical hybrid and two balanced receivers. Although a single-branch receiver of Fig. 3.1 can be used in Fig. 3.4, a balanced receiver has better performance, especially in the presence of LO laser noise.

The 90° optical hybrid composites of a 3-dB coupler and two polarization beam splitters (PBS). Using the two PBS as the reference polarization, the received signal must be controlled to be linearly polarized with a direction 45° from the PBS reference polarization. The received signal excluding noise is

$$E_r = \frac{1}{\sqrt{2}}(\mathbf{x} + \mathbf{y})A_s(t)e^{j\phi_s(t)+j\omega_c t}. \tag{3.45}$$

The LO laser must be circular polarized with an electric field of

$$E_{\mathrm{LO}} = \frac{1}{\sqrt{2}}(\mathbf{x} + e^{j\pi/2}\mathbf{y})A_L e^{j\omega_{\mathrm{LO}} t}. \tag{3.46}$$

At the output of the 3-dB coupler, the two electric fields before the PBS are

$$\begin{aligned} E_1 &= \frac{1}{\sqrt{2}}(E_r + E_{\mathrm{LO}}) \\ &= \frac{\mathbf{x}}{2}\underbrace{\left[A_s(t)e^{j\phi_s(t)+j\omega_c t} + A_L e^{j\omega_{\mathrm{LO}} t}\right]}_{0^\circ} \\ &\quad + \frac{\mathbf{y}}{2}\underbrace{\left[A_s(t)e^{j\phi_s(t)+j\omega_c t} + A_L e^{j\omega_{\mathrm{LO}} t + j\pi/2}\right]}_{90^\circ}, \end{aligned} \tag{3.47}$$

and

$$\begin{aligned} E_2 &= \frac{1}{\sqrt{2}}(E_r - E_{\mathrm{LO}}) \\ &= \frac{\mathbf{x}}{2} \underbrace{\left[A_s(t)e^{j\phi_s(t)+j\omega_c t} - A_L e^{j\omega_{\mathrm{LO}} t}\right]}_{180^\circ} \\ &\quad + \frac{\mathbf{y}}{2} \underbrace{\left[A_s(t)e^{j\phi_s(t)+j\omega_c t} - A_L e^{j\omega_{\mathrm{LO}} t + j\pi/2}\right]}_{270^\circ}. \end{aligned} \tag{3.48}$$

The two PBS separate the above two electric fields to the polarization directions of $\mathbf{x}$ and $\mathbf{y}$. The upper balanced receiver combines the photocurrents corresponding to both 0° and 180°, the received photocurrent is

$$\begin{aligned} i_I(t) &= \frac{R}{4}\left|A_s(t)e^{j\phi_s(t)+j\omega_c t} + A_L e^{j\omega_{\mathrm{LO}} t}\right|^2 \\ &\quad - \frac{R}{4}\left|A_s(t)e^{j\phi_s(t)+j\omega_c t} - A_L e^{j\omega_{\mathrm{LO}} t}\right|^2 \\ &= R A_s(t) A_L \cos[\omega_{\mathrm{IF}} t + \phi_s(t)]. \end{aligned} \tag{3.49}$$

The lower balanced receiver combines the photocurrents corresponding to both 90° and 270°, the received photocurrent is

$$\begin{aligned} i_Q(t) &= \frac{R}{4}\left|A_s(t)e^{j\phi_s(t)+j\omega_c t} + A_L e^{j\omega_{\mathrm{LO}} t + j\pi/2}\right|^2 \\ &\quad - \frac{R}{4}\left|A_s(t)e^{j\phi_s(t)+j\omega_c t} - A_L e^{j\omega_{\mathrm{LO}} t + j\pi/2}\right|^2 \\ &= R A_s(t) A_L \sin[\omega_{\mathrm{IF}} t + \phi_s(t)]. \end{aligned} \tag{3.50}$$

In both photocurrents of Eqs. (3.49) and (3.50), an nonzero IF frequency of ω_{IF} is assumed for a general quadrature receiver. Homodyne quadrature receiver has $\omega_{\mathrm{IF}} = 0$.

Mathematically, a 2 × 2 90° optical hybrid has an input and output relationship of

$$S = \frac{1}{2}\begin{bmatrix} 1 & 1 \\ 1 & j \end{bmatrix}. \tag{3.51}$$

and a 2 × 4 90° optical hybrid has an input and output relationship of

$$S = \frac{1}{2}\begin{bmatrix} 1 & 1 \\ 1 & j \\ 1 & -1 \\ 1 & -j \end{bmatrix}. \tag{3.52}$$

For a homodyne receiver with a PLL of Fig. 3.4, $\omega_{\mathrm{IF}} = 0$ and $i_I(t)$ gives the in-phase component of $A_s(t)\cos\phi_s(t)$, and $i_Q(t)$ provides the quadrature-phase component of $A_s(t)\sin\phi_s(t)$. The quadrature homodyne receiver can be used as the receiver of both M-ary PSK and quadrature-amplitude modulation (QAM) signals.

In the homodyne quadrature receiver of Fig. 3.4, the shot-noise limited SNR is the same as that in Eq. (3.35) and the amplifier-noise limited SNR is that of Eq. (3.40). The noises in the in- and quadrature-phase components are independent of each other. The SNR for the photocurrents of Eqs. (3.49) and (3.50) is identical.

For a heterodyne quadrature receiver with $\omega_{\mathrm{IF}} \neq 0$, $i_I(t)$ and $i_Q(t)$ are quadrature components with a phase difference of 90°. With the condition of Eq. (3.12), the amplifier-noise limited SNR is that of Eq. (3.40). The shot-noise limited SNR is that of Eq. (3.34).

In heterodyne receiver, it may be more convenient to use electrical PLL to obtain both in- and quadrature-phase components. In another application as shown later, a heterodyne quadrature receiver similar to Fig. 3.4 can be used to provide image-rejection.

In- and quadrature-phase detection was first used to simultaneously measure phase and amplitude of an optical electric field (Walker and Carroll, 1984). The first application to coherent optical communications was by Hodgkinson et al. (1985). The 90° optical hybrid of Fig. 3.4 is designed according to Kazovsky et al. (1987). Other implementations of 90° optical hybrid were proposed by Delavaux and Riggs (1990), Delavaux et al. (1990), Hoffman et al. (1989), and Langenhorst et al. (1991). Recently, Cho et al. (2004c) shown an integrated $LiNbO_3$ optical 90° hybrid. As shown later, both image-rejection and phase-diversity receivers are similar to the quadrature receiver of Fig. 3.4. Recently, without phase-locking, quadrature receivers similar to Fig. 3.4 were used for measurement purpose (Dorrer et al., 2003, 2005).

1.4 Image-Rejection Heterodyne Receiver

In a densely space coherent wavelength-division-multiplexed (WDM) systems, the most important issue is to improve the spectral efficiency by allotting more channels within the amplifier passband of the system. In the heterodyne receiver with $\omega_{\mathrm{IF}} \neq 0$ in both Figs. 3.1 and 3.3, the signals at both ω_{c1} and ω_{c2} fall to the same IF band if $\omega_{\mathrm{IF}} = \omega_{c1} - \omega_{\mathrm{LO}} = \omega_{\mathrm{LO}} - \omega_{c2}$. In a regular balanced heterodyne receiver of Fig. 3.3, a large guard-band of about $2\omega_{\mathrm{IF}}$ is required such that the real- and image-band signals do not interfere with each other. The requirement of large guard-band limits the spectral efficiency of the system.

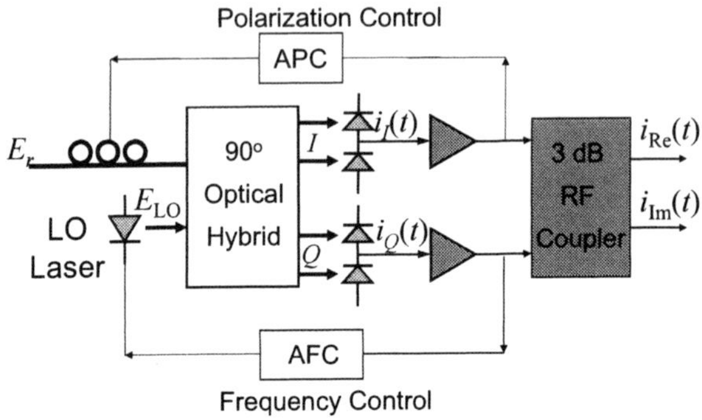

Figure 3.5. A heterodyne image-rejection receiver.

Figure 3.5 shows an image-rejection receiver with an optical front-end similar to the quadrature receiver of Fig. 3.4 with an 90° optical hybrid and two balanced receivers. The image-rejection receiver of Fig. 3.5 is for heterodyne receiver with $\omega_{\text{IF}} > 0$. Assume that there are two signals at ω_{c1} and ω_{c2} with a relationship of

$$\omega_{\text{IF}} = \omega_{c1} - \omega_{\text{LO}} = \omega_{\text{LO}} - \omega_{c2}. \tag{3.53}$$

If the two signals are

$$E_{c1}(t) = A_{s1}(t)e^{j\omega_{c1}t + j\phi_{s1}(t)}, \text{ and } E_{c2}(t) = A_{s2}(t)e^{j\omega_{c2}t + j\phi_{s2}(t)}, \tag{3.54}$$

similar to Eqs. (3.49) and (3.50) and without going into details, we obtain two photocurrents of

$$\begin{aligned} i_I(t) &= RA_{s1}(t)A_L\cos[(\omega_{c1} - \omega_{\text{LO}})t + \phi_{s1}(t)] \\ &\quad + RA_{s2}(t)A_L\cos[(\omega_{\text{LO}} - \omega_{c2})t - \phi_{s2}(t)], \end{aligned} \tag{3.55}$$

and

$$\begin{aligned} i_Q(t) &= RA_{s1}(t)A_L\sin[(\omega_{c1} - \omega_{\text{LO}})t + \phi_{s1}(t)] \\ &\quad - RA_{s2}(t)A_L\sin[(\omega_{\text{LO}} - \omega_{c2})t - \phi_{s2}(t)]. \end{aligned} \tag{3.56}$$

After a 90° microwave hybrid[2], we obtain

$$i_{\text{Re}}(t) = \sqrt{2}RA_{s1}(t)A_L\cos[(\omega_{c1} - \omega_{\text{LO}})t + \phi_{s1}(t)], \tag{3.57}$$

$$i_{\text{Im}}(t) = \sqrt{2}RA_{s2}(t)A_L\cos[(\omega_{\text{LO}} - \omega_{c2})t - \phi_{s2}(t)] \tag{3.58}$$

---

[2]The transfer matrix must be

$$S = \frac{1}{\sqrt{2}}\begin{bmatrix} 1 & j \\ 1 & -j \end{bmatrix}$$

with a 90° shift from Eq. (3.1). The transfer matrix is called 90° microwave hybrid but similar matrix is for 180° optical hybrid.

for the real and image frequency band, respectively.

In an image-rejection heterodyne receiver of Fig. 3.5, even without an optical filter to filter-out the image-band amplified spontaneous emission, the amplifier-noise limited SNR is $\rho_s = N_s/n_{\text{eq}}$ from Eq. (3.40). Not only reject the signal from the image band, the image-rejection receiver also rejects the noise from the image band.

In another application of the image-rejection receiver of Fig. 3.5, the LO laser can have a frequency between two adjacent WDM channels. The real and image bands can be processed to receive two channels having larger or smaller frequency than the LO laser.

Used by Chikama et al. (1990a) and Lachs et al. (1990) for WDM systems, heterodyne image-rejection receiver was proposed by Darcie and Glance (1986), Glance (1986b), and Westphah and Strebel (1988). Even without optical filter, as indicated in Glance et al. (1988), Walker et al. (1990) and Jørgensen et al. (1992) showed that the SNR is improved by 3 dB using image-rejection receiver. The output SNR of image-rejection receiver is the same as that of homodyne receiver.

1.5 SNR of Basic Coherent Receivers

Table 3.1 summaries the SNR of system with different receiver structures. Single-branch receiver is more likely to be limited by the noise from LO laser.

For system limited by amplifier noise, without optical filter, heterodyne receiver is 3-dB worse than homodyne receiver. Image-rejection receiver eliminates the effects of the amplifier noise from the image frequency band even for the system without optical filtering. Homodyne and heterodyne receivers have the same SNR for both cases of having image-rejection or optical filtering. The optical filter of a heterodyne receiver must have a bandwidth conformed to the relationship of Eq. (3.12).

For system limited by shot-noise, the performance of heterodyne receiver is always 3-dB worse than homodyne receiver. This 3-dB difference was given in all standard textbooks (Agrawal, 2002, Betti et al., 1995, Okoshi and Kikuchi, 1988).

In later parts of this book, system performance is analyzed based on the representation of a received signal by

$$r(t) = A_s(t)\cos[\omega_{\text{IF}} + \phi_s(t)] + n(t) \tag{3.59}$$

with the SNR from Table 3.1. The noise in the receiver is considered to be within the narrow receiver bandwidth of the receiver and with a band-pass representation of

$$n(t) = n_1(t)\cos\omega_{\text{IF}}t - n_2(t)\sin\omega_{\text{IF}}t \tag{3.60}$$

Table 3.1. Comparison of the SNR of Different Receiver Structures.

Receiver Types	Limited Noise Sources	
	Shot Noise	Amplifier Noise
Single-branch heterodyne w/ optical filter†	ηN_s	$\dfrac{N_s}{n_{\text{eq}}}$
Single-branch heterodyne w/o optical filter†	ηN_s	$\dfrac{N_s}{2n_{\text{eq}}}$
Single-branch homodyne†	$2\eta N_s$	$\dfrac{N_s}{n_{\text{eq}}}$
Balanced received heterodyne w/ optical filter	ηN_s	$\dfrac{N_s}{n_{\text{eq}}}$
Balanced received heterodyne w/o optical filter	ηN_s	$\dfrac{N_s}{2n_{\text{eq}}}$
Balanced received homodyne	$2\eta N_s$	$\dfrac{N_s}{n_{\text{eq}}}$
Quadrature homodyne	$2\eta N_s$	$\dfrac{N_s}{n_{\text{eq}}}$
Image-rejection heterodyne receiver	ηN_s	$\dfrac{N_s}{n_{\text{eq}}}$

†Single-branch receiver is more likely to be limited by LO noise.

with

$$E\{n^2(t)\} = E\{n_1^2(t)\} = E\{n_2^2(t)\} = \sigma_n^2. \tag{3.61}$$

As discussed earlier, a quantum-limited system has the limit of a quantum efficiency of $\eta = 1$ or equivalent spontaneous emission of $n_{\text{eq}} = 1$. The quantum-limited SNR is equal to the average number of photons per bit (or per symbol) for heterodyne system limited by either shot or amplifier noise.

2. Performance of Synchronous Receivers

When an optical PLL is used to track the phase of the LO laser for homodyne systems or an electrical PLL is used to track the phase of an IF oscillator for heterodyne systems, the system is called a coherent system according to the terminology of conventional digital communications (Proakis, 2000). In coherent optical communications, system with phase tracking is called synchronous detection system. This section calculates the error probability of synchronous detection systems as a function of SNR. With the same SNR of ρ_s, homodyne and heterodyne systems have the same performance. This section considers only heterodyne systems with balanced receiver.

2.1 Amplitude-Shift Keying

When the optical carrier is ASK modulated, the signal current of Eq. (3.24) in a heterodyne receiver can be expressed as

$$s_1(t) = A\cos\omega_{\mathrm{IF}}t, \tag{3.62}$$

$$s_2(t) = 0. \tag{3.63}$$

The above binary ASK signal can be received by the heterodyne receiver of Fig. 1.3(b).

Including noise, the overall received signal is

$$r(t) = s(t) + n(t) = \begin{cases} [A + n_1(t)]\cos\omega_{\mathrm{IF}}t - n_2(t)\sin\omega_{\mathrm{IF}}t & \text{for } s_1(t) \\ n_1(t)\cos\omega_{\mathrm{IF}}t - n_2(t)\sin\omega_{\mathrm{IF}}t & \text{for } s_2(t) \end{cases}. \tag{3.64}$$

At the output of an ASK receiver with PLL, the decision variable is $r_d(t) = A + n_1(t)$ and $n_1(t)$ for $s_1(t)$ and $s_2(t)$, respectively. With decision threshold of $\frac{1}{2}A$, the error probability is

$$p_e = \frac{1}{2}\int_{-\infty}^{\frac{A}{2}} p_1(x)\mathrm{d}x + \frac{1}{2}\int_{\frac{A}{2}}^{+\infty} p_2(x)\mathrm{d}x, \tag{3.65}$$

where

$$p_1(x) = \frac{1}{\sqrt{2\pi}\sigma_n}e^{-\frac{(x-A)^2}{2\sigma_n^2}}, \tag{3.66}$$

$$p_2(x) = \frac{1}{\sqrt{2\pi}\sigma_n}e^{-\frac{x^2}{2\sigma_n^2}} \tag{3.67}$$

are the Gaussian probability density function (p.d.f.) of the decision random variables.

We obtain

$$p_e = \frac{1}{2}\mathrm{erfc}\left(\frac{A}{2\sqrt{2}\sigma_n}\right) \tag{3.68}$$

$$= \frac{1}{2}\mathrm{erfc}\left(\sqrt{\frac{\rho_s}{2}}\right), \tag{3.69}$$

where the SNR is , $\mathrm{erfc}(x) = 2/\sqrt{\pi}\int_x^\infty e^{-t^2}\mathrm{d}t$ is the complementary error function, and

$$\rho_s = \frac{A^2/4}{\sigma_n^2}. \tag{3.70}$$

with signal power equal to $\frac{1}{2}\frac{A^2}{2}+0=\frac{A^2}{4}$.

To achieve an error probability of 10^{-9}, a SNR of $\rho_s = 36$ (15.6 dB) is required. The quantum-limited binary ASK signal requires 36 photons/bit.

2.2 Phase-Shift Keying

When the light is PSK modulated, the signal of Eq. (3.24) in a heterodyne receiver can be expressed as

$$s_1(t) = A\cos\omega_{\mathrm{IF}}t, \quad (3.71)$$

$$s_2(t) = -A\cos\omega_{\mathrm{IF}}t. \quad (3.72)$$

The above binary PSK signal can be received by the heterodyne receiver of Fig. 1.3(b). Including noise, the overall received signal is

$$r(t) = s(t)+n(t) = \begin{cases} [A+n_1(t)]\cos\omega_{\mathrm{IF}}t - n_2(t)\sin\omega_{\mathrm{IF}}t & \text{for } s_1(t) \\ [-A+n_1(t)]\cos\omega_{\mathrm{IF}}t - n_2(t)\sin\omega_{\mathrm{IF}}t & \text{for } s_2(t) \end{cases} \quad (3.73)$$

At the output of the PLL, the decision random variable is $r_d(t) = \pm A + n_1(t)$. With a decision threshold of zero, the error probability is

$$p_e = \frac{1}{2}\int_{-\infty}^{0} p_1(x)\mathrm{d}x + \frac{1}{2}\int_{0}^{+\infty} p_2(x)\mathrm{d}x, \quad (3.74)$$

where

$$p_1(x) = \frac{1}{\sqrt{2\pi}\sigma_n}e^{-\frac{(x-A)^2}{2\sigma_n^2}}, \quad (3.75)$$

$$p_2(x) = \frac{1}{\sqrt{2\pi}\sigma_n}e^{-\frac{(x+A)^2}{2\sigma_n^2}} \quad (3.76)$$

are the Gaussian p.d.f. of the decision random random variables.

We obtain

$$p_e = \frac{1}{2}\mathrm{erfc}\left(\frac{A}{\sqrt{2}\sigma_n}\right) \quad (3.77)$$

$$= \frac{1}{2}\mathrm{erfc}\left(\sqrt{\rho_s}\right), \quad (3.78)$$

where the SNR is

$$\rho_s = \frac{A^2/2}{\sigma_n^2}. \quad (3.79)$$

To achieve an error probability of 10^{-9}, a SNR of $\rho_s = 18$ (12.5 dB) is required for an improvement of 3-dB over ASK signal. The quantum-limited binary PSK signal requires 18 photons/bit.

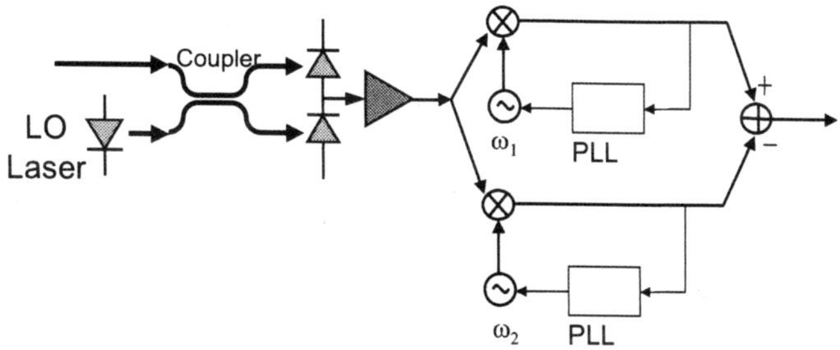

Figure 3.6. A heterodyne synchronous receiver for FSK signal.

Synchronous receivers had been demonstrated mostly for PSK signal for its superior performance (Kahn et al., 1990, Kazovsky and Atlas, 1990, Kazovsky et al., 1990, Norimatsu et al., 1990, Schöpflin et al., 1990, Watanabe et al., 1989). There were other experiments to transmit some reference carriers without phase locking (Cheng and Okoshi, 1989, Wandernoth, 1992, Watanabe et al., 1992). There is recently interest to conduct PSK experiment (Cho et al., 2004c, Taylor, 2004).

2.3 Frequency-Shift Keying

When the optical carrier is frequency-shift keying (FSK) modulated, the signal of a heterodyne receiver can be expressed as

$$s_1(t) = A\cos\omega_1 t, \tag{3.80}$$
$$s_2(t) = A\cos\omega_2 t, \tag{3.81}$$

where ω_1 and ω_2 are two angular frequencies with orthogonal condition of

$$\int_0^T s_1(t)s_2(t)\mathrm{d}t = 0, \tag{3.82}$$

where T is the symbol interval. The two frequencies should be separated by $\omega_1 - \omega_2 = k\pi/T$, $k = \pm1, \pm2, \ldots$, for the orthogonal condition. In a synchronous heterodyne FSK receiver of Fig. 3.6, two electrical PLL are required to phase lock the two oscillators with frequencies of either ω_1 or ω_2 for the two FSK signals. Equivalently, two matched filters are used with filter response matching to $s_1(t)$ and $s_2(t)$, respectively. The difference of the two outputs of Fig. 3.6 decides whether $s_1(t)$ or $s_2(t)$ is transmitted.

With the orthogonal condition, the overall received signal including noise is

$$r(t) = s(t) + n(t) = \begin{cases} [A + n_{11}(t)]\cos\omega_1 t - n_{12}(t)\sin\omega_1 t & \text{for } s_1(t) \\ [A + n_{21}(t)]\cos\omega_2 t - n_{22}(t)\sin\omega_2 t & \text{for } s_2(t) \end{cases}. \tag{3.83}$$

where $n_{11}(t)$, $n_{21}(t)$, $n_{12}(t)$, and $n_{22}(t)$ are independent of each other with the same variance of σ_n^2. When $s_1(t)$ is transmitted, the correct decision is $A + n_{11}(t) > n_{21}(t)$ or $A + n_{11}(t) - n_{21}(t) > 0$. The noise of $n_{11}(t) - n_{21}(t)$ has a variance of $2\sigma_n^2$. The error probability is the same as that of PSK signal but with a noise variance of $2\sigma_n^2$, we obtain

$$p_e = \frac{1}{2}\text{erfc}\left(\frac{A}{2\sigma_n}\right) \tag{3.84}$$

$$= \frac{1}{2}\text{erfc}\left(\sqrt{\frac{\rho_s}{2}}\right), \tag{3.85}$$

where the SNR is

$$\rho_s = \frac{A^2/2}{\sigma_n^2}. \tag{3.86}$$

A FSK signal has the same performance as ASK signal. The same as ASK signal, to achieve an error probability of 10^{-9}, a SNR of $\rho_s = 36$ (15.6 dB) is required. The quantum-limited binary FSK signal requires 36 photons/bit.

The performance of synchronous receiver, the same as digital signal with a matched filter provided by phase tracking, is analyzed in Proakis (2000) or the early paper of Yamamoto (1980).

3. Performance of Asynchronous Receivers

All of ASK, DPSK, and FSK signals can be detected without phase tracking. The detection is based on the comparison of the power or envelope of the signal. This type of detection is called noncoherent detection in conventional digital communications (Proakis, 2000) and asynchronous receiver in coherent optical communications. In this section, we consider asynchronous heterodyne receiver with signal processing by electrical circuits in the IF.

3.1 Envelope Detection of Heterodyne ASK Signal

Figure 3.7 shows the receiver for envelope detection of heterodyne binary ASK signal. The signal first passes through a band-pass filter (BPF) to limit the amount of noise. After the BPF, the signal is the

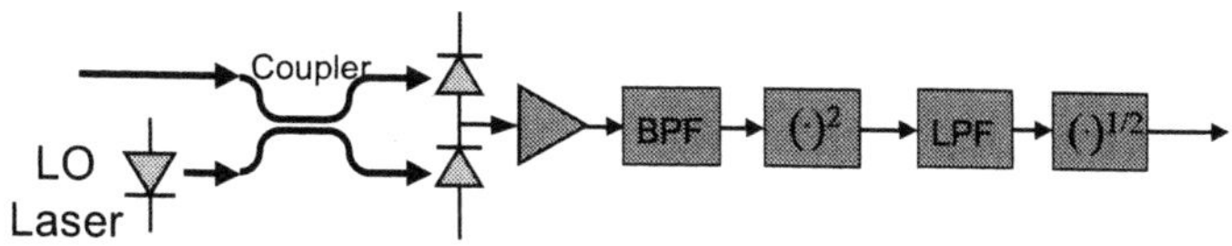

Figure 3.7. A heterodyne asynchronous receiver for ASK signal.

same as that of Eq. (3.64). The signal of Eq. (3.64) is squared, low-pass filtered (LPF), square-rooted, to obtain the envelope of

$$r_d(t) = \begin{cases} \sqrt{[A + n_1(t)]^2 + n_2^2(t)} & \text{for } s_1(t) \\ \sqrt{n_1^2(t) + n_2^2(t)} & \text{for } s_2(t) \end{cases} \tag{3.87}$$

with p.d.f. of

$$p_1(r) = \frac{r}{\sigma_n^2} I_0\left(\frac{rA}{\sigma_n^2}\right) \exp\left(-\frac{r^2 + A^2}{2\sigma_n^2}\right), \tag{3.88}$$

$$p_2(r) = \frac{r}{\sigma_n^2} \exp\left(-\frac{r^2}{2\sigma_n^2}\right), \tag{3.89}$$

as Rice and Rayleigh distribution, respectively, where $I_0(\cdot)$ is the zeroth-order modified Bessel function of the first kind. In the decision random variable of Eq. (3.87), some constant factors related to gain and loss of the squarer, LPF, and the square-root components are ignored. The envelope of Eq. (3.87) is the same as the amplitude of the signal. The error probability of the signal is

$$p_e = \frac{1}{2}\left[1 - \int_{r_{\text{th}}}^{\infty} p_1(r)\mathrm{d}r\right] + \frac{1}{2}\int_{r_{\text{th}}}^{\infty} p_2(r)\mathrm{d}r, \tag{3.90}$$

where r_{th} is the threshold. Using the Marcum's Q-function defined by Eq. (3.A.3) from Appendix 3.A, we obtain

$$p_e = \frac{1}{2}\left[1 - Q\left(\frac{A}{\sigma_n}, \frac{r_{\text{th}}}{\sigma_n}\right)\right] + \frac{1}{2}\exp\left(-\frac{r_{\text{th}}^2}{2\sigma_n^2}\right). \tag{3.91}$$

The optimal threshold can be found by $dp_e/dr_{\text{th}} = 0$ as

$$I_0\left(\frac{Ar_{\text{th}}}{\sigma_n^2}\right) \exp\left(-\frac{A^2}{2\sigma_n^2}\right) = 1. \tag{3.92}$$

The optimal decision threshold is difficult to find analytically, we may use the approximation of $r_{\text{th}} = A/2$. With this threshold, the second

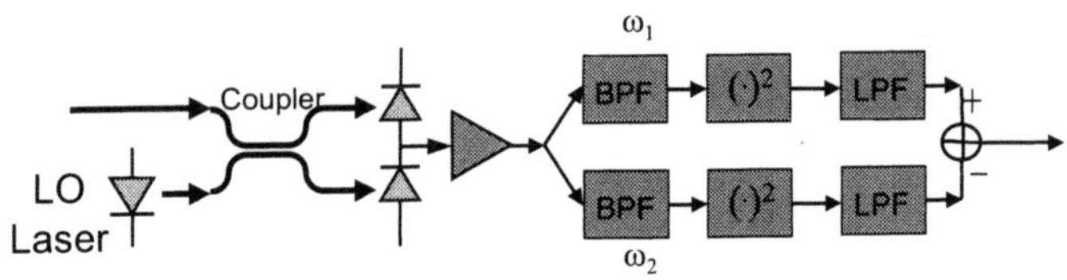

Figure 3.8. A heterodyne asynchronous receiver for FSK signal.

term of Eq. (3.91) is larger than the first term and the error probability is approximately equal to

$$\begin{aligned} p_e &\approx \frac{1}{2}\exp\left(-\frac{A^2}{8\sigma_n^2}\right) \\ &= \frac{1}{2}\exp\left(-\frac{\rho_s}{2}\right). \end{aligned} \tag{3.93}$$

Based on the approximation of Eq. (3.93), the required SNR for an error probability of 10^{-9} is $\rho_s = 40$ (16 dB). The quantum limit is 40 photons/bit. The asynchronous receiver is about 0.4 dB worse than the synchronous receiver in Sec. 3.2.1.

3.2 Dual-Filter Detection of FSK Signal

Figure 3.8 shows an asynchronous heterodyne receiver for FSK signal based on two BPF matched to the FSK signals of $s_1(t)$ and $s_2(t)$ with center angular frequencies of ω_1 and ω_2, respectively. Based on Eq. (3.83), when $s_1(t)$ is transmitted, the received signal at the first filter centered at ω_1 is the same as that of Eq. (3.87). Although not necessary, a square root is assumed for the signal of Fig. 3.8 for convenience. When $s_1(t)$ is transmitted, we obtain

$$r_1(t) = \sqrt{[A + n_{11}(t)]^2 + n_{12}^2(t)}, \tag{3.94}$$

and

$$r_2(t) = \sqrt{n_{21}^2(t) + n_{22}^2(t)} \tag{3.95}$$

with p.d.f. of

$$p_1(r_1) = \frac{r_1}{\sigma_n^2} I_0\left(\frac{r_1 A}{\sigma_n^2}\right)\exp\left(-\frac{r_1^2 + A^2}{2\sigma_n^2}\right), \tag{3.96}$$

$$p_2(r_2) = \frac{r_2}{\sigma_n^2}\exp\left(-\frac{r_2^2}{2\sigma_n^2}\right) \tag{3.97}$$

as Rice and Rayleigh distribution, respectively.

Bit error occurs when $r_1 < r_2$, or

$$p_e = \Pr\{r_1 < r_2\} = \Pr\{r_1^2 < r_2^2\}, \tag{3.98}$$

because both $r_1 \geq 0$ and $r_2 \geq 0$. From Eq. (3.98), the square-root component in the receiver of Fig. 3.8 is optional. The error probability is

$$\begin{aligned} p_e &= \int_0^\infty p_1(r_1) \int_{r_1}^\infty p_2(r_2) \mathrm{d}r_2 \mathrm{d}r_1 \\ &= \int_0^\infty \frac{r_1}{\sigma_n^2} I_0\left(\frac{r_1 A}{\sigma_n^2}\right) \exp\left(-\frac{2r_1^2 + A^2}{2\sigma_n^2}\right) \mathrm{d}r_1 \\ &= \frac{1}{2} \exp\left(-\frac{A^2}{4\sigma_n^2}\right) \int_0^\infty \frac{2r_1}{\sigma_n^2} I_0\left(\frac{2r_1 A}{2\sigma_n^2}\right) \exp\left(-\frac{4r_1^2 + A^2}{4\sigma_n^2}\right) \mathrm{d}r_1 \\ &= \frac{1}{2} \exp\left(-\frac{A^2}{4\sigma_n^2}\right) \\ &= \frac{1}{2} \exp\left(-\frac{\rho_s}{2}\right). \end{aligned} \tag{3.99}$$

The above error probability is valid only if the two signals are orthogonal. The performance of FSK signal is similar to that of ASK signal The required SNR for an error probability of 10^{-9} is $\rho_s = 40$ (16 dB). The quantum limit is 40 photons/bit.

A FSK signal can also be detected based on a single filter. The signal after the filter is the same as an ASK signal. The performance of FSK signal with a single filter is the same as that of ASK signal with an error probability of $p_e = \frac{1}{2} \exp(-\rho_s/4)$. While an ASK signal has no power at the "0" level, the power of FSK signal is the same at both "0" and "1" levels. The performance of single-filter detected FSK signal is 3-dB worse than the equivalent ASK signal. Intuitively, using a single filter, the FSK signal is also 3-dB worse than a dual-filter receiver.

Single filter FSK experiment was conducted by Emura et al. (1984) and Park et al. (1990), and dual-filter FSK experiment was conducted by Emura et al. (1990b).

3.3 Heterodyne Differential Detection of DPSK Signal

In another representation for a heterodyne DPSK system, the received signal is

$$r(t) = \Re\left\{\left[Ae^{j\phi_s(t)} + n(t)\right] e^{j\omega_{\mathrm{IF}} t}\right\}, \tag{3.100}$$

or

$$r(t) = \Re\left\{\tilde{r}(t) e^{j\omega_{\mathrm{IF}} t}\right\}, \tag{3.101}$$

where $\tilde{r}(t) = Ae^{j\phi_s(t)} + n(t)$ is the baseband representation of the IF signal. The DPSK signal is demodulated by delay-and-multiplier circuits of Fig. 1.4(b). If $\tilde{r}(t)$ is expressed by the polar representation of $\tilde{r}(t) = \tilde{A}_1(t)e^{j\tilde{\phi}_1(t)}$ with $\tilde{\phi}_1(t)$ include the noisy phase from the contribution of $n(t)$, the demodulated signal after a low-pass filter is

$$\begin{aligned} r_d(t) &= \tilde{A}_1(t)\tilde{A}_1(t-T)\cos\left[\omega_{\mathrm{IF}}\tau + \tilde{\phi}_1(t) - \tilde{\phi}_1(t-T)\right] \\ &= \Re\left\{\tilde{r}(t)\tilde{r}^*(t-T)\right\}. \end{aligned} \tag{3.102}$$

In Eq. (3.102), two symbols of τ and T are used to represent the delay of the delay-and-multiplier circuits. The delay of τ must be approximated equal to T and $\omega_{\mathrm{IF}}\tau$ must be an integer multiple of 2π for a noiseless decision variable of $r_d(t) = \pm A^2$ with $\phi_s(t) - \phi_s(t-T) = 0, \pi$, respectively. For a large frequency of ω_{IF}, a small variation of τ around the time interval T may give $\omega_{\mathrm{IF}}\tau$ equal to an integer multiple of 2π. Later in this book assumes that $\tau = T$ and $\omega_{\mathrm{IF}}T$ is equal to an integer multiple of 2π at the same time.

Some simple algebra gives

$$r_d(t) = \left|\frac{\tilde{r}(t) + \tilde{r}(t-T)}{2}\right|^2 - \left|\frac{\tilde{r}(t) - \tilde{r}(t-T)}{2}\right|^2. \tag{3.103}$$

The signals of $\tilde{r}(t) + \tilde{r}(t-T)$ and $\tilde{r}(t) - \tilde{r}(t-T)$ are independent of each other. With $r_1 = |\tilde{r}(t) + \tilde{r}(t-T)|/2$ and $r_2 = |\tilde{r}(t) + \tilde{r}(t-T)|/2$, assume that $\phi_s(t) = \phi_s(t-T)$, the error probability is similar to Eq. (3.98) with

$$p_e = \Pr\{r_1 < r_2\} = \Pr\{r_1^2 < r_2^2\} = \Pr\{r_d < 0\}. \tag{3.104}$$

If $\phi_s(t) = \phi_s(t-T)$, $r_1^2 = |A + n(t)/2 + n(t-T)/2|^2$ as the first term of Eq. (3.103) is the square of a Gaussian random variable with mean of A and variance of σ_n^2 and $r_2^2 = |n(t) - n(t-T)|^2/4$ as the second term of Eq. (3.103) is the square of a Gaussian random variable with zero mean and variance of σ_n^2. The error probability is the same as that for FSK signal of Eq. (3.85) with equivalent SNR of A^2/σ_n^2, we obtain

$$\begin{aligned} p_e &= \frac{1}{2}\exp\left(-\frac{A^2}{2\sigma_n^2}\right) \\ &= \frac{1}{2}\exp\left(-\rho_s\right). \end{aligned} \tag{3.105}$$

The required SNR for an error probability of 10^{-9} is $\rho_s = 20$ (13 dB) for DPSK signal. The quantum limit is 20 photons/bit. DPSK signal is about 3-dB better than ASK signal.

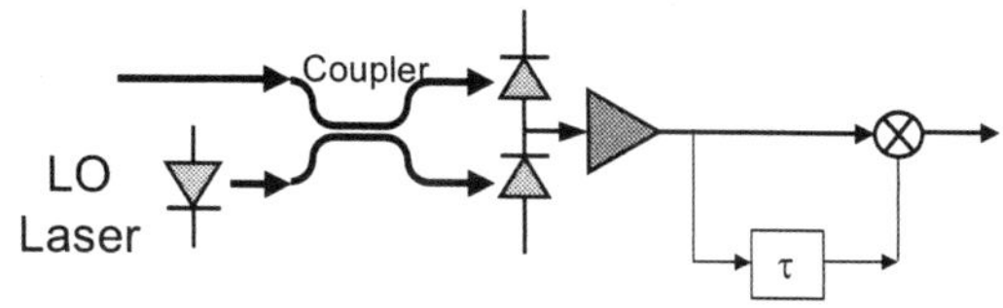

Figure 3.9. A heterodyne asynchronous receiver for CPFSK signal based on delay-and-multiplier circuits.

The error probability of Eq. (3.105) was derived by Cahn (1959) using approximation. DPSK signal was analyzed the same as orthogonal or FSK signal in Stein (1964). Using two time intervals, the performance of DPSK signal is 3-dB better than that of FSK signal. In Eq. (3.103), although the mean of the signal is the same as that of the FSK signal, the noise variance is reduced by half because the noise is the average over two time intervals.

Heterodyne DPSK signal was demonstrated in Chikama et al. (1988), Creaner et al. (1988), Meissner (1989), Naito et al. (1990), and Gnauck et al. (1990).

3.4 Heterodyne Receiver for CPFSK Signal

In binary continuous-phase frequency-shift keying (CPFSK) transmission, the signal in each time interval is equal to

$$s(t) = A\cos\left[\omega_{\mathrm{IF}}t \pm \pi\Delta f t + \phi_0\right], \tag{3.106}$$

where ϕ_0 depends the phase of previous symbols to ensure continuous phase operation and Δf is the frequency deviation between the "0" and "1" states. With a received signal of $r(t) = s(t) + n(t)$, we may define $\tilde{r}(t) = Ae^{\pm j\pi\Delta f t + j\phi_0} + n(t) = \tilde{A}_1(t)e^{j\tilde{\phi}_1(t)}$ similar to that for DPSK signal. The CPFSK signal can be demodulated using the asynchronous receiver of Fig. 3.9 based on the delay-and-multiplier circuits similar to that of Fig. 1.4(b) for DPSK signals. In the CPFSK receiver of Fig. 3.9, the delay is τ instead of the one-bit delay of T in Fig. 1.4(b). Similar to the case of DPSK signal of Eq. (3.103), the decision random variable is

$$\begin{aligned} r_d &= \tilde{A}_1(t)\tilde{A}_1(t-\tau)\cos\left[\omega_{\mathrm{IF}}\tau + \tilde{\phi}_1(t) - \tilde{\phi}_1(t-\tau)\right] \\ &= \Re\left\{\tilde{r}(t)\tilde{r}^*(t-\tau)e^{j\omega_{\mathrm{IF}}\tau}\right\}, \end{aligned} \tag{3.107}$$

where $\tilde{\phi}_1(t) = \pm\pi\Delta f t + \phi_0 + \phi_n(t)$ with $\phi_n(t)$ as the noisy phase from $n(t)$. The differential phase is $\tilde{\phi}_1(t) - \tilde{\phi}_1(t-\tau) = \pm\pi\Delta f\tau + \phi_n(t) - \phi_n(t-\tau)$. The receiver achieves its optimal performance for a design of $\omega_{\mathrm{IF}}\tau =$

$k\pi + \frac{1}{2}\pi$ with k as an integer. Excluding noise, the decision variable is $r_d(t) = \pm A^2 \sin(\Delta f \pi \tau)$. For the best system efficiency, $r_d(t) = \pm A^2$ if $\Delta f \tau = 1/2$.

With the design of $\Delta f \tau = 1/2$, assume that the phase of noise of $\phi_n(t)$ and $\phi_n(t-\tau)$ are independent of each other, the performance of CPFSK signal with differential detection is the same as DPSK signal with error probability of

$$p_e = \frac{1}{2} \exp\left(-\rho_s\right). \tag{3.108}$$

The most interesting case is $\Delta f = 1/2T$ when $\tau = T$. The frequency separation of $\Delta f = 1/2T$ is the minimum frequency separation for two orthogonal frequencies of Eq. (3.82). The CPFSK signal with $\Delta f = 1/2T$ is called minimum shift keying (MSK) modulation. Similar to DPSK signal, MSK signal may be demodulated using a one-bit delay and multiplier. The performance of MSK signal is also the same as DPSK signal. From Proakis (2000), MSK signal has more compact spectrum than DPSK signal.

Because CPFSK signal can be generated directly modulated a semiconductor laser from Sec. 2.4, it was the most popular scheme for coherent optical communications (Emura et al., 1990a, Iwashita and Matsumoto, 1987, Iwashita and Takachio, 1988, Park et al., 1991, Takachio et al., 1989).

3.5 Frequency Discriminator for FSK Signal

FSK signal can also be detected asynchronously by frequency discriminator of Fig. 3.10. The frequency discriminator may be the most popular receiver for analog frequency modulation (FM) signal for its simplicity. The system must have high SNR to ensure the correct operation of the frequency discriminator.

The frequency discriminator of Fig. 3.10 consists of two band-pass filters having frequency responses of

$$H_1(f) = \begin{cases} 2\pi K(f - f_{\mathrm{IF}}) & f_{\mathrm{IF}} < f < f + \Delta f/2 \\ 0 & \text{otherwise} \end{cases} \tag{3.109}$$

and

$$H_2(f) = \begin{cases} 2\pi K(f_{\mathrm{IF}} - f) & f - \Delta f/2 < f < f_{\mathrm{IF}} \\ 0 & \text{otherwise} \end{cases}, \tag{3.110}$$

where K is the slope of the frequency discriminator and Δf is the bandwidth of the frequency discriminator and the frequency separation of

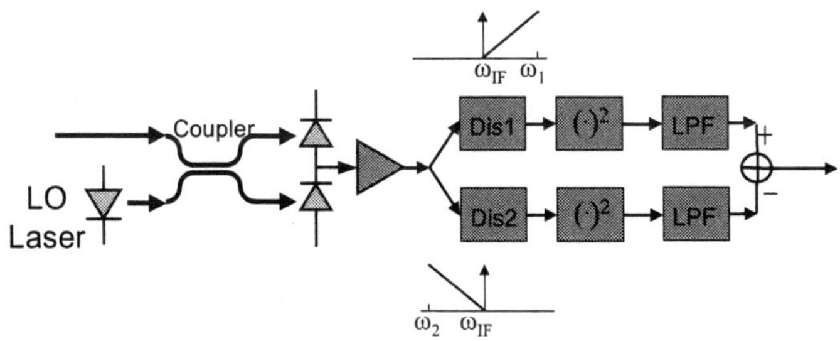

Figure 3.10. A heterodyne asynchronous receiver for FSK signal using frequency discriminator.

two FSK signals of $2\pi\Delta f = \omega_1 - \omega_2$. In time domain, the operation of the frequency discriminator is equivalent to a linear operation of $K\frac{d}{dt}$.

Assume that $s_1(t)$ is transmitted with a signal before the discriminator as

$$A\cos\omega_1 t + n_1(t)\cos\omega_{\text{IF}}t - n_2(t)\sin\omega_{\text{IF}}t, \tag{3.111}$$

the output of $H_1(f)$ is

$$\begin{aligned} r_1(t) = AK(\omega_1 - \omega_{\text{IF}})\sin\omega_1 t + K\frac{dn_{11}(t)}{dt}\cos\left(\omega_{\text{IF}}t + \pi\Delta f t/2\right) \\ - K\frac{dn_{12}(t)}{dt}\sin\left(\omega_{\text{IF}}t + \pi\Delta f t/2\right), \end{aligned} \tag{3.112}$$

where $\omega_1 - \omega_{\text{IF}} = \pi\Delta f$, $n_{11}(t)$ and $n_{12}(t)$ are the part of $n_1(t)$ and $n_2(t)$ in the frequency between f_{IF} and $f_{\text{IF}} + \Delta f/2$. The output of $H_2(f)$ is

$$r_2(t) = K\frac{dn_{21}(t)}{dt}\cos\left(\omega_{\text{IF}}t + \pi\Delta f t/2\right) - K\frac{dn_{22}(t)}{dt}\sin\left(\omega_{\text{IF}}t + \pi\Delta f t/2\right). \tag{3.113}$$

where $n_{21}(t)$ and $n_{22}(t)$ are the part of $n_1(t)$ in the frequency smaller than the IF of f_{IF}. The signal of $r_1(t)$ is similar to an amplitude-modulated signal with a power of $\pi^2[AK\Delta f]^2/2$. With the derivative operation as a linear filter, the noise of $K dn_{11}(t)/dt$ has a variance of

$$\begin{aligned} K^2 E\left\{\left|\frac{dn_{11}}{dt}\right|^2\right\} &= (2\pi K)^2 \int_0^{B_d} N_n f^2 \mathrm{d}f \\ &= (2\pi K)^2 \frac{1}{3} B_d^3 N_n = \frac{4}{3}\pi^2 K^2 B_d^2 \sigma_n^2. \end{aligned} \tag{3.114}$$

where N_n is the spectral density of $n_{11}(t)$ and $\sigma_n^2 = N_0 B_d$. At the output of the frequency discriminator, the signal is similar to that case of orthogonal signals demonstrated by two BPF followed by envelope detection.

The equivalent SNR of the discriminator output signal is

$$\rho_{\mathrm{eq}} = \frac{3\rho_s}{8}\left(\frac{\Delta f}{B_d}\right)^2. \tag{3.115}$$

Similar to the error probability of Eq. (3.99), we obtain

$$p_e = \frac{1}{2}\exp\left[-\frac{3\rho_s}{16}\left(\frac{\Delta f}{B_d}\right)^2\right]. \tag{3.116}$$

For the case that $\Delta f = 2B_d$, we obtain $p_e = \frac{1}{2}\exp(-3\rho_s/4)$, 1.76 dB improvement over the asynchronous receiver of Eq. (3.99) but twice the spectral bandwidth. Using the method based on frequency discriminator, the system performance is improved with the frequency expansion of $\Delta f/B_d$. Similar to analog FM, detection based on frequency discriminator expands the signal bandwidth to obtain improved system performance. However, the error probability of Eq. (3.116) should be considered as an approximation. While the noise outside B_d in Eq. (3.114) is ignored in the deviation, those noise gives noise-to-noise beating thought the squarer of Fig. 3.10 and degrades the performance of the system, especially at low SNR.

3.6 Envelope Detection of Correlated Binary Signals

FSK, DPSK, and MSK signals are special cases of signal modulation formats based on two orthogonal signals. The dual-filter detection of FSK signal and differential detection of both DPSK and MSK signals are asynchronous or noncoherent detection of two orthogonal signals. For envelope detection of binary signals, the receiver is the same as that of Fig. 3.8 with the band-pass filters representing two matched filters. If the two signals are correlated with a correlation coefficient of $|\rho|$, the p.d.f. of the outputs of two matched filters are

$$p_1(r_1) = \frac{r_1}{\sigma_n^2} I_0\left(\frac{r_1 A}{\sigma_n^2}\right)\exp\left(-\frac{r_1^2 + A^2}{2\sigma_n^2}\right), \tag{3.117}$$

$$p_2(r_2) = \frac{r_2}{\sigma_n^2} I_0\left(\frac{r_2|\rho| A}{\sigma_n^2}\right)\exp\left(-\frac{r_2^2 + |\rho|^2 A^2}{2\sigma_n^2}\right). \tag{3.118}$$

The random variables of R_1 and R_2 are correlated with each other. In order to derive the error probability of $p_e = \Pr\{R_2 > R_1\}$, a transform is required such that $R_2^2 - R_1^2 = \Gamma_2^2 - \Gamma_1^2$ in which the random variables of

Γ_2 and Γ_1 are independent of each other and also have Rice distribution. While the amplitude parameter of R_1 and R_2 are A and $|\rho|A$, respectively, the amplitude parameters of Γ_1 and Γ_2 are $A\left[1 \pm \sqrt{1-|\rho|^2}\right]^{1/2}$, respectively. From Appendix 3.A, the error probability of correlated binary signal with envelope detection is

$$p_e = Q(a, b) - \frac{1}{2} e^{-(a^2+b^2)/2} I_0(ab), \tag{3.119}$$

where

$$a = \sqrt{\frac{\rho_s}{2}\left(1 - \sqrt{1-|\rho|^2}\right)},$$
$$b = \sqrt{\frac{\rho_s}{2}\left(1 + \sqrt{1-|\rho|^2}\right)}. \tag{3.120}$$

For orthogonal signal, $|\rho| = 0$, $a = 0$, and $b = \sqrt{\rho_s}$, $p_e = Q(0, b) - \frac{1}{2}e^{-b^2/2} = \frac{1}{2}e^{-b^2/2} = \frac{1}{2}e^{-\rho_s/2}$, the same as the error probability of Eq. (3.85).

The error probability of Eq. (3.119) was first derived by Helstrom (1955) based on direct integration. The method to find the error probability here is based on Stein (1964) and Schwartz et al. (1966). Proakis (2000, Appendix B) also derived the error probability of Eq. (3.119). While the error probability of Eq. (3.119) is not useful if the two binary signals are well-designed without correlation, further degradation that induces correlation can be analyzed based on Eq. (3.119).

4. Performance of Direct-Detection Receivers

Other than the intensity-modulation/direct-detection (IMDD) systems of Fig. 1.1, both DPSK and FSK signals can be directly detected without mixing with an LO laser. DPSK and FSK signals can be detected using interferometer or optical filter. Direct-detection receiver is simpler than both homodyne and heterodyne receivers that require the mixing with an LO laser. This section analyzes the performance of typical direct-detection receivers for ASK or on-off keying, DPSK, and FSK signals.

4.1 Intensity-Modulation/Direct-Detection Receiver

IMDD systems of Fig. 1.1 are the simplest optical communication schemes to converse information with the presence or absence of light. The receiver is just a photodiode that converts the optical intensity to photocurrent. For system limited by amplifier noises, the performance is analyzed in, for example, Agrawal (2002), Desurvire (1994), and Becker et al. (1999).

Quantum Limit for Systems without Amplifiers

In a system without optical amplifiers, due to the quantum nature of photons, the number of photons has a Poisson distribution. At the on-state, the probability of having k photons is

$$p_k = \frac{N_b^k}{k!} e^{-N_b}, \tag{3.121}$$

where N_b is the number of photons in the on state of a binary on-off keying signal. If the detection is based on the presence or absence of photons, the error probability is

$$p_e = \frac{1}{2} p_0 = \frac{1}{2} e^{-N_b} \tag{3.122}$$

with no photon, where the factor of $1/2$ is the probability of the on state. With no photon to send, there is no error for the off state. The average number of photons is $N_s = N_b/2$. In order to achieve an error probability of 10^{-9}, an average of 10 photons/bit are required. If the quantum efficiency of η is less than unity, the required number of photons increases by the factor of η^{-1}.

Practical on-off keying receivers require thousands of photons per bit, mostly due to the contribution from the thermal noise at the receiver circuitry. As an example, assume a thermal noise density of $i_{\text{th}} = 5$ pA/$\sqrt{\text{Hz}}$, corresponding to a receiver sensitivity of -25.2 and -28.2 dBm for 10- and 2.5-Gb/s systems, respectively, if shot noise is ignored and a photodiode responsivity of $R = 1$ is assumed. For systems operating in 1.55 μm, the number of photons per bit is 4.7×10^4 and 2.3×10^4 for 10- and 2.5-Gb/s systems for an error probability of 10^{-9}. Without the usage of optical amplification, practical receiver always requires the number of photons per bit many order larger than the quantum limit of 10 photons/bit, even for receiver with very good sensitivity.

Amplifier-Noise Limited System

If the on-off keying system is limited by amplifier noise, the received electric field is

$$E_r(r) = [A_s(t) + n_x(t)] e^{j\omega_c t} \mathbf{x} + n_y(t) e^{j\omega_c t} \mathbf{y}, \tag{3.123}$$

where the transmitted data are contained in the amplitude of $A_s(t) \in \{0, A\}$ for on-off keying signal. In the direct-detection receiver, the above electric field is converted by a photodiode to photocurrent of

$$i(t) = R|A_s(t) + n_x(t)|^2 + R|n_y(t)|^2, \tag{3.124}$$

where shot noise is ignored by assuming that the amplifier noise is the major degradation. If the noise from orthogonal polarization of $n_y(t)$ is ignored, the system is the same as that using envelope detection for heterodyne ASK receiver in Sec. 3.3.1.

The common factor of photodiode responsivity of $R = 1$ is assumed in Eq. (3.124) without loss of generality. Further assumed an optical matched filter preceding the photodiode, in the on state with $A_s(t) = A$, we obtain

$$r_{\text{on}}(t) = [A + n_{x1}(t)]^2 + n_{x2}^2(t) + n_{y1}^2(t) + n_{y2}^2(t) \tag{3.125}$$

with p.d.f. of

$$p_1(y) = \frac{1}{2\sigma_n^2}\sqrt{\frac{y}{A^2}}e^{-(A^2+y)/2\sigma_n^2} I_1\left(\sqrt{y}\frac{A}{\sigma_n^2}\right), \qquad y \geq 0 \tag{3.126}$$

as the noncentral chi-square (χ^2) p.d.f. with four degrees of freedom.

In the off state with $A_s(t) = 0$, we obtain

$$r_{\text{off}}(t) = n_{x1}^2(t) + n_{x2}^2(t) + n_{y1}^2(t) + n_{y2}^2(t) \tag{3.127}$$

with p.d.f. of

$$p_2(y) = \frac{1}{4\sigma_n^4} y e^{-y/2\sigma_n^2}, \qquad y \geq 0 \tag{3.128}$$

as the χ^2 p.d.f. with four degrees of freedom, also called Gamma distribution.

With a threshold of y_{th}, the error probability of the system is

$$\begin{aligned} p_e &= \frac{1}{2}\int_0^{y_{\text{th}}} p_1(y)dy + \frac{1}{2}\int_{y_{\text{th}}}^{\infty} p_2(y)dy \\ &= \frac{1}{2}\left[1 - Q_2\left(\frac{A}{\sigma_n}, \frac{\sqrt{y_{\text{th}}}}{\sigma_n}\right)\right] + \frac{1}{2}\exp\left(-\frac{y_{\text{th}}}{2\sigma_n^2}\right)\left(1 + \frac{y_{\text{th}}}{2\sigma_n^2}\right), \end{aligned} \tag{3.129}$$

where

$$Q_2(a, b) = Q(a, b) + \frac{b}{a}e^{-(a^2+b^2)/2} I_1(ab) \tag{3.130}$$

is the second-order generalized Marcum Q function.

The optimal threshold can be determined by

$$\sqrt{y_{\text{th}}}\frac{A}{2\sigma_n^2} = e^{-A^2/2\sigma_n^2} I_1\left(\sqrt{y_{\text{th}}}\frac{A}{\sigma_n^2}\right), \qquad y \geq 0. \tag{3.131}$$

In a simplified analysis, we can approximate the decision threshold as $y_{\text{th}} = A^2/4$, the error probability is approximately equal to

$$\begin{aligned} p_e &\approx \frac{1}{2}\int_{y_{\text{th}}}^{\infty} p_2(y)\mathrm{d}y \\ &= \frac{1}{2}\exp\left(-\frac{A^2}{8\sigma_n^2}\right)\left(1+\frac{A^2}{8\sigma_n^2}\right) \\ &= \frac{1}{2}\exp\left(-\frac{\rho_s}{2}\right)\left(1+\frac{\rho_s}{2}\right). \end{aligned} \tag{3.132}$$

With the inclusion of the amplified noise from orthogonal polarization of $n_y(t)$, the error probability is increased by a factor of $1 + \rho_s/2$. Using a direct-detection receiver, based on the error probability of Eq. (3.132), a SNR of $\rho_s = 46.4$ (16.7 dB) is required to achieve an error probability of 10^{-9}. The quantum-limited receiver requires 46.4 photons/bit. The inclusion of the noise from orthogonal polarization degrades the receiver by about 0.7 dB.

Error Probability Based on Gaussian Approximation

The error probability can be analyzed based on Gaussian approximation with the assumption of two received signals of

$$r_{\text{on}}(t) = I_1 + n_1(t), \tag{3.133}$$

$$r_{\text{off}}(t) = I_0 + n_0(t), \tag{3.134}$$

where $n_0(t)$ and $n_1(t)$ are assumed zero-mean Gaussian noise with variances of $E\{n_1^2(t)\} = \sigma_1^2$ and $E\{n_0^2(t)\} = \sigma_0^2$, respectively, I_1 and I_0 are the mean photocurrents for the on and off states. In general, $I_1 = RP_r$ and $I_0 = 0$.

The p.d.f. of the on and off states are

$$p_1(x) = \frac{1}{\sqrt{2\pi}\sigma_1} e^{-(x-I_1)^2/2\sigma_1^2}, \tag{3.135}$$

$$p_0(x) = \frac{1}{\sqrt{2\pi}\sigma_0} e^{-(x-I_0)^2/2\sigma_0^2}. \tag{3.136}$$

The optimal decision threshold can be determined by

$$p_1(x_{\text{th}}) = p_0(x_{\text{th}}), \qquad I_0 < x_{\text{th}} < I_1 \tag{3.137}$$

of

$$x_{\text{th}} = \frac{\sigma_0 I_1 + \sigma_1 I_0}{\sigma_0 + \sigma_1}. \tag{3.138}$$

In the derivation of the optimal threshold, we assume that $\sigma_0 \neq \sigma_1$ but $\log(\sigma_1/\sigma_0) = 0$ as an approximation. Without the approximation of $\log(\sigma_1/\sigma_0) = 0$, the optimal threshold is a difficult to calculate but without providing further accuracy.

Defined a Q-factor of

$$Q = \frac{I_1 - x_{\text{th}}}{\sigma_1} = \frac{x_{\text{th}} - I_0}{\sigma_0} = \frac{I_1 - I_0}{\sigma_1 + \sigma_0}, \tag{3.139}$$

the error probability is equal to

$$p_e = \frac{1}{2}\text{erfc}\left(\frac{Q}{\sqrt{2}}\right). \tag{3.140}$$

With the Gaussian approximation, the noise variances are equal to

$$\sigma_1^2 = \sigma_{A_s-n_s}^2 + \sigma_{n_s-n_s}^2 + \sigma_{\text{sh}}^2 + \sigma_{\text{th}}^2, \tag{3.141}$$

$$\sigma_0^2 = \sigma_{n_s-n_s}^2 + \sigma_{\text{sh}}^2 + \sigma_{\text{th}}^2, \tag{3.142}$$

where

$$\sigma_{A_s-n_s}^2 = 4R^2 P_r S_{\text{sps}} B_d, \tag{3.143}$$

$$\sigma_{n_s-n_s}^2 = 4R^2 S_{\text{sps}}^2 \Delta\nu_{\text{opt}} B_d, \tag{3.144}$$

$$\sigma_{\text{sh}}^2 = 2eR(P_r + 2S_{\text{sps}}\Delta\nu_{\text{opt}})B_d, \tag{3.145}$$

$$\sigma_{\text{th}}^2 = \frac{4kT}{R_L}. \tag{3.146}$$

Thermal noise is included in the above equations for the case that the received signal is very small or for system without amplifier noises. A direct-detection on-off keying system is potentially limited by thermal instead of amplifier noise.

For the specific case of the signals of Eqs. (3.125) and (3.127) with amplifier noise from orthogonal polarization, we obtain

$$I_0 = 0, \qquad I_1 = A^2, \tag{3.147}$$

$$\sigma_0^2 = 8\sigma_n^4, \qquad \sigma_1^2 = 4\sigma_n^2 A^2 + 8\sigma_n^4, \tag{3.148}$$

and

$$Q = \frac{A^2}{2\sqrt{2}\sigma_n^2 + 2\sigma_n\sqrt{A^2 + 2\sigma_n^2}} = \frac{\sqrt{2}\rho_s}{1 + \sqrt{2\rho_s + 1}}. \tag{3.149}$$

Based on the Gaussian approximation, the required SNR to achieve an error probability of 10^{-9} is $\rho_s = 36 + 6\sqrt{2} = 44.5$ (16.5 dB).

When the optical filter preceding the receiver has a wide bandwidth, direct-detection on-off keying system can be approximately analyzed by adding identical and independent noise terms to both Eqs. (3.125) and (3.127) (Humblet and Azizoğlu, 1991, Marcuse, 1990, 1991). The number of noise terms is equal to twice the ratio of the optical bandwidth to data rates, the factor of two taking into account the two polarizations of amplifier noise. While Marcuse (1990, 1991) and Humblet and Azizoğlu (1991) sum independent and identical random variables to model amplifier noises, Lee and Shim (1994) sums independent random variables with variance depending on the combined effects of electrical and optical filter responses. The optical filter may also model further linear effects of chromatic and polarization-mode dispersion. These two methods are widely used for performance evaluation in direct-detection on-off keying systems (Bosco et al., 2001, Chan and Conradi, 1997, Forestieri, 2000, Holzlohner et al., 2002, Roudas et al., 2002).

Comparison of Different Models

Figure 3.11 shows the error probability of ASK signal detected using various types of receiver that are also analyzed based on different assumptions. The error probability of Eq. (3.69) with synchronous receiver has the lowest error probability for the same SNR. The error probability of Eq. (3.93) for envelope detection is also shown for comparison together as the error probability calculated with the optimal threshold of Eq. (3.92) as dashed-lines. The error probability of Eq. (3.132) for direct-detection is shown as solid line with the corresponding error probability calculated with the optimal threshold of Eq. (3.131) as dashed-lines. The error probability with optimal threshold of Eq. (3.131) almost overlaps with Eq. (3.93). The Gaussian approximation of Eq. (3.140) using Q factor is shown in Fig. 3.11 as dotted-line.

From Fig. 3.11, the performance of ASK signals with envelope detection can be evaluated using Eq. (3.93). Compared with the error probability with optimal threshold, the approximation of Eq. (3.93) is just about 0.1 dB worse at the error probability of 10^{-9}.

For direct-detection receiver, the Gaussian approximation overestimates the error probability and gives an approximated SNR penalty of about 0.45 dB comparing with the one with optimal threshold. Because the Gaussian approximation also uses an optimal threshold according to its own model, the performance with Gaussian approximation is actually better than the error probability with sub-optimal threshold of Eq. (3.132) by 0.2 dB. In practical system, for either direct- or envelope-detection, the error probability of Eq. (3.93) can be used.

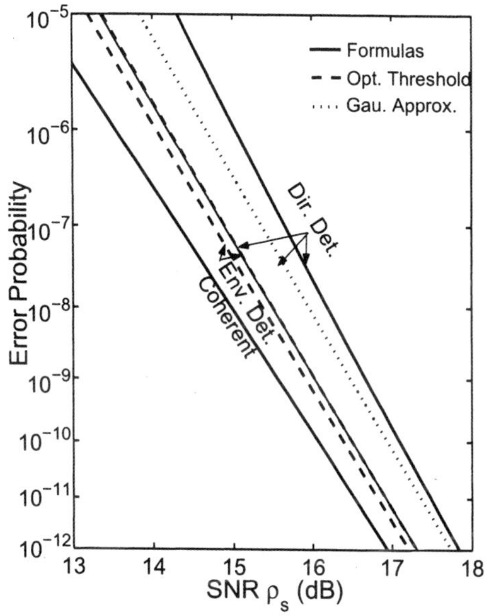

Figure 3.11. The error probability of amplitude-modulated signals as a function of SNR ρ_s.

If the optical filter before the receiver is an optical matched filter, direct-detection receiver for on-off keying signal has very good receiver sensitivity as shown in both Atia and Bondurant (1999) and Caplan and Atia (2001).

4.2 Direct-Detection DPSK Receiver

Figure 3.12 redraws the direct-detection receiver for DPSK signal of Fig. 1.4(c). The DPSK receiver uses an asymmetric Mach-Zehnder interferometer in which the signal is splitted into two paths and combined after a path difference of an one-bit delay of T. In practice, the path difference of $\tau \approx T$ must be chosen such that $\exp(j\omega_0\tau) = 1$, where ω_0 is the angular frequency of the signal. Ideally, the optical filter before the interferometer is assumed to be an optical matched filter for the transmitted signal. A balanced receiver similar to that of Fig. 3.3 is used to obtain the photocurrent. A low-pass filter reduces the receiver noise. We assume that the low-pass filter has a wide bandwidth and does not distort the received signal. With the assumption of matched filter, the analysis is applicable to both non-return-to-zero (NRZ) and RZ signals.

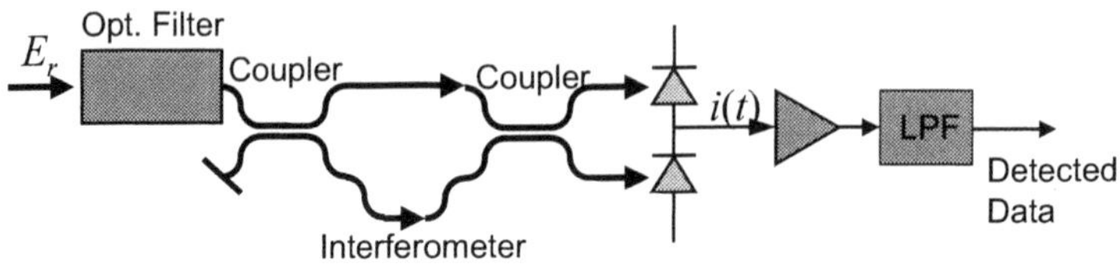

Figure 3.12. Direct-detection DPSK receiver using an unpolarized asymmetric Mach-Zehnder interferometer.

At the output of the unpolarized asymmetric Mach-Zehnder interferometer, the two output signals are

$$E_1(t) = \frac{\mathbf{x}}{2}[Ae^{-j\phi_s(t)} + n_x(t)] + \frac{\mathbf{y}}{2}n_y(t) \\ + \frac{\mathbf{x}}{2}[Ae^{-j\phi_s(t-T)} + n_x(t-T)] + \frac{\mathbf{y}}{2}n_y(t-T), \quad (3.150)$$

and

$$E_2(t) = \frac{\mathbf{x}}{2}[Ae^{-j\phi_s(t)} + n_x(t)] + \frac{\mathbf{y}}{2}n_y(t) \\ - \frac{\mathbf{x}}{2}[Ae^{-j\phi_s(t-T)} + n_x(t-T)] - \frac{\mathbf{y}}{2}n_y(t-T). \quad (3.151)$$

In the electric fields of Eqs. (3.150) and (3.151), the path difference of the interferometer is assumed exactly as the symbol time T and $\exp(j\omega_c T) = 1$. The amplifier noises of $n_x(t), n_y(t), n_x(t-T)$, and $n_y(t-T)$ are independent identically distributed complex zero-mean circular Gaussian random variables. The noise variance is $E\{|n_x(t)|^2\} = E\{|n_y(t)|^2\} = E\{|n_x(t-T)|^2\} = E\{|n_y(t-T)|^2\} = 2\sigma_n^2$, where σ_n^2 is the noise variance per dimension. In a polarized receiver, $n_y(t) = n_y(t-T) = 0$ and the error probability is the same as that for heterodyne DPSK system in Sec. 3.3.3.

In Eqs. (3.150) and (3.151), the loss in the interferometer is ignored. If the amplifier noise is the dominant noise source, both the interferometer loss and the photodiode responsivity does not affect the system performance.

Without loss of generality, we assume that $\phi_s(t) = \phi_s(t-T) = 0$ when the consecutive transmitted phases are the same. Assume an unity photodiode responsivity, similar to that of Fig. 3.3, the photocurrent at the output of the balanced receiver is

$$i(t) = |E_1(t)|^2 - |E_2(t)|^2, \quad (3.152)$$

where

$$|E_1(t)|^2 = \left| A + \frac{1}{2}[n_x(t) + n_x(t-T)] \right|^2 + \frac{1}{4}\left| n_y(t) + n_y(t-T) \right|^2, \tag{3.153}$$

and

$$|E_2(t)|^2 = \frac{1}{4}\left| n_x(t) - n_x(t-T) \right|^2 + \frac{1}{4}\left| n_y(t) - n_y(t-T) \right|^2. \tag{3.154}$$

The error probability is equal to

$$p_e = \Pr\{i(t) < 0\} = \Pr\left\{|E_2(t)|^2 > |E_1(t)|^2\right\}. \tag{3.155}$$

Because $n_x(t)+n_x(t-T)$ is independent of $n_x(t)-n_x(t-T)$ and $n_y(t)+n_y(t-T)$ is independent of $n_y(t)-n_y(t-T)$, $|E_1(t)|^2$ and $|E_2(t)|^2$ are independent of each other. From the error probability of Eq. (3.155), similar to heterodyne DPSK signal in Sec. 3.3.3, DPSK signal can be analyzed as noncoherent detection of an orthogonal binary signal.

The p.d.f. of $|E_1(t)|^2$ of Eq. (3.153) is

$$p_{|E_1|^2}(y) = \frac{1}{2\sigma^2 A}\sqrt{y}\exp\left(-\frac{A^2+y}{2\sigma^2}\right) I_1\left(\sqrt{y}\frac{A}{\sigma^2}\right), \quad y \ge 0, \tag{3.156}$$

where $I_1(\cdot)$ is the first-order modified Bessel function of the first kind and the variance parameter $\sigma^2 = \sigma_n^2/2$. The p.d.f. of $p_{|E_1|^2}(y)$ is noncentral χ^2 distribution with four degrees of freedom with a variance parameter of $\sigma^2 = \sigma_n^2/2$ and noncentrality parameter of A^2. The variance of $\sigma^2 = \sigma_n^2/2$ is the variance per dimension of the random variables of $[n_x(t) \pm n_x(t-T)]/2$ and $[n_y(t) \pm n_y(t-T)]/2$ in Eqs. (3.153) and (3.154).

The p.d.f. of $|E_2(t)|^2$ of Eq. (3.154) is

$$p_{|E_2|^2}(y) = \frac{1}{4\sigma^4} y \exp\left(-\frac{y}{2\sigma^2}\right), \quad y \ge 0. \tag{3.157}$$

The p.d.f. of $p_{|E_2|^2}(y)$ is the χ^2 distribution with four degrees of freedom.

First, we need to find the probability of (Gradshteyn and Ryzhik, 1980, §3.351)

$$\begin{aligned}\Pr\left\{|E_2(t)|^2 > y\right\} &= \int_y^{+\infty} p_{|E_2|^2}(y)\mathrm{d}y \\ &= \exp\left(-\frac{y}{2\sigma^2}\right)\left[1 + \frac{y}{2\sigma^2}\right].\end{aligned} \tag{3.158}$$

The error probability of Eq. (3.155) is

$$p_e = \int_0^{+\infty} \Pr\left\{|E_2(t)|^2 > y\right\} p_{|E_1|^2}(y)\mathrm{d}y. \tag{3.159}$$

Using the p.d.f. of Eq. (3.156) and the probability of Eq. (3.158), after some simplifications, we get

$$p_e = \frac{e^{-2\rho_s}}{\sqrt{2\rho_s}} \int_0^{+\infty} (1+x)\sqrt{x}e^{-2x} I_1\left(2\sqrt{2\rho_s x}\right) \mathrm{d}x, \tag{3.160}$$

where $x = y/(2\sigma^2)$, and $\rho_s = E_0^2/\sigma^2 = E_0^2/(2\sigma_n^2)$ is the SNR. As the special case of Gradshteyn and Ryzhik (1980, §6.631), we get

$$\int_0^\infty x^{1/2} e^{-2x} I_1(a\sqrt{x})\mathrm{d}x = (a/8)e^{a^2/8}, \tag{3.161}$$

$$\int_0^\infty x^{3/2} e^{-2x} I_1(a\sqrt{x})\mathrm{d}x = [a(16+a^2)/128]e^{a^2/8}. \tag{3.162}$$

The integration of Eq. (3.160) gives the error probability of

$$p_e = \frac{\exp(-\rho_s)}{2}\left(1 + \frac{\rho_s}{4}\right). \tag{3.163}$$

Comparing the heterodyne error probability in Sec. 3.3.3, with the amplifier noise from the orthogonal polarization, the error probability is increased by a factor of $1 + \rho_s/4$. The increase of the error probability is the similar to that for direct-detection ASK signals of Eq. (3.132).

Figure 3.13 shows the error probability of phase-modulated signal as a function of SNR ρ_s. The error probability of synchronous detection of $\frac{1}{2}\mathrm{erfc}\sqrt{\rho_s}$ from Eq. (3.78), the error probability of asynchronous heterodyne differential detection of $\frac{1}{2}e^{-\rho_s}$ from Eq. (3.105), and the error probability of direct-detection of Eq. (3.163) are also shown for comparison. For an error probability of 10^{-9}, asynchronous heterodyne detection is about 0.45 dB worse than synchronous detection, and direct-detection is about 0.40 dB worse than asynchronous differential detection. The quantum-limited receivers require 18.0, 20.0, and 21.9 photons/bit.

The degradation of direct-detection is due to the inclusion of amplifier noise from orthogonal polarization. If a lossless polarizer precedes the detector, an improvement of 0.4 dB can be expected. Tonguz and Wagner (1991) shown that direct-detection DPSK receiver performs the same as heterodyne differential detection if the amplifier noise from orthogonal direction is ignored. The error probability of Eq. (3.163) was first derived by Okoshi et al. (1988) for DPSK signals with similar noise characteristics.

If the direct-detection receiver has a noise bandwidth far larger than the signal bandwidth, the system was analyzed in Humblet and Azizoğlu (1991), Jacobsen (1993), and Chinn et al. (1996). The DPSK error probability of Eq. (3.105) assumes that there are two noise sources of

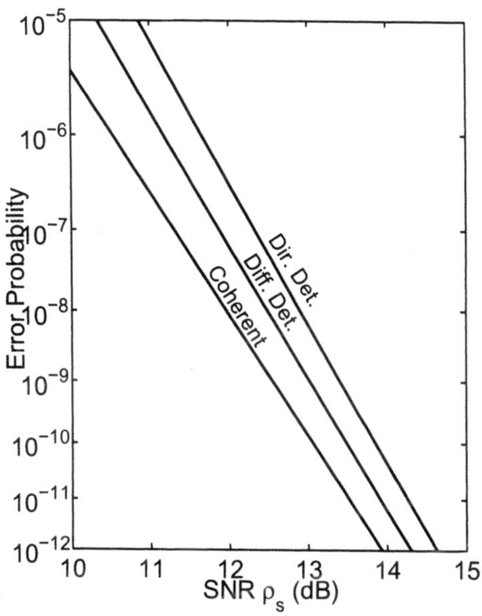

Figure 3.13. The error probability of phase-modulated signals as a function of SNR ρ_s.

$n_1(t)$ and $n_2(t)$. The error probability of Eq. (3.163) assumes that there are four noise sources of $n_{x1}(t)$, $n_{x2}(t)$, $n_{y1}(t)$, and $n_{y2}(t)$ for both real and imaginary parts noise. With the assumption of both Marcuse (1990) and Humblet and Azizoğlu (1991) that there are $2k$ independent noise sources affects the DPSK signals, the error probability are

$$p_e = \frac{e^{-\rho_s}}{2} \sum_{m=1}^{k} h_k \frac{\rho_s^m}{m!}, \tag{3.164}$$

where

$$h_k = \frac{1}{2^{2(k-1)}} \sum_{l=0}^{k-1} \begin{pmatrix} 2k - l - 2 \\ k - 1 - m \end{pmatrix}. \tag{3.165}$$

Direct-detection DPSK signal is unquestionable the most popular phase-modulated optical communication scheme as shown in Table 1.2. DPSK receivers with very good receiver sensitivity were developed by Atia and Bondurant (1999), Gnauck et al. (2003a), and Sinsky et al. (2003). Both Gnauck and Winzer (2005) and Xu et al. (2004) reviewed the activities of direct-detection DPSK systems. DPSK signal can also

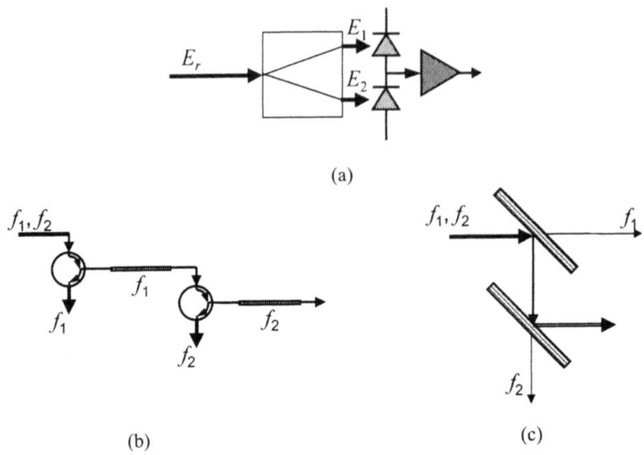

Figure 3.14. Direct-detection FSK receiver using two optical filters. (a) The schematic of the receivers. (b) Optical filter using fiber Bragg gratings. (c) Optical filter using multilayer dielectric filters

be detected using an optical filter similar to frequency discriminator (Lyubomirsky and Chien, 2005).

Single-branch direct-detection DPSK receiver converts the DPSK signal to an equivalent on-off keying signal. Comparing the receiver sensitivity of on-off keying with DPSK signal, single-branch direct-detection DPSK receiver has a receiver sensitivity 3-dB worse than the balanced receiver and has a performance the same as on-off keying signal.

4.3 Dual-Filter Direct-Detection of FSK Receiver

Figure 3.14 shows a dual-filter direct-detection FSK receiver. Figure 3.14(a) is the schematic of the receiver in which a balanced receiver is used with one detector connected to the output of each optical filter. The optical filters can be implemented using fiber Bragg grating or multilayer dielectric filters as shown in Figs. 3.14(b) and (c), respectively. The optical filters center at the optical frequencies of f_1 and f_2, corresponding to the two angular frequencies of ω_1 and ω_2 for binary FSK signal, respectively.

If the two optical filters are matched filter and the two FSK signals are orthogonal with each other, for lossless optical filter without loss of generality,

$$E_1(t) = [A \cos \omega_1 t + n_{x1}(t)]\, \mathbf{x} + n_{y1}(t)\mathbf{y} \tag{3.166}$$

if $s_1(t)$ is transmitted, where $n_{x1}(t)$ and $n_{y1}(t)$ are the amplifier noises in the polarization parallel and orthogonal to the signal, respectively.

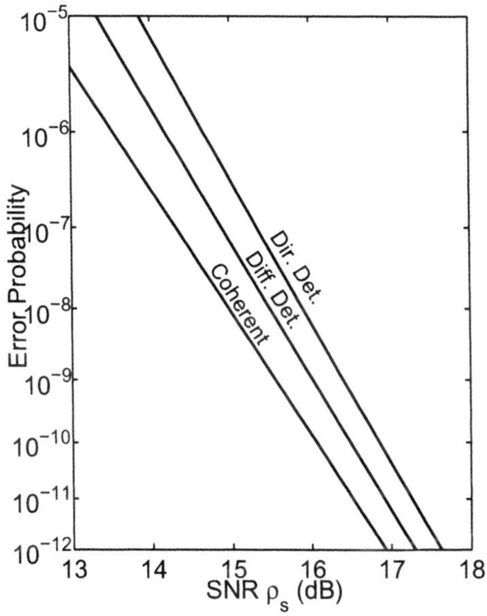

Figure 3.15. The error probability of frequency-modulated signals as a function of SNR ρ_s.

The noise variance is $E\{|n_{x1}(t)|^2\} = E\{|n_{y1}(t)|^2\} = \sigma_n^2$ and the SNR is $\rho_s = A^2/2\sigma_n^2$. If $s_1(t)$ is transmitted, the electric field at the output of the optical filter centered at f_2 is

$$E_2(t) = n_{x2}(t)\mathbf{x} + n_{y2}(t)\mathbf{y}. \tag{3.167}$$

With a photocurrent of $i(t) = R|E_1(t)|^2 - R|E_2(t)|^2$ and an error probability of $p_e = \Pr\{i(t) < 0\}$, the error probability is the same as that for DPSK signal of Eq. (3.163) but half the SNR. For dual-filter direct-detection FSK receiver, the error probability is

$$p_e = \frac{1}{2}\exp\left(-\frac{\rho_s}{2}\right)\left(1 + \frac{\rho_s}{8}\right). \tag{3.168}$$

Direct-detection FSK signal is 3-dB worse than direct-detection DPSK signal. However, using the same receiver of Fig. 3.12, direct-detection MSK receiver has the same performance as DPSK signal.

Figure 3.15 shows the error probability of FSK signal demodulated using a synchronous receiver, asynchronous heterodyne receiver, and direct-detection dual-filter receiver. Compared with Fig. 3.13, frequency-modulated signal is 3-dB worse than phase-modulated signal. For an error probability of 10^{-9}, asynchronous heterodyne detection is about 0.45 dB worse than synchronous detection, and direct-detection is about 0.40 dB worse than asynchronous differential detection. The quantum-limited receivers require 36.0, 40.0, and 43.8 photons/bit.

Practical FSK receiver may use a single filter with a performance similar to ASK signal with 3-dB worse receiver sensitivity. If the two optical filters have crosstalk, the outputs of $E_1(t)$ and $E_2(t)$ have correlation and the error probability is given by Eq. (3.119) with a correlation coefficient depending on the filter crosstalk. In order to improve the performance, FSK signal with large frequency deviation can be used with discriminator-based detector for better performance. The performance of FSK signal with frequency discriminator is the same as heterodyne system analyzed in Sec. 3.3.2. However, high frequency-deviation FSK system has very small spectral efficiency.

Direct-detection receiver for frequency-modulated signal was used for a long time by Saito and Kimura (1964), Saito et al. (1983), and Olsson and Tang (1979). Single-filter direct-detection FSK receiver can use Fabry-Perot resonator (Chraplyvy et al., 1989, Kaminow, 1990, Kaminow et al., 1988, Malyon and Stallard, 1990, Willner, 1990, Willner et al., 1990) or ring resonator (Oda et al., 1991, 1994). Using the interferometer of Fig. 3.12, CPFSK signal was directly detected by Idler et al. (2004), Malyon and Stallard (1989) and Toba et al. (1990, 1991). When an optical filter is used to demodulate the FSK signal, it can also function as a demultiplexer to select the corresponding WDM channel.

5. Phase-Diversity Receiver

Phase-diversity receiver is another type of asynchronous detector for homodyne receiver. The phase-diversity receiver is based on the quadrature receiver of Fig. 3.4. From the photocurrent of Eqs. (3.49) and (3.50) with $\omega_{\mathrm{IF}} = 0$, including a random phase of θ_0 from either the received signal or the LO signal, we obtain

$$\begin{aligned} r_I(t) &= A_s(t)\cos[\phi_s(t) + \theta_0] + n_I(t), \\ r_Q(t) &= A_s(t)\sin[\phi_s(t) + \theta_0] + n_Q(t), \end{aligned} \tag{3.169}$$

where $A_s(t)$ is due to amplitude modulation, $\phi_s(t)$ from phase modulation, and $n_I(t)$ and $n_Q(t)$ are the identical independently distributed additive Gaussian noise. The random phase of θ_0 in Eq. (3.169) is used to model a receiver without phase locking. In Eq. (3.169), the random phase of θ_0 is a constant over a bit interval of T but can be changed slowly from bit to bit.

5.1 Phase-Diversity ASK Receiver

If the signal is amplitude-modulated with $\phi_s(t) = 0$ in the signals of Eq. (3.169), amplitude modulated signal with $A(t) = \{0, A\}$ may be demodulated by the received envelope of

$$r_d(t) = \sqrt{r_I^2(t) + r_Q^2(t)}. \tag{3.170}$$

Note that the phase-diversity ASK receiver is similar to heterodyne envelope-detection receiver of Sec. 3.3.1. The error probability of phase-diversity ASK receiver is the same as Eq. (3.93) of $p_e = \frac{1}{2}\exp(-\rho_s/2)$ if the threshold is chosen as the $A/2$. As a homodyne phase-diversity ASK receiver is mathematically the same as a heterodyne ASK receiver based on envelope detection, other aspects of a homodyne phase-diversity ASK receiver can also be analyzed the same as the corresponding receivers in Sec. 3.3.1 or Fig. 3.11.

The linear optical sampling scheme of Dorrer et al. (2003) is functionally a phase-diversity ASK receiver using LO laser with short optical pulse train.

5.2 Phase-Diversity DPSK Receiver

If the data in encode in the phase difference of $\phi_s(t) - \phi_s(t-T)$ using DPSK modulation, the amplitude of $A(t) = A$ is a constant. The phase difference can be demodulated using

$$\begin{aligned} r_d(t) &= r_I(t)r_I(t-T) + r_Q(t)r_Q(t-T) \\ &= A^2\cos\left[\phi_s(t) - \phi_s(t-T)\right] + \text{noise terms}. \end{aligned} \tag{3.171}$$

Without noise, $r_d(t)$ is proportional to $\cos\left[\phi_s(t) - \phi_s(t-T)\right]$ and $r_d(t) = \pm A^2$ when $\phi_s(t) - \phi_s(t-T) = 0$ or π, respectively. The phase-diversity receiver for DPSK signal has the same performance as a DPSK heterodyne receiver using differential detection of Sec. 3.3.3 or Fig. 3.13 with an error probability of

$$p_e = \frac{1}{2}\exp(-\rho_s). \tag{3.172}$$

5.3 Phase-Diversity Receiver for Frequency-Modulated Signals

For FSK signal, the received signals of Eq. (3.169) at the output of the quadrature homodyne receiver of Fig. 3.4 are

$$r_I(t) = A\cos(\pm\pi\Delta f t) + n_I(t), \tag{3.173}$$
$$r_Q(t) = A\sin(\pm\pi\Delta f t) + n_Q(t), \tag{3.174}$$

where $\Delta f = (\omega_1 - \omega_2)/(2\pi)$ is the frequency difference between the binary FSK signal. The demodulated signal can be

$$\begin{aligned} r_d(t) &= r_I(t)\frac{dr_Q(t)}{dt} + r_Q(t)\frac{dr_I(t)}{dt} \\ &= \mp\pi\Delta f A^2 + \text{noise terms}. \end{aligned} \tag{3.175}$$

The receiver sensitivity increases with the frequency difference of Δf. The performance of the system is similar to that of Sec. 3.3.5 using frequency discriminator.

For CPFSK signal, the output from the receiver is similar to the two signals of Eq. (3.174)

$$r_I(t) = A\cos(\pm\pi\Delta f t + \phi_0 + \theta_0) + n_I(t), \tag{3.176}$$

$$r_Q(t) = A\sin(\pm\pi\Delta f t + \phi_0 + \theta_0) + n_Q(t). \tag{3.177}$$

The demodulated signal is

$$\begin{aligned} r_d(t) &= r_I(t)r_Q(t-\tau) + r_Q(t)r_I(t-\tau) \\ &= \pm A^2\sin(\pi\Delta f\tau) + \text{noise terms}. \end{aligned} \tag{3.178}$$

For MSK signal, $\tau = T$ and $\Delta f = 1/2T$, the receiver sensitivity is the same as differential detection DPSK signals of Eq. (3.105) or Fig. 3.13.

Homodyne phase-diversity receiver was mostly for ASK and DPSK signals (Cheng et al., 1989, Davis et al., 1987, Davis and Wright, 1986, Hodgkinson et al., 1985, 1988, Kazvosky et al., 1987, Okoshi and Cheng, 1987, Smith, 1987). Phase-diversity receiver for FSK and CPFSK modulation is not as popular (Davis et al., 1987, Noé et al., 1988b, Siuzdak and van Etten, 1991, Tsao et al., 1990, 1992). Reviewed by Kazovsky (1989), phase-diversity receiver was also analyzed by Hao and Wicker (1995), Nicholson and Stephens (1989), Siuzdak and van Etten (1989), and Ho and Wang (1995), especially those using 120° optical hybrid.

When the system is limited by optical amplifier noise, phase-diversity receiver performs about 0.4 dB better than direct-detection receiver for an error probability of 10^{-9}. The main advantage of phase-diversity receiver is to provide narrow channel spacing for WDM systems or reduce the bandwidth requirement of the receiver. The same as the image-rejection heterodyne receiver of Sec. 3.1.4, phase-diversity receiver also has the advantage to reduce the channel spacing of a WDM system. Both phase-diversity homodyne and image-rejection heterodyne receivers use the same optical front-end of Fig. 3.4 with a 90° optical hybrid.

6. Polarization-Diversity Receiver

The single-branch receiver of Fig. 3.1 and the balanced receiver of Fig. 3.3 all require polarization control to match the polarization of

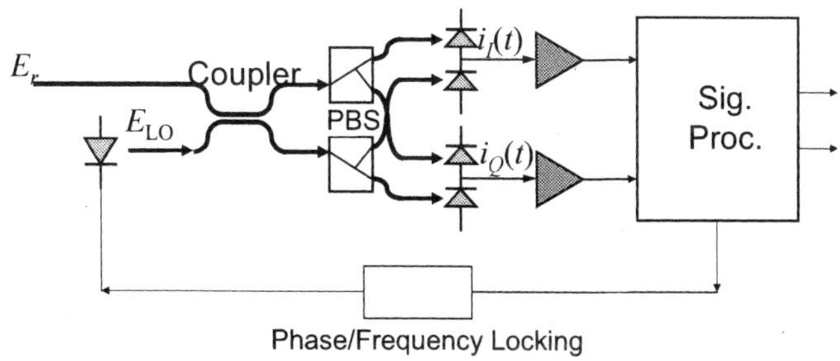

Figure 3.16. A polarization-diversity receiver.

the received signal with that of the LO laser. The image-rejection receiver of Fig. 3.5 requires polarization control such that the polarization of the received signal is 45° linearly polarized with respect to the receiver polarization. System without polarization control is possible using polarization-diversity techniques.

Figure 3.16 shows a polarization-diversity receiver with an optical front end similar to the 90° optical hybrid of Fig. 3.4. However, unlike Fig. 3.4 in which the 90° optical hybrid is operated with linearly polarized received signal, the received optical field of the polarization-diversity receiver is generally elliptically polarized and uncontrolled. The LO laser is linearly polarized at 45° with respect to the receiver polarization. The received signal is mixed with the LO signal using a 3-dB coupler and forwards to two separated PBS.

With random polarization without APC, the received signal is assumed to have an electric field of

$$E_r(t) = A_s(t)\left[\cos\varphi\mathbf{x} + \sin\varphi e^{i\theta}\mathbf{x}\right]e^{j\omega_c t + j\phi_s(t)}, \tag{3.179}$$

where the angles of φ and θ are relative to the receiver polarization. Linearly 45° polarized to the receiver polarization, the LO laser has an electric field of

$$E_{\mathrm{LO}}(t) = \frac{A_L}{\sqrt{2}}(\mathbf{x}+\mathbf{y})e^{j\omega_{\mathrm{LO}}t}. \tag{3.180}$$

The electric fields at the outputs of the 3-dB coupler are

$$\begin{aligned} E_1(t) &= \frac{1}{\sqrt{2}}\left[E_r(t) + E_{\mathrm{LO}}(t)\right] \\ &= \frac{\mathbf{x}}{2}\left[A_s(t)\cos\varphi e^{j\omega_c t + j\phi_s(t)} + A_L e^{j\omega_{\mathrm{LO}}t}\right] \\ &\quad + \frac{\mathbf{y}}{2}\left[A_s(t)\sin\varphi e^{j\omega_c t + j\phi_s(t) + j\theta} + A_L e^{j\omega_{\mathrm{LO}}t}\right], \end{aligned} \tag{3.181}$$

and

$$\begin{aligned} E_2(t) &= \frac{1}{\sqrt{2}}\left[E_r(t) - E_{\mathrm{LO}}(t)\right] \\ &= \frac{\mathbf{x}}{2}\left[A_s(t)\cos\varphi e^{j\omega_c t + j\phi_s(t)} - A_L e^{j\omega_{\mathrm{LO}} t}\right] \\ &\quad + \frac{\mathbf{y}}{2}\left[A_s(t)\sin\varphi e^{j\omega_c t + j\phi_s(t) + j\theta} - A_L e^{j\omega_{\mathrm{LO}} t}\right]. \end{aligned} \tag{3.182}$$

After the two PBS, similar to the quadrature receiver of Fig. 3.4, the photocurrents at the output of the two balanced receivers are

$$i_I(t) = R\cos\varphi A_s(t) A_L \cos[\omega_{\mathrm{IF}} t + \phi_s(t)], \tag{3.183}$$

and

$$i_Q(t) = R\sin\varphi A_s(t) A_L \cos[\omega_{\mathrm{IF}} t + \phi_s(t) + \theta]. \tag{3.184}$$

with a phase difference of θ.

In both photocurrents of Eqs. (3.183) and (3.184), the additive noise has the same variance and independent of each other. Including noise, the received signal is $i_I(t) + n_I(t)$ and $i_Q(t) + n_Q(t)$ where $E\{n_I^2(t)\} = E\{n_Q^2(t)\} = \sigma_n^2$.

6.1 Combination in Polarization-Diversity Receiver

The polarization-diversity scheme is applicable to most modulation formats. Data are demodulated by combining the information from two polarization branches. The photocurrents can be processed in either the IF or the baseband. If the signal are combined in the IF stage, the carrier phase must be matched to cancel the phase difference of θ between Eqs. (3.183) and (3.184). When the phase difference of θ changes due to external disturbance, the phases of two signals must be adjusted adaptively before the combination process. In baseband combination, the signal of either $A_s(t)$ or $\phi_s(t)$ is demodulated independently for each polarization component. If the demodulation process tracks out the phase fluctuation of the IF signals, phase matching is not necessary. In practice, baseband combining is more practical with simple implementation.

Without loss of generality with $R = 1$, we assume that the signals after phase matching for $\theta = 0$ are

$$r_i(t) = \cos\varphi A_s(t)\cos\phi_s(t) + n_i(t), \tag{3.185}$$

and

$$r_q(t) = \sin\varphi A_s(t)\cos\phi_s(t) + n_q(t), \tag{3.186}$$

corresponding to the in- and quadrature-phase components.

Among all methods to combine the two polarization components, maximum ratio is the best that maximizes the output SNR. The in-phase component of Eq. (3.183) has a gain of $\cos\varphi$ and the quadrature-phase component of Eq. (3.184) has a gain of $\sin\varphi$. With maximum-ratio combination, the combined signal is

$$\begin{aligned} r_c(t) &= \cos\varphi r_i(t) + \sin\varphi r_q(t), \\ &= A_s(t)\cos\phi_s(t) + \cos\varphi n_i(t) + \sin\varphi n_q(t), \end{aligned} \tag{3.187}$$

where the combined signal is the same as that with polarization control and the noise is also Gaussian noise having the same variance as $n_i(t)$ or $n_q(t)$. With maximum-ratio combination, there is no penalty using polarization diversity.

The simplest combining scheme may be equal-gain combining with a combined signal of

$$\begin{aligned} r_c(t) &= r_i(t) + r_q(t) \\ &= (\cos\varphi + \sin\varphi)A_s(t)\cos\phi_s(t) + n_i(t) + n_q(t). \end{aligned} \tag{3.188}$$

The SNR penalty due to equal-gain combining is

$$\delta_p = \frac{(\cos\varphi + \sin\varphi)^2}{2} = \frac{1 + \sin(2\varphi)}{2}, \qquad 0 < \varphi < \frac{\pi}{2} \tag{3.189}$$

Selection-combining scheme chooses the polarization component with the largest power. The penalty due to selection combining is

$$\delta_p = \max(\cos^2\varphi, \sin^2\varphi) = \frac{1}{2}\left(1 + |\cos(2\varphi)|\right), \qquad 0 < \varphi < \frac{\pi}{2} \tag{3.190}$$

where the factor of 2 for noise enhancement is due to the addition of two identical and independent noise sources.

Another combining scheme can be used for either heterodyne or homodyne ASK signal to square and combine the two signals. The combined signal is

$$\begin{aligned} r_c(t) &= r_i^2(t) + r_q^2(t) \\ &= A_s^2(t)\cos^2\phi_s(t) + 2A_s(t)\cos\phi_s(t)[n_i(t) + n_q(t)]n_i^2(t) + n_q^2(t). \end{aligned} \tag{3.191}$$

In homodyne detection, square-combining is only possible for ASK signal in which $\phi_s(t) = 0$ and $A_s(t) \in \{A, 0\}$. In heterodyne scheme, square combination is also possible for both DPSK and FSK signals with envelope detection.

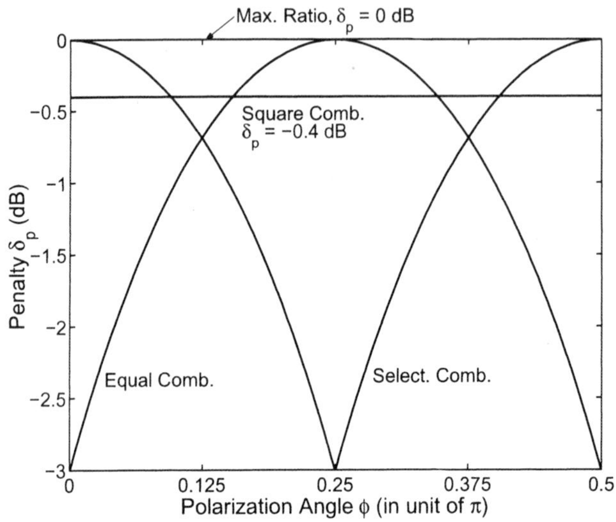

Figure 3.17. The penalty using various schemes to combine the signal from two polarization-diversified signals.

Figure 3.17 shows the SNR penalty for various signal combination schemes. The SNR penalty is shown as a function of the polarization angle of φ. Square combination has a penalty of 0.4 dB and is usually better than either equal or selected combining in most of the cases. Equal combining is the best when the polarization angle is around $\varphi = \pi/4$ and selected combining is the best when the polarization angle is around $\phi = 0$ and $\pi/2$.

First suggested by Okoshi (1985, 1986) for combination in IF, the above polarization-diversity schemes were discussed in detail in Ryu (1995). Figure 3.17 is almost identical to similar figure in Ryu (1995).

The two independent channels in polarization-division multiplexing (PDM) can be separated with maximum-ratio combination. If the orthogonal signals are $A_{s1}(t)e^{j\phi_{s1}(t)}$ and $A_{s2}(t)e^{j\phi_{s2}(t)}$, after IF processing, the two output signals are

$$r_i(t) = \cos\varphi A_{s1}(t)\cos\phi_{s1}(t) + \sin\varphi A_{s2}(t)\cos\phi_{s2}(t) + n_i(t), \quad (3.192)$$

and

$$r_q(t) = \sin\varphi A_{s1}(t)\cos\phi_{s1}(t) - \cos\varphi A_{s2}(t)\cos\phi_{s2}(t) + n_q(t). \quad (3.193)$$

The two independent signals can be recovered based on the combiner of

$$\cos\varphi r_i(t) + \sin\varphi r_q(t) \quad (3.194)$$

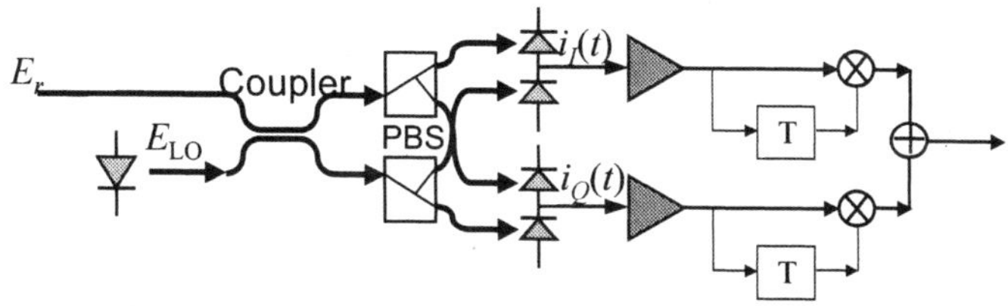

Figure 3.18. A polarization-diversity receiver for DPSK and MSK signals.

and

$$\sin\varphi r_i(t) - \cos\varphi r_q(t), \tag{3.195}$$

respectively.

6.2 Heterodyne Differential Detection with Polarization Diversity

Figure 3.18 shows a polarization-diversity receiver for both DPSK and MSK signals using electrical delay-and-multiplier circuits. Without loss of generality and assumes a DPSK signal, after the delay-and-multiplier circuitry, the upper branch of Fig. 3.18 has a signal of

$$r_1(t) = \cos^2\varphi A^2 \cos[\phi_s(t) - \phi_s(t-T)] + \text{noise terms}, \tag{3.196}$$

and the lower branch of Fig. 3.18 has a signal of

$$r_2(t) = \sin^2\varphi A^2 \cos[\phi_s(t) - \phi_s(t-T)] + \text{noise terms}, \tag{3.197}$$

where both factors $\cos^2\varphi$ and $\sin^2\varphi$ are given by the multiplexer of Fig. 3.18.

When the above two signals are combined, the decision variable becomes

$$r_d(t) = A^2 \cos[\phi_s(t) - \phi_s(t-T)] + \text{noise terms}. \tag{3.198}$$

In the above two equations, the common factor of $R^2/2$ is ignored for simplicity. If all noise terms are written down, the noise statistics is the same as that of direct-detection DPSK signal in Sec. 3.4.2 with the error probability of Eq. (3.163). The error probability of Eq. (3.163) was first derived by Okoshi et al. (1988) for phase-diversity DPSK signals. MSK signal should also have the same performance. From Fig. 3.13, the degradation of phase-diversity DPSK or MSK signal is about 0.40 dB compared with heterodyne DPSK signal, similar to the degradation of the square combination of Fig. 3.17.

When FSK signal is detected through envelope detection of Fig. 3.8 based on two filters, the signal passes through a squarer. For envelope detection of FSK signal, the decision variable is not sensitive to the phase difference of θ in Eq. (3.184). When the detected signals from two branches of Fig. 3.18 are combined together, the combination is actually based on square combination and has a degradation of 0.40 dB as from Fig. 3.13.

Polarization diversity has been considered mostly for DPSK and FSK signals using baseband combination (Cline et al., 1990, Glance, 1987, Imai et al., 1991, Kavehrad and Glance, 1988, Okoshi and Cheng, 1987, Ryu et al., 1987, 1991a, Shibutani and Yamazaki, 1989). Most early field trials of coherent optical communications used polarization diversity with frequency modulated signal (Bødker et al., 1991, Imai et al., 1990a, Ryu et al., 1988, 1992). In additional to polarization alignment using APC and polarization-diversity, polarization scrambling provides random polarization within the bit interval (Caponio et al., 1991, Cimini Jr. et al., 1988, Habbab and Cimini Jr., 1988, Hodgkinson et al., 1987, Meada and Smith, 1990). Not compatible with PDM, polarization scrambling has a power penalty of 3 dB due to the loss of signal to the polarization orthogonal to the receiver polarization.

7. Polarization-Shift Keying Modulation

A single-mode optical fiber can support two polarizations and the electric field in an optical fiber can generally be expressed as

$$E_r(t) = \left[E_x(t)\mathbf{x} + E_y(t)\mathbf{y}\right] e^{j\omega_c t}. \tag{3.199}$$

The above electric field of Eq. (3.199) is the same as the signal of Eq. (3.179) but using different notation. In the single-branch receiver of Fig. 3.1, we assume that the signal of Eq. (3.2) has a single polarization and aligned with the reference polarization of the receiver of $\mathbf{x}$.

Comparing the electric fields of Eq. (3.199) with Eq. (3.1), both $E_x(t)$ and $E_y(t)$ can be used independently to transmit two data streams using polarization-division multiplexing (PDM). In PDM system, a PBS is used to separate $E_x(t)$ and $E_y(t)$. APC is required to align the signal to the PBS. After the PBS, the two data streams encoded in $E_x(t)$ and $E_y(t)$ are then demodulated independently.

The electric fields of $E_x(t)$ and $E_y(t)$ of Eq. (3.199) can be used together to converse information by polarization-shift keying (PolSK). The simplest PolSK scheme is to transmit

$$s_1(t) = A\mathbf{x}e^{j\omega_c t}, \tag{3.200}$$

and

$$s_2(t) = A\mathbf{y}e^{j\omega_c t} \tag{3.201}$$

with $E_x(t) = E_y(t) = A$ but used alternatively to carry either "0" or "1". The performance of this simple PolSK scheme is identical to that of FSK signal with two orthogonal binary signals. Similar to FSK signal, a PolSK signal can be directly demodulated using a PBS, followed by a balanced receiver. The error probability of direct-detection PolSK is that of Eq. (3.85) of $p_e = \frac{1}{2}\exp(-\rho_s/2)$. When a heterodyne receiver is used, the LO laser can have a polarization that is 45° to both $\mathbf{x}$ and $\mathbf{y}$. The performance of heterodyne receiver is also the same as that of Eq. (3.85).

Direct-detection polarization modulation has a long history (Daino et al., 1974, Pratt, 1966). For coherent optical communications, PolSK was proposed mainly to overcome laser phase noise (Benedetto and Poggiolino, 1990, Betti et al., 1988, Calvani et al., 1988, Dietrich et al., 1987, Imai et al., 1990b). When PolSK is designed based on the Stokes parameters, Dietrich et al. (1987) used the s_1 parameter, Calvani et al. (1988) detected the s_2 parameter, Imai et al. (1990b) based on the differential s_1 parameter, and Betti et al. (1988) and Benedetto and Poggiolino (1990) detected all Stokes parameters without optical polarization control and used signal processing to find the correct polarization states of the received signal. While PolSK systems usually use for heterodyne receiver, homodyne PolSK systems were described in Betti et al. (1991). PolSK systems are analyzed in details by Benedetto and Poggiolini (1992) and Benedetto et al. (1995a,b).

8. Comparison of Optical Receivers

Table 3.2 shows the performance of a quantum-limited optical receiver for various types of signal. An optimal receiver is assumed in Table 3.2 with, for example, optimal threshold setting and optical matched filter. The SNR penalty is also calculated in Table 3.2 compared with a homodyne or heterodyne PSK receiver limited by amplifier noise. The performance of Table 3.2 includes the performance of shot- and amplifier-noise limited signals.

Synchronous receivers for phase-modulated signals have best receiver sensitivity. Asynchronous receivers for phase-modulated signals have a degradation less than 1 dB compared to the best synchronous receiver. The performance of phase-diversity receiver is the same as asynchronous heterodyne receiver.

Based on the square combination, polarization-diversity receiver has the performance the same as direct-detection receiver. If maximum-

Table 3.2. Comparison of Different Optical Receivers for an Error Probability of 10^{-9}.

Modulation Format	Sensitivity (photons/bit) Shot-Noise	Amplifier-Noise	Penalty (dB)
Homodyne PSK	9	18	0
Heterodyne PSK	18	18	0
Heterodyne DPSK, MSK	20	20	0.45
Direct-Detection DPSK	–	22	0.85
Homodyne ASK	18	36	3
Heterodyne ASK, FSK	36	36	3
Envelope Detection Heterodyne ASK	39	39	3.4
Direct-Detection ASK, FSK	–	40	3.5
Dual-Filter FSK, PolSK	40	40	3.5
Single-Filter FSK	80	80	6.5

ratio combination is used, polarization-diversity receiver has the same performance the corresponding homodyne or heterodyne receiver.

This chapter just analyzes the receiver with only the dominant amplifier noise. In next chapter, in linear regime, other degradations to the signal is studied.

APPENDIX 3.A: Marcum Q Function

For a Gaussian random variable of $A + x_1$ and $x_2(t)$, the amplitude of $R = \sqrt{[A+x_1]^2 + x_2^2}$ has a Rice distribution of

$$p(r) = \frac{r}{\sigma_n^2} I_0\left(\frac{rA}{\sigma_n^2}\right) \exp\left(-\frac{r^2 + A^2}{2\sigma_n^2}\right). \tag{3.A.1}$$

The cumulative distribution function of Rice distribution is

$$\int_x^{+\infty} p(r)\mathrm{d}r = Q\left(\frac{A}{\sigma_n}, \frac{x}{\sigma_n}\right), \tag{3.A.2}$$

where the Marcum Q function was first used for radar theory (Marcum, 1960) and is very useful in the analysis of noncoherent or asynchronous detection of binary signals. This appendix presented some important properties of Marcum Q functions.

The Marcum Q function is a real function of

$$Q(a,b) = \int_b^\infty x I_0(ax) \exp\left(-\frac{x^2+a^2}{2}\right) \mathrm{d}x, \tag{3.A.3}$$

or

$$Q\left(\sqrt{2a}, \sqrt{2b}\right) = \int_b^\infty e^{-(a+x)} I_0\left(2\sqrt{ax}\right) \mathrm{d}x, \tag{3.A.4}$$

where $I_0(x)$ is the zeroth-order modified Bessel function of the first kind. From the definition, we have

$$Q(0,b) = e^{-b^2/2}, \tag{3.A.5}$$

$$Q(a,0) = 1. \tag{3.A.6}$$

The modified Bessel function of $I_0(x)$ can be represented as inverse Laplace transform of

$$I_0(2\sqrt{ab}) = \frac{1}{2\pi j} \int_{c-j\infty}^{c+j\infty} \frac{1}{p} \exp\left(ap + \frac{b}{p}\right) \mathrm{d}p, \qquad c > 0 \tag{3.A.7}$$

where c is a real positive number. Substitute Eq. (3.A.7) into the Marcum Q function of Eq. (3.A.3) and exchange the order of integration, we get

$$\begin{aligned} Q\left(\sqrt{2a}, \sqrt{2b}\right) &= \frac{e^{-a}}{2\pi j} \int_{c-j\infty}^{c+j\infty} \frac{1}{p} \int_b^\infty e^{-x} \exp\left(ap + \frac{x}{p}\right) \mathrm{d}x \mathrm{d}p \\ &= e^{-(a+b)} \frac{1}{2\pi j} \int_{c-j\infty}^{c+j\infty} \frac{\exp\left(ap + \frac{b}{p}\right)}{p-1} \mathrm{d}p, \qquad c > 1 \end{aligned} \tag{3.A.8}$$

or

$$Q\left(\sqrt{2a}, \sqrt{2b}\right) = e^{-(a+b)} \frac{1}{2\pi j} \int_{c-j\infty}^{c+j\infty} \frac{\exp\left(bp + \frac{a}{p}\right)}{p(p-1)} \mathrm{d}p, \qquad 0 < c < 1 \tag{3.A.9}$$

and

$$Q(a,b) = -e^{-(a^2+b^2)/2} \frac{1}{2\pi j} \int_{c-j\infty}^{c+j\infty} \frac{\exp\left(\frac{b^2 p}{2} + \frac{a^2}{2p}\right)}{p(p-1)} \mathrm{d}p, \qquad 0 < c < 1 \tag{3.A.10}$$

By straightforward residue calculation involving shifting of the path, one immediately also derives the useful symmetry relationships

$$Q(a,b) + Q(b,a) = 1 + e^{-(a^2+b^2)/2} I_0(ab), \tag{3.A.11}$$

$$Q(a,a) = \frac{1}{2} + \frac{1}{2} e^{-a^2} I_0(a^2). \tag{3.A.12}$$

The Marcum Q function can be calculated using a function series of

$$Q(a,b) = e^{-(a^2+b^2)/2} \sum_{m=0}^\infty \left(\frac{a}{b}\right)^m I_m(ab), \qquad b > a \tag{3.A.13}$$

$$= 1 - e^{-(a^2+b^2)/2} \sum_{m=0}^\infty \left(\frac{b}{a}\right)^m I_m(ab), \qquad a > b. \tag{3.A.14}$$

or using the method in Cantrell and Ojha (1987).

In the derivation of the error probability of orthogonal modulation with correlation, we need to evaluate the probability that a Rice distributed random variable exceeds another. If the envelope of two Gaussian processes, R_1 and R_2, are independently distributed, with the well-known Rice p.d.f. of

$$p_{R_1}(r_1) = \frac{r_1}{\sigma_1^2} \exp\left(-\frac{A_1^2 + r_1^2}{2\sigma_1^2}\right) I_0\left(\frac{A_1 r_1}{\sigma_1^2}\right), \tag{3.A.15}$$

$$p_{R_2}(r) = \frac{r_2}{\sigma_2^2} \exp\left(-\frac{A_2^2 + r_2^2}{2\sigma_2^2}\right) I_0\left(\frac{A_2 r_2}{\sigma_2^2}\right), \tag{3.A.16}$$

the error probability is

$$\begin{aligned} p_e &= \Pr\{R_1^2 < R_2^2\} = \Pr\{R_1 < R_2\} \\ &= \int_0^\infty p_{R_1}(r_1) \int_{r_1}^\infty p_{R_2}(r_2) \mathrm{d}r_2 \mathrm{d}r_1 \\ &= \int_0^\infty p_{R_1}(r_1) Q\left(\frac{A_2}{\sigma_2}, \frac{r_1}{\sigma_2}\right) \mathrm{d}r_1. \end{aligned} \tag{3.A.17}$$

By using Eq. (3.A.8) in (3.A.17), interchanging orders of integrations, we find that

$$\begin{aligned} p_e = \exp\left[-\frac{1}{2}\left(\frac{A_1^2}{\sigma_1^2} + \frac{A_2^2}{\sigma_2^2}\right)\right] \cdot \frac{1}{2\pi j} \int_{c-j\infty}^{c+j\infty} \frac{\exp\left(\frac{A_2^2}{2\sigma_2^2} p\right)}{p-1} \\ \times \frac{1}{1 + \frac{\sigma_1^2}{\sigma_2^2}\left(1 - \frac{1}{p}\right)} \exp\left[\frac{A_1^2}{2\sigma_1^2} \frac{1}{1 + \frac{\sigma_1^2}{\sigma_2^2}\left(1 - \frac{1}{p}\right)}\right] \mathrm{d}p, \quad c > 1. \end{aligned} \tag{3.A.18}$$

Using the obvious partial fraction expansion, after some algebra via Eqs. (3.A.7) and (3.A.8), the error probability is

$$p_e = Q(a, b) - \frac{\sigma_1^2}{\sigma_1^2 + \sigma_2^2} e^{-(a^2+b^2)/2} I_0(ab), \tag{3.A.19}$$

where

$$a^2 = \frac{A_2^2}{\sigma_1^2 + \sigma_2^2}, \qquad b^2 = \frac{A_1^2}{\sigma_1^2 + \sigma_2^2}. \tag{3.A.20}$$

The results of Eq. (3.A.19) can also be written in more symmetric forms of

$$p_e = \frac{\sigma_1^2}{\sigma_1^2 + \sigma_2^2} [1 - Q(b, a)] + \frac{\sigma_2^2}{\sigma_1^2 + \sigma_2^2} Q(a, b), \tag{3.A.21}$$

and

$$p_e = \frac{1}{2} [1 - Q(b, a) + Q(a, b)] - \frac{\sigma_1^2 - \sigma_2^2}{\sigma_1^2 + \sigma_2^2} e^{-(a^2+b^2)/2} I_0(ab). \tag{3.A.22}$$

This Appendix follows the approaches of Stein (1964) for Marcum Q function. Both Schwartz et al. (1966) and Betti et al. (1995) also had similar Appendix. Higher-order Marcum Q functions are considered in Proakis (2000, Appendix B). In this book, only the second-order Marcum Q function of Eq. (3.130) is used.

Chapter 4

IMPAIRMENT TO OPTICAL SIGNAL

The last chapter is about the performance of optical receivers in the ideal case limited only by amplifier and shot noises for system with and without optical amplifiers, respectively. In practice, there are many other noise sources or distortions that degrade the performance of the optical receiver.

If the local oscillator (LO) laser is very noisy with high relative intensity noise (RIN), using the single branch receiver of Sec. 3.1.1, the optical signal is very likely to be limited by the LO noise. One of the main advantage of balanced receiver is to reduce the impact of LO intensity noise. In phase-modulated optical communications, the laser sources must be coherent with low phase noise. In early studies of phase-modulated optical communications, laser phase noise was considered the major limitation of phase-modulated signals (Betti et al., 1995, Okoshi and Kikuchi, 1988, Ryu, 1995). Because early systems typically had a low data rate and the laser phase must remain the same over multiple bit intervals, the ratio of laser linewidth to the data rate for early systems is very significant compared with contemporary 10 and 40-Gb/s systems.

When phase-modulated optical signals are transmitted in the optical fiber, the signal is further distorted by the linear effects of fiber chromatic and polarization-mode dispersion (PMD). When different parts of an optical spectrum travel in slightly different speed due to chromatic dispersion, the maximum transmission distance is limited before the usage of dispersion compensation. If the fiber core is not perfectly circular, the two polarizations in a single-mode fiber does not has the same speed. With the dependence of propagating speed on launched polarization, PMD also broadens the optical signal and limits the transmission distance. This chapter considers only linear fiber effects, signal distortion due to fiber nonlinearities is the main topic in following chapters.

1. Relative Intensity Noise

In the analysis of the single-branch receiver of Sec. 3.1.1, the RIN is included as $n_L(t)$ in Eq. (3.4). If the power of LO laser of P_{LO} is much larger than the signal power of P_r, in the photocurrent of Eq. (3.7), the two mean noise sources are LO-spontaneous beat noise and the RIN from $|A_L + n(t)|^2$. Including only LO-spontaneous beat noise and RIN, the signal-to-noise ratio (SNR) of single-branch receiver is

$$\rho_s = \frac{\frac{1}{2}R^2 P_{\mathrm{LO}} P_r}{\frac{1}{2}R^2 P_{\mathrm{LO}} S_{n_s} B_d + \frac{1}{2}R^2 P_{\mathrm{LO}}^2 \sigma_{\mathrm{RIN}}^2}, \tag{4.1}$$

where the variance due to RIN is equal to

$$\sigma_{\mathrm{RIN}}^2 = \int_{f_{\mathrm{IF}}-B_d}^{f_{\mathrm{IF}}+B_d} \mathrm{RIN}(f)\mathrm{d}f, \tag{4.2}$$

where the RIN(f) is the RIN spectrum. For a required SNR of ρ_{s0}, the SNR penalty due to RIN is equal to

$$\delta_p = \frac{1}{1 - \frac{P_{\mathrm{LO}}}{P_r}\rho_{s0}\sigma_{\mathrm{RIN}}^2} \tag{4.3}$$

that depends on the power ratio of P_{LO}/P_r.

Figure 4.1 shows the SNR penalty as a function of RIN for single-branch receiver of Fig. 3.1. Figure 4.1 assumes a required SNR of 18 (12.5 dB), corresponding to an error probability of 10^{-9} for synchronous detected phase-shift keying (PSK) signal. With a large ratio of LO power of P_{LO} to received power P_r, the impact of RIN is very significant. The SNR penalty of Eq. (4.3) also increases with the required SNR of ρ_{s0}. For example, the RIN that can be tolerated is reduced by 3 dB if the required SNR increases by 3 dB.

An expression similar to Eq. (4.3) can also be derived for balanced receiver, the impact of RIN is drastically reduced with a SNR of

$$\rho_s = \frac{P_r}{S_{n_s} B_d + \frac{1}{2}P_r \hat{\sigma}_{\mathrm{RIN}}^2} \tag{4.4}$$

where the RIN is

$$\hat{\sigma}_{\mathrm{RIN}}^2 = \int_{-B_d}^{+B_d} \mathrm{RIN}(f)\mathrm{d}f. \tag{4.5}$$

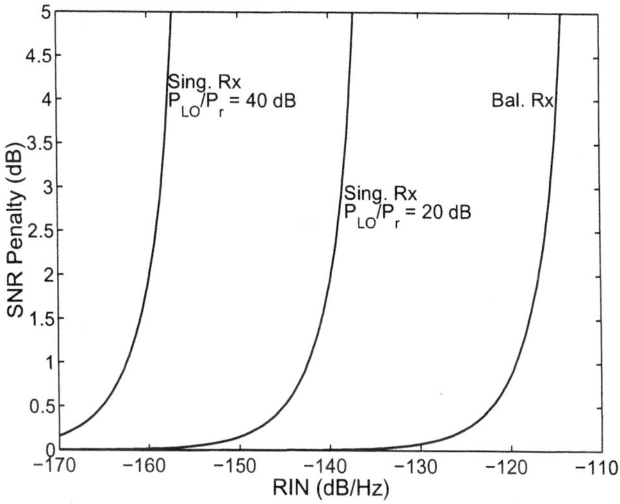

Figure 4.1. The SNR penalty as a function of the RIN of LO laser.

For a required SNR of ρ_{s0}, the SNR penalty due to RIN is equal to

$$\delta_p = \frac{1}{1 - \frac{1}{2}\rho_{s0}\hat{\sigma}^2_{\mathrm{RIN}}} \tag{4.6}$$

that is independent of the LO power of P_{LO}.

Comparing the SNR of Eq. (4.3) with (4.6), it is obvious that the impact of RIN to single branch receiver is much larger than that to balanced receiver in the normal case of $P_{\mathrm{LO}} \gg P_r$. Figure 4.1 also shows the SNR penalty as a function of RIN for balanced receiver of Fig. 3.3.

From Fig. 4.1, the SNR penalty for single-branch receiver is far higher than that for balanced receiver. For the ratio of $P_{\mathrm{LO}}/P_r = 40$ dB, even a RIN of -160 dB/Hz gives 1.5-dB penalty to a single-branch receiver. For low-speed system without optical amplifier, the ratio of P_{LO}/P_r is typically larger than 40 dB. For high-speed system with optical amplifiers, the ratio of P_{LO}/P_r is typically around 10 to 20 dB.

In direct-detection receiver without LO laser, the RIN from source laser should be less than amplifier noise.

An analysis of the impact of LO noise to single-branch and balanced receiver can be found at Abbas et al. (1985), Hodgkinson (1987), Yuen and Chan (1983), and Alexander (1987). If the two branches of the balanced receiver is not symmetric, the LO noise degrades the SNR of the system more than that of Eq. (4.6) (Abbas et al., 1985).

2. Phase Error for Differentially Detected Signals

In the heterodyne receiver of differential phase-shift keying (DPSK) signal in Fig. 1.4(b) and the direct-detection DPSK receiver of Figs. 1.4(c) or 3.12, the phase delay of the delay-and-multiplier circuits or asymmetric Mach-Zehnder interferometer must be equal to an integer multiple of 2π. In the heterodyne DPSK receiver of Fig. 1.4(b), the requirement is zero phase error of $\exp(j\omega_{\mathrm{IF}}T) = 1$, where T is the delay of the delay-and-multiplier circuits and approximately equal to the symbol time of the DPSK signal. In the direction-detection DPSK receiver of Fig. 3.12, the requirement is zero phase error of $\exp(j\omega_c T) = 1$. A phase error is induced into the system if either $\exp(j\omega_{\mathrm{IF}}T) = e^{j\theta_e}$ or $\exp(j\omega_c T) = e^{j\theta_e}$ for heterodyne or direct-detection DPSK receiver, respectively, where $-\pi < \theta_e < +\pi$ is the phase error of the DPSK receiver.

Alternatively, the interferometer or delay-and-multiplier circuits has its operation frequency. If the operation frequency does not match to the frequency of the signal, the phase error is equal to $\theta_e = 2\pi\delta f T$, where δf is the frequency mismatch.

2.1 Delay Phase Error for DPSK Signals

Using the series expansion of Appendix 4.A, almost directly from Eq. (4.A.19), the error probability for a DPSK signal with phase error is

$$p_e = \frac{1}{2} - \frac{\rho_s e^{-\rho_s}}{2} \sum_{k=0}^{\infty} \frac{(-1)^k}{2k+1} \left[I_k\left(\frac{\rho_s}{2}\right) + I_{k+1}\left(\frac{\rho_s}{2}\right) \right]^2 \cos[(2k+1)\theta_e], \tag{4.7}$$

where $I_k(\cdot)$ is the kth-order modified Bessel function of the first kind. The mth Fourier coefficient of the phase error is obviously equal to $\cos m\theta_e$.

There is simpler formula for the error probability with phase error than that of Eq. (4.7). The polarized direct-detection DPSK receiver is equivalent to heterodyne DPSK receiver. At the output photocurrent of the balanced receiver, ignoring the constant factor of coupler loss, photodiode responsivity, and receiver gain, the signal is similar to the difference of Eqs. (3.150) and (3.151) but without the noise from orthogonal polarization. The photocurrent is

$$i(t) = \left| E(t) + e^{j\theta_e} E(t-T) \right|^2 - \left| E(t) - e^{j\theta_e} E(t-T) \right|^2 \tag{4.8}$$

with phase error of θ_e.

As the signal of Eq. (4.8) is the same as that of the decision variable for envelope detection of correlated binary signals of Sec. 3.3.6, the equivalent correlation coefficient is (Proakis, 2000, Eq. 4.2-44)

$$\frac{\Re\left\{\left(1+e^{j\theta_e}\right)\left(1-e^{-j\theta_e}\right)^*\right\}}{\left|1+e^{j\theta_e}\right|\left|1-e^{-j\theta_e}\right|} = \sin\theta_e. \tag{4.9}$$

Using Eq. (3.119) for correlated binary signals, the error probability is

$$p_e = Q(a,b) - \frac{1}{2}e^{-(a^2+b^2)/2}I_0(ab),$$
$$a = \sqrt{2\rho_s}\left|\sin\frac{\theta_e}{2}\right|, \quad b = \sqrt{2\rho_s}\left|\cos\frac{\theta_e}{2}\right|, \tag{4.10}$$

where $Q(\cdot,\cdot)$ is the well-known first-order Marcum's Q function from Appendix 3.A.

The error probability of Eq. (4.10) has a SNR twice that in Eq. (3.119) as the correlated binary signal is $E(t) \pm e^{j\theta_e}E(t-T)$, uses two time slots, and doubles the energy per symbol. While difficult to prove analytically, from numerical results, the error probability of Eqs. (4.7) and (4.10) are identical.

For direct-detection DPSK receiver with amplifier noise from orthogonal polarization, without going into detail, from Proakis (2000) and Savory and Hadjifotiou (2004), the error probability of Eq. (4.10) becomes

$$p_e = Q(a,b) - \frac{1}{2}e^{-(a^2+b^2)/2}I_0(ab) + \frac{I_1(ab)}{8}e^{-(a^2+b^2)/2}\left(\frac{b}{a}-\frac{a}{b}\right), \tag{4.11}$$

where the last term arises due to the amplifiers noise from the orthogonal polarization.

From Eqs. (4.10) and (4.11), the error probability is independent of the sign of the phase error. In later parts of this section, only the results of positive phase error are shown. Figure 4.2 shows the error probability as a function of SNR ρ_s for a DPSK signal with phase errors of 10°, 20°, 30°, and 40°. Figure 4.2 also shows the error probability with no phase error of Eqs. (3.105) and (3.163), the same as the corresponding curves in Fig. 3.13. The error probability for DPSK signal with polarized direct-detection receiver is the same as that with heterodyne receiver.

Figure 4.3 shows the SNR penalty as a function of phase error for DPSK signal. The SNR penalty is calculated for an error probability of 10^{-9}, corresponding to a required SNR of 20 (13 dB) for heterodyne DPSK signal. The SNR penalty is calculated using both Eqs. (4.10) and (4.11) for direct-detection DPSK receiver without and with amplifier noise from orthogonal polarization. The SNR penalty for direct-

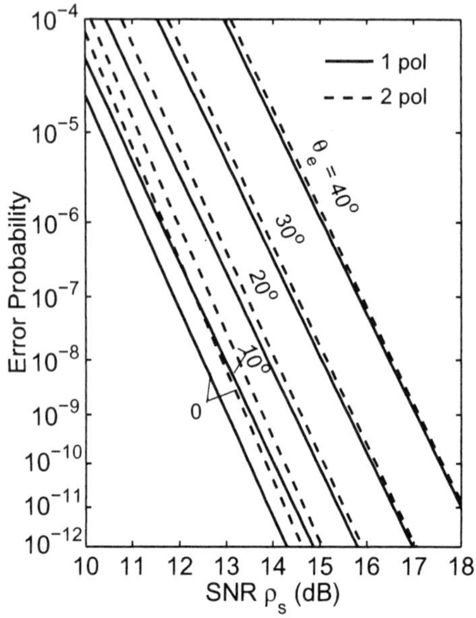

Figure 4.2. The error probability as a function of SNR ρ_s for DPSK with phase error.

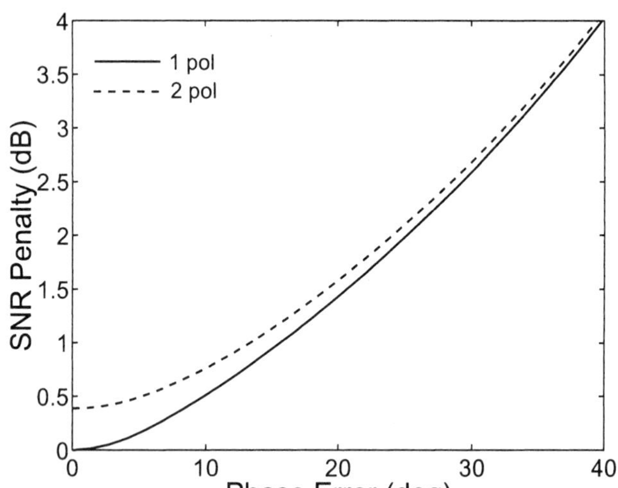

Figure 4.3. The SNR penalty as a function of interferometer phase error for DPSK signals.

detection DPSK receiver without amplifier noise from orthogonal polarization is also applicable to that for heterodyne DPSK receiver.

The curve of SNR penalty of a DPSK signal in Fig. 4.2 has insignificant difference with the corresponding curves in Kim and Winzer (2003, Fig. 3), Bosco and Poggiolini (2003, Fig. 2), and Winzer and Kim (2003, Fig.5) (required the adjusting of x-axis). The phase error for a SNR penalty of 1 dB is about 16° (or 4.5% of 360°) for a DPSK signal. In all of Bosco and Poggiolini (2003), Kim and Winzer (2003), and Winzer and Kim (2003), the phase error of 1-dB SNR penalty is about 4 to 5% from simulation or analysis. With narrow bandwidth and from Bosco and Poggiolini (2003), the SNR penalty due to phase error is more or less independent of the optical filter before the interferometer. For SNR penalty less than 2 dB and from Winzer and Kim (2003), the SNR penalty due to phase error is more or less independent of the electrical filtering after the balanced receiver. The SNR penalty of Fig. 4.3 is smaller than that from measurement (Kim and Winzer, 2003, Winzer and Kim, 2003). As explained in Winzer and Kim (2003), this discrepancy is probably due to the nonideal signal source used in the experiment. The 10% (or 36°) mismatch of Rohde et al. (2000) gives a penalty of about 3.5 dB.

The error probability of Eq. (4.10) assumes an optical matched filter, the typical cases of Bosco and Poggiolini (2003) and Winzer and Kim (2003). Within the matched filter, in additional to the amplifier noise in the same polarization as the signal (Chinn et al., 1996, Humblet and Azizoğlu, 1991, Pires and de Rocha, 1992), we also consider that amplifier noise from orthogonal polarization. When optical matched filter is assumed, the results here are applicable to systems with both return-to-zero (RZ) and non-return-to-zero (NRZ) linecodes.

2.2 Phase Error in CPFSK and MSK Signals

In differential detection of CPFSK and MSK signals of Fig. 3.9, without noise, the phase difference at the output of the delay-and-multiplier circuits is

$$\Delta\phi_{\pm} = \omega_{\mathrm{IF}}\tau \pm \pi\Delta f\tau. \tag{4.12}$$

In the phase difference of Eq. (4.12), there are two types of equivalent phase error of $\exp(\omega_{\mathrm{IF}}\tau) \neq \pm j$ or $\Delta f\tau \neq 1/2$. The consequence of either case provides a phase error as the phase difference to the nominal phases of $\pm\pi/2$. The error probability is the same as that of Eq. (4.10) with, for example,

$$\theta_e = \pi\left(\frac{1}{2} - \Delta f\tau\right) \tag{4.13}$$

or $\exp(j\omega_{\mathrm{IF}}T) = \pm j e^{j\theta_e}$. In either case, we have $\theta_e = \phi_\pm \bmod 2\pi$, $-\pi < \theta_e < \pi$.

When there are phase errors of both $\exp(\omega_{\mathrm{IF}}T) \neq \pm j$ and $\Delta f T \neq 1/2$, there are two phase errors of $\theta_{e,\pm} = \Delta\phi_\pm \bmod 2\pi$ with $-\pi < \theta_{e,\pm} < \pi$. The error probability of CPFSK signal is equal to the average of the two error probabilities given by Eq. (4.10) with two phase errors of $\theta_{e,\pm}$.

After minor modification, both Figs. 4.2 and 4.3 are applicable to CPFSK and MSK signal with phase errors in the receiver.

Other than phase error, other degradation in a direct-detection DPSK receiver was analyzed in Winzer and Kim (2003). If the two photodiodes of the balanced receiver is not the same, the SNR penalty is from 0 to 3 dB. The 3-dB SNR penalty is the degenerated case of a single-branch receiver. Direct-detection DPSK receiver is not sensitive to the extinction ratio of the asymmetric Mach-Zehnder interferometer (Winzer and Kim, 2003).

3. Laser Phase Noise

For phase-modulated or coherent optical communications, especially with synchronous receiver with phase tracking, the laser sources for both transmitter and LO must be "coherent". The phase fluctuation or incoherence of a laser limits the sensitivity of a coherent receiver. Semiconductor lasers widely used in lightwave communication systems are not as coherent as other types of laser source. From the semiclassical theory of semiconductor laser noise of Sec. 2.2.3, given by Eq. (2.41), the laser linewidth is broadened by a factor of $1 + \alpha^2$ compared with the Schawlow-Townes formula of Eq. (2.40), where α is the linewidth enhancement factor of the laser, typically in the range of 3 to 4. With low facet reflectivity and short laser cavity, even without the linewidth enhancement, the Schawlow-Townes formula also gives a large linewidth for semiconductor laser. The impact of laser phase noise to phase-modulated optical signal will be analyzed here in details.

Ignored the effect of laser relaxation oscillation, the instantaneous phase of a laser is a random walk or Brownian motion. The frequency noise of the laser, the same as Eq. (2.39), has a spectral density of

$$S_{\dot{\phi}_n}(f) = \frac{\Delta f_L}{2\pi f^2}, \tag{4.14}$$

where Δf_L is the full-width-half-maximum (FWHM) linewidth of the Lorentzian line-shape of Eq. (2.37)

$$S_{\phi_n}(f) = \frac{\Delta f_L^2}{\Delta f_L^2 + 4f^2}. \tag{4.15}$$

From Sec. 2.2.3, the phase of the semiconductor laser can be modeled as a Wiener process with a variance parameter of

$$\sigma_{n_\phi}^2 = 2\pi\Delta f_L, \tag{4.16}$$

and autocorrelation function of

$$E\left\{\phi_n(t_1)\phi_n(t_2)\right\} = \sigma_{n_\phi}^2 \min(t_1, t_2). \tag{4.17}$$

Both the spectral density of Eq. (4.14) and the line-shape of Eq. (4.15) are for a single laser. When two lasers are beating together in either homodyne or heterodyne system, the sum of the linewidth of both lasers becomes the linewidth at the intermediate frequency (IF). For homodyne and heterodyne systems, without changing the notation, the linewidth of Δf_L is the sum of the linewidth of both transmitter and LO lasers. For direct-detection system, the linewidth of Δf_L is assumed to be that of the transmitter laser only.

The potential limitation of laser linewidth to coherent systems was first studied by Yamamoto and Kimura (1981) and observed by Favre and LeGuen (1982), Kikuchi et al. (1983), and Tamburrini et al. (1983). There were many follow-up studies as outlined later.

3.1 Impact to PSK Signals

A PSK signal is demodulated using the receiver schematically shown in Fig. 1.3, either using a homodyne or heterodyne receiver. The phase-locked loop (PLL) of Fig. 1.3 comes with many different architectures. The simplest PLL may be the square loop for binary PSK in which a square device is used to generate a carrier frequency at $2f_{\text{IF}}$. However, the squaring operation leads to noise enhancement that increases the noise power level at the input to the PLL and results in an increase of the variance of phase error (Proakis, 2000).

Figure 4.4 shows a decision-feedback (or decision-directed, decision-driven) PLL. A quadrature receiver of Fig. 3.4 is used in Fig. 4.4 to provide both the data and the phase error. For binary PSK signal, the input to the decision circuit is

$$\frac{1}{2}[A(t) + n_1(t)]\cos\phi_e - \frac{1}{2}n_2(t)\sin\phi_e, \tag{4.18}$$

where ϕ_e is the phase error of the PLL. Without decision error, the output of the upper branch is $A_s(t)$ after the decision circuit with a delay of T. The output of the lower branch is (Proakis, 2000)

$$\frac{1}{2}[A_s(t) + n_1(t)]\sin\phi_e - \frac{1}{2}n_2(t)\cos\phi_e. \tag{4.19}$$

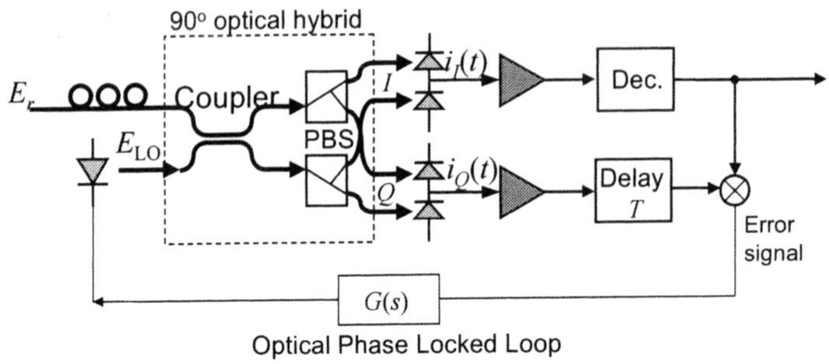

Figure 4.4. A decision-feedback PLL to demodulate binary PSK signal.

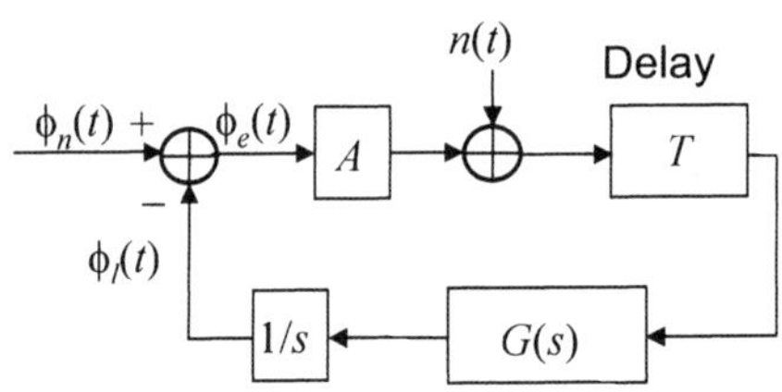

Figure 4.5. A linearized model of the PLL of Fig. 4.4.

The error signal is thus

$$e(t) = \frac{1}{2}A_s^2(t)\sin\phi_e + \frac{1}{2}A_s(t)[n_1(t)\sin\phi_e - n_2(t)\cos\phi_e]. \tag{4.20}$$

Other than the noise term, the phase error is contained in $\frac{1}{2}A_s^2(t)\sin\phi_e$. With small phase error of $\phi_e \to 0$, only the noise term of $n_2(t)$ is important in the error signal of Eq. (4.20).

Figure 4.5 shows a linearized model of the PLL of Fig. 4.4. The parameters of Fig. 4.5 are $A = E\{A_s^2(t)/2\}$ and $n(t) = A_s(t)n_2(t)/2$ with a SNR of $A^2/2\sigma_n^2$. For simplicity without changing the SNR, we can assume that $A = |A_s(t)|$ for $A_s(t) = \pm A$ and $n(t) = n_2(t)$. In the linearized model of Fig. 4.4, the LO laser is modeled as an integrator of a frequency response of $1/s$ with $s = j2\pi f$. For a second-order PLL, the loop filter is $G(s) = K(s+a)/s$, where K includes the gain of the LO laser, multiplier, and other circuits. The transfer function of the loop is defined as $\phi_l(s)/\phi_n(s)$ that is the ratio of the response of the phases of $\phi_l(t)$ to $\phi_n(t)$. The closed-loop transfer function, expressed in terms of the loop filter $G(s)$ is

$$H(s) = \frac{\phi_l(s)}{\phi_n(s)} = \frac{AKG(s)e^{-sT}}{s + AKG(s)e^{-sT}}. \tag{4.21}$$

For small T such that the term of $\exp(-sT)$ is equal to unity, we obtain the second-order transform function of

$$H(s) = \frac{AK(s+a)}{s^2 + AKs + AKa} = \frac{\omega_n^2 + 2\zeta\omega_n s}{\omega_n^2 + 2\zeta\omega_n s + s^2}. \tag{4.22}$$

where $\omega_n^2 = AKa$ and $\zeta = \frac{1}{2}\sqrt{AK/a}$, corresponding to the cut-off frequency and damping factor of a second-order response, respectively. With $\exp(-sT) \neq 1$, we obtain

$$H(s) = \frac{2\zeta\omega_n s + \omega_n^2}{s^2 e^{sT} + 2\zeta\omega_n s + \omega_n^2} \tag{4.23}$$

instead.

The phase error of $\phi_e(t)$ has two components arising from $n(t)$ and the phase noise of $\phi_n(t)$, respectively. In the linearized analysis, we obtain

$$\phi_e(t) = \frac{1}{A} n(t) \otimes h(t) + \phi_n(t) \otimes [1 - h(t)], \tag{4.24}$$

where $\otimes$ denotes convolution and $h(t)$ is the impulse response corresponding to the frequency response of $H(s)$. The spectral density of $\phi_e(t)$ is

$$S_{\phi_e}(f) = \frac{S_N(f)}{A^2} |H(j2\pi f)|^2 + S_{\phi_n}(s)|1 - H(j2\pi f)|^2. \tag{4.25}$$

The phase error variance is

$$\sigma_{\phi_e}^2 = \frac{N_0}{A^2} \int_0^\infty |H(j2\pi f)|^2 \mathrm{d}f + 2 \int_0^\infty S_{\phi_n}(f)|1 - H(j2\pi f)|^2 \mathrm{d}f, \tag{4.26}$$

where $N_0 = 2\sigma_n^2 T$ is the spectral density of $n(t)$. The above phase error variance of Eq. (4.26) is based only on the linearized model of the PLL. The transfer function of $H(s)$ can be from either Eq. (4.22) or (4.23), or other more general loop transform functions.

The short-delay approximation of $\exp(-sT) \approx 1$

When $e^{-sT} \approx 1$, we obtain

$$B_n = \int_0^\infty |H(j2\pi f)|^2 \mathrm{d}f = \frac{\omega_n}{8\zeta}(1 + 4\zeta)^2. \tag{4.27}$$

In the usual case of $\zeta = 1/\sqrt{2}$, we obtain

$$B_n = \frac{3\omega_n}{4\sqrt{2}} = 0.53\omega_n. \tag{4.28}$$

The tracking error function is

$$|1 - H(j2\pi f)|^2 = \frac{(2\pi f/\omega_n)^4}{[1 + (2\pi f/\omega_n)^2]^2 + (4\zeta\pi f/\omega_n)^2}. \tag{4.29}$$

With the phase error spectrum of Eq. (4.14), we obtain

$$\begin{aligned}&\int_0^\infty S_{\phi_n}(f)|1 - H(j2\pi f)|^2 \mathrm{d}f \\ &= \frac{\Delta f_L}{2\omega_n}\int_0^\infty \frac{x^2}{1 + 2x^2(2\zeta^2 - 1) + x^4}\mathrm{d}x = \frac{\pi}{4\zeta}\frac{\Delta f_L}{\omega_n}.\end{aligned} \tag{4.30}$$

For $\zeta = 1/\sqrt{2}$, the overall phase error variance is

$$\sigma^2_{\phi_e} = \frac{0.53\omega_n T}{\rho_s} + 2.22\frac{\Delta f_L}{\omega_n}. \tag{4.31}$$

The optimal bandwidth is

$$\omega_{n,\mathrm{opt}} = 2.04\sqrt{\frac{\Delta f_L \rho_s}{T}} \tag{4.32}$$

to give a minimum phase error variance of

$$\sigma^2_{\phi_e,\mathrm{min}} = 2.17\sqrt{\frac{\Delta f_L T}{\rho_s}}. \tag{4.33}$$

For general case, the minimum phase error variance is

$$\sigma^2_{\phi_e,\mathrm{min}} = \sqrt{\pi\frac{\Delta f_L T}{\rho_s}\left(1 + \frac{1}{4\zeta^2}\right)}. \tag{4.34}$$

With phase error, the error probability of a PSK signal is

$$p_e = \frac{1}{2}\int_{-\pi}^{\pi} \mathrm{erfc}\left(\sqrt{\rho_s}\cos\phi_e\right) p_{\Phi_e}(\phi_e)\mathrm{d}\phi_e, \tag{4.35}$$

where $p_{\Phi_e}(\phi_e)$ is the probability density of the phase error. Alternatively, using the series expansion of Appendix 4.A, similar to Eq. (4.A.16), we

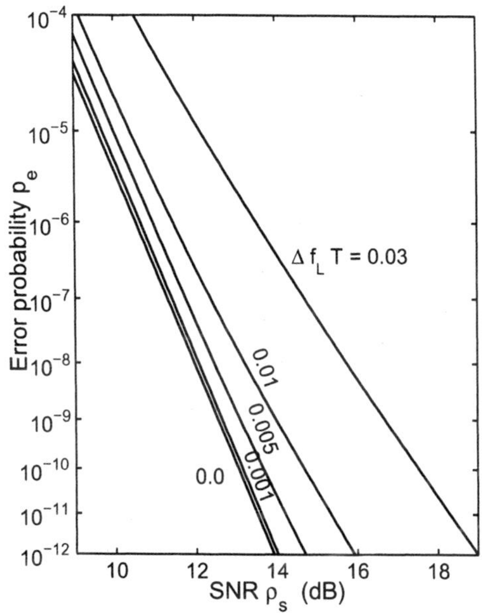

Figure 4.6. The error probability of binary PSK signal with laser phase noise.

can obtain

$$p_e = \frac{1}{2} - \exp\left(-\frac{\rho_s}{2}\right)\sqrt{\frac{\rho_s}{\pi}}\sum_{k=0}^{\infty}\frac{(-1)^k}{2k+1}\left[I_k\left(\frac{\rho_s}{2}\right) + I_{k+1}\left(\frac{\rho_s}{2}\right)\right] \times \exp\left[-\frac{(2k+1)^2}{2}\sigma^2_{\phi_e}\right], \tag{4.36}$$

if the phase error is assumed to be Gaussian distributed with a variance of $\sigma^2_{\phi_e}$.

Using the minimum phase error variance of Eq. (4.34), Figure 4.6 shows the error probability of binary PSK signal of Eq. (4.36). The error probability of Fig. 4.6 is shown for several different values of normalized combined laser linewidth of $\Delta f_L T$. Figure 4.7 shows the SNR penalty of binary PSK signal as a function of the normalized combined laser linewidth of $\Delta f_L T$. In additional to use the optimal bandwidth of Eq. (4.32), the SNR penalty of Fig. 4.7 is also calculated when the normalized PLL bandwidth of $\omega_n T$ is equal to 0.1 and 0.05 of 2π.

For an 1-dB SNR penalty, from Fig. 4.7, the normalized laser linewidth of $\Delta f_L T$ must be less 7.7×10^{-3}. The same as Kazovsky (1986b), a

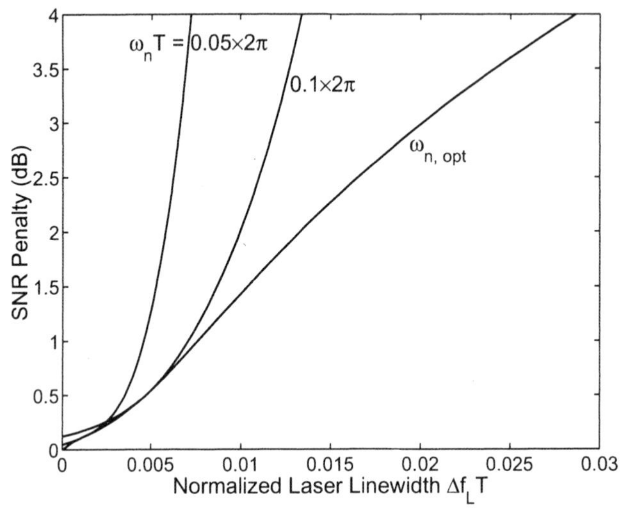

Figure 4.7. The SNR penalty of binary PSK signal as a function of laser linewidth for PLL without delay.

normalized laser linewidth of $\Delta f_L T$ of about 5×10^{-3} gives a SNR penalty of 0.5 dB.

Accurate Results of $\exp(-sT) \neq 1$

For large normalized laser linewidth of $\Delta f_L T$, the PLL must have a bandwidth comparable to the data rate of $1/T$. With a large PLL bandwidth, the approximation of $\exp(-sT) \approx 1$ is not valid. Without the approximation of $\exp(-sT) \neq 1$, the variance of phase error is equal to

$$\sigma_{\phi_e}^2 = \frac{\omega_n}{2\pi\rho_s}\int_0^\infty \frac{1+(2\zeta x)^2}{D(x)}\mathrm{d}x + \frac{2\Delta f_L}{\omega_n}\int_0^\infty \frac{x^2}{D(x)}\mathrm{d}x, \tag{4.37}$$

where $D(x) = [1 - x^2\cos(\omega_n T x)]^2 + [2\zeta x - x^2 \sin(\omega_n T x)]^2$. The phase error variance of Eq. (4.37) is very complicated and difficult to find the parameters to optimize the system analytically.

Figure 4.8 shows the SNR penalty for binary PSK signal calculated using Eq. (4.37) with normalized PLL bandwidth of $\omega_n T/2\pi$ equal to 0.01 to 0.07 and $\zeta = 1/\sqrt{2}$. With minimum PLL delay of T, the maximum normalized linewidth of $\Delta f_L T$ must be less than 2×10^{-3} for a SNR penalty less than 1 dB. This is substantially less than the 5×10^{-3} requirement of Fig. 4.7 for PLL without delay. The optimal PLL bandwidth is $\omega_n T/2\pi \approx 0.04$ when $\Delta f_L T = 2 \times 10^{-3}$. In both Figs. 4.7 and

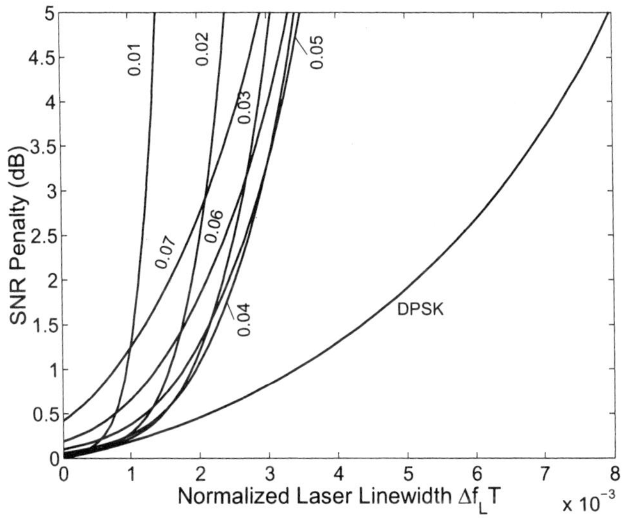

Figure 4.8. The SNR penalty of binary PSK signal as a function of laser linewidth for PLL with delay of T. The labels of each curve are $\omega_n T/2\pi$.

4.8, the standard deviation (STD) of phase error is about 11.5° for a SNR penalty of 1 dB.

For a 10-Gb/s system, the maximum normalized linewidth of $\Delta f_L T$ of 2×10^{-3} gives a combined laser linewidth of 20 MHz or 10 MHz linewidth for either transmitter or LO laser. Common DFB lasers have a linewidth in the order of few MHz. The performance of high-speed PSK signal is not likely to be limited by the laser linewidth.

For system limited by amplifier noise, the effects of laser phase noise to heterodyne and homodyne receivers are the same. For system limited by shot noise, the analysis here is also applicable to heterodyne PSK system (Kazovsky, 1986b, Norimatsu and Iwashita, 1992). The analysis here followed that of Spilker, Jr. (1977) and Kazovsky (1986b) but also taken into account the loop delay of T (Grant et al., 1987, Norimatsu and Iwashita, 1992).

The impact of phase error variance to PSK signal was first analyzed by Prabhu (1976) and further studied by Glance (1986a), Kikuchi et al. (1984), and Hodgkinson (1987). The procedure of optimizing the PLL, especially decision-feedback PLL, is the same as that in Kazovsky (1985, 1986b), Norimatsu and Iwashita (1991, 1992), Spilker, Jr. (1977), and Norimatsu and Ishida (1994). Note that the propagating delay of the PLL cannot be ignored.

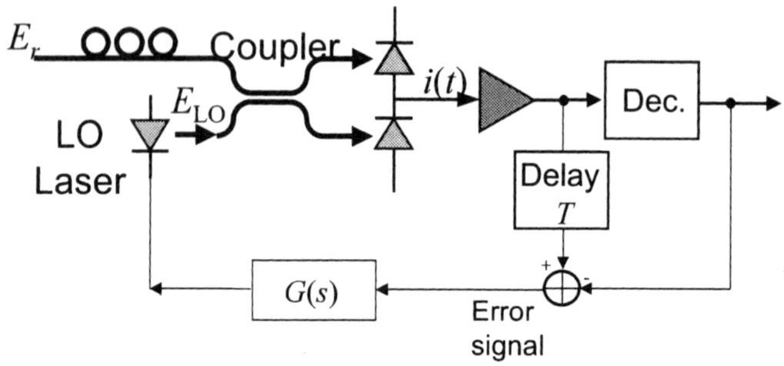

Figure 4.9. The balanced PLL to demodulate binary PSK signal.

Without decision feedback, the PLL of Fig. 4.4 is similar to a Costa loop. Costa loop was implemented for PSK signal by Schöpflin et al. (1990) and Norimatsu et al. (1990). The performance of Costa loop is the same as square-loop with noise enhancement due to the square operation (Proakis, 2000). Most implementation of homodyne PLL receiver using the balanced PLL similar to Fig. 4.9 with or without decision feedback. Other than Sun and Ye (1990) and Kahn (1990), most implementation of the PLL of Fig. 4.9 does not remove the data to phase error crosstalk (Altas and Kazovsky, 1990, Kahn, 1989, Kahn et al., 1990). Other than using a square or Costa loop, the implementation of the PLL of Fig. 4.9 required the usage of part of the signal to generate the error signal (Kahn, 1990, Kahn et al., 1990, Kazovsky, 1986a, Sun and Ye, 1990).

For shot-noise limited homodyne systems, because the signal splitting ratio of the coupler preceding the balanced receiver also affects the performance of the systems, the optimization of the PLL becomes very complicated (Kazovsky, 1985, 1986a). The balanced PLL of Fig. 4.9 is further complicated by the requirement to split the signal for data and phase locking (Huang and Wang, 1995, 1996). Homodyne system limited by shot noise has a very small linewidth requirement down to 10^{-5} the date rate.

In heterodyne PSK system, the effect of laser phase noise can be reduced by the recovery of a carrier with the same phase noise (Cheng and Okoshi, 1989, Chikama et al., 1990b, Watanabe et al., 1989, 1992). When carrier is recovered using a band-pass filter, the bandwidth of the band-pass filter needs to pass the carrier with phase noise. A large filter bandwidth translates to larger phase error variance due to receiver noise than those based on PLL. Heterodyne PSK signal was also demodulated using electrical PLL in Kazovsky et al. (1990).

3.2 Impact to DPSK Signals

For a DPSK signal demodulated by delay-and-multiplier circuits, the phase error variance due to laser phase noise in two consecutive symbols is equal to

$$\sigma^2_{\phi_e} = 2\pi\Delta f_L T. \tag{4.38}$$

There are two methods to calculate the error probability due to laser phase noise, the first method using an integration similar to Eq. (4.35) of

$$p_e = \int_{-\pi}^{+\pi} p_e(\phi_e) p_{\Phi_e}(\phi_e) \mathrm{d}\phi_e, \tag{4.39}$$

where $p_e(\phi_e)$ is the error probability of DPSK signal having a phase error of ϕ_e, the same as that of Eq. (4.10) or (4.11) with $\theta_e = \phi_e$. Similar to the series of Eqs. (4.A.19) and (4.36), with Gaussian distributed phase error, the error probability of a DPSK signal is equal to

$$\begin{aligned} p_e &= \frac{1}{2} - \frac{\rho_s e^{-\rho_s}}{2} \sum_{k=0}^{\infty} \frac{(-1)^k}{2k+1} \left[I_k\left(\frac{\rho_s}{2}\right) + I_{k+1}\left(\frac{\rho_s}{2}\right)\right]^2 \\ &\quad \times \exp\left[-(2k+1)^2 \pi \Delta f_L T\right]. \end{aligned} \tag{4.40}$$

Figure 4.10 shows the error probability of DPSK signal with laser phase noise. The SNR penalty of DPSK signal with laser phase noise is already shown in Fig. 4.8 for comparison. From Fig. 4.8, a normalized linewidth $\Delta f_L T = 3.4 \times 10^{-3}$ gives a SNR penalty of 1 dB for DPSK signal. The linewidth requirement of DPSK signals is 1.7 times that of PSK signals.

In heterodyne DPSK receiver, the linewidth of Δf_L is the overall linewidth from both transmitter and LO laser. In direct-detection DPSK receiver, the linewidth of Δf_L is the linewidth from the transmitter alone. For 10-Gb/s direct-detection DPSK signal, the linewidth must be less than 34 MHz. For a 10-Gb/s heterodyne system, the linewidth of either transmitter and LO laser must be less than 17 MHz individually.

The analysis here for DPSK signals is the same as that of Nicholson (1984) by assuming that the filter after the delay-and-multiplier circuits does not further distort the signal and the usage of matched IF filter. Jacobsen and Garrett (1987) and Jacobsen (1993) further assume that the IF filter is not a matched filter. The statistics of the filtered light with phase noise is found by Foschini and Vannucci (1988) and Foschini et al. (1989) and used by Azizoğlu and Humblet (1991) and Kaiser et al. (1993). Methods to analyze DPSK signal with laser phase noise was reviewed by Smith et al. (1995).

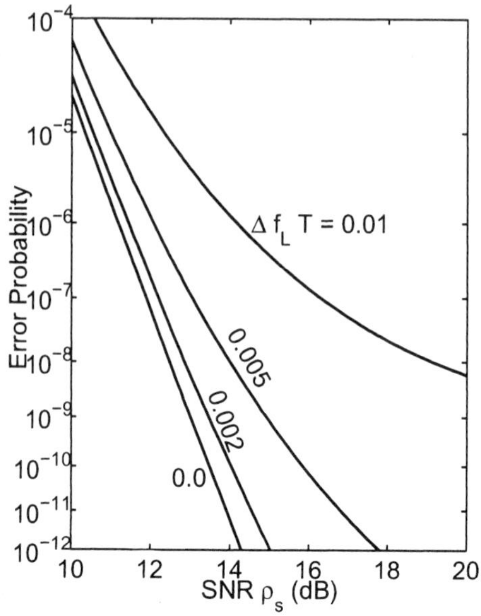

Figure 4.10. The error probability of DPSK signal with laser phase noise

3.3 Impact to Other Signal Formats

The impact of laser phase noise to MSK signal is identical to that to DPSK signal if differential demodulator with a delay of the symbol time of T is used. The error probability of MSK signal with laser phase noise is the same as that of Eq. (4.40). For the general case of CPFSK signal, the error probability is the combination of Eqs. (4.10) and (4.40) of

$$p_e = \frac{1}{2} - \frac{\rho_s e^{-\rho_s}}{2} \sum_{k=0}^{\infty} \frac{(-1)^k}{2k+1} \left[I_k\left(\frac{\rho_s}{2}\right) + I_{k+1}\left(\frac{\rho_s}{2}\right) \right]^2 \times \exp\left[-(2k+1)^2 \pi \Delta f_L \tau\right] \cos[(2k+1)\theta_e], \quad (4.41)$$

where θ_e is that of Eq. (4.13) and τ is the delay for the differential demodulator of the CPFSK receiver of Fig. 3.9. The performance of CPFSK signal was analyzed by both Iwashita and Matsumoto (1987) and Emura et al. (1990a).

Phase noise also degrades dual-filter detected FSK signal or envelope detected ASK signal. However, most analysis showed that those systems can tolerate a laser linewidth up to at least 10% of the data rate (Corvaja et al., 1992, Foschini et al., 1988, Garrett et al., 1990, Greenstein et al., 1989, Kazovsky and Tonguz, 1990, Marti and Capmany, 1996,

Siuzdak and van Etten, 1991, Tsao et al., 1990, 1992). Currently, most communication systems use semiconductor DFB lasers with a linewidth of few MHz. For high-speed transmission with over Gb/s of data rate, the linewidth of few MHz does not affect the performance of both ASK and FSK signals.

4. Fiber Chromatic Dispersion

In most dielectric materials, the speed of light depends slightly on frequency. The glass material also has refractive index as a function of optical frequency, as an example, the glass prism to separate sun light into rainbow. Chromatic dispersion is especially serious when an on-off keying signal is generated by directly modulated a semiconductor laser due to the frequency modulation accompany the amplitude modulation of the laser (Corvini and Koch, 1987). Even when external modulator is used without chirp, the modulated optical carrier occupies an optical spectral bandwidth. Signal from different part of the spectrum propagates with different speed due to chromatic dispersion. Without dispersion compensation, the transmission distance is limited.

In frequency domain, different spectral components propagate inside the fiber according to the simple relationship of

$$B(z, \omega) = B(0, \omega) \exp\left[j\beta(\omega) z\right], \tag{4.42}$$

where z is the fiber location, and $B(z, \omega)$ and $B(0, \omega)$ are the low-pass representation of the signal spectra at the locations of 0 or z, respectively, and $\beta(\omega)$ is the propagation constant as a function of optical frequency. The propagation constant of $\beta(\omega)$ may be expanded as a Taylor series around the carrier frequency of ω_c and retains terms up to third order, we obtain

$$\beta(\omega) = n_r(\omega)\frac{\omega}{c} \approx \beta_0 + \beta_1(\omega - \omega_c) + \frac{1}{2}\beta_2(\omega - \omega_c)^2 + \frac{1}{6}\beta_3(\omega - \omega_c)^3, \tag{4.43}$$

where $\beta_k = d^k\beta/d\omega^k\big|_{\omega=\omega_c}$. The group delay of the optical signal is

$$v_g = \beta_1^{-1} = \left(\frac{d\beta}{d\omega}\right)^{-1}\Bigg|_{\omega=\omega_c}. \tag{4.44}$$

The difference in arrival time is the dispersion coefficient in the unit of ps/km/nm of

$$D = \frac{d}{d\lambda}\left(\frac{1}{v_g}\right) = -\frac{2\pi c}{\lambda^2}\beta_2. \tag{4.45}$$

Standard single-mode fiber has a dispersion coefficient from 16 to 19 ps/km/nm at the low-loss window of 1.55 μm[1]. Fiber dispersion can be controlled by balancing material dispersion with waveguide dispersion. Long-haul system uses nonzero dispersion-shifted fiber (NZDSF) with dispersion in the range of 3 to 5 ps/km/nm from the wavelength of 1.530 to 1.565 μm[2]. Some nonzero dispersion fibers have larger dispersion coefficient, typically from 3 to 10 ps/km/nm in the wavelength from 1.460 to 1.565 μm[3].

With a dispersion coefficient of D, two signals with wavelength separation of $\Delta\lambda$ walk-off by a time of $D\Delta\lambda L$ after a distance of L. Except for the simplest case with Gaussian pulses, the propagation equation of Eq. (4.42) does not have analytical solution. As a linear system, the equation of Eq. (4.42) has a simple numerical solution based on Fourier analysis.

For a fiber distance of L, including only the second-order term of β_2, the equation of Eq. (4.42) can be simplified to

$$B(L, \omega) = B(0, \omega) \exp\left[j \frac{1}{2} \beta_2 \omega^2 L\right]. \tag{4.46}$$

The system performance depends solely on the parameter of $\beta_2 L B_d^2$ or DLB_d^2, where B_d is the data or symbol rate of the signal.

Figure 4.11 shows the eye penalty as a function of the dispersion parameter of DLB_d^2 for ASK, PSK, DPSK, and MSK signals. Using numerical simulation, the eye-penalty is calculated as the ratio of eye-closure to the nominal eye opening without dispersion and other distortions. ASK or on-off keying signal is assumed to be directly detected using only a photodiode. PSK signal is demodulated using a synchronous homodyne receiver with phase tracking. Both DPSK and MSK signals are demodulated using an interferometer with a path delay of T. Figure 4.11 also shows the transmission distance for a 10-Gb/s signal in a standard single-mode fiber with dispersion coefficient of $D = 17$ ps/km/nm.

For ASK, PSK, and DPSK signals, the transmitters for Fig. 4.11 use amplitude modulator. The drive signal is a rectangular pulse passed through a fifth-order Bessel filter with a bandwidth of $0.75B_d$. The receiver also has the same Bessel filter to shape the waveform.

---

[1] ITU G.652 (2003): Characteristics of a single-mode optical fibre and cable

[2] ITU G.655 (2003): Characteristics of a non-zero dispersion-shifted single-mode optical fibre and cable.

[3] ITU G.656 (2004): Characteristics of a fibre and cable with non-zero dispersion for wideband optical transport.

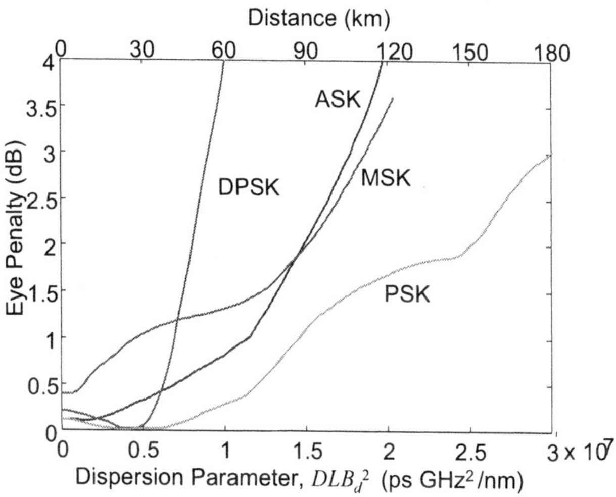

Figure 4.11. Eye penalty as a function of dispersion parameter of DLB_d^2 for several modulation formats. Also shown is the corresponding transmission distance for a 10-Gb/s signals in standard single-mode fiber with dispersion coefficient of $D = 17$ ps/km/nm.

The MSK signal of Fig. 4.11 is assumed to be an ideal MSK signal in the transmitter. The receiver has a fifth-order Bessel filter with a bandwidth the same as the data-rate.

With phase tracking, PSK signal has a very large tolerance to fiber dispersion. The dispersion tolerance of DPSK signal is the worst among all modulation formats. For PSK signal, the receiver for Figure 4.11 also tracks out the constant phase shift due to fiber dispersion.

Because RZ-DPSK signal is currently the most popular modulation format, Figure 4.12 shows the eye penalty for three types of RZ-DPSK signal with the RZ pulses from Fig. 2.18. The transmitter is the same as that in Fig. 2.17 with the phase modulator there replaced by a zero-chirp amplitude modulator driven with a $2V_\pi$ peak-to-peak voltage, the same as that in Fig. 4.11.

Figure 4.12 shows the dispersion tolerance decrease with duty-cycle. The smaller is the duty-cycle, or the shorter is the pulse, the smaller is the dispersion tolerance.

The effects of chromatic dispersion to coherent lightwave systems were studied in Elrefaie et al. (1988), Winters and Gitlin (1990), and Elrefaie and Wagner (1991). The book of Betti et al. (1995) also has section about the effect of chromatic dispersion to coherent optical signal. The broadening of Gaussian pulses with fiber dispersion was considered in

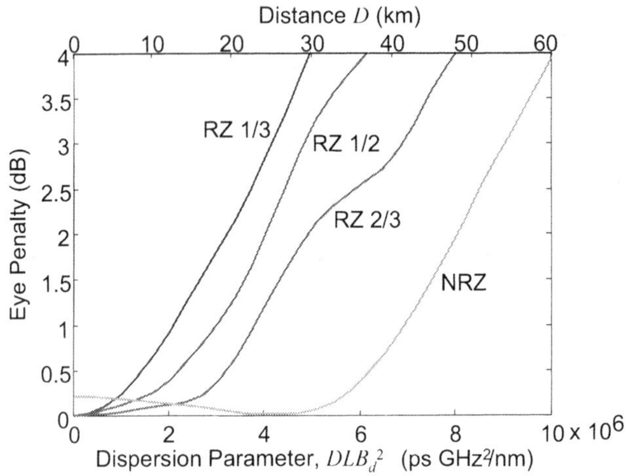

Figure 4.12. Eye penalty as a function of dispersion parameter of DLB_d^2 for various types of direct-detection DPSK signal. Also shown is the corresponding transmission distance for a 10-Gb/s signals in standard single-mode fiber with dispersion coefficient of $D = 17$ ps/km/nm.

Miyagi and Nishida (1979), and Marcuse (1980, 1981). DPSK signal was studied recently by Pan et al. (2003) and Wang and Kahn (2004a,b).

Applicable for both phase-modulated and on-off keying signals, typical WDM systems use optical fiber or optical components with a dispersion opposite to that of the fiber link for dispersion management (Bakhshi et al., 2004, Cai et al., 2002, Murakami et al., 2000, Suzuki and Edagawa, 2003, Willner and Hoanca, 2002). Dispersion compensated fiber (Grüner-Nielsen et al., 1998, Lin et al., 1980, Marcuse and Lin, 1981) may be the most popular technique for dispersion management. Electrical dispersion compensation is also very effective for coherent optical signals, especially those using heterodyne receiver (Iwashita and Takachio, 1990, Winters, 1990, Winters and Gitlin, 1990, Yamazaki et al., 1993). For homodyne receiver, dispersion compensation requires the usage of quadrature receiver (Taylor, 2004). For 40-Gb/s signals, tunable dispersion compensator may be required (Lunardi et al., 2002, Pan et al., 2002, Sandel et al., 2004, Yagi et al., 2005).

In additional to the fiber, optical filter also has dispersion that gives the same effects as chromatic dispersion (Lenz et al., 1998, Nyoklak et al., 1998). Optical filter also induces amplitude filtering to the optical signal (Ho et al., 2000a, Roudas et al., 2002, Testa et al., 1994).

5. Polarization-Mode Dispersion

A single-mode optical fiber can support two polarization modes, these two polarization modes are used for polarization-division multiplexing (PDM) or polarization-shift keying (PolSK) as shown in Sec. 3.7. If the core of a single-mode fiber is perfectly circular, the two polarization modes propagate with the same speed. However, due to manufacturing tolerance, the core of the fiber varies slightly from prefect circle. The propagation speeds of the two polarization modes have small difference, leading to polarization-mode dispersion (PMD). The optical signal transmitted through the two polarizations arrives with a timing offset at the receiver, called differential group delay (DGD).

For a short piece of fiber less than the birefringence correlation length of a single mode fiber, the DGD increases with fiber length. The birefringence correlation length of a fiber is typically less than 20 m (Galtarossa and Palmieri, 2004, Galtarossa et al., 2001, Huttner et al., 1998). Within the birefringence correlation length, the timing offset between two polarization modes is $\delta\tau = (\beta_{1x} - \beta_{1y})\delta L$, where β_{1x} and β_{1y} are the first-order propagation coefficients of the two polarization modes and δL is the fiber length.

Even if the fiber is perfectly circular, external mechanical or thermal stresses cause small asymmetric to the fiber core. The DGD changes with time due to external stress. Other than polarization-maintained fiber (PMF), beyond the birefringence correlation length, the two polarization axes of the fiber are uncorrelated. A long optical fiber can model as the cascade of many pieces of polarized components, each polarized component corresponds to a short piece of fiber with a length about the birefringence correlation length. When many short pieces of polarized fiber are randomly oriented in a long fiber, the whole fiber can still be modeled as a single polarized component with two principle states of polarization (PSP) and a group delay of DGD. The two PSPs are randomly oriented and the DGD is a random variable, depending on the alignment of individual polarized fiber pieces. Time varying external stress changes both the PSP and DGD.

With a mean DGD of $\langle \Delta\tau \rangle$, the instantaneous DGD has a Maxwellian distribution of

$$p(\Delta\tau) = \frac{32\Delta\tau^2}{\pi^2 \langle \Delta\tau \rangle^3} \exp\left(-\frac{4\Delta\tau^2}{\pi \langle \Delta\tau \rangle^2} \right), \quad \Delta\tau > 0, \tag{4.47}$$

with a DGD second-order moment of $E\{\Delta\tau^2\} = 3\pi \langle \Delta\tau \rangle^2 / 8$. For a long optical fiber, all statistical PMD properties of the fiber depend solely on the mean DGD of $\langle \Delta\tau \rangle$.

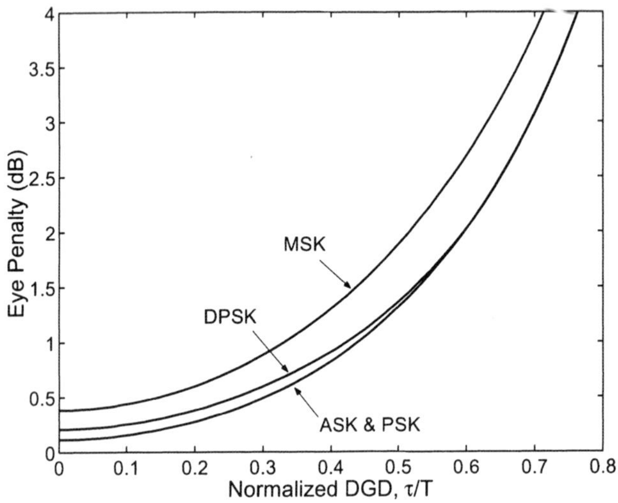

Figure 4.13. Eye penalty as a function of normalized instantaneous DGD of $\Delta\tau/T$ for different modulation formats.

With random coupling of light between the polarization modes, the mean DGD of $\langle \Delta\tau \rangle$ increases with $\sqrt{L}$, where L is the fiber length. Before early 90's, the PMD coefficient of fiber may be as large as 1 ps/$\sqrt{\text{km}}$. Currently, optical fiber has a PMD coefficient commonly around 0.1 ps/$\sqrt{\text{km}}$, with some commercially available fiber types have PMD coefficient even approaching 0.05 ps/$\sqrt{\text{km}}$.

Figure 4.13 shows the eye-penalty as a function of normalized instantaneous DGD of $\Delta\tau/T$ for different modulation formats. The transmitter and receiver is the same as that for Fig. 4.11. ASK and PSK modulation formats have the same tolerance to DGD. MSK modulation format, due to its pulse shape, has the least tolerance to DGD. The simulation of Fig. 4.13 assumes that the two PSP has equal optical power, the worst case for pulse spreading.

Figure 4.14 shows the eye-penalty for the popular RZ-DPSK format. The DGD tolerance is the smallest for RZ format with a duty-cycle of 1/3. The NRZ DPSK scheme has the largest tolerance to DGD.

A system with PMD is usually studied for the outage probability larger than certain instantaneous DGD. For example, for NRZ-DPSK format, the instantaneous DGD of $\Delta\tau = 0.42T$ gives 1-dB power penalty. If more than 1-dB power penalty is defined as outage, based on the

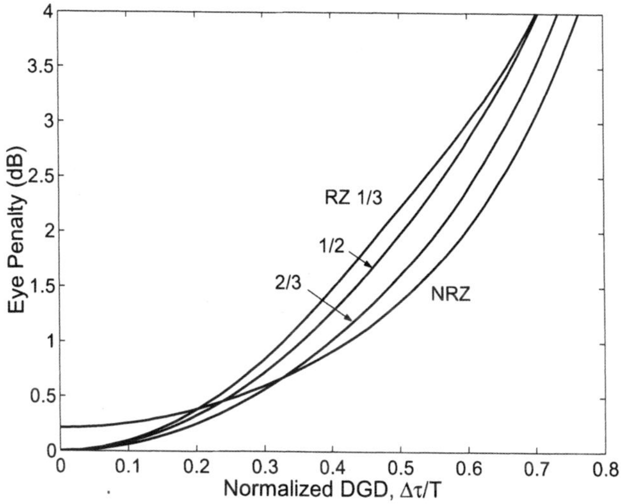

Figure 4.14. Eye penalty as a function of normalized instantaneous DGD of $\Delta\tau/T$ for direct-detection DPSK signals.

probability density of Eq. (4.47), the outage probability is

$$\Pr\{\Delta\tau > \kappa \langle \Delta\tau \rangle\} = \frac{4\kappa}{\pi} \exp\left(-\frac{4\kappa^2}{\pi}\right) + \operatorname{erfc}\left(\frac{4\kappa}{\pi}\right), \tag{4.48}$$

as shown in Fig. 4.15. From Fig. 4.15, with $\Delta\tau = 0.42T$, the mean DGD must be less than $\langle \Delta\tau \rangle < 0.17T$ or $\langle \Delta\tau \rangle < 0.12T$ for an outage probability less than 10^{-3} and 10^{-6}, respectively. In additional to the dependence on modulation formats from Figs. 4.13 and 4.14, the tolerance to PMD is also a function of the allowance of eye-penalty and the outage probability. For example, the eye-penalty may be 2 dB instead of 1 dB for outage.

The PMD of single-mode fiber had been studied for a long time by Rashleigh and Ulrich (1978) and Kaminow (1981), mainly focused on polarization fluctuation in an optical fiber and the requirement of automatic polarization control (APC) for coherent optical communications (Noé et al., 1988a, Okoshi, 1985, Walker and Walker, 1990). The statistical model for PMD was developed by Poole and Wagner (1986), Poole et al. (1991), and Foschini and Poole (1991) that gave the Maxwellian distribution of Eq. (4.47). Gordon and Kogelnik (2000) and Poole and Nagel (1997) provided a review of PMD issues in optical fiber.

The simulation of Figs. 4.13 and 4.14, even for the same modulation format, depends on the transmitter and receiver (Garcia et al., 1996,

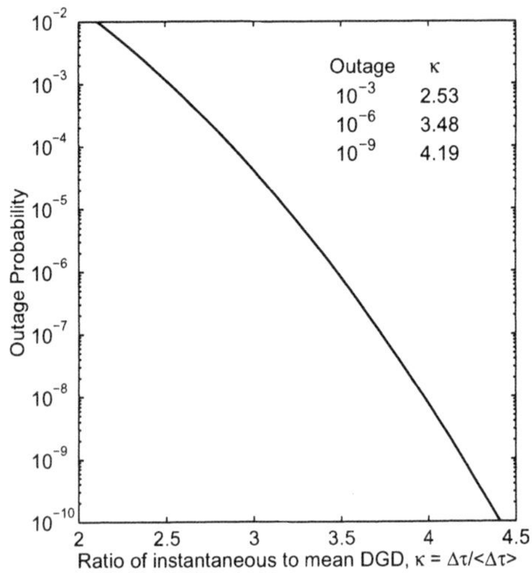

Figure 4.15. The outage probability as a function of the ratio of instantaneous to mean DGD of $\kappa = \Delta\tau / \langle\Delta\tau\rangle$. Also list the ratio of $\kappa = \Delta\tau / \langle\Delta\tau\rangle$ for some typical outage probabilities.

Sunnerud et al., 2003, Winzer et al., 2003). The eye-penalty for RZ pulses in Fig. 4.14 can be reduced using a receiver with smaller bandwidth than $0.75B_d$. Practical RZ-DPSK signal can have a large tolerance to PMD if the receiver is optimized accordingly.

With some improvements in Tomizawa et al. (2002), forward-error correction cannot in principle improve a system with PMD (Ho and Lin, 1997). Polarization scrambling may be used with forward-error correction to improve the system performance (Liu et al., 2004b, Wedding and Haslach, 2001).

In additional to first-order PMD, higher-order PMD also degrades the performance of a system (Bruyère, 1996, Bülow, 1998, Francia et al., 1998). The statistical properties of second-order PMD were studied in Foschini et al. (1999, 2001).

PMD can be compensated by both optical and electrical techniques (Buchali and Bülow, 2004, Hakki, 1997, Haunstein et al., 2004, Lanne and Corbel, 2004, Noé et al., 2004, Sunnerud et al., 2002, Takahashi et al., 1994, Winters and Gitlin, 1990). Typical optical based PMD compensators eliminate all first-order PMD and part of the second-order PMD.

The impact of PMD to coherent optical signal was first studied in Winters and Gitlin (1990). The impact of PMD to DPSK signal was studied in Kim et al. (2002), Pan et al. (2003), Wang and Kahn (2004a,b), and Xie et al. (2003).

Small amount of polarization-dependent loss exists in most optical components (Giles et al., 1991, Yamamoto et al., 1993). In a long-haul network, polarization-dependent loss is induced by the cascade of many those polarized components. When fiber between those polarized components also has PMD to randomize the input polarization into those components, the output optical power is fluctuated due to polarization-dependent loss (Huttner et al., 2000, Willner et al., 2004, Xie and Mollenauer, 2003). The statistics of polarization-dependent loss with PMD is studied in Fukada (2002), Lu et al. (2001), Yu et al. (2002), and Mecozzi and Shtaif (2004).

For single channel system, the EDFA gain in the polarization of the signal is reduced due to polarization hole-burning. Polarization-dependent gain increases the noise in the polarization orthogonal to the signal (Bruyère and Audouin, 1994, Lichtman, 1995, Mazurczyk and Zyskind, 1994, Taylor, 1993). Polarization-dependent gain is not an issue for WDM system with many channels that have random polarization. Even if all channels are launched to the fiber co-polarized, the polarization is randomized after a short piece of fiber due to PMD effects. For WDM system with many channels, common counter-pump Raman amplifier does not have large polarization-dependent gain, even stimulated Raman scattering is polarization dependent (Stolen, 1979).

6. Summary

The impact of impairment other than amplifier noises is studied in this chapter. Using balanced receiver, RIN from LO laser does not affect the performance of a well-designed systems. The laser phase noise was used to be a limitation for low-speed phase-modulated optical communication systems, especially homodyne or heterodyne PSK signals with phase tracking. For high-speed systems limited by amplifier noises, laser phase noise is not a major degradation. Typical high-speed direct-detection DPSK systems will not be affected by laser phase noise.

Both chromatic and polarization dispersion of a single-mode fiber distort the waveform of a phase-modulated optical signal. When chromatic dispersion is compensated by dispersion-compensated fiber, the tolerance to fiber dispersion gives the allowance residual chromatic dispersion. PMD compensation is still not widely used and the major degradation will be first-order PMD.

APPENDIX 4.A: Phase Distribution of Gaussian Random Variables

With only additive Gaussian noise, the baseband representation of the received signal is

$$s(t) = A + n(t), \tag{4.A.1}$$

where A is a real-number representing the transmitted signal and $n(t)$ is complex Gaussian noise of $n(t) = n_1(t)+jn_2(t)$. For a noise variance of $E\{n_1^2(t)\} = E\{n_2^2(t)\} = \sigma_n^2$, the p.d.f. of the received signal is

$$p(x_1, x_2) = \frac{1}{2\pi\sigma_n^2} \exp\left[-\frac{(x_1 - A)^2 + x_2^2}{2\sigma_n^2}\right], \tag{4.A.2}$$

with characteristic function of

$$\Psi(\omega) = \exp\left(j\omega_1 A - \frac{1}{2}\sigma_n^2|\omega|^2\right). \tag{4.A.3}$$

where $\omega = \omega_1 + j\omega_2$ as a complex number.

The received signal can convert to polar representation of $s(t) = r(t)e^{j\theta_n(t)}$ with $x_1 = r(t)\cos\theta_n(t)$ and $x_2 = r(t)\sin\theta_n(t)$, we obtain the distribution of $r(t)$ and $\theta_n(t)$ as

$$\begin{aligned} p_{R,\Theta_n}(r,\theta) &= \frac{r}{2\pi\sigma_n^2} \exp\left(-\frac{(r\cos\theta - A)^2 + r^2\sin^2\theta}{2\sigma_n^2}\right) \\ &= \frac{r}{2\pi\sigma_n^2} \exp\left(-\frac{r^2 + A^2 + 2r\cos\theta}{2\sigma_n^2}\right). \end{aligned} \tag{4.A.4}$$

The distribution of the received amplitude R is (Rice, 1944, 1948)

$$p_R(r) = \int_{-\pi}^{+\pi} p_{R,\Theta_n}(r,\theta)\mathrm{d}\theta = \frac{r}{\sigma_n^2} \exp\left(-\frac{r^2 + A^2}{2\sigma_n^2}\right) I_0\left(\frac{rA}{\sigma_n^2}\right) \tag{4.A.5}$$

as the Rice distribution, where $I_0(\cdot)$ is zero-order modified Bessel function of first kind given by $I_0(z) = \frac{1}{\pi}\int_0^\pi e^{z\cos\theta}\mathrm{d}\theta$ (Gradshteyn and Ryzhik, 1980, §8.431).

The received intensity is $Y = R^2$ with a noncentral chi-square (χ^2) distribution of

$$p_Y(y) = \frac{1}{2\sigma_n^2} \exp\left(-\frac{y + A^2}{2\sigma_n^2}\right) I_0\left(\frac{\sqrt{y}A}{\sigma_n^2}\right). \tag{4.A.6}$$

For phase modulation, the phase component of the received electric field has a distribution of

$$\begin{aligned} p_{\Theta_n}(\theta) &= \int_0^\infty p_{R,\Theta_n}(r,\theta)\mathrm{d}r \\ &= \frac{1}{2\pi}e^{-\rho_s} + \sqrt{\frac{\rho_s}{4\pi}}\cos\theta e^{-\rho_s\sin^2\theta}\mathrm{erfc}\left(-\sqrt{\rho_s}\cos\theta\right). \end{aligned} \tag{4.A.7}$$

The distribution of Eq. (4.A.7) is a function of $\cos\theta$ and thus an even periodic function with a period of 2π. We can expand the p.d.f. of Eq. (4.A.7) as a Fourier series of

$$p_{\Theta_n}(\theta) = \frac{1}{2\pi}\sum_{m=-\infty}^{\infty} c_m \exp(-jm\theta). \tag{4.A.8}$$

It is difficult to find the coefficients of c_m directly from the p.d.f. of Eq. (4.A.7), we can calculate the coefficients according to

$$\begin{aligned} c_m &= \int_{-\pi}^{\pi} p_{\Theta_n}(\theta) e^{jm\theta} \mathrm{d}\theta \\ &= \int_0^{\infty} \int_{-\pi}^{\pi} p_{R,\Theta_n}(r,\theta) e^{jm\theta} \mathrm{d}\theta \mathrm{d}r, \quad m \geq 0, \end{aligned} \tag{4.A.9}$$

and $c_{-m} = c_m$. Using Gradshteyn and Ryzhik (1980, §8.431), we obtain

$$\int_{-\pi}^{\pi} p_{R,\Theta_n}(r,\theta) e^{jm\theta} \mathrm{d}\theta = \frac{r}{\sigma_n^2} \exp\left(-\frac{r^2 + A^2}{2\sigma_n^2}\right) I_m\left(\frac{rA}{\sigma_n^2}\right). \tag{4.A.10}$$

Using Gradshteyn and Ryzhik (1980, §6.614, §9.220), the integration of Eq. (4.A.10) over amplitude r is equal to

$$\begin{aligned} c_m &= \frac{\rho_s^{\frac{m}{2}}}{m!} \Gamma\left(\frac{m}{2}+1\right) {}_1F_1\left(\frac{m}{2}; m+1; -\rho_s\right) \\ &= \frac{\sqrt{\pi \rho_s}}{2} e^{-\rho_s/2} \left[I_{\frac{m-1}{2}}\left(\frac{\rho_s}{2}\right) + I_{\frac{m+1}{2}}\left(\frac{\rho_s}{2}\right)\right], \end{aligned} \tag{4.A.11}$$

where $\Gamma(\cdot)$ is the Gamma function, ${}_1F_1(a; b; \cdot)$ is the confluent hypergeometric function of the first kind , and $I_k(\cdot)$ is the k-th order modified Bessel function of the first kind. The p.d.f. of Eq. (4.A.7) is equal to

$$p_{\Theta_n}(\theta) = \frac{1}{2\pi} + \frac{e^{-\frac{\rho_s}{2}}}{2} \sqrt{\frac{\rho_s}{\pi}} \sum_{m=1}^{\infty} \left[I_{\frac{m-1}{2}}\left(\frac{\rho_s}{2}\right) + I_{\frac{m+1}{2}}\left(\frac{\rho_s}{2}\right)\right] \cos(m\theta). \tag{4.A.12}$$

The series representation of the phase distribution Eq. (4.A.12) can be found in Middleton (1960, §9.2-2), Jain (1974), Jain and Blachman (1973), Prabhu (1969), and Blachman (1981, 1988). The conversion from confluent hypergeometric function to modified Bessel function is given by Jain (1974), Jain and Blachman (1973), Prabhu (1969), and Blachman (1981, 1988).

The variance of the phase of amplifier noise is equal to

$$\begin{aligned} \sigma_{\Theta_n}^2 &= 2\int_0^{\pi} \theta^2 p_{\Theta_n}(\theta) \mathrm{d}\theta \\ &= \frac{\pi^2}{3} + 2e^{-\frac{\rho_s}{2}} \sqrt{\pi \rho_s} \sum_{m=1}^{\infty} \frac{(-1)^m}{m^2} \left[I_{\frac{m-1}{2}}\left(\frac{\rho_s}{2}\right) + I_{\frac{m+1}{2}}\left(\frac{\rho_s}{2}\right)\right]. \end{aligned} \tag{4.A.13}$$

For high SNR, the p.d.f. of Eq. (4.A.7) can be approximated as

$$p_{\Theta_n}(\theta) \approx \sqrt{\frac{\rho_s}{\pi}} \cos\theta e^{-\rho_s \sin^2\theta} \tag{4.A.14}$$

with variance of

$$\sigma_{\Theta_n}^2 \approx \frac{1}{2\rho_s}. \tag{4.A.15}$$

Figure 4.A.1 shows the exact phase variance of Eq. (4.A.13) and the approximated phase variance of Eq. (4.A.15) as a function of SNR ρ_s. In high SNR of $\rho_s > 10$ dB,

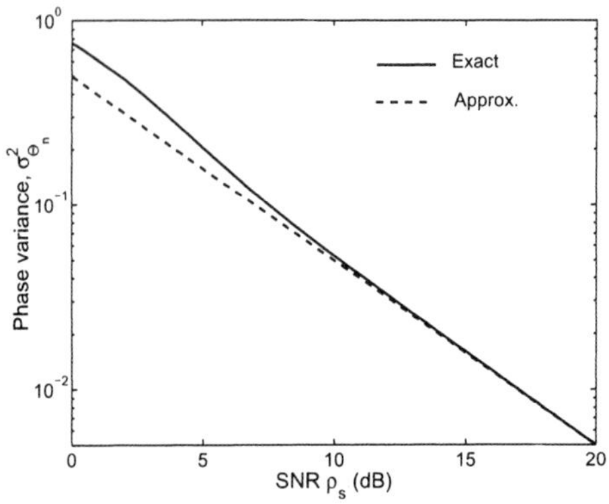

Figure 4.A.1. The phase variance of $\sigma^2_{\Theta_n}$ as a function of SNR ρ_s.

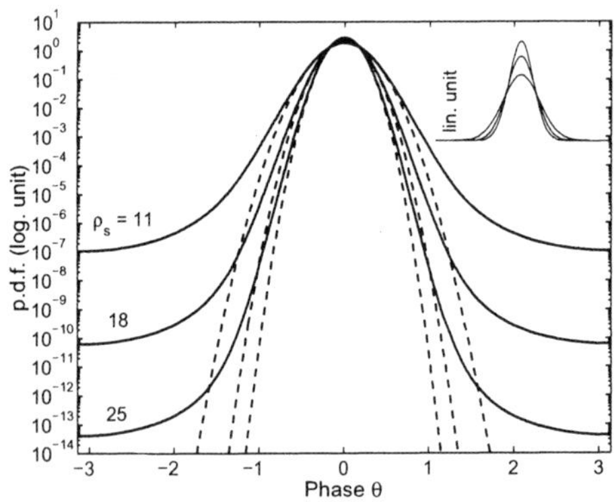

Figure 4.A.2. The p.d.f. of $p_{\Theta_n}(\theta)$ as compared with the Gaussian approximation. Solid and dashed lines are the exact and Gaussian approximated p.d.f.

the exact [Eq. (4.A.13)] and approximated [Eq. (4.A.15)] phase variance are almost the same.

Figure 4.A.2 shows the phase distribution $p_{\Theta_n}(\theta)$ of either Eq. (4.A.7) or (4.A.12) for a SNR of $\rho_s = 11, 18, 25$ (10.4 12.6 14.0 dB). The zero-mean Gaussian approximation with a variance of Eq. (4.A.15) is also plotted in Fig. 4.A.2 for comparison. Figure 4.A.2 plots in logarithmic scale to show the difference in the tail between the exact p.d.f. and Gaussian approximation. An inset plots the p.d.f. in linear scale. In the linear scale inset, the exact and approximated p.d.f. overlaps with each other and has not observable difference. While the phase distribution of $p_{\Theta_n}(\theta)$ is the same as Gaussian distribution in linear scale, Gaussian distribution cannot be used if a tail probability is interested. For optical communications interest in an error probability of 10^{-9} or lower, the tail probability is essential to evaluate the error probability.

Based on the series expansion of Eq. (4.A.12), the error probability of binary PSK signal [see Eq. (3.78)] is also equal to

$$\begin{aligned} p_e &= 1 - \int_{-\frac{\pi}{2}}^{\frac{\pi}{2}} p_{\Theta_n}(\theta)\mathrm{d}\theta \\ &= \frac{1}{2} - \frac{e^{-\frac{\rho_s}{2}}}{2}\sqrt{\frac{\rho_s}{\pi}} \sum_{m=1}^{\infty} \frac{2\sin(m\pi/2)}{m} \left[I_{\frac{m-1}{2}}\left(\frac{\rho_s}{2}\right) + I_{\frac{m+1}{2}}\left(\frac{\rho_s}{2}\right)\right] \\ &= \frac{1}{2} - \exp\left(-\frac{\rho_s}{2}\right)\sqrt{\frac{\rho_s}{\pi}} \sum_{k=0}^{\infty} \frac{(-1)^k}{2k+1} \left[I_k\left(\frac{\rho_s}{2}\right) + I_{k+1}\left(\frac{\rho_s}{2}\right)\right]. \end{aligned} \tag{4.A.16}$$

Because $\sin(m\pi/2) = 0$ for m as even number, the error probability of Eq. (4.A.16) is simplified.

From the factor of $(-1)^k$, the terms of Eq. (4.A.16) oscillate between positive and negative values. Although the summation of Eq. (4.A.16) converges, the calculation is numerically challenging for small error probability. Note that the multiplication factor of the summation is a small value in the order of $e^{-\frac{\rho_s}{2}}\sqrt{\rho_s}$, the summation has a value in the order of $e^{\frac{\rho_s}{2}}/\sqrt{\rho_s}$ for small error probability. For large SNR ρ_s, the summation has very large terms although the error probability is small. The error probability is the difference between 1/2 and a value a little bit smaller than 1/2. The series summation of Eq. (4.A.16) can be calculated to an error probability of 10^{-13} to 10^{-14} with an accuracy of three to four significant digits. Symbolic mathematical software can provide better accuracy by using variable precision arithmetic in the calculation of low error probability.

A DPSK signal can be demodulated using the differential phase of

$$\Delta\Theta_n = \Theta_n(t) - \Theta_n(t-T), \tag{4.A.17}$$

where $\Theta_n(\cdot)$ is the phase of amplifier noise as a function of time and T is the symbol interval. The phases of $\Theta_n(t)$ and $\Theta_n(t-T)$ are two identical independently distributed random variables with p.d.f. given by $p_{\Theta_n}(\theta)$ of either Eq. (4.A.7) and Eq. (4.A.12).

When two independently distributed random variables are summed (or subtracted) together, the characteristic function of the sum (or difference) is equal to the product of its individual characteristic functions. For the series expansion like Eq. (4.A.12), the sum of two random variables has a Fourier coefficient that is the product of the corresponding Fourier coefficients. Based on the series expansion of Eq. (4.A.12), the differential phase of Eq. (4.A.17) has a p.d.f. of

$$p_{\Delta\Theta_n}(\theta) = \frac{1}{2\pi} + \frac{\rho_s e^{-\rho_s}}{4} \sum_{m=1}^{\infty} \left[I_{\frac{m-1}{2}}\left(\frac{\rho_s}{2}\right) + I_{\frac{m+1}{2}}\left(\frac{\rho_s}{2}\right)\right]^2 \cos(m\theta). \tag{4.A.18}$$

The error probability corresponding to Eq. (4.A.16) for DPSK signal [see Eq. (3.105) is

$$p_e = \frac{1}{2} - \frac{\rho_s e^{-\rho_s}}{2} \sum_{k=0}^{\infty} \frac{(-1)^k}{2k+1} \left[I_k\left(\frac{\rho_s}{2}\right) + I_{k+1}\left(\frac{\rho_s}{2}\right)\right]^2. \tag{4.A.19}$$

Comparing the error probability of Eq. (3.78) with (4.A.16) for PSK signal and Eq. (3.105) with Eq. (4.A.19) for DPSK signal, the series of Eq. (4.A.12) is not very useful for performance analysis of phase-modulated signals with additive Gaussian noise only. However, when the system has additive phase noise that is independent of the additive Gaussian noise, the series of Eq. (4.A.12) becomes useful. The summed phase noise has Fourier coefficients that are the multiplication of the corresponding coefficients of each individual phase noise components. Using the series of Eq. (4.A.12), the error probability of PSK or DPSK signals was derived by Jain (1974), Jain and Blachman (1973), Lindsey and Simon (1973), and Prabhu (1976) with PLL noise, Blachman (1981) with phase error, Nicholson (1984) with laser phase noise, Jacobsen and Garrett (1987) and Iwashita and Matsumoto (1987) with phase error and laser phase noise, and Ho (2003b, 2004b) with nonlinear phase noise. In later chapter, followed Prabhu (1969), the series of Eq. (4.A.12) is used to derive the error probability of multilevel phase-modulated signals.

Chapter 5

NONLINEAR PHASE NOISE

The response of all dielectric materials to light becomes nonlinear under strong optical intensity (Boyd, 2003), and optical fiber has no exception. Due to fiber Kerr effect, the refractive index of optical fiber increases with optical intensity to slightly slow down the propagation speed, inducing intensity depending nonlinear phase shift. With optical amplifier noises, the optical intensity has a noisy component and the nonlinear phase shift includes nonlinear phase noise.

Gordon and Mollenauer (1990) first showed that when optical amplifiers are used to periodically compensate for fiber loss, the interaction of amplifier noises and the fiber Kerr effect causes phase noise, often called the Gordon-Mollenauer effect, or more precisely, self-phase modulation induced nonlinear phase noise. Phase-modulated optical signals, both phase-shift keying (PSK) and differential phase-shift keying (DPSK), carry information by the phase of an optical carrier. Added directly to the phase of a signal, nonlinear phase noise degrades both PSK and DPSK signals and limits the maximum transmission distance.

Early literatures studied the spectral broadening induced by nonlinear phase noise (Ryu, 1991, 1992, Saito et al., 1993). The performance degradation due to nonlinear phase noise is assumed the same as that due to laser phase noise in Sec. 4.3. However, the statistical properties of nonlinear phase noise are not the same as laser phase noise as shown later. The probability density function (p.d.f.) of nonlinear phase noise is required for performance evaluation of a phase-modulated signal with nonlinear phase noise.

This chapter investigates nonlinear phase noise based on either discrete or distributed assumption for finite or infinite number of fiber spans. When the optical signal is periodically amplified by optical amplifiers, amplifier noise is unavoidably added to the optical signal. Nonlinear phase noise is accumulated span after span. When the number of fiber spans is very large, the accumulation of nonlinear phase noise can be modeled as a distributed process asymptotically. For small number of fiber spans, the accumulation of nonlinear phase noise is the summation of the contribution from each individual span.

The exact error probability of a signal with nonlinear phase noise is derived when the dependence between linear and nonlinear phase noise is taken into account. The dependence between linear and nonlinear phase noise increases the error probability of the signal. Simulation is conducted to verify theoretical results.

1. Nonlinear Phase Noise for Finite Number of Fiber Spans

In a lightwave system, nonlinear phase noise is induced by the interaction of fiber Kerr effect and optical amplifier noise. Here, nonlinear phase noise is induced by self-phase modulation through the amplifier noise in the same polarization as the signal and within an optical bandwidth matched to the signal. The phase noise induced by cross-phase modulation from adjacent channels of a WDM system is first ignored in this chapter.

1.1 Self-Phase Modulation Induced Nonlinear Phase Noise

At high optical power of P, the refractive index of silica must include the nonlinear contribution of (Agrawal, 2001, Boyd, 2003)

$$n_r' = n_{r0} + \bar{n}_2(P/A_{\text{eff}}), \tag{5.1}$$

where n_{r0} is the refractive index at small optical power, n_r' is the refractive index depending on optical power, $\bar{n}_2$ is the nonlinear-index coefficient, and A_{eff} is the effective core area. The nonlinear-index coefficient is $\bar{n}_2 \approx 3.2 \times 10^{-20}$ m^2/W for silica fibers (Boyd, 2003). Typically, the nonlinear contribution to the refractive index is quite small (less than 10^{-7}). Compared with other materials, the fiber material of fused silica also has very small nonlinear-index coefficient. Because optical fiber has very small loss and thus a long interaction length, the effect of nonlinear refractive index becomes significant[1], especially when optical amplifiers are used to maintain high optical power in the fiber link. The

[1] Most experiments in nonlinear optics use a crystal with a length in the order of several centimeters compared with an effective length of about 20 km in typical optical fiber.

propagation constant becomes power dependent and can be written as $\beta' = \beta_0 + \gamma P$ (Agrawal, 2001) where

$$\gamma = \frac{\omega_0 \bar{n}_2}{A_{\text{eff}} c} \tag{5.2}$$

is the fiber nonlinear coefficient with ω_0 as the angular frequency and c as the speed of light in free space.

In each fiber span, the overall nonlinear phase shift is equal to

$$\Phi_{\text{NL}} = \int_0^L \gamma P(z) \mathrm{d}z = \gamma L_{\text{eff}} P, \tag{5.3}$$

where P is assumed to be the launched power of $P = P(0)$. For a fiber span length of L and attenuation coefficient of α, $P(z) = Pe^{-\alpha z}$ and

$$L_{\text{eff}} = \frac{1 - e^{-\alpha L}}{\alpha} \tag{5.4}$$

is the effective nonlinear length.

If the electric field is E and amplifier noise is n, both as complex number for the baseband representation of the electric field, with proper unit, we have $P = |E + n|^2$. For the amplifier noise within the bandwidth of the signal, self-phase modulation causes a mean nonlinear phase shift[2] of about $\gamma L_{\text{eff}} |E|^2$ and phase noise of $\gamma L_{\text{eff}} \left[2\Re\{E \cdot n\} + |n|^2 \right]$. For high signal-to-noise ratio (SNR), the first term of $2\Re\{E \cdot n\}$ is much larger than the second term of $|n|^2$.

In the refractive index of Eq. (5.1), the actual electric field in the fiber is $\sqrt{P/A_{\text{eff}}}$. In practice, a proportional constant does not change the physical meaning of the equations. The electric field in the fiber is also not uniformly distributed as implied by the simple division of P/A_{eff}.

For an N_A-span fiber system, the overall nonlinear phase noise is

$$\Phi_{\text{NL}} = \gamma L_{\text{eff}} \Big\{ |E_0 + n_1|^2 + |E_0 + n_1 + n_2|^2 + \cdots + |E_0 + n_1 + \cdots + n_{N_A}|^2 \Big\}, \tag{5.5}$$

where E_0 is the baseband representation of the transmitted electric field, $n_k, k = 1, \ldots, N_A$, are independent identically distributed zero-mean circular Gaussian random complex number as the optical amplifier noise introduced into the system at the kth fiber span. The variance of n_k is $E\{|n_k|^2\} = 2\sigma_0^2$, $k = 1, \ldots, N_A$, where σ_0^2 is the noise variance per span per dimension. In the linear regime, ignoring the fiber loss of the

[2] For a simplified discussion, we ignore the mean of $|n|^2$.

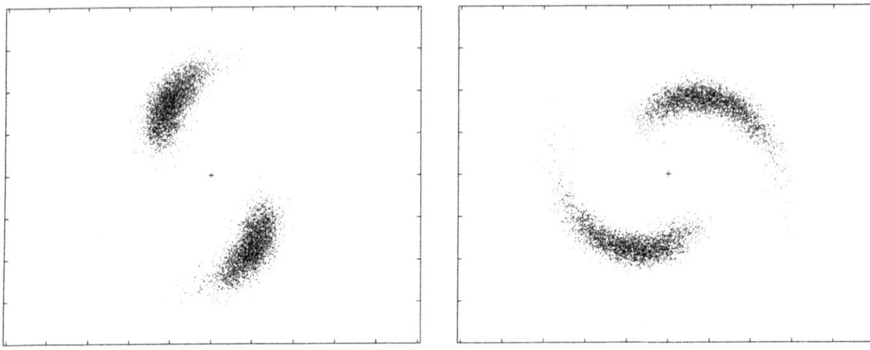

(a) $\langle\Phi_{\mathrm{NL}}\rangle = 1$ rad (b) $\langle\Phi_{\mathrm{NL}}\rangle = 2$ rad

Figure 5.1. Simulated distribution of the received electric field for mean nonlinear phase shift of (a) $\langle\Phi_{\mathrm{NL}}\rangle = 1$ rad and (b) $\langle\Phi_{\mathrm{NL}}\rangle = 2$ rad.

last span and the amplifier gain required to compensate it, the signal received after N_A spans is

$$E_N = E_0 + n_1 + n_2 + \cdots + n_{N_A} \tag{5.6}$$

with a power of $P_N = |E_N|^2$ and SNR of $\rho_s = P_N^2/(2N_A\sigma_0^2)$. In Eq. (5.5), the configuration of each fiber spans is assumed to be identical with the same length and launched power.

Figures 5.1 show the simulated distribution of the received electric field including the contribution from nonlinear phase noise of $E_r = E_N \exp(-j\Phi_{\mathrm{NL}})$. The mean nonlinear phase shifts $\langle\Phi_{\mathrm{NL}}\rangle$ are 1 and 2 rad for Figs. 5.1(a) and (b), respectively. The mean nonlinear phase shift of $\langle\Phi_{\mathrm{NL}}\rangle = 1$ rad corresponds to the limitation estimated by Gordon and Mollenauer (1990). The limitation of $\langle\Phi_{\mathrm{NL}}\rangle = 2$ rad is when the standard deviation (STD) of nonlinear phase noise is halved using a linear compensator. We will discuss nonlinear phase noise compensation in detail in next chapter.

Figures 5.1 are plotted for the case that the SNR $\rho_s = 18$ (12.6 dB), corresponding to an error probability of 10^{-9} if the amplifier noise is the sole impairment for PSK signal as from Fig. 3.13. The number of fiber spans is $N_A = 32$. The transmitted signal is $E_0 = \pm A$ for binary PSK signal. The distribution of Figs. 5.1 has 5000 points for different noise combinations.

In practice, the signal distribution of Figs. 5.1 can be measured using a quadrature optical phase-locked loop (PLL) of Fig. 3.4. Note that although the optical PLL actually tracks out the mean nonlinear phase shift of $\langle\Phi_{\mathrm{NL}}\rangle$, nonzero values of $\langle\Phi_{\mathrm{NL}}\rangle$ have been preserved in plotting Figs. 5.1 to better illustrate the nonlinear phase noise.

Early this section considers a non-return-to-zero (NRZ) signal or a continuous-wave (cw) optical signal with noise. In practice, the optical signal may be a short return-to-zero (RZ) pulse of, for example, a Gaussian pulse of $u_0(t) = A_0 \exp\left(-t^2/2T_0^2\right)$ with $1/e$-pulse width of T_0. Assume a single span system and the pulse does not have distortion in the fiber, the nonlinear phase noise is time dependent and proportional to $\gamma L_{\text{eff}}|u_0(t) + n(t)|^2$. When the nonlinear phase noise is weighted and averaged using the pulse shape of $u_0(t)$, the nonlinear phase noise is

$$\Phi_{\text{NL}} = \frac{\gamma L_{\text{eff}} \int_{-\infty}^{+\infty} |u_0(t) + n(t)|^2 u_0^2(t)\mathrm{d}t}{\int_{-\infty}^{+\infty} u_0^2(t)\mathrm{d}t} \tag{5.7}$$

with mean nonlinear phase shift of

$$\langle \Phi_{\text{NL}} \rangle = \frac{\gamma L_{\text{eff}}|A_0|^2}{\sqrt{2}} + 2\sigma_0^2. \tag{5.8}$$

If the noise is constant over the pulse period, the noise is given by

$$2A_0\sqrt{\frac{2}{3}}n(t) \tag{5.9}$$

with a small increase of about $\sqrt{4/3}$ more than the mean of Eq. (5.8), i.e., 33% in variance.

If the optical pulse has a period of T, the average channel power is $P_0 = |A_0|^2\sqrt{\pi}T_0/T$. The mean nonlinear phase shift is increased by a factor of about $\frac{1}{\sqrt{2\pi}}\frac{T}{T_0}$ to Eq. (5.3) with $P = P_0$. The nonlinear phase noise is increased by a factor of

$$\frac{2}{3\sqrt{\pi}}\frac{T}{T_0}. \tag{5.10}$$

With proper scaling, the same expression of nonlinear phase noise of Eq. (5.5) may be used for system using short pulse. However, the above analysis may consider as a first-order approximation. A more rigorous deviation is given in later chapter based on better model.

The impact of nonlinear phase noise to a phase-modulated system was first studied by Gordon and Mollenauer (1990). Early works measured the linewidth broadening due to nonlinear phase noise (Ryu, 1991, 1992, Saito et al., 1993). Recent measurement of nonlinear phase noise includes Kim and Gnauck (2003), Mizuochi et al. (2003), Xu et al. (2002), and Kim (2003). As shown in next chapter, Liu et al. (2002b), Xu and Liu (2002), and Ho and Kahn (2004a), nonlinear phase noise can be compensated using a scale version of the received intensity.

1.2 Probability Density

Equivalent to the p.d.f., the characteristic function of the nonlinear phase noise of Φ_{NL} is derived here. For simplicity, we ignore the product of fiber nonlinear coefficient and the effective length per span of γL_{eff}. A constant factor does not change the properties of a random variable. When the nonlinear phase noise is normalized with respect to the mean nonlinear phase shift of $\langle \Phi_{\text{NL}} \rangle$, the value of γL_{eff} is not essential to the probability density of the nonlinear phase noise of Eq. (5.5).

Characteristic Function

With a transmitted electrical field of $E_0 = A$ as a real number, we consider the random variable of

$$\varphi_1 = |A + x_1|^2 + |A + x_1 + x_2|^2 + \cdots + |A + x_1 + \cdots + x_{N_A}|^2, \quad (5.11)$$

where $n_k = x_k + j y_k$, $k = 1, \ldots, N_A$, with x_k and y_k as the real and imaginary parts of n_k, respectively. The random variable of Eq. (5.11) is similar to noncentral chi-square (χ^2) distribution but the variance of the Gaussian random variables of $A + x_1 + \cdots + x_k$ are not the same. The overall nonlinear phase noise of Eq. (5.5) is $\Phi_{\text{NL}} = \varphi_1 + \varphi_2$, where

$$\varphi_2 = y_1^2 + |y_1 + y_2|^2 + \cdots + |y_1 + \cdots + y_{N_A}|^2 \quad (5.12)$$

is independent of φ_1 and has a p.d.f. equal to that of φ_1 when $A = 0$. In matrix format, the random variable of Eq. (5.11) is

$$\varphi_1 = N_A A^2 + 2A \vec{w}^T \vec{x} + \vec{x}^T \mathcal{C} \vec{x}, \quad (5.13)$$

where $\vec{w} = (N_A, N_A - 1, \ldots, 2, 1)^T$, $\vec{x} = (x_1, x_2, \ldots, x_{N_A})^T$, and the covariance matrix is

$$\mathcal{C} = \mathcal{M}^T \mathcal{M} \quad (5.14)$$

with

$$\mathcal{M} = \begin{bmatrix} 1 & 0 & 0 & \cdots & 0 \\ 1 & 1 & 0 & \cdots & 0 \\ 1 & 1 & 1 & \cdots & 0 \\ \vdots & \vdots & \vdots & \ddots & \vdots \\ 1 & 1 & 1 & \cdots & 1 \end{bmatrix}. \quad (5.15)$$

The p.d.f. of the vector $\vec{x}$ is a multi-dimensional Gaussian distribution of

$$\frac{1}{(2\pi\sigma_0^2)^{\frac{N_A}{2}}} \exp\left(-\frac{\vec{x}^T \vec{x}}{2\sigma_0^2} \right). \quad (5.16)$$

While the p.d.f. is difficult to find directly, the characteristic function of a random variable is the Fourier transform of the p.d.f. The characteristic function of φ_1, $\Psi_{\varphi_1}(\nu) = E\{\exp(j\nu\varphi_1)\}$, is

$$\Psi_{\varphi_1}(\nu) = \frac{\exp(j\nu N_A A^2)}{(2\pi\sigma_0^2)^{\frac{N_A}{2}}} \int \exp\left[2j\nu A\vec{w}^T\vec{x} - \vec{x}^T\Gamma\vec{x}\right] \mathrm{d}\vec{x}, \tag{5.17}$$

where $\Gamma = \mathcal{I}/(2\sigma_0^2) - j\nu\mathcal{C}$ and $\mathcal{I}$ is an $N_A \times N_A$ identity matrix. Using the relationship of

$$\vec{x}^T\Gamma\vec{x} - 2j\nu A\vec{w}^T\vec{x} = \left(\vec{x} - j\nu A\Gamma^{-1}\vec{w}\right)^T\Gamma\left(\vec{x} - j\nu A\Gamma^{-1}\vec{w}\right) + \nu^2A^2\vec{w}^T\Gamma^{-1}\vec{w}, \tag{5.18}$$

with some algebra, the characteristic function of Eq. (5.17) is

$$\Psi_{\varphi_1}(\nu) = \frac{\exp\left[j\nu N_A A^2 - \nu^2A^2\vec{w}^T\Gamma^{-1}\vec{w}\right]}{(2\sigma_0^2)^{\frac{N_A}{2}}\det[\Gamma]^{1/2}}, \tag{5.19}$$

where $\det[\cdot]$ is the determinant of a matrix. The characteristic function of Eq. (5.19) is rewritten as

$$\Psi_{\varphi_1}(\nu) = \frac{\exp\left[j\nu N_A A^2 - 2\sigma_0^2\nu^2A^2\vec{w}^T(\mathcal{I} - 2j\nu\sigma_0^2\mathcal{C})^{-1}\vec{w}\right]}{\det\left[\mathcal{I} - 2j\nu\sigma_0^2\mathcal{C}\right]^{\frac{1}{2}}}. \tag{5.20}$$

Substitute $A = 0$ into Eq. (5.20), the characteristic function of φ_2 is

$$\Psi_{\varphi_2}(\nu) = \frac{1}{\det\left[\mathcal{I} - 2j\nu\sigma_0^2\mathcal{C}\right]^{\frac{1}{2}}}. \tag{5.21}$$

The characteristic function of Φ_{NL} is $\Psi_{\Phi_{\mathrm{NL}}}(\nu) = \Psi_{\varphi_1}(\nu)\Psi_{\varphi_2}(\nu)$, or

$$\Psi_{\Phi_{\mathrm{NL}}}(\nu) = \frac{\exp\left[j\nu N_A A^2 - 2\sigma_0^2\nu^2A^2\vec{w}^T(\mathcal{I} - 2j\nu\sigma_0^2\mathcal{C})^{-1}\vec{w}\right]}{\det\left[\mathcal{I} - 2j\nu\sigma_0^2\mathcal{C}\right]}. \tag{5.22}$$

If the covariance matrix $\mathcal{C}$ has eigenvalues and eigenvectors of λ_k, $\vec{v}_k$, $k = 1, 2, \ldots, N_A$, respectively, the characteristic function of Eq. (5.22) becomes

$$\Psi_{\Phi_{\mathrm{NL}}}(\nu) = \frac{\exp\left[j\nu N_A A^2 - 2\sigma_0^2\nu^2A^2\sum_{k=1}^{N_A}\frac{(\vec{v}_k^T\vec{w})^2}{1 - 2j\nu\sigma_0^2\lambda_k}\right]}{\prod_{k=1}^{N_A}\left(1 - 2j\nu\sigma_0^2\lambda_k\right)}, \tag{5.23}$$

or

$$\Psi_{\Phi_{\rm NL}}(\nu) = \prod_{k=1}^{N_A} \frac{1}{1 - 2j\nu\sigma_0^2\lambda_k} \exp\left[\frac{j\nu A^2(\vec{v}_k^T \vec{w})^2/\lambda_k}{1 - 2j\nu\sigma_0^2\lambda_k}\right]. \tag{5.24}$$

From the characteristic function of Eq. (5.24), the random variable $\Phi_{\rm NL}$ of Eq. (5.5) is the summation of N_A independently distributed noncentral χ^2 random variables with two degrees of freedom. The characteristic function in the form of Eq. (5.24) based on eigenvalues and eigenvectors had been known for a long time (Turin, 1960). As a positive define matrix, the eigenvalues of the covariance matrix of $\mathcal{C}$ are all positive and multiply to unity because of

$$\prod_{k=1}^{N_A} \lambda_k = \det[\mathcal{C}] = \det[\mathcal{M}]^2 = 1. \tag{5.25}$$

Without going into detail, the matrix

$$\mathcal{C}^{-1} = \begin{bmatrix} 1 & -1 & 0 & \cdots & 0 & 0 \\ -1 & 2 & -1 & \cdots & 0 & 0 \\ 0 & -1 & 2 & \cdots & 0 & 0 \\ \vdots & \vdots & \vdots & \ddots & \vdots & \vdots \\ 0 & 0 & 0 & \cdots & -1 & 2 \end{bmatrix} \tag{5.26}$$

is approximately a Toeplitz matrix for the series of $2, -1, 0, \ldots$ For large number of spans of N_A, the eigenvalues of the covariance matrix of $\mathcal{C}$ are asymptotically equal to (Gray, 1973)

$$\frac{1}{\lambda_k} \approx 2\left[1 - \cos\left(\frac{(2k+1)\pi}{2N_A}\right)\right] = 4\sin^2\left(\frac{(2k-1)\pi}{4N_A}\right), \quad k = 1, \ldots, N_A. \tag{5.27}$$

The values of Eq. (5.27) are the discrete Fourier transform of each row of the matrix $\mathcal{C}^{-1}$, i.e., that of $2, -1, 0, \ldots$ The approximation of Eq. (5.27) can be used to understand the behavior of the characteristic function of Eq. (5.24) when the number of fiber spans is very large.

Numerical Results

The p.d.f. of nonlinear phase noise of Eq. (5.5) can be calculated by taking the inverse Fourier transforms of the corresponding characteristic functions $\Psi_{\Phi_{\rm NL}}(\nu)$ of Eq. (5.24). Figure 5.2 shows the p.d.f. of $\Phi_{\rm NL}$ of Eq. (5.5). Figure 5.2 is plotted for the case that the SNR of $\rho_s = A^2/(2N_A\sigma_0^2) = 18$, corresponding to an error probability of 10^{-9} for PSK signals if the amplifier noise is the only impairment as shown in Fig. 3.13.

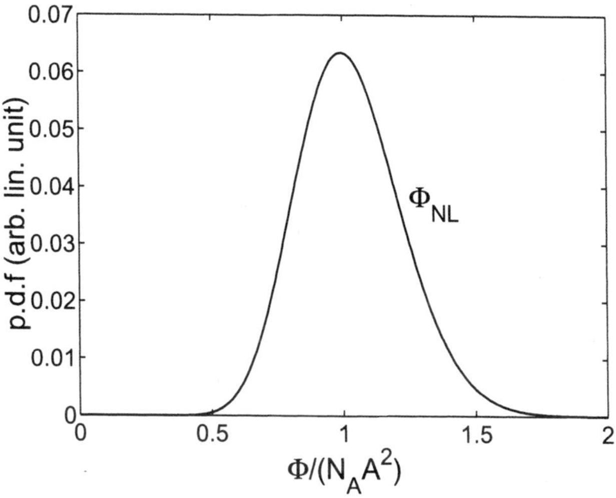

Figure 5.2. The p.d.f. of nonlinear phase noise of $\Phi_{\rm NL}$.

The number of fiber spans is $N_A = 32$. The x-axis is normalized with respect to $N_A A^2$, approximately equal to the mean nonlinear phase shift.

From the characteristic function of Eq. (5.24), the random variables of $\Phi_{\rm NL}$ can be modeled as the combination of $N_A = 32$ independently distributed noncentral χ^2-random variables. Some studies implicitly assume a Gaussian distribution by using the Q-factor to characterize the random variables. When many independently distributed random variables with more or less the same variance are summed together, the summed random variable approaches the Gaussian distribution from central limit theorem. For the characteristic function of Eq. (5.24), the Gaussian assumption is valid only if the eigenvalues λ_k are more or less the same. From Eq. (5.27), the largest eigenvalue λ_1 of the covariance matrix C is about nine times larger than the second largest eigenvalue λ_2. Numerical results show that the approximation of Eq. (5.27) is accurate within 3.2% for $N_A = 32$.

While the Gaussian assumption for $\Phi_{\rm NL}$ may not be valid, other than the noncentral χ^2-random variables corresponds to the largest eigenvalue, the other random variables should sum to Gaussian distribution. By modeling the summation of random variables with smaller eigenvalues as Gaussian distribution, the nonlinear phase noise of Eq. (5.24) can be modeled as a summation of two instead of $N_A = 32$ independently distributed random variables.

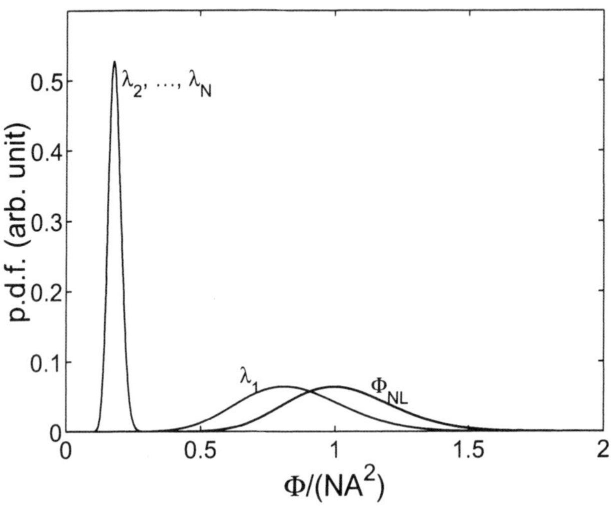

Figure 5.3. The p.d.f. of Φ_{NL} is the convolution of a Gaussian p.d.f. and a noncentral χ^2-p.d.f. with two degrees of freedom. [Adapted from Ho (2003f)]

Note that the variance of the noncentral χ^2-random variables with two degrees of freedom in Eq. (5.24) is $4\sigma_0^4\lambda_k^2 + 4A^2(\vec{v}_k^T\vec{w})^2$ (Proakis, 2000). While the above reasoning just takes into account the contribution from the eigenvalue of λ_k but ignores the contribution from the eigenvector $\vec{v}_k$, numerical results show that the variance of each individual noncentral χ^2-random variable increases with the corresponding eigenvalue of λ_k.

From Fig. 5.2, the p.d.f. of Φ_{NL} has significant difference with that of a Gaussian distribution. Figure 5.3 divides the p.d.f. of Φ_{NL} into the convolution of two parts. The first part has no observable difference with a Gaussian p.d.f. and corresponds to the second largest to the smallest eigenvalues, $\lambda_k, k = 2, \ldots, N_A$, of the characteristic function of Eq. (5.24). The second part is a noncentral χ^2-p.d.f. and corresponds to the largest eigenvalue λ_1, where $\sigma_0^2\lambda_1 \approx 2/(\pi^2\rho_s)\cdot N_A A^2$. The p.d.f. of Φ_{NL} in Fig. 5.2 is also plotted in Fig. 5.3 for comparison. The mean and variance of the Gaussian random variable are

$$\sum_{k=2}^{N_A} \frac{A^2(\vec{v}_k^T\vec{w})^2}{\lambda_k} + 2\sigma_0^2\lambda_k \qquad (5.28)$$

and

$$4\sum_{k=2}^{N_A} \sigma_0^4\lambda_k^2 + A^2(\vec{v}_k^T\vec{w})^2, \qquad (5.29)$$

respectively. The second part noncentral χ^2-p.d.f. with two degrees of freedom has a variance parameter of $\sigma_0^2\lambda_1$ and noncentrality parameter of $A^2(\vec{v}_1^T\vec{w})^2/\lambda_1$.

Traditionally, the performance of the system with nonlinear phase noise is evaluated based on the variance of the nonlinear phase noise (Gordon and Mollenauer, 1990, Ho and Kahn, 2004a, Liu et al., 2002b, McKinstrie and Xie, 2002, McKinstrie et al., 2002, Mecozzi, 1994a, Xu and Liu, 2002, Xu et al., 2003). However, it is found that nonlinear phase noise is not Gaussian-distributed both experimentally (Kim and Gnauck, 2003) and analytically (Ho, 2003a,f, Mecozzi, 1994a). For non-Gaussian noise, neither the variance nor the Q-factor (Hiew et al., 2004, Wei et al., 2003a,b) is sufficient to characterize the performance of the system. The p.d.f. is necessary to better understand the noise properties and evaluates the system performance.

This section mainly studies the nonlinear phase noise for finite number of fiber spans. The p.d.f. of nonlinear phase noise is derived analytically based on the method of Kac and Siegert (1947) and Turin (1960). These classes of random variable may be called a generalized noncentral χ^2 random variable (Middleton, 1960).

Nonlinear phase noise can be particularly modeled as the summation of a χ^2-random variable and a Gaussian random variable. Ho (2003f) also calculated the tail probability from different models for the nonlinear phase noise to confirm the model here.

The characteristic function of $\Phi_{\Phi_{\mathrm{NL}}}(\nu)$ can be used to approximately evaluate the error probability of a phase-modulated signal with nonlinear phase noise based on the assumption that nonlinear phase noise is independent of the phase of amplifier noise (Ho, 2003b). This section finds that the nonlinear phase noise is not Gaussian distributed, confirming the experimental measurement of Kim and Gnauck (2003).

2. Asymptotic Nonlinear Phase Noise

In previous section, nonlinear phase noise is given by a summation from the contribution of many fiber spans. If the number of fiber spans is very large, the summation can be replaced by integration. This distributed model of nonlinear phase noise enables us to model the nonlinear phase noise as a transform of Wiener process. The joint statistics of nonlinear phase noise with received electric field can be derived accordingly.

Later parts of this section first find a series representation of the nonlinear phase noise after a convenient normalization. The characteristic function of nonlinear phase noise is derived afterward. Similarly, the joint characteristic function of nonlinear phase noise and received electric field can also be derived.

2.1 Statistics of Nonlinear Phase Noise

The characteristic function of nonlinear phase noise is derived in this section after normalization to simplify the problem. Nonlinear phase noise is found to be the summation of infinitely many independently distributed noncentral χ^2-distributed random variables. The joint statistics of nonlinear phase noise with received electric field depends on only two parameters: the SNR of ρ_s and the mean nonlinear phase shift of $\langle \Phi_{\mathrm{NL}} \rangle$.

Normalization

With large number of fiber spans, the summation of Eq. (5.5) can be replaced by integration as

$$\Phi_{\mathrm{NL}} = \frac{\gamma L_{\mathrm{eff}}}{L} \int_0^{L_T} |E_0 + n(z)|^2 \mathrm{d}z, \tag{5.30}$$

where $L_T = N_A L$ is the overall fiber length, $\gamma L_{\mathrm{eff}}/L$ is the average nonlinear coefficient per unit length, and $n(z)$ is a zero-mean complex value Wiener process with autocorrelation of

$$E\{n(z_1) \cdot n^*(z_2)\} = \sigma_s^2 \min(z_1, z_2). \tag{5.31}$$

The variance of $\sigma_s^2 = 2\sigma_0^2/L$ is the noise variance per unit length where $E\{|n_k|^2\} = 2\sigma_0^2, k = 1, \ldots, N_A$ is noise variance per amplifier per polarization in the optical bandwidth matched to the signal.

We investigate the joint statistical properties of the normalized electric field and normalized nonlinear phase noise

$$e_N = \xi_0 + b(1), \quad \Phi = \int_0^1 |\xi_0 + b(t)|^2 \mathrm{d}t, \tag{5.32}$$

where $b(t)$ is a zero-mean complex Wiener process with an autocorrelation function of

$$R_b(t, s) = E\{b(s) \cdot b^*(t)\} = \min(t, s). \tag{5.33}$$

Comparing the phase noise of Eqs. (5.30) and (5.32), the normalized nonlinear phase noise of Eq. (5.32) is scaled by $\Phi = L^2 \sigma_s^2 \Phi_{\mathrm{NL}}/(\gamma L_{\mathrm{eff}})$, $t = z/L$ is the normalized distance, $b(t) = n(tL_T)/\sigma_s/\sqrt{L_T}$ is the normalized amplifier noise, $\xi_0 = E_0/\sigma_s/\sqrt{L_T}$ is the normalized transmitted vector. Compared with Eq. (5.6), the normalized electric field of e_N is scaled by the inverse of the noise variance. The SNR is

$$\rho_s = |\xi_0|^2 = \frac{|E_0|^2}{L_T \sigma_s^2} = \frac{|E_0|^2}{2N_A \sigma_0^2}. \tag{5.34}$$

In Eq. (5.32), the normalized electric field e_N is the normalized received electric field without nonlinear phase noise. The actual normalized received electric field, corresponding to Fig. 5.1, is $e_r = e_N \exp(-j\Phi)$. The actual normalized received electric field has the same intensity as that of the normalized electric field e_N, i.e., $|e_r|^2 = |e_N|^2$. The values of $Y = |e_N|^2$ and $R = |e_N|$ are called normalized received intensity and amplitude, respectively.

Series Expansion

The complex Wiener process of $b(t)$ can be expanded using the standard Karhunen-Loéve expansion of

$$b(t) = \sum_{k=1}^{\infty} \lambda_k x_k \psi_k(t), \tag{5.35}$$

where x_k are identical independently distributed complex Gaussian random variable with zero mean and unity variance, λ_k^2 are the eigenvalues, and the functions of $\psi_k(t), 0 \leq t \leq 1$, are orthonormal functions of

$$\int_0^1 \psi_k(t)\psi_l(t)dt = \begin{cases} 1 & k = l \\ 0 & k \neq l \end{cases}. \tag{5.36}$$

The autocorrelation function is equal to

$$R_b(t, s) = E\{b(t) \cdot b^*(s)\} = \sum_{k=1}^{+\infty} \lambda_k^2 \psi_k(t)\psi_k(s), \tag{5.37}$$

and $\lambda_k^2, \psi_k(t), 0 \leq t \leq 1$ are the eigenvalues and eigenfunctions, respectively, of the following integral equation

$$\lambda_k^2 \psi_k(t) = \int_0^1 R_b(t, s)\psi_k(s)\mathrm{d}s. \tag{5.38}$$

Substitute the correlation function of Eq. (5.33) into the integral equation of Eq. (5.38), we have

$$\lambda_k^2 \psi_k(t) = \int_0^t s\psi_k(s)\mathrm{d}s + t\int_t^1 \psi_k(s)\mathrm{d}s. \tag{5.39}$$

Take the second derivative of both sides of Eq. (5.39) with respect to t, we get

$$\lambda_k^2 \frac{d^2\psi_k(t)}{dt^2} = -\psi_k(t) \tag{5.40}$$

with solution of $\psi(t) = \sqrt{2}\sin(t/\lambda_k)$. Substitute into Eq. (5.38) or Eq. (5.39), we find that

$$\lambda_k = \frac{2}{(2k-1)\pi}, \quad \psi_k(t) = \sqrt{2}\sin\left[\frac{(2k-1)\pi}{2}t\right]. \tag{5.41}$$

Karhunen-Loéve expansion of Wiener process was a standard exercise in random process (Papoulis, 1984, §10-6). The orthogonal process of Sec. 5.1.2 are equivalent to the Karhunen-Loéve transform of finite number of random variables of Eq. (5.5) based on numerical calculation. While the eigenvalues of the covariance matrix of Eq. (5.27) correspond approximately to λ_k^2 of Eq. (5.41), the eigenvectors in Eq. (5.24) always require numerical calculations. The assumption of a distributed process of Eq. (5.32) can derive both eigenvalues and eigenfunctions of Eq. (5.41) analytically.

Substitute Eq. (5.35) with Eq. (5.41) to the normalized phase of Eq. (5.32), because $\int_0^1 \sin(t/\lambda_k)\mathrm{d}t = \lambda_k$, we obtain

$$\Phi = |\xi_0|^2 + 2\sqrt{2}\sum_{k=1}^{\infty} \lambda_k^2 \Re\{\xi_0 \cdot x_k^*\} + \sum_{k=1}^{\infty} \lambda_k^2 |x_k|^2. \tag{5.42}$$

Because $\sum_{k=1}^{\infty} \lambda_k^2 = 1/2$ [see Gradshteyn and Ryzhik (1980, §0.234)], we obtain

$$\Phi = \sum_{k=1}^{\infty} \lambda_k^2 |\sqrt{2}\xi_0 + x_k|^2. \tag{5.43}$$

The random variable $|\sqrt{2}\xi_0 + x_k|^2$ is a noncentral χ^2 random variable with two degrees of freedom with a noncentrality parameter of $2\rho_s$ and a variance parameter of $1/2$. The normalized nonlinear phase noise is the summation of infinitely many independently distributed noncentral χ^2-random variables with two degrees of freedom with noncentrality parameters of $2\lambda_k^2\rho_s$ and variance parameters of $\lambda_k^4/2$. The mean and standard deviation (STD) of the random variables are both proportional to the square of the reciprocal of all odd natural numbers.

Characteristic Function

While it may be difficult to find the p.d.f. of the normalized nonlinear phase noise of Eq. (5.43) directly, its characteristic function has a very simple expression. Because x_k is a zero-mean and unit-variance complex

Gaussian random variable, the characteristic function of $|\sqrt{2}\xi_0 + x_k|^2$ is

$$\begin{aligned}\Psi_{|\sqrt{2}\xi_0+x_k|^2}(j\nu) &= \frac{1}{\pi}\int_{-\infty}^{+\infty}\int_{-\infty}^{+\infty} e^{j\nu(\sqrt{2\rho_s}+\zeta_1)^2+j\nu\zeta_2^2-\zeta_1^2-\zeta_2^2}\mathrm{d}\zeta_1\mathrm{d}\zeta_2 \\ &= \frac{1}{1-j\nu}\exp\left(\frac{2j\nu\rho_s}{1-j\nu}\right), \end{aligned} \tag{5.44}$$

with $\rho_s = |\xi_0|^2$. As the summation of many independent χ^2 random variables, the characteristic function of the normalized phase Φ of Eq. (5.32) is

$$\Psi_\Phi(j\nu) = \prod_{k=1}^{\infty}\frac{1}{1-j\nu\lambda_k^2}\exp\left(\frac{2j\nu\rho_s\lambda_k^2}{1-j\nu\lambda_k^2}\right). \tag{5.45}$$

Using the expressions of Gradshteyn and Ryzhik (1980, §1.431, §1.421)

$$\cos x = \prod_{k=1}^{\infty}\left(1-\frac{4x^2}{(2k-1)^2\pi^2}\right), \tag{5.46}$$

$$\tan\frac{\pi x}{2} = \frac{4x}{\pi}\sum_{k=1}^{\infty}\frac{1}{(2k-1)^2-x^2}, \tag{5.47}$$

the characteristic function of Eq. (5.45) can be simplified to

$$\Psi_\Phi(j\nu) = \sec\sqrt{j\nu}\exp\left[\rho_s\sqrt{j\nu}\tan\sqrt{j\nu}\right]. \tag{5.48}$$

The trigonometric function with complex argument is calculated by, for example,

$$\left(\sec\sqrt{j\nu}\right)^{-1} = \cos\sqrt{\frac{\nu}{2}}\cosh\sqrt{\frac{\nu}{2}} - j\sin\sqrt{\frac{\nu}{2}}\sinh\sqrt{\frac{\nu}{2}}.$$

From the characteristic function of Eq. (5.48), the mean normalized nonlinear phase shift is

$$\langle\Phi\rangle = -j\left.\frac{d}{d\nu}\Psi_\Phi(\nu)\right|_{\nu=0} = \rho_s + \frac{1}{2}. \tag{5.49}$$

Note that the differentiation or partial differentiation operation can be handled by most symbolic mathematical software. The scaling from normalized nonlinear phase noise to the nonlinear phase noise of Eq. (5.30) is

$$\Phi_{\mathrm{NL}} = \frac{\langle\Phi_{\mathrm{NL}}\rangle}{\rho_s+\frac{1}{2}}\Phi, \tag{5.50}$$

depending on the mean nonlinear phase shift of $\langle\Phi_{\mathrm{NL}}\rangle$ and SNR of ρ_s.

The second moment of the nonlinear phase noise is

$$\langle \Phi^2 \rangle = -\left.\frac{d^2}{d\nu^2}\Psi_\Phi(\nu)\right|_{\nu=0} = \frac{2}{3}\rho_s + \frac{1}{6} + \left(\rho_s + \frac{1}{2}\right)^2, \tag{5.51}$$

that gives the variance of normalized nonlinear phase noise as

$$\sigma_\Phi^2 = \frac{2}{3}\rho_s + \frac{1}{6}. \tag{5.52}$$

Using the scale factor of Eq. (5.50) with the variance of Eq. (5.52), the variance of nonlinear phase noise can be found.

The first eigenvalue of Eq. (5.41) is much larger than other eigenvalues. The normalized phase of Eq. (5.42) is dominated by the noncentral χ^2 random variable corresponding to the first eigenvalue because of

$$\frac{\lambda_1^2}{\lambda_2^2 + \lambda_3^2 + \cdots} = \frac{(2/\pi)^2}{1/2 - (2/\pi)^2} = 4.27, \tag{5.53}$$

and

$$\frac{\lambda_1^4}{\lambda_2^4 + \lambda_3^4 + \cdots} = \frac{(2/\pi)^4}{1/6 - (2/\pi)^4} = 68.12. \tag{5.54}$$

The relationship of $\sum_{k=1}^{\infty} \lambda_k^4 = 1/6$ is based on Gradshteyn and Ryzhik (1980, §0.234).

Beside the noncentral χ^2 random variable corresponding to the largest eigenvalue of λ_1, the other χ^2 random variables of $\lambda_k^2|\sqrt{2}\xi_0 + x_k|^2$, $k > 1$, have more or less than same variance. From the central limit theorem, the summation of many random variables with more or less the same variance approaches a Gaussian random variable. The characteristic function of Eq. (5.45) can be accurately approximated by

$$\Psi_\Phi(j\nu) \approx \frac{1}{1 - 4j\nu/\pi^2} \exp\left(\frac{8j\nu\rho_s/\pi^2}{1 - 4j\nu/\pi^2}\right)$$
$$\times \exp\left[j\nu(2\rho_s + 1)\left(\frac{1}{2} - \frac{4}{\pi^2}\right) - \frac{1}{2}\nu^2(4\rho_s + 1)\left(\frac{1}{6} - \frac{16}{\pi^4}\right)\right] \tag{5.55}$$

as a summation of a noncentral χ^2 random variable with two degrees of freedom and a Gaussian random variable.

The p.d.f. of the normalized phase noise of Eq. (5.32) can be calculated by taking the inverse Fourier transform of either the exact [Eq. (5.48)] or the approximated [Eq. (5.55)] characteristic functions. Figure 5.4 shows the p.d.f. of the normalized nonlinear phase noise for three different SNR of $\rho_s = 11, 18,$ and 25, corresponding to about an error probability of

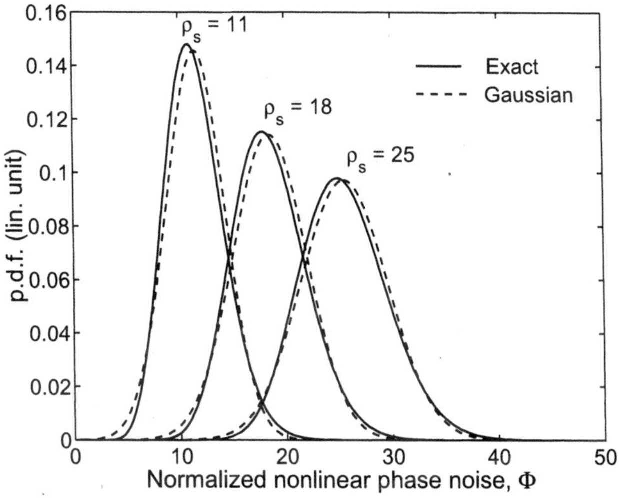

Figure 5.4. The p.d.f. of the normalized nonlinear phase noise Φ for SNR of $\rho_s = 11, 18$, and 25. [Adapted from Ho (2003a)]

10^{-6}, 10^{-9}, and 10^{-12} for binary PSK signal, respectively, when amplifier noise is the sole impairment. Figure 5.4 shows that the p.d.f. using the exact or the approximated characteristic function, and the Gaussian approximation with mean and variance of Eqs. (5.49) and (5.52), respectively. The exact and approximated p.d.f. overlap and cannot be distinguished with each other.

The p.d.f. for finite number of fiber spans was derived base on the orthogonalization of Eq. (5.5) by N_A independently distributed random variables in Sec. 5.1.2. Figure 5.5 shows a comparison of the p.d.f. for $N_A = 4, 8, 16, 32$, and 64 of fiber spans with the distributed case of Eq. (5.48). Using the SNR of $\rho_s = 18$, Figure 5.5 is plotted in logarithmic scale to show the difference in the tail. Figure 5.5 also provides an inset in linear scale of the same p.d.f. to show the difference around the mean. The asymptotic p.d.f. of Eq. (5.48) with distributed noise has the smallest spread in the tail as compared with those p.d.f. with N_A discrete noise sources. The asymptotic p.d.f. is very accurate for $N_A \geq 32$ fiber spans.

The method to find the characteristic function of nonlinear phase noise is similar to Foschini and Poole (1991) for polarization-mode dispersion. The method of Cameron and Martin (1945) and Mecozzi (1994a,b) gave the analytical characteristic function of Eq. (5.48) almost directly. The summation of Eq. (5.43) shows that the nonlinear phase noise is a gener-

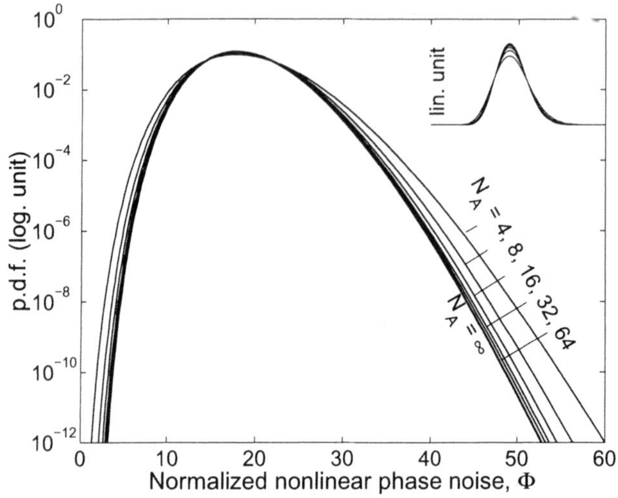

Figure 5.5. The asymptotic p.d.f. of normalized nonlinear phase noise of Φ as compared with the p.d.f. of $N_A = 4, 8, 16, 32$, and 64 fiber spans. The p.d.f. in linear scale is shown in the inset. [From Ho (2003a)]

alized χ^2 random variable. While the characteristic function of Eq. (5.48) is a simpler expression than that of the approximation of Eq. (5.55) and can be derived easily (Cameron and Martin, 1945, Mecozzi, 1994a), the physical meaning of Eq. (5.55) is more obvious.

The analysis here assumes dispersionless fiber. With fiber chromatic dispersion, if the nonlinear phase noise is confined to that induced by the amplifier noise having a bandwidth matched to the signal, the analysis here should be a very good approximation. Having the same wavelength, both signal and amplifier noise propagate in the same speed. The analysis here should be very accurate even for dispersive fiber. For RZ-DPSK signal, later chapter will derive the variance of self-phase modulation induced nonlinear phase noise in highly dispersive fiber.

2.2 Cross-Phase Modulation Induced Nonlinear Phase Noise

The nonlinear phase noise here is induced by self-phase modulation. The effects of amplifier noise outside the signal bandwidth and the amplifier noise from orthogonal polarization are all ignored for simplicity.

For the case of the nonlinear phase noise from wide-band amplifier noise, the marginal characteristic function of the normalized nonlinear phase noise of Eq. (5.48) becomes

$$\sec^m \sqrt{j\nu} \exp\left[\rho_s \sqrt{j\nu} \tan\sqrt{j\nu}\right]. \tag{5.56}$$

where m is product of the ratio of the amplifier noise bandwidth to the signal bandwidth and the number of polarizations of the amplifier noises. If only the amplifier noise from same polarization as signal is included, $m = 1$ gives the characteristic function of Eq. (5.48). If the amplifier noise from orthogonal polarization matched to signal bandwidth is also considered, $m = 2$ for two polarizations. The characteristic function of Eq. (5.56) does not include the nonlinear phase noise induced from other WDM channels. The nonlinear phase noise from other WDM channels through cross-phase modulation will be considered in one of the later chapters.

With cross-phase modulation induced nonlinear phase noise through amplifier noise only, the mean and variance of the nonlinear phase noise increase slightly to $\rho_s + \frac{1}{2}m$ and $\frac{2}{3}\rho_s + \frac{1}{6}m$, respectively. The nonlinear phase noise is induced mainly by the beating of the signal and amplifier noise from the same polarization as the signal, similar to the case of signal-spontaneous beat noise in an amplified IMDD receiver. For high SNR of ρ_s, it is obvious that the signal-amplifier noise beating is the major contribution to nonlinear phase noise. The parameter of m can equal to $1/2$ for the case if the amplifier noise from another dimension is ignored by confining to single-dimensional signal and noise. The characteristic function of Eq. (5.48) can be changed to Eq. (5.56) if necessary.

The characteristic function of Eq. (5.56) assumes a dispersionless fiber. With fiber dispersion, due to walk-off effect, the nonlinear phase noise caused by cross-phase modulation should approximately Gaussian distributed. Methods similar to Chiang et al. (1996) and Ho (2000) can be used to find the variance of the nonlinear phase noise due to cross-phase modulation in dispersive fiber. This approach is used in later of this book to find the nonlinear phase noise from other WDM channels.

For DPSK signal, the cross-phase modulation induced nonlinear phase noises in adjacent symbols are correlated to each other. The characteristic function of the differential phase due to cross-phase modulation can be found using the power spectral density similar to that in Chiang et al. (1996), taking the inverse Fourier transform to get the autocorrelation function, and getting the correlation coefficient as the autocorrelation with a time difference of the symbol interval. The characteristic function of the differential phase decreases by the correlation coefficient.

All the derivations here assume NRZ pulses (or continuous-wave signal) but most experiments in Table 1.2 use RZ pulses. For flat-top RZ pulse, the mean nonlinear phase shift of $\langle\Phi_{\mathrm{NL}}\rangle$ should be the mean nonlinear phase shift when the peak amplitude is transmitted. Usually, the mean nonlinear phase shift of $\langle \Phi_{\mathrm{NL}} \rangle$ is increased with the inverse of the duty cycle. However, for soliton and dispersion-managed soliton, based on soliton perturbation (Georges, 1995, Iannone et al., 1998, Kaup, 1990, Kivshar and Malomed, 1989) or variational principle (McKinstrie and Xie, 2002, McKinstrie et al., 2002), the mean nonlinear phase shift of $\langle \Phi_{\mathrm{NL}} \rangle$ is reduced by a factor of 2 when dispersion and self-phase modulation balance each other. The nonlinear phase noise of RZ or soliton signal will be considered later.

2.3 Dependence between Nonlinear Phase Noise and Received Electric Field

The joint characteristic function of the normalized nonlinear phase noise and electric field of Eq. (5.32) is presented here analytically. Using the series expansion of Eq. (5.35), the normalized electric field of Eq. (5.32) is

$$e_N = \xi_0 + \sqrt{2}\sum_{k=1}^{\infty}(-1)^{k+1}\lambda_k x_k = e_{N1} + j e_{N2}. \tag{5.57}$$

where e_{N1} and e_{N2} are the real and imaginary parts of the electric field e_N, respectively. Using Gradshteyn and Ryzhik (1980, §0.232), we get $\sum_{k=1}^{\infty}(-1)^{k+1}\lambda_k = 1/2$ and

$$e_N = e_{N1} + j e_{N2} = \sqrt{2}\sum_{k=1}^{\infty}(-1)^{k+1}\lambda_k(\sqrt{2}\xi_0 + x_k). \tag{5.58}$$

The normalized electric field of Eq. (5.58) has a complex Gaussian distribution with a mean of of ξ_0 and unity variance.

The joint characteristic function of the normalized nonlinear phase noise and the electric field of Eq. (5.32) is

$$\Psi_{\Phi, e_N}(\nu, \omega) = E\left\{\exp\left(j\nu\Phi + j\omega_1 e_{N1} + j\omega_2 e_{N2}\right)\right\}, \tag{5.59}$$

where $\omega = \omega_1 + j\omega_2$.

From Appendix 5.A, the joint characteristic of normalized nonlinear phase noise and electric field is

$$\Psi_{\Phi, e_N}(\nu, \omega) = \sec\sqrt{j\nu}\exp\left[\left(|\xi_0|^2\sqrt{j\nu} - \frac{|\omega|^2}{4\sqrt{j\nu}}\right)\tan\sqrt{j\nu}\right.$$
$$\left. + j\sec(\sqrt{j\nu})\Re\{\xi_0 \cdot \omega^*\}\right]. \tag{5.60}$$

The marginal characteristic function of

$$\Psi_{e_N}(\omega) = \Psi_{\Phi,e_N}(0,\omega) = \exp\left(j\Re\{\xi_0 \cdot \omega^*\} - \frac{|\omega|^2}{4}\right) \tag{5.61}$$

is the characteristic function of a two-dimensional Gaussian distribution for the normalized electric field of Eq. (5.58). Comparing joint characteristic function of Eq. (5.60) with the marginal characteristic functions of $\Psi_\Phi(\nu)$ and $\Psi_{e_N}(\omega)$ of Eqs. (5.48) and (5.61), respectively, $\Psi_{\Phi,e_N}(\nu,\omega) \neq \Psi_\Phi(\nu)\Psi_{e_N}(\omega)$ due to some very weak dependence between nonlinear phase noise of Φ with the received electric field of e_N.

In the received signal of $e_r = e_N \exp(-j\Phi)$, the nonlinear phase noise is added directly to the phase of the electric field of e_N. The joint characteristic function of nonlinear phase noise with the phase of e_N is a more interesting topic. For the phase, as a periodic function with a period of 2π, the p.d.f. can be expanded by a Fourier series with coefficients as the value of the characteristic function at integer "angular frequency".

From Eq. (5.A.13) of Appendix 5.A, the Fourier coefficients are

$$\Psi_{\Phi,\Theta_n}(\nu,m) = \frac{\sqrt{\pi}}{2}\Psi_\Phi(\nu)\gamma_\nu^{1/2}e^{-\gamma_\nu/2}\left[I_{\frac{m-1}{2}}\left(\frac{\gamma_\nu}{2}\right) + I_{\frac{m-1}{2}}\left(\frac{\gamma_\nu}{2}\right)\right], \quad m \geq 0 \tag{5.62}$$

where γ_ν from Eq. (5.A.12) is the angular depending SNR. If $\gamma_\nu = \gamma_0 = \rho_s$, the joint coefficient of Eq. (5.62) is equal to the product of $\Psi_\Phi(\nu)$ and the coefficient of Eq. (4.A.11).

The statistics of nonlinear phase noise given here is mostly based on Appendix 5.A. The joint characteristic function of nonlinear phase noise and the received electric field is also given. The joint characteristic function of nonlinear phase noise with the p.d.f. of received electric field, as shown in Eq. (5.A.8), resembles a Gaussian distribution with both mean and variance as a complex number depending on $j\nu$. This "Gaussian" property is used later, mainly to find the error probability of phase-modulated signal with and without compensation.

3. Exact Error Probability for Distributed Systems

In performance assessment, the ultimate goal is to investigate the impact of nonlinear phase noise to phase-modulated signals. The error probability of the system is the most important parameter to characterize the system performance. The characteristic function in previous section can be used to approximately evaluate the error probability based on the assumption that nonlinear phase noise is independent of the phase of amplifier noise. Although it is obvious that nonlinear phase noise is uncorrelated with the phase of amplifier noise, as non-Gaussian random variables, they are weakly depending on each other. In this section, the error probability is derived by taking into account the dependence between the nonlinear phase noise and the phase of amplifier noise. Even with the assumption of independence between nonlinear phase noise and the phase of amplifier noise, inferred from Figs. 5.2 and 5.5, the received phase does not distribute symmetrically with respect to the mean nonlinear phase shift. The decision regions of PSK signal with nonlinear phase noise do not center with respect to the mean nonlinear phase shift. The error probability is also verified by Monte-Carlo simulation.

3.1 Distribution of Received Phase

The overall received phase of the signal is the summation of transmitted phase, nonlinear phase noise, and the phase of amplifier noise,

$$\Phi_r = \theta_0 + \Theta_n - \Phi_{\mathrm{NL}} = \theta_0 + \Theta_n - \frac{\langle \Phi_{\mathrm{NL}} \rangle}{\rho_s + 1/2} \Phi, \tag{5.63}$$

where θ_0 is the transmitted phase, Θ_n is the phase of amplifier noise, Φ_{NL} is the nonlinear phase noise, $\langle \Phi_{\mathrm{NL}} \rangle$ is the mean nonlinear phase shift, Φ is the normalized nonlinear phase noise defined in Sec. 5.2, $\langle \Phi \rangle = \rho_s + 1/2$ [Eq. (5.49)] is the mean normalized nonlinear phase noise, and ρ_s is the SNR of the signal. Without the loss of generality, we assume that the transmitted phase is $\theta_0 = 0$ in later parts of this section. The linear phase noise term of Θ_n is solely contributed by the additive amplifier noise. Without changing the results, the nonlinear phase noise of Φ_{NL} may be added or subtracted to the received phase depending on whether the transmitted signal is represented as $\exp(\pm j\theta_0)$.

In order to find the p.d.f. of Φ_r of Eq. (5.63), we need to find the joint characteristic function of nonlinear phase noise with the phase of amplifier noise. The p.d.f. of the phase of amplifier noise can be expanded as a Fourier series as shown in Appendix 4.A. If the nonlinear phase noise is assumed to be Gaussian distributed and independent of the phase of amplifier noise, the analysis of error probability is the same as a phase-modulated signal with laser phase noise of Eq. (4.40).

Comparing with the assumption of independence, the error probability is increased due to the dependence between the nonlinear phase noise and the phase of amplifier noise. The optimal operating point of the system is estimated by Gordon and Mollenauer (1990) using the insight that the variance of linear and nonlinear phase noise should be approximately the same. With the exact error probability, the system can be optimized rigorously by the condition that the increase in SNR penalty is less than the increase of launched power.

The received phase of Eq. (5.63) is confined to the range of $[-\pi, +\pi)$. The p.d.f. of the received phase is a periodic function with a period of 2π. If the characteristic function of the received phase is $\Psi_{\Phi_r}(\nu)$, the p.d.f. of the received phase has a Fourier series expansion of

$$p_{\Phi_r}(\theta) = \frac{1}{2\pi} \sum_{m=-\infty}^{+\infty} \Psi_{\Phi_r}(m) \exp(jm\theta). \tag{5.64}$$

Because the characteristic function has the property of $\Psi_{\Phi_r}(-\nu) = \Psi^*_{\Phi_r}(\nu)$, we get

$$p_{\Phi_r}(\theta) = \frac{1}{2\pi} + \frac{1}{\pi} \sum_{m=1}^{+\infty} \Re\left\{\Psi_{\Phi_r}(m) \exp(jm\theta)\right\}, \tag{5.65}$$

where $\Re\{\cdot\}$ denotes the real part of a complex number.

For the received phase of Eq. (5.63) with $\theta_0 = 0$, using Eq. (5.62), the Fourier series coefficients are

$$\Psi_{\Phi_r}(m) = \Psi_{\Phi,\Theta_n}\left(-m\frac{\langle\Phi_{\mathrm{NL}}\rangle}{\rho_s + 1/2}, m\right). \tag{5.66}$$

The characteristic function with an expression of Eq. (5.66) is due to the dependence between nonlinear phase noise and the phase of amplifier noise. If nonlinear phase noise is assumed independent to the phase of amplifier noise, the characteristic function of Eq. (5.66) can be separated to the product of two parts that depend only on Φ and Θ_n, respectively. Due to the dependence, the characteristic function of Eq. (5.66) cannot be separated into two independent parts.

Figure 5.6 shows the p.d.f. of the received phase of Eq. (5.65) with mean nonlinear phase shift of $\langle\Phi_{\mathrm{NL}}\rangle = 0, 0.5, 1.0, 1.5$, and 2.0 rad. Shifted by the mean nonlinear phase shift $\langle\Phi_{\mathrm{NL}}\rangle$, the p.d.f. is plotted in logarithmic scale to show the difference in the tail. Not shifted by $\langle\Phi_{\mathrm{NL}}\rangle$, the same p.d.f. is plotted in linear scale in the inset. Figure 5.6 is plotted for the case that the SNR is equal to $\rho_s = 18$ (12.6 dB), corresponding to an error probability of 10^{-9} for binary PSK signal if amplifier noise is the sole impairment from Fig. 3.13. Without nonlinear phase noise of $\langle\Phi_{\mathrm{NL}}\rangle = 0$, the p.d.f. is the same as that in Fig. 4.A.2 and symmetrical with respect to the zero phase.

From Fig. 5.6, when the p.d.f. is broadened by the nonlinear phase noise, the broadening is not symmetrical with respect to the mean nonlinear phase shift of $\langle\Phi_{\mathrm{NL}}\rangle$. With small mean nonlinear phase shift

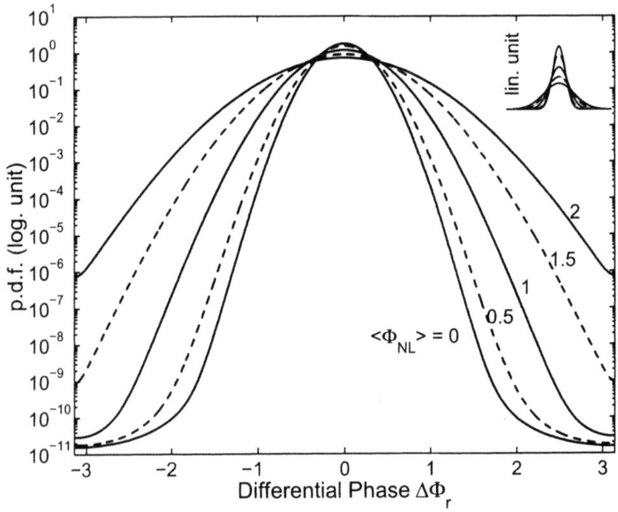

Figure 5.6. The p.d.f. of the received phase $p_{\Phi_r}(\theta + \langle\Phi_{\rm NL}\rangle)$ in logarithmic scale. The inset is the p.d.f. of $p_{\Phi_r}(\theta)$ in linear scale.

of $\langle\Phi_{\rm NL}\rangle = 0.5$ rad, the received phase spreads further in the positive phase than the negative phase. With large mean nonlinear phase shift of $\langle\Phi_{\rm NL}\rangle = 2$ rad, the received phase spreads further in the negative phase than the positive phase. The difference in the spreading for small and large mean nonlinear phase shift is due to the dependence between nonlinear phase noise and the phase of amplifier noise. After normalization, the p.d.f. of nonlinear phase noise depends solely on the SNR. If nonlinear phase noise is independent of the phase of amplifier noise, the spreading of the received phase noise is independent of the mean nonlinear phase shift.

3.2 PSK Signals

If the p.d.f. of Eq. (5.65) were symmetrical with respect to the mean nonlinear phase shift of $\langle\Phi_{\rm NL}\rangle$, the decision region would center at $\langle\Phi_{\rm NL}\rangle$ and the decision angles for binary PSK signals should be $\pm\pi/2 - \langle\Phi_{\rm NL}\rangle$. From Fig. 5.6, because the p.d.f. is not symmetrical with respect to the mean nonlinear phase shift, assume that the decision angles are $\pm\pi/2 - \theta_c$ with the center phase of θ_c, the error probability is

$$p_e = 1 - \int_{-\pi/2-\theta_c}^{\pi/2-\theta_c} p_{\Phi_r}(\theta)\mathrm{d}\theta, \qquad (5.67)$$

or

$$p_e = \frac{1}{2} - \frac{1}{\pi}\sum_{m=1}^{+\infty} \frac{2\sin(m\pi/2)}{m} \Re\left\{\Psi_{\Phi_r}(m) e^{jm\theta_c}\right\}. \tag{5.68}$$

After some simplifications for $\sin(m\pi/2) = 0$ when m are even numbers, we get

$$p_e = \frac{1}{2} - \frac{2}{\pi}\sum_{k=0}^{+\infty} \frac{(-1)^k}{2k+1} \Re\left\{\Psi^*_{\Phi_r}(2k+1) e^{-j(2k+1)\theta_c}\right\}. \tag{5.69}$$

From both Eqs. (5.62) and (5.66), the coefficients for the error probability Eq. (5.69) are

$$\begin{aligned}\Psi^*_{\Phi_r}(2k+1) = \frac{\sqrt{\pi r_k}}{2} e^{-r_k/2} \left[I_k\left(\frac{r_k}{2}\right) + I_{k+1}\left(\frac{r_k}{2}\right)\right] \\ \times \Psi_\Phi\left[\frac{(2k+1)\langle\Phi_{\mathrm{NL}}\rangle}{\rho_s + 1/2}\right], \qquad k \geq 0,\end{aligned} \tag{5.70}$$

where, using Eq. (5.A.12),

$$r_k = \frac{2\left[\dfrac{j(2k+1)\langle\Phi_{\mathrm{NL}}\rangle}{\rho_s + 1/2}\right]^{1/2}}{\sin\left\{2\left[\dfrac{j(2k+1)\langle\Phi_{\mathrm{NL}}\rangle}{\rho_s + 1/2}\right]^{1/2}\right\}}\rho_s, \quad k \geq 0, \tag{5.71}$$

are equivalent to the angular frequency depending SNR parameters, and $\Psi_\Phi(\nu)$ is the marginal characteristic function of nonlinear phase noise of Eq. (5.48). From Eq. (5.48), the shape of the p.d.f. of nonlinear phase noise depends solely on the signal SNR.

If the nonlinear phase noise is assumed to be independent to the phase of amplifier noise, similar to the approaches in Chapter 4 in which the extra phase noise is independent of the signal phase, the error probability can be approximated as

$$\begin{aligned}p_e \approx \frac{1}{2} - \exp\left(-\frac{\rho_s}{2}\right)\sqrt{\frac{\rho_s}{\pi}} \sum_{k=0}^{\infty} \frac{(-1)^k}{2k+1} \left[I_k\left(\frac{\rho_s}{2}\right) + I_{k+1}\left(\frac{\rho_s}{2}\right)\right] \\ \times \Re\left\{\Psi_\Phi\left[\frac{(2k+1)\langle\Phi_{\mathrm{NL}}\rangle}{\rho_s + 1/2}\right] e^{-j(2k+1)\theta_c}\right\}.\end{aligned} \tag{5.72}$$

The center phase of θ_c of Eq. (5.72) may be assumed as the mean nonlinear phase shift of $\theta_c = \langle\Phi_{\mathrm{NL}}\rangle$.

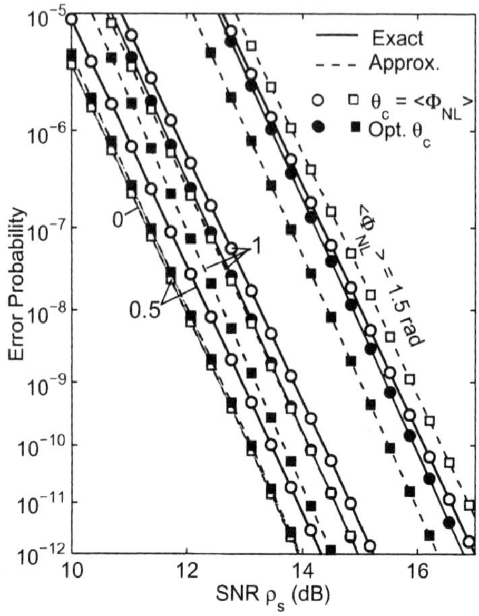

Figure 5.7. The error probability of PSK signal as a function of SNR ρ_s.

Figure 5.7 shows the exact [Eq. (5.69)] and approximated [Eq. (5.72)] error probabilities as a function of SNR ρ_s. Figure 5.7 also plots the error probability without nonlinear phase noise of Eq. (3.78) and Fig. 3.13. Figure 5.7 plots the error probability for both the center phase equal to the mean nonlinear phase shift $\theta_c = \langle \Phi_{\rm NL} \rangle$ (empty symbol) and optimized to minimize the error probability (solid symbol). From Fig. 5.7, the approximated error probability of Eq. (5.72) always underestimates the error probability for signal with optimized center phase.

Figure 5.8 shows the SNR penalty of PSK signal for an error probability of 10^{-9} calculated by the exact and approximated error probability formulae. Figure 5.8 is plotted for both cases of the center phase equal to the mean nonlinear phase shift $\theta_c = \langle \Phi_{\rm NL} \rangle$ or optimized to minimize the error probability. The corresponding optimal center phase is shown in Fig. 5.9.

The discrepancy between the exact and approximated error probability is smaller for small and large nonlinear phase shift. With the optimal center phase, the largest discrepancy between the exact and approximated SNR penalty is about 0.49 dB at a mean nonlinear phase shift of $\langle \Phi_{\rm NL} \rangle$ around 1.25 rad. When the center phase is equal to the mean nonlinear phase shift $\theta_c = \langle \Phi_{\rm NL} \rangle$, the largest discrepancy between

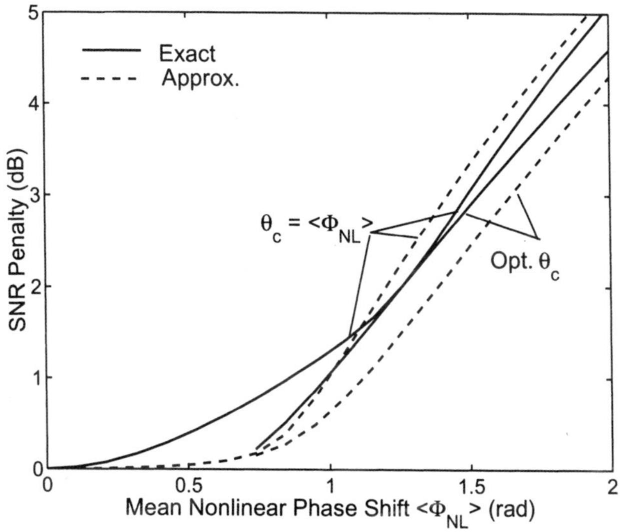

Figure 5.8. The SNR penalty of PSK signal as a function of mean nonlinear phase shift $\langle \Phi_{\mathrm{NL}} \rangle$.

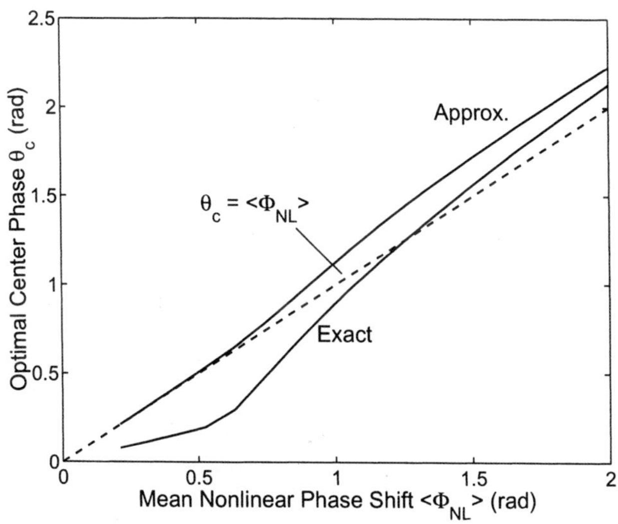

Figure 5.9. The optimal center phase corresponding to the operating point of Fig. 5.8 as a function of mean nonlinear phase shift $\langle \Phi_{\mathrm{NL}} \rangle$.

the exact and approximated SNR penalty is about 0.6 dB at a mean nonlinear phase shift of $\langle \Phi_{\mathrm{NL}} \rangle$ about 0.75 rad. For PSK signal, the approximated error probability of Eq. (5.72) may not have sufficient accuracy for practical applications.

Using the exact error probability of Eq. (5.69) with optimal center phase, the mean nonlinear phase shift must be less than 1 rad for a SNR penalty less than 1 dB. The optimal operating level is that the increase of mean nonlinear phase shift, proportional to the increase of launched power and SNR, does not decrease the system performance. In Fig. 5.8, the optimal operating point can be found by

$$\frac{d\rho_s}{d\,\langle \Phi_{\mathrm{NL}} \rangle} \to 1 \tag{5.73}$$

when both the required SNR ρ_s and mean nonlinear phase shift $\langle \Phi_{\mathrm{NL}} \rangle$ are expressed in decibel unit. The optimal operating level is for the mean nonlinear phase noise $\langle \Phi_{\mathrm{NL}} \rangle = 1.25$ rad, close to the estimation of Mecozzi (1994a) when the center phase is assumed to be $\langle \Phi_{\mathrm{NL}} \rangle$.

From the optimal center phase of Fig. 5.9 with the exact error probability of Eq. (5.69), the optimal center phase is less than the mean nonlinear phase shift of $\langle \Phi_{\mathrm{NL}} \rangle$ when the mean nonlinear phase shift is less than about 1.25 rad. At small mean nonlinear phase shift, from Fig. 5.6, the p.d.f. of the received phase spreads further to positive phase such that the optimal center phase is smaller that the mean nonlinear phase shift. At large mean nonlinear phase shift, the received phase is dominated by the nonlinear phase noise. Because the p.d.f. of nonlinear phase noise spreads further to the negative phase as from Fig. 10.3, the optimal center phase is larger than the mean nonlinear phase shift for large mean nonlinear phase shift. For the same reason, when the nonlinear phase noise is assumed to be independent of the phase of amplifier noise, the optimal center phase is always larger than the mean nonlinear phase shift. From Fig. 5.9, the approximated error probability of Eq. (5.72) is not useful to find the optimal center phase.

Comparing the exact [Eq. (5.69)] and approximated [Eq. (5.72)] error probability, the approximated error probability of Eq. (5.72) is evaluated when the angular SNR of r_k of Eq. (5.71) is approximated by the SNR of ρ_s. The parameters of r_k are complex numbers. Because $|r_k|$ are always less than ρ_s, with optimized center phase and from Figs. 5.7 and 5.8, the approximated error probability of Eq. (5.72) always gives an error probability smaller than the exact error probability.

To verify the accuracy of the error probability in Fig. 5.7, Figure 5.10 compares the theoretical and simulated error probability as a function of SNR for a typical PSK system having mean nonlinear phase shift of

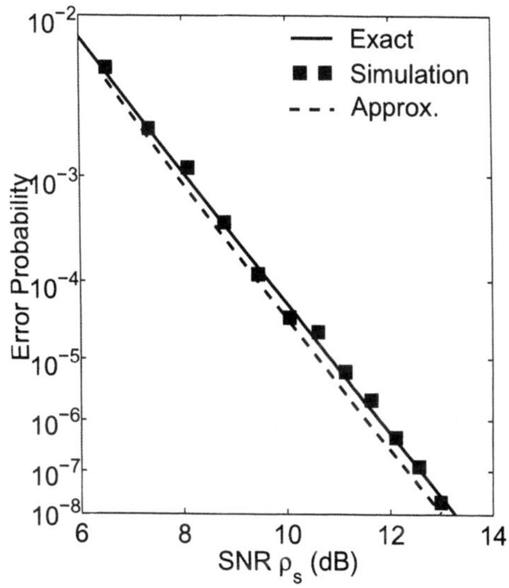

Figure 5.10. Calculated and simulated error probability for a PSK system with mean nonlinear phase shift of $\langle \Phi_{\rm NL} \rangle = 1$ rad.

$\langle \Phi_{\rm NL} \rangle = 1$ rad. The simulation is conducted for $N_A = 32$ fiber spans based on Monte-Carlo error counting. Equivalently speaking, the distribution of the received electric field of Fig. 5.1 is found and the error probability is equal to the ratio of points outside the decision region. The number of error counts is more than 10 for a good confident interval (Jeruchim, 1984). In the simulation of Fig. 5.10, the decision regions are centered at the mean nonlinear phase shift of $\langle \Phi_{\rm NL} \rangle$ for simplicity. Including both exact and approximated error probability, the theoretical results are the same as that in Fig. 5.7 but extend to high error probability. Figure 5.10 shows that the approximated and simulated results have an insignificant difference of about 0.15 dB and the exact and simulated results are virtually identical. From Fig. 5.10, we may conclude that the exact error probability of Eq. (5.69) is very accurate to evaluate the error probability of PSK signals with nonlinear phase noise.

Note that the exact error probability Eq. (5.69) is very similar to that in Mecozzi (1994a)[3]. The major difference between the exact error probability Eq. (5.69) and that in Mecozzi (1994a) is the observation that the center phase is not equal to the mean nonlinear phase shift.

[3]The error probability of Mecozzi (1994a, eq. 71) is for PSK instead of DPSK signal

When the center phase is equal to the mean nonlinear phase shift, the results using the exact error probability of Eq. (5.69) should be the same as that of Mecozzi (1994a). When the center phase is equal to the mean nonlinear phase shift $\theta_c = \langle \Phi_{\rm NL} \rangle$, the SNR penalty given by the approximated error probability is the same as that in Ho (2003e) but calculated by a simple formula of Eq. (5.72).

Using the Fourier series of Eq. (4.A.12), the error probability was derived for DPSK signals with a noisy reference (Jain, 1974), phase error (Blachman, 1981), and laser phase noise (Nicholson, 1984). In those studies, the extra phase noise is independent to the phase of the signal. Because of the dependence between the nonlinear phase noise and the linear phase noise (Ho, 2003g, Mecozzi, 1994a,b), the error probability here is far more complicated then those early works.

3.3 DPSK Signals

Direct-detection DPSK signal is the most popular signal format for phase-modulated optical communications. Equivalently, the asymmetric Mach-Zehnder interferometer of Fig. 1.4(c) gives the differential phase of

$$\begin{aligned}\Delta\Phi_r &= \Phi_r(t) - \Phi_r(t-T)\\ &= \Theta_n(t) - \Phi_{\rm NL}(t) - \Theta_n(t-T) + \Phi_{\rm NL}(t-T),\end{aligned} \tag{5.74}$$

where $\Phi_r(\cdot)$, $\Theta_n(\cdot)$, and $\Phi_{\rm NL}(\cdot)$ are the received phase, the phase of amplifier noise, and the nonlinear phase noise as a function of time, and T is the symbol interval. The phases at t and $t-T$ are independent of each other but are identically distributed random variables similar to that of Eq. (5.63). The differential phase of Eq. (5.74) assumes that the transmitted phases at t and $t-T$ are the same.

When two independent random variables are added (or subtracted) together, the sum has a characteristic function that is the product of the corresponding individual characteristic functions. The p.d.f. of the sum of the two random variables has Fourier series coefficients that are the product of the corresponding Fourier series coefficients. From the p.d.f. of Eq. (5.65), the p.d.f. of the differential phase is

$$p_{\Delta\Phi_r}(\theta) = \frac{1}{2\pi} + \frac{1}{\pi}\sum_{m=1}^{+\infty} |\Psi_{\Phi_r}(m)|^2 \cos(m\theta). \tag{5.75}$$

As the difference of two independent identically distributed random variables, with the same transmitted phase between two consecutive symbols, the p.d.f. of the differential phase $\Delta\Phi_r$ is symmetrical with respect to the zero phase.

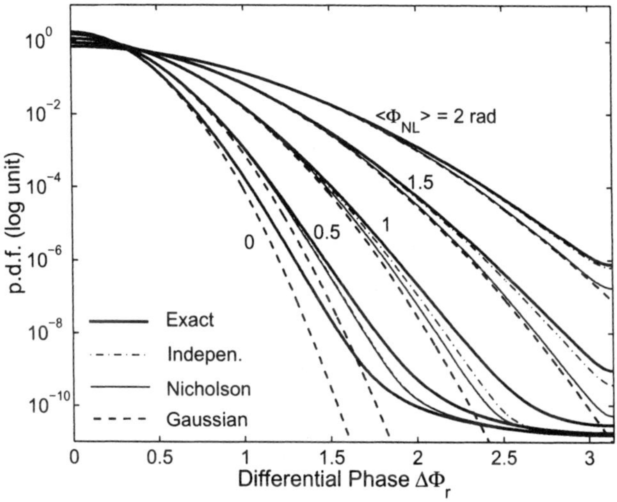

Figure 5.11. The probability density of differential phase $\Delta\Phi_r$ based on four different models. [Adapted from Ho (2004c)]

Figure 5.11 shows the p.d.f. of Eq. (5.75) for the differential received phase with mean nonlinear phase shift of $\langle\Phi_{\rm NL}\rangle = 0, 0.5, 1.0, 1.5$, and 2.0 rad. In additional to the p.d.f. from Eq. (5.75), Figure 5.11 also shows the p.d.f. obtain from other models that will be discussed later. The p.d.f. is plotted in logarithmic scale to show the difference in the tail. Because the p.d.f. is symmetrical with respect to the zero phase, only the p.d.f. with positive differential received phase is shown in Fig. 5.11. Figure 5.11 is plotted for the case that the SNR is equal to $\rho_s = 20$ (13 dB), corresponding to an error probability of 10^{-9} for DPSK signal if amplifier noise is the sole impairment from Fig. 3.13.

Interferometer based receiver gives an output proportional to $\cos(\Delta\Phi_r)$ from Sec. 3.4.2. The detector makes a decision on whether $\cos(\Delta\Phi_r)$ is positive or negative that is equivalent to whether the differential phase $\Delta\Phi_r$ is within or without the angles of $\pm\pi/2$. Similar to that for PSK signal of Eq. (5.69), the error probability for DPSK signal is

$$p_e = 1 - \int_{-\frac{\pi}{2}}^{\frac{\pi}{2}} p_{\Delta\Phi_r}(\theta)\mathrm{d}\theta \tag{5.76}$$

or

$$p_e = \frac{1}{2} - \frac{2}{\pi}\sum_{k=0}^{+\infty}\frac{(-1)^k}{2k+1}\left|\Psi_{\Phi_r}(2k+1)\right|^2, \tag{5.77}$$

where the coefficients of $\Psi_{\Phi_r}(2k+1)$ are given by Eq. (5.70). The formula of Eq. (5.77) is a generally valid formula for various models. When other models for nonlinear phase noise or the phase of amplifier noise are used, only different coefficients of $\Psi_{\Phi_r}(2k+1)$ are used in Fig. 5.77.

Similar to the approximation for PSK signal Eq. (5.72), if the nonlinear phase noise is assumed to be independent to the phase of amplifier noise, the error probability of Eq. (5.77) can be approximated as

$$p_e \approx \frac{1}{2} - \frac{\rho_s e^{-\rho_s}}{2} \sum_{k=0}^{\infty} \frac{(-1)^k}{2k+1} \left[I_k\!\left(\frac{\rho_s}{2}\right) + I_{k+1}\!\left(\frac{\rho_s}{2}\right) \right]^2 \times \left| \Psi_\Phi \left[\frac{(2k+1)\langle \Phi_{\rm NL} \rangle}{\rho_s + 1/2} \right] \right|^2. \tag{5.78}$$

The corresponding p.d.f. for Eq. (5.78) by independence assumption is also shown in Fig. 5.11 for comparison.

Comparing the exact [Eq. (5.77)] and approximated [Eq. (5.78)] error probability, the approximated error probability of Eq. (5.78) is evaluated when the angular SNR of r_k is approximated by the SNR ρ_s. Because $|r_k|$ is always less than ρ_s, the approximated error probability of Eq. (5.78) always gives an error probability smaller than the exact error probability of Eq. (5.77). From Fig. 5.11, the approximation of Eq. (5.78) also underestimates the spreading of the tail.

Figure 5.12 shows the exact [Eq. (5.77)] and approximated [Eq. (5.78)] error probabilities as a function of SNR ρ_s. Figure 5.12 also plots the error probability without nonlinear phase noise of Eq. (3.105) or Fig. 3.13. From Fig. 5.12, the approximated error probability Eq. (5.78) always underestimates the error probability.

Figure 5.13 shows the SNR penalty of DPSK signal for an error probability of 10^{-9} calculated by the exact and approximated error probability formulae. The discrepancy between the exact and approximated error probability is very small for small and large mean nonlinear phase shift. The largest discrepancy between the exact and approximated SNR penalty is about 0.23 dB at a mean nonlinear phase shift of about 0.53 rad.

For a power penalty less than 1 dB, the mean nonlinear phase shift must be less than 0.57 rad. The optimal level of the mean nonlinear phase shift is about 1 rad such that the increase of SNR penalty is always less than the increase of mean nonlinear phase shift, similar to the estimation of Gordon and Mollenauer (1990) as the limitation of the mean nonlinear phase shift.

To verify the accuracy of the error probability in Fig. 5.12, Figure 5.14 compares the theoretical and simulated error probability as a function

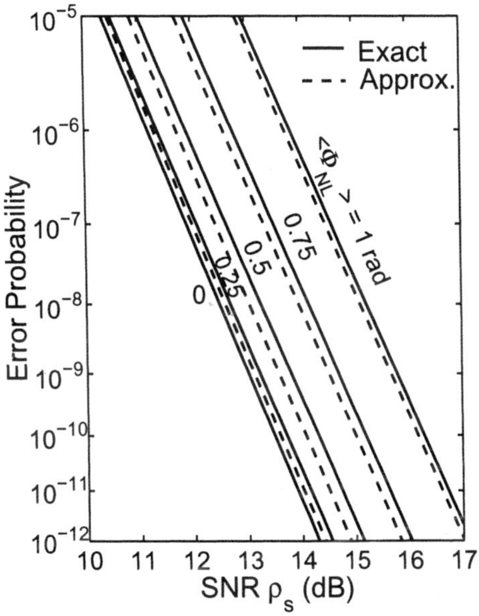

Figure 5.12. The error probability of DPSK signal as a function of SNR ρ_s.

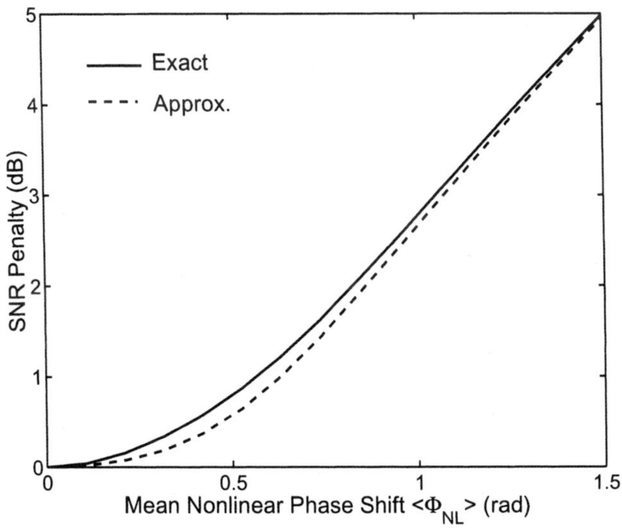

Figure 5.13. The SNR penalty of DPSK signal as a function of mean nonlinear phase shift $\langle\Phi_{NL}\rangle$.

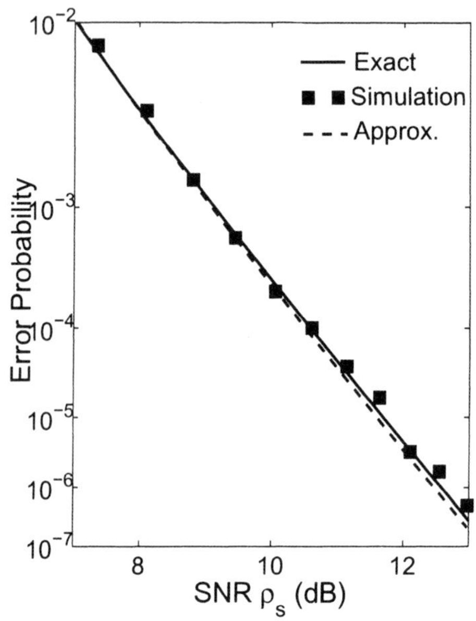

Figure 5.14. Calculated and simulated error probability for a DPSK system for a mean nonlinear phase shift of $\langle \Phi_{\text{NL}} \rangle = 1/\sqrt{2}$ rad.

of SNR for a typical DPSK system having mean nonlinear phase shift of $\langle \Phi_{\text{NL}} \rangle = 1/\sqrt{2}$ rad. The simulation is similar to that of Fig. 5.10 for PSK systems with $N_A = 32$ fiber spans. Including both exact [Eq. (5.77)] and approximated [Eq. (5.78)] error probability, the theoretical results are the same as that in Fig. 5.12 but extend to larger error probability. Figure 5.14 shows that the approximated, exact and simulated results have insignificant difference. From Fig. 5.14, we may conclude that the exact error probability of Eq. (5.77) is very accurate to evaluate the error probability of DPSK signal with nonlinear phase noise.

3.4 Comparison of Different Models

DPSK signal is by far the most popular modulation format for phase-modulated optical communications. There are many methods in the literatures to find the error probability of DPSK signal with nonlinear phase noise.

In additional to the models in this section with the exact p.d.f. of nonlinear phase noise given by Eq. (5.48) and numerical results in Figs. 5.12 and 5.13, the error probability was approximated mostly by Gaussian assumption in Gordon and Mollenauer (1990), Liu et al. (2002b), Xu and Liu (2002), Xu et al. (2003), and Wei et al. (2003a,b).

Gaussian Approximation Based on Q-factor

From Eq. (4.A.15) of Appendix 4.A, the variance of the phase of amplifier noise is approximately equal to

$$\sigma_{\Theta_n}^2 = \frac{1}{2\rho_s}. \tag{5.79}$$

With the combination of the variance of Eq. (5.52) with the scale relationship of Eq. (5.50), the variance of nonlinear phase noise is

$$\sigma_{\Phi_{\mathrm{NL}}}^2 = \frac{\langle \Phi_{\mathrm{NL}} \rangle^2}{(\rho_s + \frac{1}{2})^2} \left(\frac{2}{3}\rho_s + \frac{1}{6} \right) \approx \frac{2 \langle \Phi_{\mathrm{NL}} \rangle^2}{3\rho_s}. \tag{5.80}$$

Using both Eqs. (5.79) and (5.80), for DPSK signal, the Q-factor is

$$Q = \frac{\pi}{2\sqrt{2(\sigma_{\Theta_n}^2 + \sigma_{\Phi_{\mathrm{NL}}}^2)}}, \tag{5.81}$$

where $\pi/2$ is the phase difference between the constellation points and the decision threshold, and a further factor of 2 is for differential signal. Based on the Q-factor, similar to Eq. (3.140), the error probability is $p_e = \frac{1}{2}\mathrm{erfc}(Q/\sqrt{2})$.

The approximation of Eq. (5.80) was given in Gordon and Mollenauer (1990). The Q-factor based analysis of DPSK signal is first proposed by Wei et al. (2003a,b) and used in Hiew et al. (2004). The phase of amplifier noise of Θ_n is certainly non-Gaussian distributed as shown in Fig. 4.A.2. With only the phase of amplifier noise, the assumption of Gaussian distribution of the phase underestimates the error probability by $\pi/2$ ($\approx$ 4 dB) as a SNR gain for PSK signal and about 1.2 dB for DPSK signals. The usage of Q-factor is not accurate (Bosco and Poggiolini, 2004a).

Gaussian Approximation of Nonlinear Phase Noise (Nicholson Model)

When the nonlinear phase noise difference between two consecutive symbols is assumed to be Gaussian distributed, the variance of $2\sigma_{\Phi_{\mathrm{NL}}}^2$ for the differential phase is sufficient to characterize the nonlinear phase noise. The same as DPSK signals with laser phase noise in Sec. 4.3.2,

Table 5.1. Different Models for DPSK Signal with Nonlinear Phase Noise.

Model	Θ_n	Φ_{NL}	Dependence	1 dB Penalty	Optimal Point
Gaussian	Gaussian	Gaussian	Ind.	0.45 rad	0.86 rad
Nicholson	non-Gaussian	Gaussian	Ind.	0.64 rad	0.95 rad
Independent	non-Gaussian	non-Gaussian	Ind.	0.63 rad	0.92 rad
Exact	non-Gaussian	non-Gaussian	Dep.	0.56 rad	0.97 rad

the error probability is

$$p_e = \frac{1}{2} - \frac{\rho_s e^{-\rho_s}}{2} \sum_{k=0}^{\infty} \frac{(-1)^k}{2k+1} \left[I_k\!\left(\frac{\rho_s}{2}\right) + I_{k+1}\!\left(\frac{\rho_s}{2}\right) \right]^2 \exp\left[-(2k+1)^2 \sigma_{\Phi_{\mathrm{NL}}}^2\right]. \tag{5.82}$$

In Eq. (5.82), the term of $\exp\left[-(2k+1)^2\sigma_{\Phi_{\mathrm{NL}}}^2\right]$ is the characteristic function of the Gaussian distributed phase noise at "angular frequency" of $2k+1$, the same as that in Eq. (4.40). Comparing with Eq. (4.40) with laser phase noise, the error probability of Eq. (5.82) just replaces the noise variance of Eq. (4.38) by $2\sigma_{\mathrm{NL}}^2$.

In another approximation shows in Figs. 5.12 and 5.13, the nonlinear phase noise is assumed to be independent of the phase of amplifier noise.

Figure 5.11 shows the p.d.f. of the differential phase for DPSK signal according to different models. In Fig. 5.11, the transmitted phases in two consecutive timing intervals are assumed to be the same. From Fig. 5.11, all approximated models underestimate the spreading of the differential phase. Fast decreasing, the Gaussian approximation gives a very small probability density at the tail, especially for small mean nonlinear phase shift of $\langle\Phi_{\mathrm{NL}}\rangle$. With smaller p.d.f. spreading than the exact model, all approximated models underestimate the error probability of DPSK signals with nonlinear phase noise.

Figure 5.15 shows the required SNR for an error probability of 10^{-9} as a function of mean nonlinear phase shift of $\langle\Phi_{\mathrm{NL}}\rangle$. From Fig. 5.15, all approximated models underestimate the required SNR. Table 5.1 summarizes the key parameters from various models. The optimal operating point is such that the increase of mean nonlinear phase shift, proportional to SNR, is larger than the increase of required SNR. The Nicholson and independence approximation give larger (about 13%) mean nonlinear phase shift for 1-dB power penalty but smaller (within 6%) optimal operating point than the exact model.

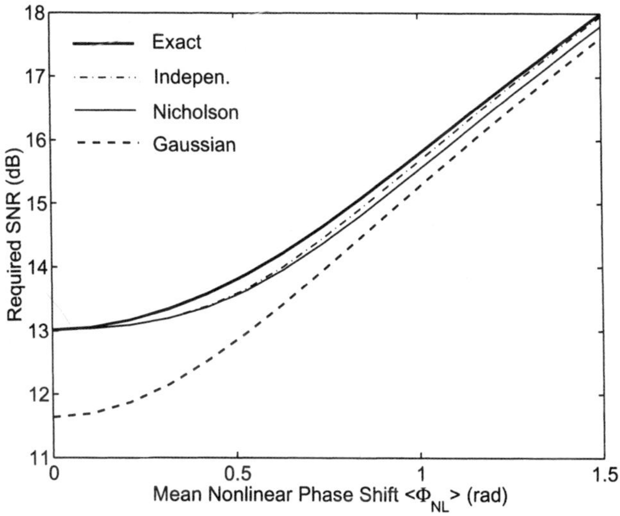

Figure 5.15. The required SNR of DPSK signal as a function of mean nonlinear phase shift $\langle \Phi_{\mathrm{NL}} \rangle$. [Adapted from Ho (2004c)]

While the error probability based on Q-factor is not able to predict the system performance except at very large nonlinear phase shift, the Nicholson and independence approximation of nonlinear phase noise underestimate the required SNR of up to 0.27 and 0.23 dB, respectively, and may not conform to the principle of conservative system design. If a prior correction of about 0.3 dB is added to both the Nicholson and independence approximation, both models can provide a conservative system design.

Direct-detection DPSK signal was analyzed by Humblet and Azizoğlu (1991), Jacobsen (1993), Pires and de Rocha (1992), Tonguz and Wagner (1991), and Chinn et al. (1996). The analysis here is also applicable to differential detection of CPFSK signal (Jacobsen, 1993).

The exact error probability is derived analytically for PSK and DPSK signals with nonlinear phase noise. The p.d.f. of the received phase is first expressed as a Fourier series. The Fourier coefficients are given by the joint characteristic function of nonlinear phase noise and the phase of amplifier noise.

For PSK signal, although the mean of the received phase is equal to the mean nonlinear phase shift $\langle \Phi_{\mathrm{NL}} \rangle$, the optimal decision region does not center around $\langle \Phi_{\mathrm{NL}} \rangle$. The SNR penalty of PSK signal increases by up to 0.49 dB due to the dependence between nonlinear phase noise and the phase of amplifier noise. With optimal decision angle, the mean nonlinear phase shift must be less than 1 rad for a SNR penalty less than 1 dB.

For DPSK signal, the differential phase has a symmetrical distribution with respect to the zero phase. The SNR penalty of DPSK signal increases by up to 0.23 dB due to the dependence between nonlinear phase noise and the phase of amplifier noise. The mean nonlinear phase shift must be less than 0.57 rad for a SNR penalty less than 1 dB. The optimal mean nonlinear phase shift is about 1 rad, similar to the estimation of Gordon and Mollenauer (1990).

The approximated formula Eq. (5.78) is the same as that in Ho (2003b) but using the asymptotic characteristic function of Eq. (5.48) instead of Eq. (5.22). The approximated error probability in Fig. 5.12 is the same as that in Ho (2003e) but calculated by a simple formula of Eq. (5.78).

4. Exact Error Probability of DPSK Signals with Finite Number of Spans

When a DPSK signal propagates in a system with less than $N_A = 32$ spans, the asymptotic model of last section may not applicable. Appendix 5.B derives the joint statistics of received intensity with the nonlinear phase noise for systems with small number of fiber spans. Here, the error probability is evaluated for DPSK signals. Without going into details, similar to Eq. (5.77), the error probability for DPSK signal is

$$p_e = \frac{1}{2} - \frac{1}{2}\sum_{k=0}^{\infty} \frac{(-1)^k \left|r_k e^{-r_k}\right|}{2k+1} \left| I_k\left(\frac{r_k}{2}\right) + I_{k+1}\left(\frac{r_k}{2}\right)\right|^2 \left|\Psi_{\Phi_{\mathrm{NL}}}(2k+1)\right|^2, \tag{5.83}$$

where

$$r_k = \frac{m_N^2(2k+1)}{2\sigma_N^2(2k+1)} \tag{5.84}$$

is analogous to the "angular frequency" depending SNR as the ratio of complex power of $\frac{1}{2}m_N^2(\nu)$ to the noise variance of $\sigma_N^2(\nu)$. Both $m_N(\nu)$ and $\sigma_N^2(\nu)$ are given by Eqs. (5.B.11) and (5.B.12) of Appendix 5.B, respectively.

If the dependence between nonlinear phase noise and the phase of amplifier noise is ignored, the error probability is approximated as

$$p_e \approx \frac{1}{2} - \frac{1}{2}\sum_{k=0}^{\infty} \frac{(-1)^k \rho_s e^{-\rho_s}}{2k+1} \left| I_k\left(\frac{\rho_s}{2}\right) + I_{k+1}\left(\frac{\rho_s}{2}\right)\right|^2 \left|\Psi_{\Phi_{\mathrm{NL}}}(2k+1)\right|^2. \tag{5.85}$$

The error probability expression of Eq. (5.83) is almost the same as that of Eq. (5.77) but with different parameters of Eq. (5.84). The

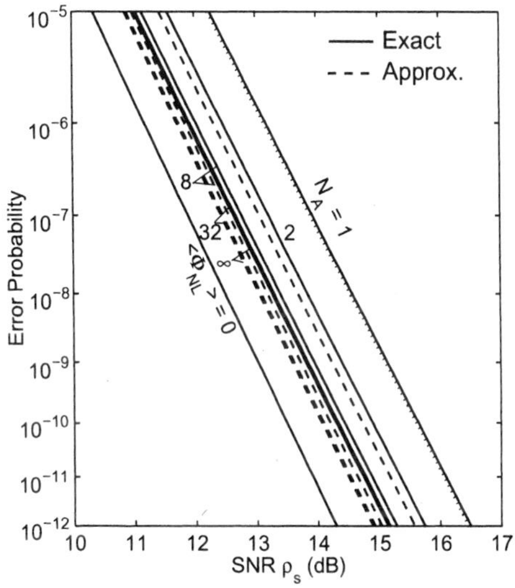

Figure 5.16. The error probability of DPSK signal as a function of SNR for N_A =1, 2, 4, 8, 32, and infinite number of fiber spans and mean nonlinear phase shift of $\langle \Phi_{\mathrm{NL}} \rangle = 0.5$ rad. [Adapted from Ho (2004d)]

approximated error probability of Eq. (5.85) is similar to the cases when additive phase noise is independent to Gaussian noise. The frequency depending SNR of Eq. (5.83) is originated from the dependence between the nonlinear phase noise and the additive Gaussian noise.

For DPSK signals with nonlinear phase noise, Figure 5.16 shows the exact error probability as a function of SNR ρ_s for mean nonlinear phase shift of $\langle \Phi_{\mathrm{NL}} \rangle = 0.5$ rad. Figure 5.17 shows the SNR penalty for an error probability of 10^{-9} as a function of mean nonlinear phase shift $\langle \Phi_{\mathrm{NL}} \rangle$. The SNR penalty is defined as the additional required SNR to achieve the same error probability of 10^{-9}. Both Figs. 5.16 and 5.17 are calculated using Eq. (5.83) and the independence approximation of Eq. (5.85). The independence approximation of Eq. (5.85) underestimates both the error probability and SNR penalty of a DPSK signal with nonlinear phase noise. Both Figs. 5.16 and 5.17 also include the exact and approximated error probability for $N_A = \infty$ that are the distributed model from Sec. 5.3. The distributed model is applicable when the number of fiber spans is larger than 32. The required SNR for systems without nonlinear phase noise of $\langle \Phi_{\mathrm{NL}} \rangle = 0$ is $\rho_s = 20$ (13 dB) for an error probability of 10^{-9} from Fig. 3.13.

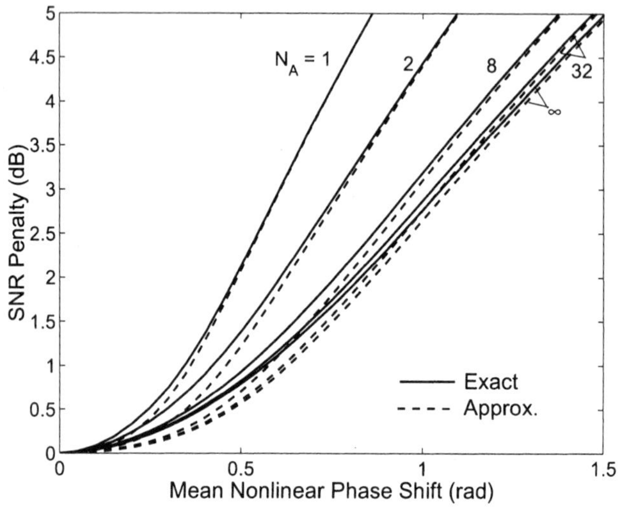

Figure 5.17. The SNR penalty as a function of mean nonlinear phase shift $\langle \Phi_{\mathrm{NL}} \rangle$ for system with finite number of fiber spans. [Adapted from Ho (2004d)]

From Figs. 5.16 and 5.17, for the same mean nonlinear phase shift of $\langle \Phi_{\mathrm{NL}} \rangle$, the SNR penalty is larger for smaller number of fiber spans. When the mean nonlinear phase shift is $\langle \Phi_{\mathrm{NL}} \rangle = 0.56$ rad, the SNR penalty is about 1 dB with large number ($N_A \geq 32$) of fiber spans but up to 3-dB SNR penalty for small number ($N_A = 1, 2$) of fiber spans. For 1-dB SNR penalty, the mean nonlinear phase shift is also reduced from 0.56 to 0.35 rad with small number of fiber spans.

In Sec. 5.3, the optimal operating point is calculated rigorously by the condition in which the increase of launched power does not further degrade the system performance. With the decrease of the number of fiber spans, the optimal operating point is reduced from 0.97 to 0.55 rad.

When the exact error probability is compared with the independence approximation of Sec. 5.3, the independence approximation is closer to the exact error probability for small number of fiber spans. In all cases, the independence assumption underestimates the error probability of the system, contradicting to the conservative principle of system design. The dependence between linear and nonlinear phase noise increases the SNR penalty up to 0.23 dB.

From the SNR penalty of Fig. 5.17, if a prior penalty of about 0.3 dB is added into the system, the independence assumption can be used to provide a conservative system design

5. Summary

The statistical properties of nonlinear phase noise are studied in detail. The p.d.f. of nonlinear phase noise is derived for finite and infinite number of fiber spans. The joint statistics of nonlinear phase noise with the phase of amplifier noises are also derived analytically. With the joint statistics, the exact error probability of phase-modulated optical signals with nonlinear phase noise is calculated for various system types, and compared with various other models based on different assumptions.

APPENDIX 5.A: Asymptotic Joint Characteristic

The characteristic function of Eq. (5.59) was derived by Mecozzi (1994a,b) based on the method of Cameron and Martin (1945). Similar to the method of Sec. 5.2.1 for nonlinear phase noise, the joint characteristic function is derived here using simpler method.

Note that the normalized nonlinear phase noise of Φ and the received electric field of e_N can be expressed as the summation of Eqs. (5.43) and (5.58), respectively, with terms determine by $|\sqrt{2}\xi_0 + x_k|^2$ and $\sqrt{2}\xi_0 + x_k$. First of all, we obtain

$$E\left\{\exp\left[j\nu|\sqrt{2}\xi_0 + x_k|^2 + j\Re\{\omega^* \cdot (\sqrt{2}\xi_0 + x_k)\}\right]\right\}$$
$$= \frac{1}{1-j\nu}\exp\left(\frac{2j\nu|\xi_0|^2 + \sqrt{2}j\Re\{\xi_0 \cdot \omega^*\} - \frac{1}{4}|\omega|^2}{1-j\nu}\right), \tag{5.A.1}$$

where $\Re\{\xi_0 \cdot \omega^*\}$ is the inner product for ξ_0 and ω when both of these two complex numbers are expressed as vector. In the above expression, if $\omega = 0$, the characteristic function of $|\sqrt{2}\xi_0 + x_k|^2$ is

$$\Psi_{|\sqrt{2}\xi_0 + x_k|^2}(\nu) = \frac{1}{1-j\nu}\exp\left(\frac{2j\nu\rho_s}{1-j\nu}\right) \tag{5.A.2}$$

for a noncentral χ^2-distribution with mean and variance of $2\rho_s + 1$ and $4\rho_s + 1$, respectively.

The joint characteristic function of Ψ_{Φ,e_N} is

$$\Psi_{\Phi,e_N}(\nu,\omega) = \prod_{k=1}^{\infty}\frac{1}{1-j\nu\lambda_k^2}\exp\left[\frac{2j\nu|\xi_0|^2\lambda_k^2 + 2j(-1)^{k+1}\lambda_k\Re\{\xi_0 \cdot \omega^*\} - \frac{1}{2}\lambda_k^2|\omega|^2}{1-j\nu\lambda_k^2}\right] \tag{5.A.3}$$

as the product of the joint characteristic function of the corresponding independently distributed random variables in the series expansion of Eqs. (5.43) and (5.58).

Using the expressions of Eqs. (5.46), (5.47) and Gradshteyn and Ryzhik (1980, §1.422)

$$\sec\frac{\pi x}{2} = \frac{4}{\pi}\sum_{k=1}^{\infty}\frac{(-1)^{k+1}(2k-1)}{(2k-1)^2 - x^2}, \tag{5.A.4}$$

the characteristic function of Eq. (5.A.3) can be simplified to

$$\Psi_{\Phi,e_N}(\nu,\omega) = \sec\sqrt{j\nu}\exp\left[\left(|\xi_0|^2\sqrt{j\nu} - \frac{|\omega|^2}{4\sqrt{j\nu}}\right)\tan\sqrt{j\nu} + j\sec(\sqrt{j\nu})\Re\{\xi_0\cdot\omega^*\}\right]. \tag{5.A.5}$$

The p.d.f. of $p_{\Phi,e_N}(\phi,z)$ is the inverse Fourier transform of the characteristic function $\Psi_{\Phi,e_N}(\nu,\omega)$ of Eq. (5.60). So far, there is no analytical expression for the p.d.f. of $p_{\Phi,e_N}(\phi,z)$.

In the field of lightwave communications, the approach here to derive the joint characteristic function of normalized nonlinear phase noise and electric field is similar to that of Foschini and Poole (1991) to find the joint characteristic function for polarization-mode dispersion (Poole et al., 1991), or that of Foschini and Vannucci (1988) for filtered phase noise. Another approach is to solve the Fokker-Planck equation of the corresponding diffusion process (Gardiner, 1985).

From the characteristic function Eq. (5.60), we can take the inverse Fourier transform with respect to ω and get

$$\mathcal{F}_\omega^{-1}\left\{\Psi_{\Phi,e_N}(\nu,\omega)\right\} = \mathcal{F}_\phi\left\{p_{\Phi,e_N}(\phi,z_1,z_2)\right\}, \tag{5.A.6}$$

where $\mathcal{F}_\omega^{-1}$ denotes the inverse Fourier transform with respect to ω_1 and ω_2, and $\mathcal{F}_\phi$ denotes the Fourier transform with respect to ϕ.

The characteristic function of Eq. (5.60) can be rewritten as

$$\Psi_{\Phi,e_N}(\nu,\omega) = \Psi_\Phi(\nu)\exp\left[-\frac{|\omega|^2\tan\sqrt{j\nu}}{4\sqrt{j\nu}} + j\sec(\sqrt{j\nu})\Re\{\xi_0\cdot\omega^*\}\right], \tag{5.A.7}$$

where $\Psi_\Phi(\nu)$ is the marginal characteristic function of nonlinear phase noise from Eq. (5.48). The inverse Fourier transform is

$$\mathcal{F}_\omega^{-1}\left\{\Psi_{\Phi,e_N}\right\} = \frac{\Psi_\Phi(\nu)}{2\pi\sigma_\nu^2}\exp\left[-\frac{(z_1-\xi_{\nu 1})^2 + (z_2-\xi_{\nu 2})^2}{2\sigma_\nu^2}\right], \tag{5.A.8}$$

where

$$\sigma_\nu^2 = \frac{\tan\sqrt{j\nu}}{2\sqrt{j\nu}} \quad \text{and} \quad (\xi_{\nu 1},\xi_{\nu 2}) = \sec(\sqrt{j\nu})(\xi_{01},\xi_{02}), \tag{5.A.9}$$

where $\xi_0 = \xi_{01} + j\xi_{02}$.

In Eq. (5.A.8), the electric field is a two-dimensional Gaussian distribution with a variance and mean of σ_ν^2 and $(\xi_{\nu 1},\xi_{\nu 2})$, respectively. The two components of the two-dimensional electric field are independent of each other. The dependence of both mean and variance on the nonlinear phase noise is given by Eq. (5.A.9) as a function of ν, the "angular frequency" of nonlinear phase noise. Properties of Gaussian random variable can be used to study the joint statistics of Eq. (5.A.8). When $\nu = 0$, the p.d.f. of $p_{e_N}(\vec{z}) = \exp(-|z-\xi_0|^2)/\pi$ is a two-dimensional Gaussian distribution with variance of $1/2$ and mean of ξ_0.

In the expression of Eq. (5.A.8), the partial characteristic function and p.d.f. is well-defined. Although σ_ν^2, $\xi_{\nu 0}$, and $\xi_{\nu 2}$ do not have a physical meaning, they are mathematically well defined as the Fourier transform of the joint p.d.f. of nonlinear phase noise and electric field with respect to the phase noise.

Using the partial p.d.f. and characteristic function of Eq. (5.A.8), similar to Appendix 4.A, change the random variables from rectangular coordinate of $e_N = e_{N1} + je_{N2}$ to polar coordinate of $e_N = Re^{j\Theta_n}$, we obtain

$$\mathcal{F}_\omega^{-1}\left\{\Psi_{\Phi,R,\Theta_n}\right\} = \frac{r\Psi_\Phi(\nu)}{2\pi\sigma_\nu^2}\exp\left[-\frac{r^2 + |\xi_\nu|^2 - 2r|\xi_\nu|\cos\theta}{2\sigma_\nu^2}\right], \tag{5.A.10}$$

where, not a standard symbol, $|\xi_\nu| = \sec(\sqrt{j\nu})|\xi_0|$ and $\Psi_\Phi(\nu)$ is the characteristic function of nonlinear phase noise of Eq. (5.48). The expression of Eq. (5.A.10) is very similar to Eq. (4.A.4) in Appendix 4.A. Take the integration over the received amplitude r of Eq. (5.A.10), we get

$$\mathcal{F}_\omega^{-1}\left\{\Psi_{\Phi,\Theta_n}\right\} = \Psi_\Phi(\nu)\left\{\frac{1}{2\pi}e^{-\gamma_\nu} + \sqrt{\frac{\gamma_\nu}{4\pi}}\cos\theta e^{-\gamma_\nu\sin^2\theta}\operatorname{erfc}\left[-\sqrt{\gamma_\nu}\cos\theta\right]\right\}, \tag{5.A.11}$$

where

$$\gamma_\nu = \frac{|\xi_\nu|^2}{2\sigma_\nu^2} = \frac{2\sqrt{j\nu}}{\sin\left(2\sqrt{j\nu}\right)}\rho_s \tag{5.A.12}$$

with σ_ν^2 and $|\xi_\nu|^2$ given by Eq. (5.A.9) can be interpreted as the angular frequency depending SNR. The "phase" distribution of Eq. (5.A.11) can be found in standard textbook (Proakis, 2000, §5.2.7).

Taking a Fourier series expansion of Eq. (5.A.11) with respect to the phase of amplifier noise Θ_n, similar to Eq. (4.A.11), the Fourier coefficients are (Middleton, 1960, §9.2-2)

$$\begin{aligned}\Psi_{\Phi,\Theta_n}(\nu, m) &= \frac{\Psi_\Phi(\nu)\gamma_\nu^{m/2}}{m!}\Gamma\left(\frac{m}{2}+1\right){}_1F_1\left(\frac{m}{2}; m+1; -\gamma_\nu\right) \\ &= \frac{\sqrt{\pi}}{2}\Psi_\Phi(\nu)\gamma_\nu^{1/2}e^{-\gamma_\nu/2}\left[I_{\frac{m-1}{2}}\left(\frac{\gamma_\nu}{2}\right) + I_{\frac{m+1}{2}}\left(\frac{\gamma_\nu}{2}\right)\right], \\ &\qquad m \geq 0,\end{aligned} \tag{5.A.13}$$

and $\Psi_{\Phi,\Theta_n}(\nu, -m) = \Psi_{\Phi,\Theta_n}(\nu, m)$, where $\Gamma(\cdot)$ is the Gamma function, ${}_1F_1(a; b; \cdot)$ is the confluent hypergeometric function of the first kind, and $I_k(\cdot)$ is the k-th order modified Bessel function of the first kind.

The Fourier coefficients of Eq. (5.62) is similar to that in Jain (1974), Jain and Blachman (1973), and Blachman (1981, 1988) but with a frequency dependent γ_ν of Eq. (5.A.12). To simplify calculation, the confluent hypergeometric function is converted to modified Bessel function of the first kind. Bessel or modified Bessel functions with complex argument are well-defined (Amos, 1986).

To derive the coefficients of Eq. (5.62), we can also find the Fourier coefficients of Eq. (5.A.10) with respect to the angle θ, and then take the integration over the received amplitude.

APPENDIX 5.B: Joint Statistics for Finite Number of Spans

In Appendix 5.A, all the analysis and subsequent calculations are based on the joint "Gaussian" distribution of Eq. (5.A.8) with mean and variance of ξ_ν and σ_ν^2, respectively. For system with finite number of fiber spans, the marginal characteristic function for nonlinear phase noise corresponding to $\Psi_\Phi(\nu)$ of Eq. (5.48) is $\Psi_{\Phi_{\rm NL}}(\nu)$ of Eq. (5.24). Comparing the simple expression of $\Psi_\Phi(\nu)$ with that of $\Psi_{\Phi_{\rm NL}}(\nu)$, the joint statistics of the nonlinear phase noise with received electric field is difficult to derive for systems with finite number of fiber spans.

When the system has finite number of fiber spans, the joint statistics of nonlinear phase noise with received field has the same expression of Eq. (5.A.8) but all parameters of $\Psi_\Phi(\nu)$, σ_ν^2, and ξ_ν are not the same as Eq. (5.48) or (5.A.9). While $\Psi_{\Phi_{\rm NL}}(\nu)$ was derived in Sec. 5.1, this appendix derives the corresponding σ_ν^2 and ξ_ν for finite number of fiber spans. After that, for example, to evaluate the error probability, the expressions of Eqs. (5.69) and (5.77) can be used with the new parameters of Eqs. (5.71) according to the angular SNR of Eq. (5.A.12).

To find the dependence between the nonlinear phase noise and the received electric field, the joint characteristic function of

$$\Psi_{\Phi_{\rm NL}, E_N}(\nu, \omega) = E\left\{\exp(j\nu\Phi_{\rm NL} + j\Re\{\omega^* \cdot E_N\}\right\} \tag{5.B.1}$$

will be derived here with $\Phi_{\rm NL}$ and E_N given by Eqs. (5.5) and (5.6), respectively.

Ignored the constant factor of $\gamma L_{\rm eff}$ for simplicity, with $\omega = \omega_1 + j\omega_2$ and $E_N = E_{N1} + jE_{N2}$, we obtain

$$j\nu\varphi_1 + j\omega_1 E_{N1} = j\nu N_A A^2 + j\omega_1 A + 2j\nu A\vec{w}^T\vec{x} + j\omega_1\vec{w}_I^T\vec{x} + j\nu\vec{x}^T C\vec{x}, \tag{5.B.2}$$

where φ_1 is given by Eq. (5.11),

$$j\omega_1 E_{N1} = j\omega_1(A + x_1 + x_2 + \cdots + x_{N_A}) = j\omega_1 A + j\omega_1\vec{w}_I^T\vec{x}, \tag{5.B.3}$$

$\vec{x} = (x_1, x_2, \ldots, x_{N_A})^T$, and $\vec{w}_I = (1, 1, \ldots, 1)^T$, $\vec{w}$ and C are the same as those in Eq. (5.13).

Similar to Sec. 5.1.2, using the N_A-dimensional Gaussian of Eq. (5.16) for $\vec{x}$, we obtain

$$\Psi_{\varphi_1, E_{N1}}(\nu, \omega_1) = \frac{e^{j\nu N_A A^2 + j\omega_1 A}}{(2\pi\sigma_0^2)^{\frac{N_A}{2}}}\int \exp\left[2j\nu A\vec{w}^T\vec{x} + j\omega_1\vec{w}_I^T\vec{x} - \vec{x}^T\Gamma\vec{x}\right] {\rm d}\vec{x}. \tag{5.B.4}$$

or

$$\Psi_{\varphi_1, E_{N1}}(\nu, \omega_1) = \frac{e^{j\nu N_A A^2 + j\omega_1 A}}{(2\sigma_0^2)^{\frac{N_A}{2}}\det[\Gamma]^{\frac{1}{2}}} \times \exp\left[-\left(\nu A\vec{w} + \frac{1}{2}\omega_1\vec{w}_I\right)^T \Gamma^{-1}\left(\nu A\vec{w} + \frac{1}{2}\omega_1\vec{w}_I\right)\right], \tag{5.B.5}$$

where Γ is the same as that in Eq. (5.17). Similarly, using $A = 0$ in Eq. (5.B.5), we have

$$\Psi_{\varphi_2, E_{N2}}(\nu, \omega_2) = \frac{\exp\left[-\frac{1}{4}\omega_2^2\vec{w}_I^T\Gamma^{-1}\vec{w}_I\right]}{(2\sigma_0^2)^{\frac{N_A}{2}}\det[\Gamma]^{\frac{1}{2}}}, \tag{5.B.6}$$

where φ_2 is that from Eq. (5.12).

The joint characteristic function of

$$\Psi_{\Phi_{\rm NL}, E_N}(\nu, \omega) = \Psi_{\varphi_1, E_{N1}}(\nu, \omega_1)\Psi_{\varphi_2, E_{N2}}(\nu, \omega_2) \tag{5.B.7}$$

becomes

$$\Psi_{\Phi_{\mathrm{NL}}, E_N}(\nu, \omega) = \Psi_{\Phi_{\mathrm{NL}}}(\nu) \exp\left[j\omega_1 m_N(\nu) - \sigma_N^2(\nu) \frac{|\omega|^2}{2} \right], \tag{5.B.8}$$

where $\Psi_{\Phi_{\mathrm{NL}}}(\nu)$ is the same as that in Eq. (5.22) and

$$m_N(\nu) = A + j\nu A \vec{w}^T \Gamma^{-1} \vec{w}_I, \tag{5.B.9}$$

$$\sigma_N^2(\nu) = \frac{1}{2} \vec{w}_I^T \Gamma^{-1} \vec{w}_I. \tag{5.B.10}$$

Based on the eigenvalues and eigenvectors of the covariance matrix $\mathcal{C}$, the characteristic function of $\Psi_{\Phi_{\mathrm{NL}}}(\nu)$ becomes that of Eq. (5.24), and

$$\begin{aligned} m_N(\nu) &= A + 2j\nu\sigma_0^2 A \sum_{k=1}^{N_A} \frac{(\vec{v}_k^T \vec{w})(\vec{v}_k^T \vec{w}_I)}{1 - 2j\nu\sigma_0^2 \lambda_k} \\ &= A \sum_{k=1}^{N_A} \frac{(\vec{v}_k^T \vec{w})(\vec{v}_k^T \vec{w}_I)/\lambda_k}{1 - 2j\nu\sigma_0^2 \lambda_k}, \end{aligned} \tag{5.B.11}$$

$$\sigma_N^2(\nu) = \sigma_0^2 \sum_{k=1}^{N_A} \frac{(\vec{v}_k^T \vec{w}_I)^2}{1 - 2j\nu\sigma_0^2 \lambda_k}. \tag{5.B.12}$$

The characteristic function of Eq. (5.B.8) is similar to the corresponding characteristic function with the distributed assumption of Eq. (5.60). If the number of spans N_A approaches infinite, the characteristic function should converge to that of Eq. (5.60).

Based on Eq. (5.B.8), we obtain

$$\mathcal{F}_\omega^{-1}\left\{ \Psi_{\Phi_{\mathrm{NL}}, E_N} \right\} = \frac{\Psi_{\Phi_{\mathrm{NL}}}(\nu)}{2\pi\sigma_N^2(\nu)} \exp\left\{ -\frac{[z_1 - m_N(\nu)]^2 + z_2^2}{2\sigma_N^2(\nu)} \right\}, \tag{5.B.13}$$

with $\mathcal{F}_\omega^{-1}\{\cdot\}$ denotes inverse Fourier transform with respect to ω. The partial characteristic function and p.d.f. of Eq. (5.B.13) is similar to a two-dimensional Gaussian p.d.f. with mean of $(m_N(\nu), 0)$ and variance of $\sigma_N^2(\nu)$. Notice that Eq. (5.B.13) is similar to Eq. (5.A.8). With the dependence between the nonlinear phase noise and the phase of amplifier noise, the variance of $\sigma_N^2(\nu)$ and the mean of $m_N(\nu)$ are both complex numbers depending on the "angular frequency" of ν. The marginal p.d.f. of the received electric field $\vec{E}_N$ is a two-dimensional Gaussian distribution with variance of $\sigma_N^2(\nu)|_{\nu=0} = N_A\sigma_0^2$ and mean of $m_N(\nu)|_{\nu=0} = A$.

In all the analysis for nonlinear phase noise in this book, the required optical amplification in the last span and its additive noise is ignored for simplicity. For systems with large number of fiber spans, the assumption is valid as the $(N_A + 1)$th amplifier is not that important. For systems with small number of fiber spans, the variance of $\sigma_N^2(\nu)$ may be modified slightly to include the amplifier noise from the last span by using

$$\sigma_N^2(\nu) = \sigma_0^2 \left[1 + \sum_{k=1}^{N_A} \frac{(\vec{v}_k^T \vec{w}_I)^2}{1 - 2j\nu\sigma_0^2 \lambda_k} \right]. \tag{5.B.14}$$

In a more general system configuration, the nonlinear phase noise of Eq. (5.5) may be equal to

$$\Phi_{\mathrm{NL}} = \sum_{k=1}^{N_A} \gamma_k L_{\mathrm{eff}k} \left| A_k + \frac{A_k}{A_1} n_1 + \cdots + \frac{A_k}{A_{k-1}} n_{k-1} + n_k \right|^2, \tag{5.B.15}$$

where γ_k and $L_{\mathrm{eff}k}$ are the fiber nonlinear coefficient and effective length of the kth fiber span, A_k, $k = 1, \ldots, N_A$, are the launched electric field of each span, and n_k, $k = 1, \ldots, N_A$, are the amplifier noises of each span with variance of $E\{|n_k|^2\} = 2\sigma_k^2$. When the signal is changed, for example, from A_1 to A_2, the noise is also changed by the factor of A_2/A_1.

Using the matrix $\mathcal{M}$ of

$$\mathcal{M} = \begin{bmatrix} 1 & 0 & 0 & \cdots & 0 \\ \frac{A_2}{A_1} & 1 & 0 & \cdots & 0 \\ \frac{A_3}{A_1} & \frac{A_3}{A_2} & 1 & \cdots & 0 \\ \vdots & \vdots & \vdots & \ddots & \vdots \\ \frac{A_{N_A}}{A_1} & \frac{A_{N_A}}{A_2} & \frac{A_{N_A}}{A_3} & \cdots & 1 \end{bmatrix} \times \begin{bmatrix} \sigma_1 & 0 & 0 & \cdots & 0 \\ 0 & \sigma_2 & 0 & \cdots & 0 \\ 0 & 0 & \sigma_3 & \cdots & 0 \\ \vdots & \vdots & \vdots & \ddots & \vdots \\ 0 & 0 & 0 & \cdots & \sigma_{N_A} \end{bmatrix}, \tag{5.B.16}$$

we can obtain a covariance matrix of

$$\mathcal{C} = \mathcal{M}^T \mathrm{diag}\left\{\gamma_1 L_{\mathrm{eff}1}, \gamma_1 L_{\mathrm{eff}2}, \cdots, \gamma_{N_A} L_{\mathrm{eff}N_A}\right\} \mathcal{M}, \tag{5.B.17}$$

and a vector of

$$\vec{w} = \mathcal{M}^T \left(\gamma_1 L_{\mathrm{eff}1} A_1, \gamma_1 L_{\mathrm{eff}2} A_2, \cdots, \gamma_{N_A} L_{\mathrm{eff}N_A} A_{N_A}\right)^T, \tag{5.B.18}$$

where $\mathrm{diag}\{\cdot\}$ is a diagonal matrix form by a vector.

Together with the vector of

$$\vec{w_I} = \left(\frac{A_{N_A}}{A_1}\sigma_1, \frac{A_{N_A}}{A_2}\sigma_2, \frac{A_{N_A}}{A_3}\sigma_3, \cdots, \frac{A_{N_A}}{A_{N_A-1}}\sigma_{N_A-1}, \sigma_{N_A}\right)^T, \tag{5.B.19}$$

we obtain, without change the symbols,

$$\Psi_{\Phi_{\mathrm{NL}}}(\nu) = \prod_{k=1}^{N_A} \frac{1}{1 - 2j\nu\lambda_k} \exp\left[\frac{j\nu(\vec{v}_k^T \vec{w})^2/\lambda_k}{1 - 2j\nu\lambda_k}\right], \tag{5.B.20}$$

$$m_N(\nu) = \sum_{k=1}^{N_A} \frac{(\vec{v}_k^T \vec{w})(\vec{v}_k^T \vec{w}_I)/\lambda_k}{1 - 2j\nu\lambda_k}, \tag{5.B.21}$$

$$\sigma_N^2(\nu) = \sum_{k=1}^{N_A} \frac{(\vec{v}_k^T \vec{w}_I)^2}{1 - 2j\nu\lambda_k}, \tag{5.B.22}$$

where $\lambda_k, \vec{v}_k, k = 1, \ldots, N_A$ are the eigenvectors and eigenvalues of the covariance matrix $\mathcal{C}$. The usage of the above three functions of ν in Eq. (5.B.13) can obtain all necessary statistics between the nonlinear phase noise and received electric field.

If the effective of the amplifier noise of that last span cannot be ignored, we can add $\sigma_{N_A+1}^2$ to $\sigma_N^2(\nu)$.

Chapter 6

COMPENSATION OF NONLINEAR PHASE NOISE

As shown in the helix shape scattergram of Figs. 5.1, nonlinear phase noise is correlated with the received intensity. The received intensity can be used to compensate for the nonlinear phase noise. Ideally, the optimal compensator should minimize the error probability of the system after compensation. However, the optimal compensator or detector given by the maximum *a posteriori* probability (MAP) criterion to minimize the error probability may be difficult to find and design. The minimum mean-square error (MMSE) compensator is always analytically simple to find and leads to practical implementation.

In this chapter, the linear compensator is optimized in term of the variance, or MMSE criterion, of the residual nonlinear phase noise. The linear MMSE compensator is derived when a single compensator preceding the receiver or in the middle of the fiber span. The MMSE compensator is generalized to the cases when many compensators are located in the fiber link for mid-span or distributed compensation.

The exact error probability with a single linear compensator preceding the receiver is derived analytically. With a simple expression to calculate the error probability, numerical optimization is used to find the linear MAP compensator to minimize the error probability. While linear MAP compensator gives improvement over MMSE compensator for phase-shift keying (PSK) signals, linear MMSE compensator performs close to linear MAP compensator for differential phase-shift keying (DPSK) signals.

When the joint distribution of received amplitude and phase is derived analytically for signal with nonlinear phase noise. The optimal MAP detector can be derived to minimize the error probability of the signal after the detector. Based on the joint distribution of received amplitude and phase, the error probability of nonlinear MMSE or MAP detector can also be evaluated numerically for PSK signals.

1. Electronic Compensator for Nonlinear Phase Noise

The received intensity $P_N = R_r^2$ may be used to compensate the nonlinear phase noise of Eq. (5.5), where the received amplitude is

$$R_r = |E_0 + n_1 + \cdots + n_N| . \tag{6.1}$$

The received amplitude of Eq. (6.1) ignores the fiber loss of the last fiber span and the required optical amplifier to compensate for it. The actual received electric field is

$$E_r = (E_0 + n_1 + \cdots + n_N) \exp(-j\Phi_{\rm NL}) \tag{6.2}$$

with nonlinear phase rotation of $\Phi_{\rm NL}$ as shown in Figs. 5.1. In the received electric field of Eq. (6.2), the nonlinear phase noise of Eq. (5.5) does not change the received amplitude R_r of Eq. (6.1). For binary PSK and DPSK signals, only the real part of the received electric field or the corresponding differential component is required to detect the transmitted data. Both real and imaginary parts of the electric field of E_r can be measured by an optical phase-locked loop (PLL) for PSK signal using the quadrature receiver of Fig. 3.4. The corresponding differential components can be measured by two interferometers for DPSK signal. In order to compensate the nonlinear phase noise, either the received phase or the complex representation of the received electric field of E_r must be measured.

In the simplest nontrivial example of two fiber spans, the received electric field in linear regime is $E_2 = E_0 + n_1 + n_2$. The nonlinear phase shift is $\Phi_{\rm NL} = \gamma L_{\rm eff}\left(|E_0 + n_1|^2 + |E_0 + n_1 + n_2|^2\right)$, and the received electric field with nonlinear phase shift is $E_2 \exp(-j\Phi_{\rm NL})$. Hence, the received phase is $\Phi_2 = \theta_2 - \Phi_{\rm NL} = \theta_2 - \gamma L_{\rm eff}\left(|E_0 + n_1|^2 + |E_0 + n_1 + n_2|^2\right)$, where θ_2 is the phase of $E_2 = |E_2|e^{j\theta_2}$. Because the received signal intensity $P_2 = |E_2|^2 = |E_0+n_1+n_2|^2$ is correlated with $\Phi_{\rm NL}$, the impact of the nonlinear phase shift can be reduced by adding a correction term of $\gamma L_{\rm eff} P_2$ to the received phase. The corrected estimation of the received phase is $\Phi_2 + \gamma L_{\rm eff} P_2 = \theta_2 - \gamma L_{\rm eff}|E_0 + n_1|^2$. In fact, the impact of the nonlinear phase noise can be reduced further, because the remaining nonlinear phase shift of $\gamma L_{\rm eff}|E_0 + n_1|^2$ is still correlated with the received intensity P_2. As shown below, the optimal correction term is about $1.5\gamma L_{\rm eff} P_2$.

Figure 6.1(a) shows an example of a homodyne optical receiver using an optical PLL to detect both in-phase and quadrature components

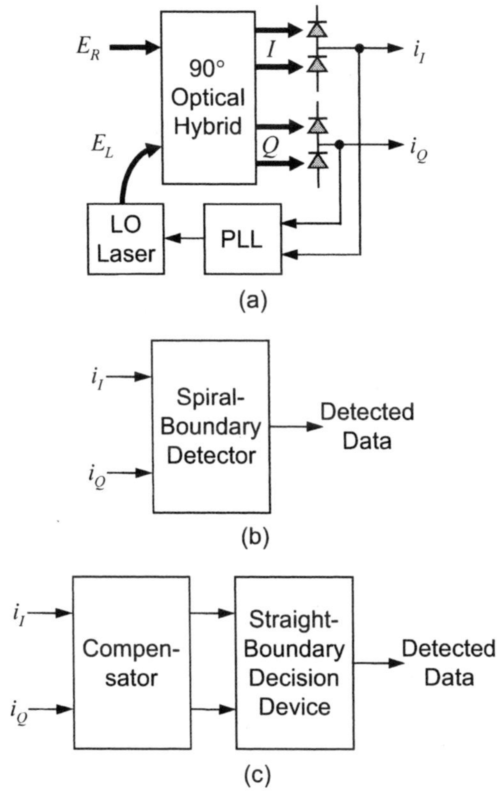

Figure 6.1. (a) Typical coherent receiver detecting both in-phase and quadrature components of the received electric field E_r (see also Fig. 3.4). (b) Nonlinear phase noise is compensated by using a spiral-boundary decision device. (c) Nonlinear phase noise is compensated by using a compensator, followed by a straight-boundary decision device. [From Ho and Kahn (2004a), © 2005 IEEE]

of the received electric field E_r, similar to the quadrature receiver of Fig. 3.4. A 90° optical hybrid is used to combine the signal with a phase-locked local oscillator (LO) laser, yielding four combinations with relative phase shifts of 0°, 180°, 90° and 270°. A pair of balanced photodetectors provides in-phase and quadrature photocurrents, i_I and i_Q, equivalently the two components of $\cos \Phi_r$ and $\sin \Phi_r$ or the whole complex number of $E_r = |E_r|e^{j\Phi_r}$. In systems where the dominant noise source is amplified spontaneous emission from optical amplifiers, a synchronous heterodyne optical receiver can also be used without loss of

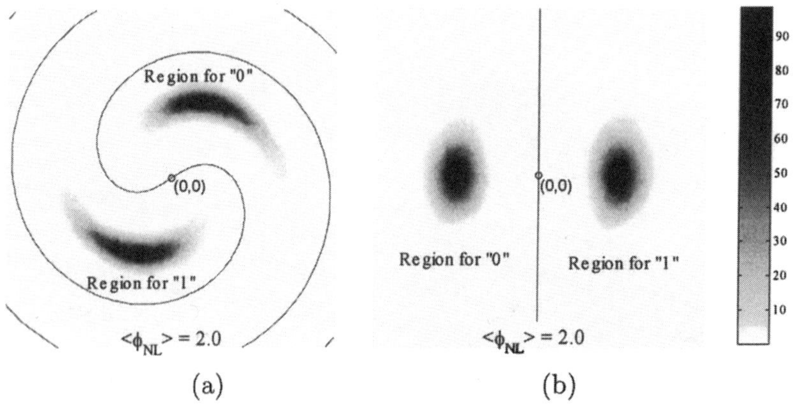

Figure 6.2. (a) The distribution of received electric field E_r together with the spiral decision boundary for binary PSK signal. (b) The distribution of the compensated electric field together with the straight line decision boundary for binary PSK signal. [From Ho and Kahn (2004a), © 2005 IEEE]

sensitivity as shown in Sec. 3.1.1 and Table 3.1. An electrical PLL is required to measure the in- and quadrature-components of $\cos \Phi_r$ and $\sin \Phi_r$ for a heterodyne receiver.

Similar to Figs. 5.1, Figure 6.2(a) shows the simulated distribution of the received electric field E_r for a binary PSK system with $N_A = 32$ spans, after detection by a coherent receiver as in Fig. 6.1. Note that although the optical PLL of Fig. 6.1(a) actually tracks out the mean nonlinear phase shift, nonzero values of $\langle \Phi_{\text{NL}} \rangle$ have been preserved in plotting Figs. 6.2 to better illustrate the nonlinear phase noise. For a mean nonlinear phase shift of $\langle \Phi_{\text{NL}} \rangle = 2$ rad, Figure 6.2(a) also shows the decision boundary when a detector with a spiral boundary of Fig. 6.1(b) is used. The spiral region of Fig. 6.1(a) resembles the Yin-Yang logo of Chinese mysterism, call the Yin-Yang detector. Once the mean nonlinear phase shift of $\langle \Phi_{\text{NL}} \rangle$ is known, one can implement the spiral-boundary decision device using a lookup table. The Yin-Yang detector can only be implemented using electronic techniques.

An alternative method, shown in Fig. 6.1(c), employs a compensator that subtracts from the received phase a correction as a function of the received intensity. The compensated phase, depending on the received intensity, is the central phase of the decision region of either zero or one of Fig. 6.2(a). The compensator is followed by a straight-boundary decision device. Figure 6.2(b) shows the distribution of the residual phase noise after the compensator, together with the straight-boundary deci-

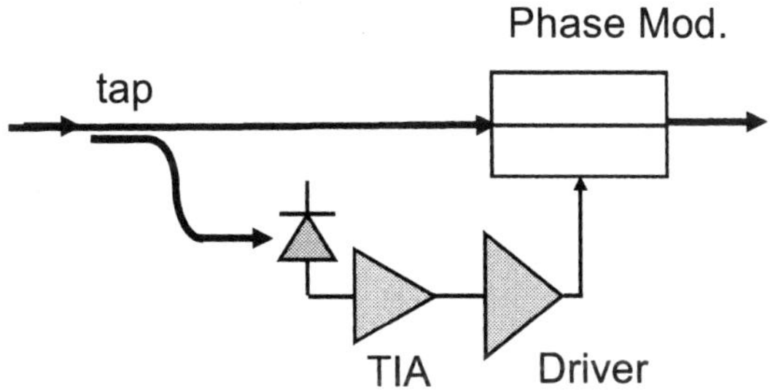

Figure 6.3. A compensator of nonlinear phase noise implemented using a phase modulator.

sion boundary as the y-axis. The compensator phase can be a linear function of the received intensity, as suggested by Liu et al. (2002b), Xu et al. (2003), and Ho and Kahn (2004a) or a nonlinear function of the received intensity, as suggested by Ho (2003d) and Appendix 6.A. Theoretically, the optimal compensator should be nonlinear compensator without linearity constraint.

Figure 6.3 shows a schematic diagram to compensate nonlinear phase noise using a phase modulator, first suggested by Xu and Liu (2002) and implemented in both Xu et al. (2002) and Hansryd et al. (2004, 2005). A passive fiber tap is used to split out part of the optical signal, amplified using both a TIA and driver amplifier to drive a phase modulator. The modulated phase of the phase modulator cancels part of the nonlinear phase noise in the fiber link. Both linear and nonlinear compensators can be implemented using a phase modulator of Fig. 6.3 using either a linear or nonlinear driver amplifier. While most external modulators are polarization sensitive, polarization independent modulator is also available (Chen et al., 1992, 1997a, Sugiyama, 2003). If polarization dependent modulator is used in Fig. 6.3, automatic polarization control (APC) is required preceding the phase modulator.

While the electronic compensator of Fig. 6.1(c) or the Yin-Yang detector of Fig. 6.1(b) is used in the receiver, the electro-optical compensator of Fig. 6.3 can be used in the middle of a fiber link for distributed compensation of nonlinear phase noise. Both electronic and electro-optical implementations of Figs. 6.1 and 6.3 can be used as both linear and nonlinear compensator, some nonlinear optical devices can be used solely as a linear compensator (Liu et al., 2002b).

2. Linear MMSE Compensator for Finite Number of Fiber Spans

The simplest compensator is a linear compensator to minimize the variance of the residual nonlinear phase noise. Although the probability density function (p.d.f.) of nonlinear phase noise was derived in Chapter 5 for finite and infinite number of fiber spans, this type of MMSE compensator can be derived without using the p.d.f. or joint characteristic between nonlinear phase noise and received intensity but the correlation between them. In this section, we find either the decision boundary in Fig. 6.2(a) or the corresponding compensated curve to minimize the variance of the residual nonlinear phase noise. In terms of the variance of residual nonlinear phase noise, linear and nonlinear MMSE compensators do not have significant difference.

2.1 Minimum Mean-Square Error Compensation

In an N_A-span system, the linear MMSE compensator can be derived by finding a scale factor α to minimize the variance of the residual nonlinear phase noise of $\Phi_{\mathrm{NL}} - \alpha P_N$. The corrected phase estimation is $\Phi_r + \alpha P_N$, where Φ_r is the phase of the received electric field E_r of Eq. (6.2).

The intensity of signal plus noise has a noncentral chi-square (χ^2) distribution. The MMSE compensator minimizes the variance that depends on the covariance between its compositions. First, we consider a simple mathematical problem to find the covariance between two χ^2 distributed random variables. For a real amplitude of $A = |E_0|$ and two independent complex circular Gaussian distributed random variables ζ_1 and ζ_2, both $|A + \zeta_1|^2$ and $|A + \zeta_1 + \zeta_2|^2$ are noncentral χ^2 distributed random variables with two degrees of freedom. The mean and variance of $|A + \zeta_1|^2$ are

$$E\{|A + \zeta_1|^2\} = A^2 + 2\sigma_{\zeta_1}^2, \tag{6.3}$$

$$\sigma^2_{|A+\zeta_1|^2} = f\left(\sigma_{\zeta_1}^2\right) = 4A^2\sigma_{\zeta_1}^2 + 4\sigma_{\zeta_1}^4, \tag{6.4}$$

where $E\{|\zeta_1|^2\} = 2\sigma_{\zeta_1}^2$ is the variance of ζ_1. In Eq. (6.4), the variance of the random variable $|A+\zeta_1|^2$ is defined as a function of $f(\sigma_{\zeta_1}^2)$. Without going into detail, the covariance between $|A + \zeta_1|^2$ and $|A + \zeta_1 + \zeta_2|^2$ is

$$E\left\{\left(|A+\zeta_1|^2 - E\{|A+\zeta_1|^2\}\right)\left(|A+\zeta_1+\zeta_2|^2 - E\{|A+\zeta_1+\zeta_2|^2\}\right)\right\} \\ = 4A^2\sigma_{\zeta_1}^2 + 4\sigma_1^4 = f\left(\sigma_{\zeta_1}^2\right), \tag{6.5}$$

The covariance relationship of Eq. (6.5) is obvious, the random variable of $|A + \zeta_1|^2$ does not depend on, and is not correlated with, the random variable of ζ_2. Note that $\sigma^2_{\zeta_1}$ is the covariance between ζ_1 and $\zeta_1 + \zeta_2$.

Using the expressions of Eqs. (6.4) and (6.5), the variance of the nonlinear phase noise of Eq. (5.5) is found to be

$$\sigma^2_{\Phi_{\rm NL}} = (\gamma L_{\rm eff})^2 \left[\sum_{k=1}^{N_A} f\left(k\sigma^2\right) + 2\sum_{k=1}^{N_A} (N_A - k) f\left(k\sigma^2\right) \right]. \tag{6.6}$$

The first summation of Eq. (6.6) corresponds to $\zeta_1 = n_1 + \cdots + n_k$ with $\sigma^2_{\zeta_1} = k\sigma^2$ for all terms of $|E_0 + n_1 + \cdots + n_k|^2$. The second summation of Eq. (6.6) corresponds to all the covariance terms between $|E_0 + n_1 + \cdots + n_k|^2$ and $|E_0 + n_1 + \cdots + n_{k+1}|^2 + \cdots + |E_0 + n_1 + \cdots + n_{N_A}|^2$ with $\zeta_1 = n_1 + \cdots + n_k$ and $\zeta_2 = n_{k+1} + \cdots + n_{N_A}$. Substituting Eq. (6.4) to (6.6), we obtain

$$\sigma^2_{\Phi_{\rm NL}} = \frac{4}{3} N_A (N_A + 1) \left(\gamma L_{\rm eff} \sigma_0\right)^2 \left[\left(N_A + \frac{1}{2}\right) |E_0|^2 + \left(N_A^2 + N_A + 1\right) \sigma_0^2 \right]. \tag{6.7}$$

Similar to Eq. (6.6), the variance of the residual nonlinear phase noise of $\Phi_{\rm NL} - \alpha P_N$, is found using Eqs. (6.4) and (6.6) to be

$$\sigma^2_{\Phi_{\rm NL} - \alpha P_N} = (\gamma L_{\rm eff})^2 \left[\sigma^2_{\Phi_{\rm NL}} + (\alpha - 1)^2 f(N_A \sigma^2) - 2(\alpha - 1) \sum_{k=1}^{N_A - 1} f\left(k\sigma^2\right) \right]. \tag{6.8}$$

The optimal scale factor can be found by solving $d\sigma^2_{\Phi_{\rm NL} - \alpha P_N}/d\alpha = 0$ to obtain

$$\alpha_{\rm mse} = 1 + \frac{\sum_{k=1}^{N_A - 1} f\left(k\sigma^2\right)}{f\left(N_A \sigma^2\right)}. \tag{6.9}$$

Using Eq. (6.4), the optimal scale factor is found to be

$$\alpha_{\rm mse} = \gamma L_{\rm eff} \frac{N_A + 1}{2} \cdot \frac{|E_0|^2 + (2N_A + 1)\sigma_0^2/3}{|E_0|^2 + N_A \sigma_0^2}. \tag{6.10}$$

At high signal-to-noise ratio (SNR), we obtain

$$\alpha_{\rm mse} \approx \gamma L_{\rm eff} \frac{N_A + 1}{2}. \tag{6.11}$$

The variance of the residual nonlinear phase noise is reduced to

$$\begin{aligned} \sigma^2_{\Phi_{\rm NL} - \alpha_{\rm mse} P_N} &= (N_A - 1) N_A (N_A + 1) (\gamma L_{\rm eff} \sigma_0)^2 \\ &\times \frac{|E_0|^4 + 2N_A \sigma_0^2 |E_0|^2 + \left(2N_A^2 + 1\right)\sigma_0^4/3}{3(|E_0|^2 + N_A \sigma_0^2)}. \end{aligned} \tag{6.12}$$

The mean residual nonlinear phase shift is

$$\langle\Phi_{\mathrm{RES}}\rangle = \langle\Phi_{\mathrm{NL}}\rangle - \alpha_{\mathrm{mse}}(|E_0|^2 + 2N_A\sigma_0^2). \tag{6.13}$$

In high SNR, similar to Eq. (5.80), the variance of nonlinear phase noise of Eq. (6.7) can be approximated as

$$\sigma^2_{\Phi_{\mathrm{NL}}} \approx \frac{4}{3}\left(\gamma L_{\mathrm{eff}}\sigma_0|E_0|\right)^2 \approx \frac{2\left\langle\Phi_{\mathrm{NL}}\right\rangle^2}{3\rho_s}. \tag{6.14}$$

and the variance of residual nonlinear phase noise of Eq. (6.12) can be approximated as

$$\sigma^2_{\Phi_{\mathrm{NL}}-\alpha P_N} \approx \frac{1}{3}\left(\gamma L_{\mathrm{eff}}\sigma_0|E_0|\right)^2 \approx \frac{\left\langle\Phi_{\mathrm{NL}}\right\rangle^2}{6\rho_s}, \tag{6.15}$$

where the mean nonlinear phase shift is

$$\langle\Phi_{\mathrm{NL}}\rangle = N_A\gamma L_{\mathrm{eff}}\left[|E_0|^2 + (N_A+1)\,\sigma_0^2\right] \approx N_A\gamma L_{\mathrm{eff}}|E_0|^2. \tag{6.16}$$

To derive Eqs. (6.10), (6.7), and (6.12), the following relationships are used:

$$\sum_{k=1}^{N} k^2 = \frac{1}{6}N(2N+1)(N+1), \tag{6.17}$$

$$\sum_{k=1}^{N}(N-k)^2 = \frac{1}{6}N(2N-1)(N-1), \tag{6.18}$$

$$\sum_{k=1}^{N} k^2(N-k) = \sum_{k=1}^{N} k(N-k)^2 = \frac{1}{12}N^2(N^2-1), \tag{6.19}$$

$$\sum_{k=1}^{N} k(N-k) = \frac{1}{6}N(N^2-1). \tag{6.20}$$

Comparing the approximated variances of Eqs. (6.14) and (6.15), using linear compensation, the variance of nonlinear phase noise is reduced by a factor of 4.

At high SNR, the variance of the phase of amplifier noise is $\sigma^2_{\Theta_n} \approx 1/2\rho_s$ from Eq. (4.A.15). Gordon and Mollenauer (1990) gave an insight that the optimal operating point can be found by choosing $\sigma^2_{\Theta_n} = \sigma^2_{\Phi_{\mathrm{NL}}}$ or $\langle\Phi_{\mathrm{NL}}\rangle = \sqrt{0.75} = 0.866$ rad for both PSK and DPSK signals. Gordon and Mollenauer (1990) actually gave an estimation of $\langle\Phi_{\mathrm{NL}}\rangle = 1$ rad as a simple approximated result. Proportional to the mean nonlinear phase shift of $\langle \Phi_{\mathrm{NL}} \rangle$, the standard deviation (STD) of the residual nonlinear phase noise is reduced by a factor of 2 using linear compensator and enable the maximum $\langle \Phi_{\mathrm{NL}} \rangle$ increases by a factor of 2. Because the mean nonlinear phase shift of Eq. (6.16) is proportional to the number of fiber spans, as shown in Eq. (6.16), doubling the mean nonlinear phase shift doubles the number of fiber spans, and thus doubles the transmission distance, assuming that nonlinear phase noise is the primary limitation.

Linear compensator for nonlinear phase noise was first independently proposed by Liu et al. (2002b) using special nonlinear optical devices, Xu and Liu (2002) using a phase modulator, and Ho and Kahn (2004a) using electronic circuits. To use a scale version of the received intensity to compensate for nonlinear phase noise may be the simplest implementation.

While the foregoing discussion has focused on binary PSK, DPSK signals have generated much more interest recently from Table 1.2. In a DPSK system, information is encoded in phase differences between successive symbols, and is decoded using equivalently the differential phase of $\Delta\Phi_r = \Phi_r(t) - \Phi_r(t-T)$, where T is the symbol interval. When the differential phase is corrupted by the nonlinear phase noise difference of $\Phi_{\mathrm{NL}}(t+T) - \Phi_{\mathrm{NL}}(t)$, the impact of nonlinear phase noise can be compensated by the linear combination of $\Delta\Phi_r + \alpha \Delta P_N$, where $\Delta P_N = P_N(t) - P_N(t-T)$ is the power difference between consecutive symbols. The optimal scale factor for DPSK systems is precisely analogous to that for PSK systems, and optimal compensation also approximately doubles the transmission distance.

Using electronic circuits of Fig. 6.1, there are two implementations of the linear compensator (or any compensators). Using the received electric field shown in Figs. 5.1, a Yin-Yang detector can be implemented based on the spiral-boundary decision device. Alternatively, when both received phase Φ_r and intensity P_N are given, a simple linear combiner of $\Phi_r + \alpha P_N$ can be used as a compensator.

A practical quadrature receiver of Fig. 3.4 may yield the in-phase and quadrature components of $\cos\Phi_r$ and $\sin\Phi_r$. Instead of the corrected phase of $\Phi_r + \alpha P_N$, the corrected quadrature components can be calculated using electronic signal processing techniques, as $\cos(\Phi_r + \alpha P_N) = \sin\Phi_r \sin(\alpha P_N) - \cos\Phi_r \cos(\alpha P_N)$ and $\sin(\Phi_r + \alpha P_N) = \sin\Phi_r \cos(\alpha P_N) + \cos\Phi_r \sin(\alpha P_N)$. Compensator for DPSK signal can also be implemented similarly.

To compensate for nonlinear phase noise, the electro-optic implementation of Fig. 6.3 and all-electronic implementation of Figs. 6.1 can use electronic signal processing to provide arbitrary linear and nonlinear functions. To correct the nonlinear phase noise, a fundamental prob-

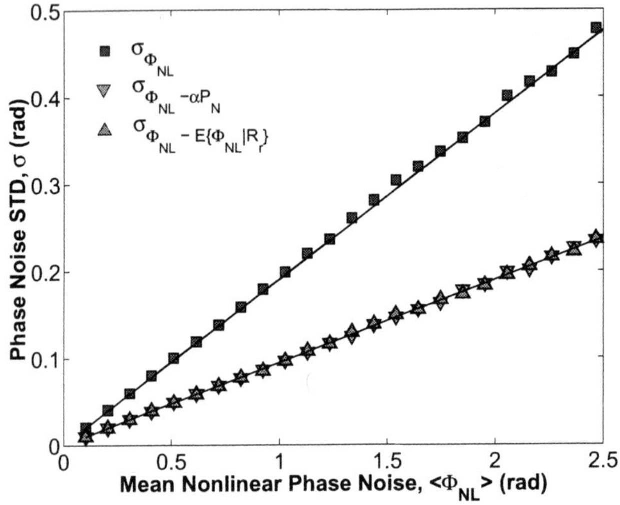

Figure 6.4. The STD of nonlinear phase noise with and without compensation. The solid lines are the approximation of Eq. (6.21). [Adapted from Ho (2003d)]

lem is to find the optimal MMSE compensator without the constraint of linearity. It is also important to verify whether the linear MMSE compensator is close to the optimal MMSE compensator. Appendix 6.A derives the optimal nonlinear compensator to minimize the variance of nonlinear phase noise after the compensator.

Figure 6.4 shows the simulated phase noise STD as a function of mean nonlinear phase shift with and without compensation. For all 5000 points in Figs. 5.1, the STD of the nonlinear phase noise Φ_{NL} of Eq. (5.5), denotes as $\sigma_{\Phi_{\mathrm{NL}}}$, is calculated. When the nonlinear phase noise is compensated using a linear compensator of $\alpha_{\mathrm{mse}} P_N$ proportional to the received intensity of P_N, the STD of the residual nonlinear phase noise of $\Phi_{\mathrm{NL}} - \alpha_{\mathrm{mse}} P_N$ is calculated and denotes as $\sigma_{\Phi_{\mathrm{NL}} - \alpha P_N}$, where α_{mse} is the scale factor of Eq. (6.10). When the nonlinear phase noise is compensated using the nonlinear MMSE compensator $E\{\Phi_{\mathrm{NL}}|R_r\}$ of Eq. (6.A.15) of Appendix 6.A, the STD of the residual nonlinear phase noise of $\Phi_{\mathrm{NL}} - E\{\Phi_{\mathrm{NL}}|R_r\}$ is calculated and denotes as $\sigma_{\Phi_{\mathrm{NL}} - E\{\Phi_{\mathrm{NL}}|R_r\}}$. Figure 6.4 also plots the approximation of

$$\sigma_{\Phi_{\mathrm{NL}}} \approx 0.1925 \langle \Phi_{\mathrm{NL}} \rangle \quad \text{and} \quad \sigma_{\Phi_{\mathrm{NL}} - \alpha R_r^2} \approx 0.0962 \langle \Phi_{\mathrm{NL}} \rangle . \tag{6.21}$$

given by Eqs. (6.14) and (6.15), respectively, as solid lines for $\rho_s = 18$ (12.6 dB).

Figure 6.4 shows that there is almost no difference between the STD of the residual nonlinear phase noise of $\sigma_{\Phi_{\text{NL}}-\alpha P_N}$ using the scale factor given by Eq. (6.10) and the STD of the residual nonlinear phase noise of $\sigma_{\Phi_{\text{NL}}-E\{\Phi_{\text{NL}}|R_r\}}$ using the nonlinear MMSE compensator of Eq. (6.A.15). In general, $\sigma_{\Phi_{\text{NL}}-E\{\Phi_{\text{NL}}|R_r\}}$ is smaller than $\sigma_{\Phi_{\text{NL}}-\alpha R_r^2}$ by about only 0.2%. Figure 6.4 also shows that the approximation of Eq. (6.21) is also very accurate. The linear compensator, if optimized, is very close to the optimal nonlinear MMSE compensator.

As shown in Fig. 6.4, the variances of the residual nonlinear phase noise of the linear and optimal MMSE compensators have little difference. When linear and optimal MMSE compensators are used, the p.d.f. of the residual nonlinear phase noise may not be the same, especially the tail probabilities. With the approximation of Eq. (6.A.11) from the Appendix 6.A, the p.d.f. of the residual nonlinear phase noise is Gaussian distributed. In practice, the residual nonlinear phase noise has a larger tail probability than that of the Gaussian distribution. While it is possible to derive the p.d.f. of the residual nonlinear phase noise for the linear compensator, that for nonlinear compensator is difficult to derive.

The usage of linear compensation to reduce nonlinear phase noise was proposed by Liu et al. (2002b), Xu and Liu (2002), and Ho and Kahn (2004a). The reduction of the variance of nonlinear phase noise by a factor of 4 was first given by the observation of Liu et al. (2002b) and Xu and Liu (2002) by simulation and Ho and Kahn (2004a) by analysis.

2.2 Probability Density of Residual Nonlinear Phase Noise

The probability density of residual nonlinear phase noise can also be derived using method similar to that in Sec. 5.1.2. Using a notation similar to Sec. 5.1.2 by ignoring the constant fact of γL_{eff}, with the linear correction of the phase noise using received intensity, the residual nonlinear phase noise is

$$\Phi_{\text{RES}} = |E_0 + n_1|^2 + |E_0 + n_1 + n_2|^2 + \cdots + |E_0 + n_1 + \cdots + n_{N_A-1}|^2 - (\alpha_{\text{mse}} - 1)|E_0 + n_1 + \cdots + n_{N_A}|^2, \tag{6.22}$$

As from Eq. (6.10), $\alpha_{\text{mse}} \approx (N_A + 1)/2$ is the optimal scale factor to correct the nonlinear phase noise of Eq. (5.5) using the received intensity of P_N. The random variable corresponding to φ_1 of Eq. (5.13) becomes

$$(N_A - \alpha_{\text{mse}})\, A^2 + 2A\vec{w}_r^{\,T}\vec{x} + \vec{x}^T C_r \vec{x}, \tag{6.23}$$

where $\vec{w_r} = \vec{w} - \alpha_{\text{mse}}\vec{w}_I$ with $\vec{w}_I = (1, 1, \ldots, 1)$ and

$$\mathcal{C}_r = (\mathcal{M} - \mathcal{L})^T(\mathcal{M} - \mathcal{L}) - (\alpha_{\text{mse}} - 1)\mathcal{L}^T\mathcal{L}, \tag{6.24}$$

where

$$\mathcal{L} = \begin{bmatrix} 0 & 0 & \cdots & 0 & 0 \\ 0 & 0 & \cdots & 0 & 0 \\ \vdots & \vdots & \ddots & \vdots & \vdots \\ 0 & 0 & \cdots & 0 & 0 \\ 1 & 1 & \cdots & 1 & 1 \end{bmatrix}. \tag{6.25}$$

Following the procedure from Eqs. (5.13) to (5.22), the characteristic function of Φ_{RES} is

$$\Psi_{\Phi_{\text{RES}}}(\nu) = \frac{\exp\left[-j\nu(N_A - \alpha_{\text{mse}})A^2 - 2\sigma_0^2\nu^2A^2\vec{w_r}^T(\mathcal{I} - 2j\nu\sigma_0^2\mathcal{C}_r)^{-1}\vec{w_r}\right]}{\det\left[\mathcal{I} - 2j\nu\sigma_0^2\mathcal{C}_r\right]}. \tag{6.26}$$

The characteristic functions of Φ_{RES} in the form of eigenvalues and eigenvectors are similar to that of Eqs. (5.23) and (5.24). The characteristic functions of Φ_{RES} has the same expression as Eq. (5.24) using a new set of eigenvalues and eigenvectors of the covariance matrix $\mathcal{C}_r$ and the vector of $\vec{w_r}$. The characteristic function of Eq. (6.26) is a special case of Eq. (5.B.20).

Except the first and last rows, the matrix $\mathcal{C}_r^{-1}$ is also approximately a Toeplitz matrix for the series of $2, -1, 0, \ldots$ For large number of fiber spans of N_A, the eigenvalues of $\mathcal{C}_r$ are asymptotically equal to

$$\frac{1}{\lambda_k} \approx 4\sin^2\left[\frac{(k-1.25)\pi}{2(N_A - 1)}\right], \qquad k = 2, \ldots, N_A, \qquad \lambda_1 \approx -\sum_{k=2}^{N_A}\lambda_k. \tag{6.27}$$

Other than the largest one in absolute value, the eigenvalues of $\mathcal{C}_r$ are all positive. All eigenvalues of the covariance matrix $\mathcal{C}_r$ sum to approximately zero and multiple to $\alpha_{\text{mse}} - 1 \approx (N_A - 1)/2$.

The p.d.f. of the residual nonlinear phase noise of Φ_{RES} of Eq. (6.22) can be calculated by taking the inverse Fourier transform of the corresponding characteristic functions of $\Psi_{\Phi_{\text{RES}}}(\nu)$ [Eq. (6.26)]. Figure 6.5 shows the p.d.f. of Φ_{RES}. The same as Fig. 5.2, Fig. 6.5 is plotted for the case that the SNR of $\rho_s = A^2/(2N_A\sigma_0^2) = 18$ (12.6 dB), corresponding to an error probability of 10^{-9} for PSK signals if the amplifier noise is the only impairment as shown in Eq. (3.78) and Fig. 3.13. The number of spans is $N_A = 32$. The x-axis is normalized with respect to $N_A A^2$, approximately equal to the mean nonlinear phase shift.

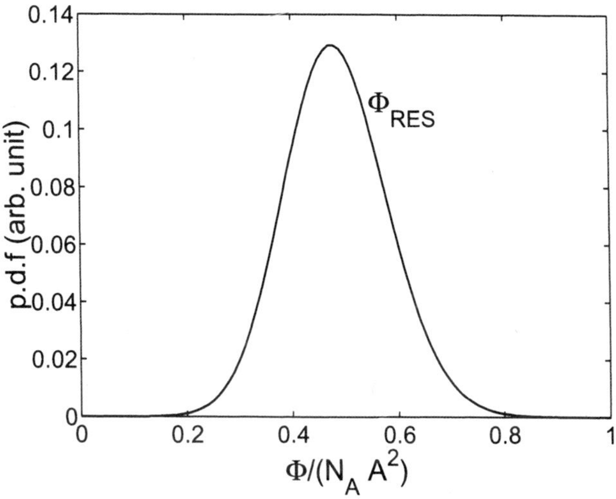

Figure 6.5. The p.d.f. of residual nonlinear phase noise of Φ_{RES}.

Similar to Φ_{NL}, the random variables of Φ_{RES} can be modeled as the combination of $N_A = 32$ independently distributed noncentral χ^2-random variables with two degrees of freedom. Some studies implicitly assume a Gaussian distribution by using the Q-factor to characterize the random variables. While the Gaussian assumption for Φ_{RES} may not be valid, other than the two noncentral χ^2-random variables corresponds to the two largest eigenvalues, the other random variables should sum to Gaussian distribution. By modeling the summation of random variables with smaller eigenvalues as Gaussian distribution, the nonlinear phase noise of Eq. (6.26) can be modeled as a summation of three instead of $N_A = 32$ independently distributed random variables.

From Fig. 6.5, the p.d.f. of Φ_{RES} has significant difference with that of a Gaussian distribution. Figure 6.6 divides the p.d.f. of Φ_{RES} into the convolution of three parts. The first part has no observable difference with a Gaussian p.d.f. and corresponds to the third largest to the smallest eigenvalues, $\lambda_k, k = 3, \ldots, N_A$, of the characteristic function in the format of Eq. (5.24). The second and third parts are noncentral χ^2-p.d.f. with two degrees of freedom and corresponds to the largest and second largest eigenvalues λ_1 and λ_2, respectively. The p.d.f. of Φ_{RES} in Fig. 6.5 is also plotted in Fig. 6.6 for comparison.

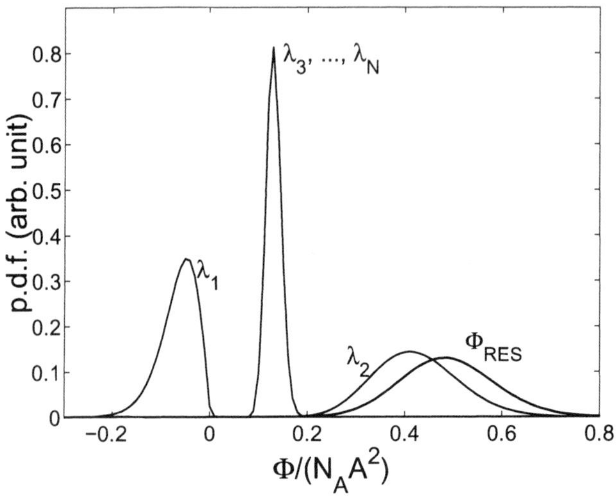

Figure 6.6. The p.d.f. of $\Phi_{\rm RES}$ is the convolution of a Gaussian p.d.f. and the difference of two noncentral χ^2-p.d.f. with two degrees of freedom. [Adapted from Ho (2003f)]

3. Linear Compensator for Infinite Number of Fiber Spans

With linear compensator, the decision variable is a linear combination of nonlinear phase noise $\Phi_{\rm NL}$ [either Eqs. (5.5) or (5.32)], the phase of amplifier noise Θ_n, and the received intensity Y. This section first finds the received phase with linear compensation. Like Sec. 5.3, the p.d.f. of decision variable is then expanded using as a Fourier series and the error probability can be found analytically as a series summation. This section is based on the distributed model of Sec. 5.2 but the results and methods are also applicable to a system with finite number of fiber spans using the model of Appendix 5.B.

3.1 Minimum Mean-Square Error Compensation

The simplest method to compensate the nonlinear phase noise is to add a scaled received intensity into the received phase. The optimal linear MMSE compensator minimizes the variance of the normalized residual phase noise of $\Phi_\alpha = \Phi - \alpha R^2 = \Phi - \alpha Y$. Using the joint characteristic function of Eq. (6.B.7) from Appendix 6.B, the characteristic function for the normalized residual nonlinear phase noise is

$$\Psi_{\Phi_\alpha}(\nu) = \Psi_{\Phi,Y}(\nu, -\alpha\nu). \tag{6.28}$$

The mean of the normalized residual nonlinear phase noise is

$$\begin{aligned}\langle \Phi_\alpha \rangle &= -j \left. \frac{d}{d\nu} \Psi_{\Phi_\alpha}(\nu) \right|_{\nu=0} \\ &= \rho_s + \frac{1}{2} - \alpha\,(\rho_s + 1)\,. \end{aligned} \tag{6.29}$$

The variance of the normalized residual nonlinear phase noise is

$$\begin{aligned}\sigma^2_{\Phi_\alpha} &= - \left. \frac{d^2}{d\nu^2} \Psi_{\Phi_\alpha}(\nu) \right|_{\nu=0} - \langle \Phi_\alpha \rangle^2 \\ &= \frac{2}{3}\rho_s + \frac{1}{6} - 2\left(\rho_s + \frac{1}{3}\right)\alpha + (2\rho_s + 1)\,\alpha^2. \end{aligned} \tag{6.30}$$

Solving $d\sigma^2_{\Phi_\alpha}/d\alpha = 0$, the optimal scale factor for linear compensator is

$$\alpha_{\text{mse}} = \frac{1}{2}\frac{\rho_s + \frac{1}{3}}{\rho_s + \frac{1}{2}}. \tag{6.31}$$

In high SNR, $\alpha_{\text{mse}} \to 1/2$. Other than the normalization, the optimal scale factor of Eq. (6.31) is the same as that from Sec. 6.2. Here, the optimal scale factor is derived by the joint characteristic function between nonlinear phase noise and the received intensity. Similar to Sec. 6.2, the results here can also be derived using the covariance properties of the received intensity and nonlinear phase noise.

With the optimal scale factor of Eq. (6.31), the mean and variance of the normalized residual nonlinear phase noise are

$$\langle \Phi_{\alpha_{\text{mse}}} \rangle = \frac{1}{2}\frac{\rho_s^2 + \frac{2}{3}\rho_s + \frac{1}{6}}{\rho_s + \frac{1}{2}}, \tag{6.32}$$

$$\sigma^2_{\Phi_{\alpha_{\text{mse}}}} = \frac{1}{6}\frac{\rho_s^2 + \rho_s + \frac{1}{6}}{\rho_s + \frac{1}{2}}. \tag{6.33}$$

The mean of the residual nonlinear phase noise is about half the mean of the nonlinear phase noise of $\langle \Phi \rangle$ from Eq. (5.49). The variance of the residual nonlinear phase noise is about a quarter of that of the variance of the nonlinear phase noise of σ^2_Φ from Eq. (5.52).

After the linear compensation, the characteristic function of the residual normalized nonlinear phase noise is

$$\Psi_{\Phi_{\alpha_{\text{mse}}}}(\nu) = \Psi_{\Phi,Y}\left(\nu, -\frac{\nu}{2}\frac{\rho_s + \frac{1}{3}}{\rho_s + \frac{1}{2}}\right). \tag{6.34}$$

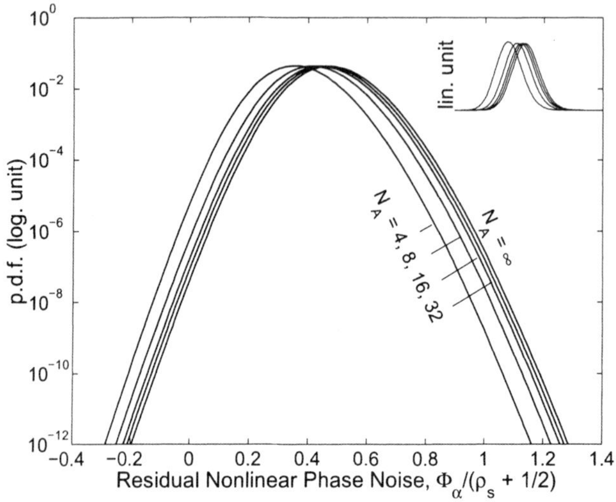

Figure 6.7. The asymptotic p.d.f. of the residual nonlinear phase noise Φ_α as compared with the p.d.f. of $N_A = 4, 8, 16$, and 32 fiber spans. The p.d.f. in linear scale is shown in the inset.

The p.d.f. of the residual normalized nonlinear phase noise is the inverse Fourier transform of $\Psi_{\Phi_{\alpha\text{mse}}}(\nu)$.

The p.d.f. of the residual nonlinear phase noise for finite number of fiber spans was derived by modeling the residual nonlinear phase noise Φ_α as the summation of N_A independently distributed random variables in Sec. 6.2.2. Figure 6.7 shows a comparison of the p.d.f. for $N_A = 4, 8, 16$, and 32 of fiber spans with the distributed case of Eq. (6.34). The residual nonlinear phase noise is scaled by the mean normalized phase shift of $\langle \Phi \rangle = \rho_s + \frac{1}{2}$ from Eq. (5.49). Using a SNR of $\rho_s = 18$, Figure 6.7 is plotted in logarithmic scale to show the difference in the tail. Figure 6.7 also provides an inset in linear scale of the same p.d.f. to show the difference around the mean. Like that of Fig. 5.5, the asymptotic p.d.f. for residual nonlinear phase noise of Fig. 6.7 is also very accurate for $N_A \geq 32$ fiber spans. Unlike that of Fig. 5.5, the asymptotic p.d.f. for residual nonlinear phase noise of Fig. 6.7 has slightly larger spreading then that of the finite cases. The mean of the residual nonlinear phase noise is about $0.5(\rho_s + \frac{1}{2})$, the same as that of Eq. (6.32). Comparing Figs. 5.5 and 6.7, with linear compensation, both the mean and STD of the residual nonlinear phase noise is about half of that of the nonlinear phase noise before compensation.

3.2 Distribution of the Linearly Compensated Received Phase

With nonlinear phase noise, received by a quadrature receiver of Fig. 3.4 with an optical PLL, before compensation, the overall received phase is that of Eq. (5.63). With linear compensation, the compensated received phase is

$$\Phi_{\text{cm}} = \Theta_n - \frac{\langle \Phi_{\text{NL}} \rangle}{\rho_s + \frac{1}{2}} (\Phi - \alpha Y). \tag{6.35}$$

The compensated received phase is confined to the range of $[-\pi, +\pi)$. The p.d.f. of the compensated received phase is a periodic function with a period of 2π. If the characteristic function of the compensated received phase is $\Psi_{\Phi_{\text{cm}}}(\nu)$, the p.d.f. of the compensated received phase has a Fourier series expansion of

$$p_{\Phi_{\text{cm}}}(\theta) = \frac{1}{2\pi} \sum_{m=-\infty}^{+\infty} \Psi_{\Phi_{\text{cm}}}(m) \exp(jm\theta), \tag{6.36}$$

or

$$p_{\Phi_{\text{cm}}}(\theta) = \frac{1}{2\pi} + \frac{1}{\pi} \sum_{m=1}^{+\infty} \Re\left\{\Psi_{\Phi_{\text{cm}}}(m) \exp(jm\theta)\right\}, \tag{6.37}$$

where $\Re\{\cdot\}$ denotes the real part of a complex number.

Using the joint characteristic function of $\Psi_{\Phi,Y,\Theta_n}(\nu, \omega, m)$ with Θ_n at integer "angular" frequency from Eq. (6.B.10) of Appendix 6.B, similar to that of Eq. (6.34), from the compensated received phase of Eq. (6.35), the Fourier series coefficients are

$$\Psi_{\Phi_{\text{cm}}}(m) = \Psi_{\Phi,Y,\Theta_n}\left(-\frac{m\langle \Phi_{\text{NL}} \rangle}{\rho_s + \frac{1}{2}}, \alpha \frac{m\langle \Phi_{\text{NL}} \rangle}{\rho_s + \frac{1}{2}}, m \right). \tag{6.38}$$

Using the scale factor of α_{mse} from Eq. (6.31), Figure 6.8 shows the p.d.f. of the compensated received phase of Eq. (6.37) with mean nonlinear phase shift of $\langle \Phi_{\text{NL}} \rangle = 0, 1, 2,$ and 3 rad. Shifted by the mean residual nonlinear phase shift of

$$\langle \Phi_{\text{RES}} \rangle = \frac{\langle \Phi_{\text{NL}} \rangle}{\rho_s + \frac{1}{2}} \langle \Phi_{\alpha_{\text{mse}}} \rangle, \tag{6.39}$$

the p.d.f. is plotted in logarithmic scale to show the difference in the tail. Not shifted by $\langle \Phi_{\text{RES}} \rangle$, the same p.d.f. is plotted in linear scale in the inset. Figure 6.8 is plotted for the case that the SNR is equal to $\rho_s = 18$ (12.6 dB), the same as that of Fig. 5.6 without compensation.

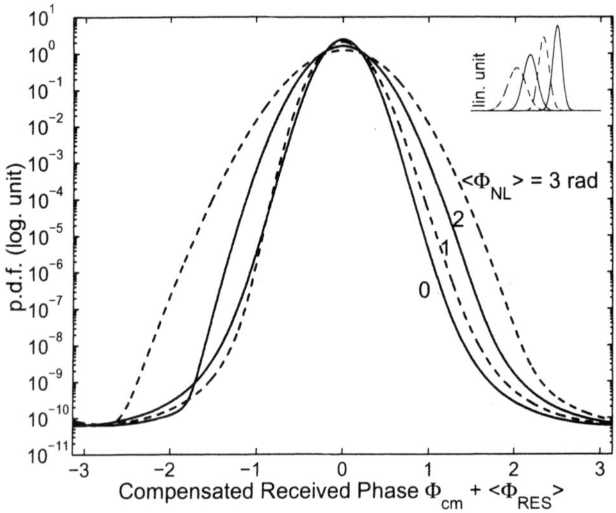

Figure 6.8. The p.d.f. of the compensated received phase $p_{\Phi_{\mathrm{cm}}}(\theta + \langle\Phi_{\mathrm{RES}}\rangle)$ in logarithmic scale. The inset is the p.d.f. of $p_{\Phi_{\mathrm{cm}}}(\theta)$ in linear scale.

From Fig. 6.8, when the p.d.f. is broadened by the nonlinear phase noise, similar to the case without compensation of Fig. 5.6, the broadening is not symmetrical with respect to the mean residual nonlinear phase shift. With small mean nonlinear phase shift of $\langle\Phi_{\mathrm{NL}}\rangle = 1$ rad, the received phase spreads further in the positive than the negative phase. With large mean nonlinear phase shift of $\langle\Phi_{\mathrm{NL}}\rangle = 3$ rad, the received phase spreads further in the negative than the positive phase. The difference in the spreading for small and large mean nonlinear phase shift is due to the dependence between the residual nonlinear phase noise and the phase of amplifier noise.

3.3 PSK Signals

For PSK signal with two antipodal phases, assume that the decision angles are $\pm\pi/2 - \theta_c$ with the center phase of θ_c, the error probability is

$$p_e = 1 - \int_{-\frac{1}{2}\pi-\theta_c}^{\frac{1}{2}\pi-\theta_c} p_{\Phi_{\mathrm{cm}}}(\theta)\mathrm{d}\theta, \tag{6.40}$$

or, similar to the error probability of Eq. (5.69),

$$p_e = \frac{1}{2} - \frac{2}{\pi}\sum_{k=0}^{+\infty}\frac{(-1)^k}{2k+1}\Re\left\{\Psi^*_{\Phi_{\mathrm{cm}}}(2k+1)e^{-j(2k+1)\theta_c}\right\}. \tag{6.41}$$

From both Eqs. (6.38) and (6.B.10), the coefficients for the error probability Eq. (6.41) are

$$\Psi^{*}_{\Phi_{\mathrm{cm}}}(2k+1) = \frac{\sqrt{\pi} r_{k,\omega}^{\frac{3}{2}}}{2r_k} e^{-r_k + \frac{r_{k,\omega}}{2}} \Psi_{\Phi}\left[\frac{(2k+1)\langle\Phi_{\mathrm{NL}}\rangle}{\rho_s + \frac{1}{2}}\right] \times \left[I_k\left(\frac{r_{k,\omega}}{2}\right) + I_{k+1}\left(\frac{r_{k,\omega}}{2}\right)\right], \quad k \geq 0, \tag{6.42}$$

where, from Eq. (5.A.12) of Appendix 5.A and Eq. (6.B.3) of Appendix 6.B,

$$r_k = \frac{2\left[\frac{j(2k+1)\langle\Phi_{\mathrm{NL}}\rangle}{\rho_s + \frac{1}{2}}\right]^{\frac{1}{2}}}{\sin\left\{2\left[\frac{j(2k+1)\langle\Phi_{\mathrm{NL}}\rangle}{\rho_s + \frac{1}{2}}\right]^{\frac{1}{2}}\right\}} \rho_s, \tag{6.43}$$

and

$$r_{k,\omega} = \frac{r_k}{1 + \alpha\left[\frac{j(2k+1)\langle\Phi_{\mathrm{NL}}\rangle}{\rho_s + \frac{1}{2}}\right]^{\frac{1}{2}} \tan\left\{\left[\frac{j(2k+1)\langle\Phi_{\mathrm{NL}}\rangle}{\rho_s + \frac{1}{2}}\right]^{\frac{1}{2}}\right\}}, \tag{6.44}$$

and $\Psi_{\Phi}(\nu)$ is the marginal characteristic function of nonlinear phase noise of Eq. (5.48) that depends solely on SNR. Note that the angular SNR of Eq. (6.43) is the same as Eq. (5.71).

Based on the MMSE criterion, the center phase in Eq. (6.41) is $\theta_c = \langle\Phi_{\mathrm{RES}}\rangle$ of Eq. (6.39) and the scale factor in Eq. (6.44) is $\alpha = \alpha_{\mathrm{mse}}$ of Eq. (6.31). Because the series summation of Eq. (6.41) is a simple expression, numerical optimization can be used to find the linear MAP compensator to minimize the error probability.

If the residual nonlinear phase noise is assumed to be independent to the phase of amplifier noise, similar to Eq. (5.72), the error probability based on MMSE criterion can be approximated as

$$p_e \approx \frac{1}{2} - e^{-\frac{\rho_s}{2}} \sqrt{\frac{\rho_s}{\pi}} \sum_{k=0}^{\infty} \frac{(-1)^k}{2k+1} \left[I_k\left(\frac{\rho_s}{2}\right) + I_{k+1}\left(\frac{\rho_s}{2}\right)\right] \times \Re\left\{\Psi_{\Phi_{\alpha_{\mathrm{mse}}}}\left[\frac{(2k+1)\langle\Phi_{\mathrm{NL}}\rangle}{\rho_s + \frac{1}{2}}\right] e^{-j(2k+1)\langle\Phi_{\mathrm{RES}}\rangle}\right\}, \tag{6.45}$$

where the characteristic function $\Psi_{\Phi_{\alpha_{\mathrm{mse}}}}(\nu)$ is the characteristic function of Eq. (6.34) with $\alpha = \alpha_{\mathrm{mse}}$ of Eq. (6.31).

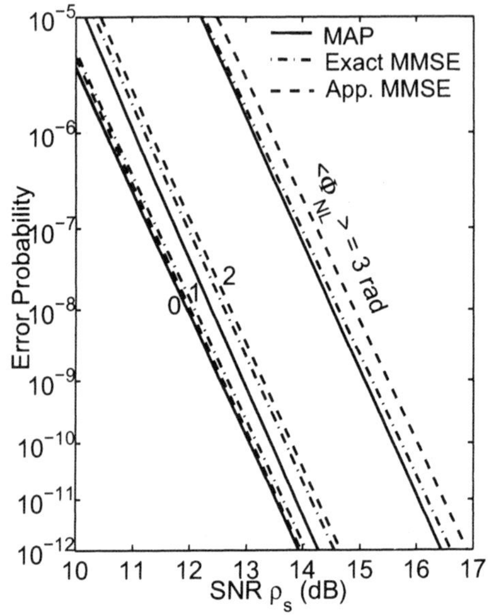

Figure 6.9. With linear compensation, the error probability of PSK signal as a function of SNR ρ_s.

Figure 6.9 shows the exact [Eq. (6.41)] and approximated [Eq. (6.45)] error probabilities as a function of SNR ρ_s when nonlinear phase noise is compensated using the linear MMSE compensator. The error probability of Eq. (6.41) is also minimized based on the MAP criterion and shown in Fig. 6.9. Figure 6.9 also plots the error probability without nonlinear phase noise of Fig. 3.13.

Figure 6.10 shows the SNR penalty of PSK signal for an error probability of 10^{-9} calculated by the exact Eq. (6.41) and approximated Eq. (6.45) error probability formulae with the linear MMSE compensator. The SNR penalty with the linear MAP compensator is also shown in Fig. 6.10. The corresponding optimal center phase and scale factor of Fig. 6.10 are shown in Fig. 6.11.

Figure 6.10 also shows the SNR penalty of PSK signal without compensation from Fig. 5.8 using the exact error probability with optimal center phase there. For the same SNR penalty, the mean nonlinear phase shift with compensation is slightly larger than twice of that without compensation from Fig. 5.8.

The discrepancy between the exact and approximated error probability is smaller for small and large mean nonlinear phase shift of $\langle\Phi_{\mathrm{NL}}\rangle$.

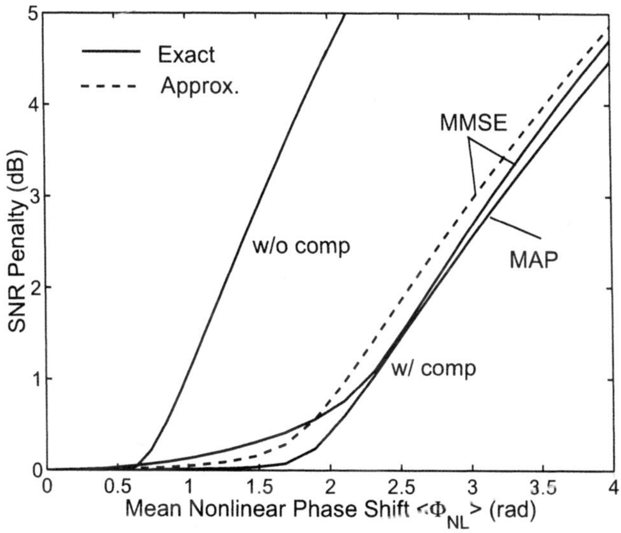

Figure 6.10. With linear compensation, the SNR penalty of PSK signal as a function of mean nonlinear phase shift $\langle\Phi_{\text{NL}}\rangle$.

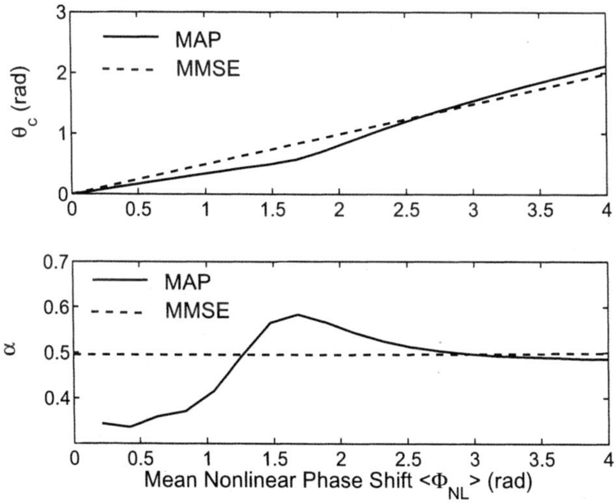

Figure 6.11. The optimal center phase corresponding to the operating point of Fig. 6.10 as a function of mean nonlinear phase shift $\langle\Phi_{\text{NL}}\rangle$.

With the MMSE criterion, the largest discrepancy between the exact and approximated SNR penalty is about 0.37 dB at a mean nonlinear phase shift around $\langle \Phi_{\mathrm{NL}} \rangle = 2.53$ rad. Like the conclusion of Sec. 5.3 without compensation, with the MMSE criterion, the approximated error probability Eq. (6.45) is not accurate enough for linearly compensated PSK signals.

Figure 6.10 also shows that the linear MMSE compensator using the optimal scale factor of Eq. (6.31) does not perform well as compared with the linear MAP compensator. The largest discrepancy is about 0.34 dB at mean nonlinear phase shift of $\langle \Phi_{\mathrm{NL}} \rangle = 1.68$ rad. The major reason of this large discrepancy is due to the non-symmetrical p.d.f. of Fig. 6.8.

Using the linear MAP compensator, the mean nonlinear phase shift $\langle \Phi_{\mathrm{NL}} \rangle$ must be less than 2.30 rad for a SNR penalty less than 1 dB, slightly more than twice that of Fig. 5.8 of 1 rad without compensation. The optimal operating level is for a mean nonlinear phase shift of $\langle \Phi_{\mathrm{NL}} \rangle = 2.15$ rad, slightly less than twice that of Fig. 5.8 of 1.25 rad without compensation.

From the optimal center phase from Fig. 6.11 for the linear MAP compensator, the optimal center phase is less than the mean nonlinear phase shift when the mean nonlinear phase shift is less than about 2.68 rad. The changing of the optimal center phase with the mean nonlinear phase shift is consistent with Figs. 5.9 and 6.8. Similar to Fig. 5.9, when the residual nonlinear phase noise is assumed to be independent of the phase of amplifier noise using Eq. (6.45), the optimal center phase is always larger than the mean residual nonlinear phase shift. The approximated error probability of Eq. (6.45) is not useful to find the optimal center phase.

The optimal scale factor from Fig. 6.11 is also not equal to the optimal MMSE scale factor from Eq. (6.31) except when the mean nonlinear phase shift is very large. From Figs. 6.10 and 6.11, the MMSE criterion is not close to the performance of the optimal linearly compensated PSK signals based on MAP criterion.

3.4 DPSK Signals

A compensated DPSK signal is demodulated using interferometer to find the compensated differential phase of

$$\begin{aligned} \Delta\Phi_{\mathrm{cm}} &= \Phi_{\mathrm{cm}}(t) - \Phi_{\mathrm{cm}}(t-T) \\ &= \Theta_n(t) - \Phi_{\mathrm{RES}}(t) - \Theta_n(t-T) + \Phi_{\mathrm{RES}}(t-T), \end{aligned} \tag{6.46}$$

where $\Phi_{\mathrm{cm}}(\cdot)$, $\Theta_n(\cdot)$, and $\Phi_{\mathrm{RES}}(\cdot)$ are the compensated received phase, the phase of amplifier noise, and the residual nonlinear phase noise as a

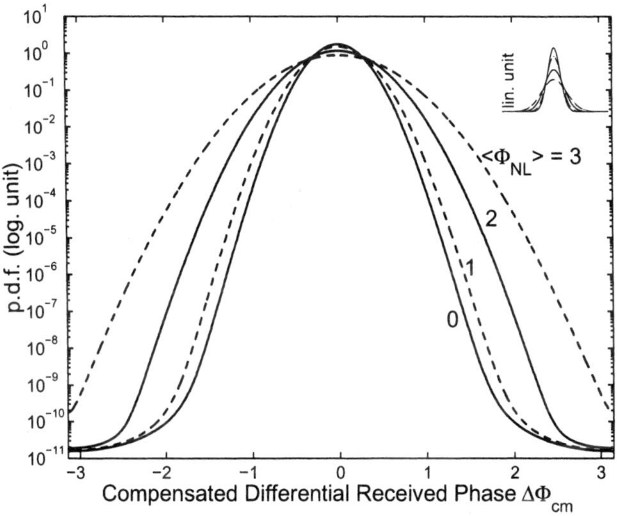

Figure 6.12. The p.d.f. of the differential compensated received phase $p_{\Delta\Phi_{\text{cm}}}(\theta)$ in logarithmic scale. The inset is the same p.d.f. in linear scale.

function of time, and T is the symbol interval. The phases at t and $t-T$ are independent of each other but are identically distributed random variables similar to that of Eqs. (5.63) and (6.35). The differential phase of Eq. (6.46) assumes that the transmitted phases at t and $t-T$ are the same.

From the p.d.f. of Eq. (6.37), similar to Eq. (5.75), the p.d.f. of the differential compensated phase Eq. (6.46) is

$$p_{\Delta\Phi_{\text{cm}}}(\theta) = \frac{1}{2\pi} + \frac{1}{\pi}\sum_{m=1}^{+\infty} |\Psi_{\Phi_{\text{cm}}}(m)|^2 \cos(m\theta), \tag{6.47}$$

that is symmetrical with respect to the zero phase.

Figure 6.12 shows the p.d.f. of the differential received phase with linear compensation of Eq. (6.47) for the mean nonlinear phase shift of $\langle\Phi_{\text{NL}}\rangle = 0, 1, 2,$ and 3 rad using the linear MMSE compensator with $\alpha = \alpha_{\text{mse}}$ of Eq. (6.31). The p.d.f. is plotted in logarithmic scale to show the difference in the tail. The same p.d.f. is plotted in linear scale in the inset. Figure 6.12 is plotted for the case that the SNR is equal to $\rho_s = 20$ (13 dB), corresponding to an error probability of 10^{-9} with only amplifier noise from Fig. 3.13. From Fig. 6.12, when the p.d.f. of differential phase is broadened by the residual nonlinear phase noise, the broadening is symmetrical with respect to the zero phase.

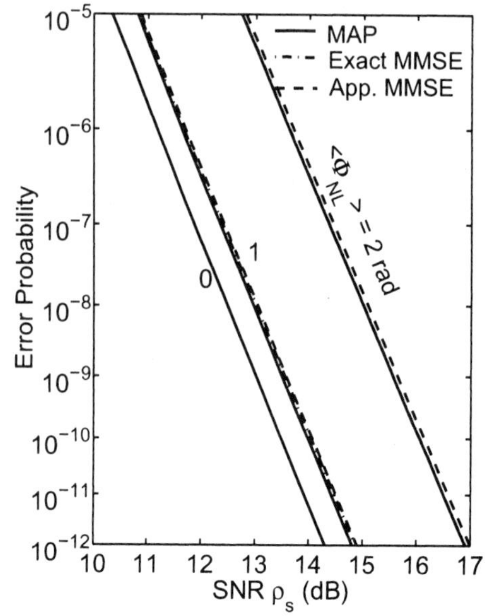

Figure 6.13. With linear compensation, the error probability of DPSK signal as a function of SNR ρ_s.

Similar to Eqs. (5.77) and (6.41), the error probability for DPSK signal is

$$p_e = \frac{1}{2} - \frac{2}{\pi} \sum_{k=0}^{+\infty} \frac{(-1)^k}{2k+1} \left| \Psi_{\Phi_{\rm cm}}(2k+1) \right|^2, \tag{6.48}$$

where the coefficients of $\Psi_{\Phi_{\rm cm}}(2k+1)$ are given by Eq. (6.42) with parameters from Eqs. (6.43) and (6.44).

Similar to the approximation for PSK signal Eq. (6.45), if the nonlinear phase noise is assumed to be independent to the phase of amplifier noise, the error probability of Eq. (6.48) can be approximated as

$$p_e \approx \frac{1}{2} - \frac{\rho_s e^{-\rho_s}}{2} \sum_{k=0}^{\infty} \frac{(-1)^k}{2k+1} \left[I_k\!\left(\frac{\rho_s}{2}\right) + I_{k+1}\!\left(\frac{\rho_s}{2}\right) \right]^2 \times \left| \Psi_{\Phi_{\alpha_{\rm mse}}} \left[\frac{(2k+1)\langle \Phi_{\rm NL} \rangle}{\rho_s + \frac{1}{2}} \right] \right|^2. \tag{6.49}$$

Figure 6.13 shows the exact [Eq. (6.48)] and approximated [Eq. (6.49)] error probabilities as a function of SNR ρ_s for DPSK signal with linear MMSE compensator. The scale factor in Eq. (6.44) is also numerically

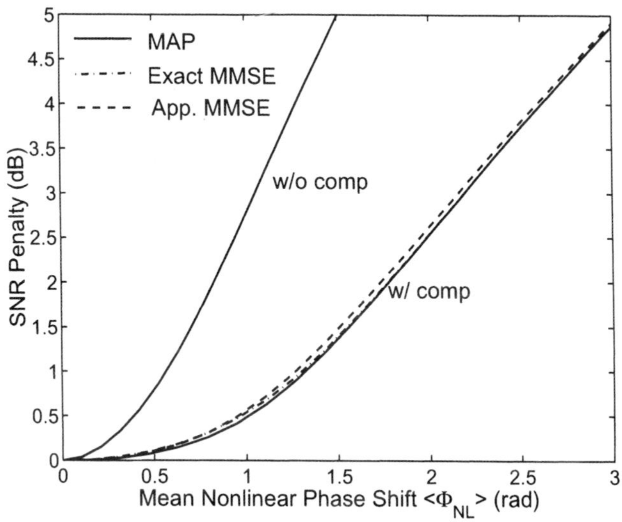

Figure 6.14. With linear compensation, the SNR penalty of DPSK signal as a function of mean nonlinear phase shift $\langle\Phi_{\mathrm{NL}}\rangle$.

optimized to find the linear MAP compensator to minimize the exact error probability of Eq. (6.48). From Fig. 6.13, the linear MMSE and MAP compensators do not have big difference for linearly compensated DPSK signal. Figure 6.13 also plots the error probability without nonlinear phase noise. From Fig. 6.13, the approximated error probability Eq. (6.49) always slightly overestimates the error probability.

Figure 6.14 shows the SNR penalty of DPSK signal for an error probability of 10^{-9} calculated by the exact and approximated error probability formulae for DPSK signal with linear MMSE and MAP compensation. The discrepancy between the exact and approximated error probability with either MMSE or MAP criteria is very small. The largest discrepancy between the exact and approximated SNR penalty with MMSE criterion is about 0.1 dB at a mean nonlinear phase shift of $\langle\Phi_{\mathrm{NL}}\rangle = 1.74$ rad. Both with the exact error probability Eq. (6.48), the linear MMSE and MAP compensators have the largest discrepancy of 0.06 dB at a mean nonlinear phase shift $\langle\Phi_{\mathrm{NL}}\rangle$ of 0.95 rad.

Figure 6.14 also shows the SNR penalty of DPSK signal without compensation from Eq. (5.13) using the exact error probability formula. For a power penalty less than 1 dB, the mean nonlinear phase shift of $\langle\Phi_{\mathrm{NL}}\rangle$ must be less than 1.31 rad, slightly larger than twice that of Fig. 5.13 of 0.57 rad without compensation. The optimal operating level of the mean nonlinear phase shift is about 1.81 rad such that the increase of

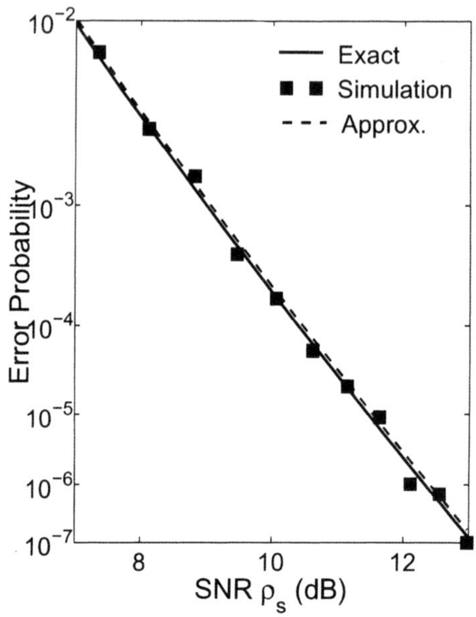

Figure 6.15. Calculated and simulated error probability for a DPSK system with linearly compensated nonlinear phase noise for a mean nonlinear phase shift of $\langle\Phi_{\rm NL}\rangle = \sqrt{2}$ rad.

power penalty is always less than the increase of mean nonlinear phase shift, slightly smaller than twice that of Fig. 5.13 of 1 rad without compensation.

To verify the accuracy of the error probability in Fig. 6.13, Figure 6.15 compares the theoretical and simulated error probability as a function of SNR for a typical linearly compensated DPSK system having mean nonlinear phase shift of $\langle\Phi_{\rm NL}\rangle = \sqrt{2}$ rad. The Monte-Carlo simulation is similar to that of Figs. 5.10 and 5.14 for $N_A = 32$ fiber spans using the optimal scale factor $\alpha_{\rm mse}$ of Eq. (6.31). Including both exact [Eq. (6.48)] and approximated [Eq. (6.49)] error probability, the theoretical results are the same as that in Fig. 6.13 but extend to high error probability. Figure 6.15 shows that the approximated, exact and simulated results have insignificant difference. From Fig. 6.15, we may conclude that the exact error probability of Eq. (6.48) is very accurate to evaluate the error probability of DPSK signal with linearly compensated nonlinear phase noise.

The SNR penalty given by the approximated error probability of Eq. (6.49) is the same as that in Ho (2003b) but for large number of fiber spans. With the assumption of independent additive phase noise, the approximated error probability of Eq. (6.49) is the same as that of Eqs. (5.78), (5.82), and (4.40) with nonlinear and linear phase noise. Similar to Ho (2003b), for the same SNR penalty, DPSK signal with linear compensator can tolerate slightly larger than twice the mean nonlinear phase noise without linear compensation.

In this section, the exact error probability is derived analytically for PSK and DPSK signals with linearly compensated nonlinear phase noise. The p.d.f. of the compensated received phase is first expressed as a Fourier series. The Fourier coefficients are given by the joint characteristic function of nonlinear phase noise, received intensity, and the phase of amplifier noise.

For PSK signal, the linear MMSE compensator is up to 0.34 dB worse than the linear MAP compensator that minimizes the error probability after compensation. For PSK signal with linear MAP compensator, the mean nonlinear phase shift must be less than 2.30 rad for a SNR penalty less than 1 dB. The optimal mean nonlinear phase shift is about 2.15 rad such that the increase of required SNR is smaller than the increase of launched optical power.

For DPSK signal with linear MAP compensator, the phase of amplifier noise can be assumed to be independent of the residual nonlinear phase noise. The mean nonlinear phase shift must be less than 1.31 rad for a SNR penalty less than 1 dB. The optimal mean nonlinear phase shift is about 1.81 rad.

The formulae to evaluate the error probability of DPSK signal with linearly compensated nonlinear phase noise are verified by Monte-Carlo simulation. Theoretical and simulation results have no significant difference.

4. Mid-Span Linear MMSE Compensation

Unlike the electronic compensator of Fig. 6.1, the phase-modulator based compensator of Fig. 6.3 can be used in the middle of the fiber link. The system performance can be greatly improved using mid-span compensation even with only a single compensator. When more than one compensator are used, the performance of the system can be further improved. With a compensator preceding the receiver, the variance of the residual nonlinear phase noise can be reduced to about a quarter of the variance of nonlinear phase noise without compensation as show earlier. With a compensator located optimally in the middle of fiber span, the variance of the residual nonlinear phase noise can be reduced to about 1/9 of the variance of nonlinear phase noise without compensation.

The compensator is independent of signal format and applicable to NRZ and RZ PSK and DPSK signals.

The optimal locations for mid-span compensation are derived analytically here. For simplicity, the model of nonlinear phase noise is according to the distributed model with infinite number of fiber spans. The optimal linear compensator is derived using the covariance between optical intensity and nonlinear phase noise.

4.1 Single Compensator

If only one compensator is used, the compensator can be located anywhere in the fiber link to optimize the system performance. Using the compensation factor of α, based on the distributed model, the optimal compensator minimizes the variance of the residual nonlinear phase noise of

$$\Phi_\alpha = \Phi - \alpha|\sqrt{\rho_s} + b(s)|^2, \tag{6.50}$$

where s is the location of the compensator. In order to derive the optimal factor of α, we need to find the covariance of

$$\kappa(s,t) = E\Big\{ \left(|\sqrt{\rho_s} + b(s)|^2 - m_{|\sqrt{\rho_s}+b(s)|^2}\right) \times \left(|\sqrt{\rho_s} + b(t)|^2 - m_{|\sqrt{\rho_s}+b(t)|^2}\right) \Big\} \tag{6.51}$$

for Wiener process of $b(s)$, where $m_{|\sqrt{\rho_s}+b(s)|^2}$ and $m_{|\sqrt{\rho_s}+b(t)|^2}$ are the mean values of $\left|\sqrt{\rho_s} + b(s)\right|^2$ and $\left|\sqrt{\rho_s} + b(t)\right|^2$, respectively. Because $|\sqrt{\rho_s} + b(t)|^2$ is noncentral χ^2 random variable with two degrees of freedom, the noncentrality parameter is $\sqrt{\rho_s}$ and the variance is $t/2$. We have $m_{|\sqrt{\rho_s}+b(s)|^2} = \rho_s + s$ and $m_{|A+b(t)|^2} = \rho_s + t$.

The covariance between $b(s)$ and $b(t)$ is $\min(s,t)$ from the properties of Wiener process. Similar to Eq. (6.5), the covariance of Eq. (6.51) is

$$\kappa(s,t) = 2\rho_s \min(s,t) + \min(s,t)^2. \tag{6.52}$$

For the normalized nonlinear phase noise Φ defined in Eq. (5.32), the covariance between Φ and the intensity of $|\sqrt{\rho_s} + b(s)|^2$ is

$$\begin{aligned}
\xi(s) &= E\left\{ \left(|\sqrt{\rho_s} + b(s)|^2 - \rho_s - s\right) \left(\int_0^1 |\sqrt{\rho_s} + b(t)|^2 \mathrm{d}t - \rho_s - \frac{1}{2} \right) \right\} \\
&= \int_0^1 \kappa(s,t) \mathrm{d}t \\
&= \rho_s s(2-s) + s^2 \left(1 - \frac{2s}{3}\right).
\end{aligned} \tag{6.53}$$

Using both Eqs. (6.52) and (6.53), the variance of the residual nonlinear phase noise is

$$\sigma^2_{\Phi_\alpha} = \sigma^2_\Phi + \alpha^2 \kappa(s,s) - 2\alpha\xi(s), \tag{6.54}$$

The variance of Eq. (6.54) is a quadratic function of α, the optimal compensation factor is given by

$$\alpha_{\text{opt}} = \frac{\xi(s)}{\kappa(s,s)} = \frac{\rho_s s(2-s) + s^2(1-2s/3)}{2\rho_s s + s^2}. \tag{6.55}$$

If $s = 1$, the optical compensation factor of α_{opt} is the same as Eq. (6.31) when the compensator precedes the receiver. If $s = 0$, there is no compensation. For high SNR of ρ_s, the optimal compensation factor is approximately equal to $\alpha_{\text{opt}} \to 1 - s/2$.

When the optimal compensation factor of Eq. (6.55) is substituted to the variance of Eq. (6.54), we obtain

$$\sigma^2_{\Phi_\alpha,\min} = \sigma^2_\Phi - \frac{\xi^2(s)}{\kappa(s,s)}. \tag{6.56}$$

The location of the compensator is optimized by solving $d\sigma^2_{\Phi_\alpha,\min}/ds = 0$ to obtain

$$s_{\text{opt}} = \frac{3}{8} + \frac{9\rho_s}{8}\left(\sqrt{1 + \frac{9+42\rho_s}{81\rho_s^2}} - 1\right) \approx \frac{2}{3} + \frac{2}{91\rho_s}. \tag{6.57}$$

In most value of SNR, the optimal location is about 2/3 from the beginning of the fiber link.

At high SNR of ρ_s, the ratio of the variance of residual nonlinear phase noise to that of nonlinear phase noise of Eq. (5.52) is equal to

$$\frac{\sigma^2_{\Phi_\alpha,\min}}{\sigma^2_\Phi} \to 1 - 3s + 3s^2 - \frac{3}{4}s^3 \tag{6.58}$$

with a minimum value of 1/9 when $s_{\text{opt}} \to 2/3$.

Figure 6.16 shows the ratio of standard deviation (STD) of residual nonlinear phase noise of σ_{RES} to the mean nonlinear phase shift $\langle\Phi_{\text{NL}}\rangle$. The STD of residual nonlinear phase noise is given by $\sigma_{\text{RES}} = \langle\Phi_{\text{NL}}\rangle \sigma_{\Phi_\alpha}/(\rho_s + 1/2)$ without normalization. In Fig. 6.16, the compensator is located at $s = 1/4, 1/2$ and $3/4$ from the beginning of fiber link, or preceding the receiver of $s = 1$. Figure 6.16 also shows $\sigma_{\text{RES}}/\langle\Phi_{\text{NL}}\rangle$ without compensation and with optimal location of $s = s_{\text{opt}}$ of Eq. (6.57).

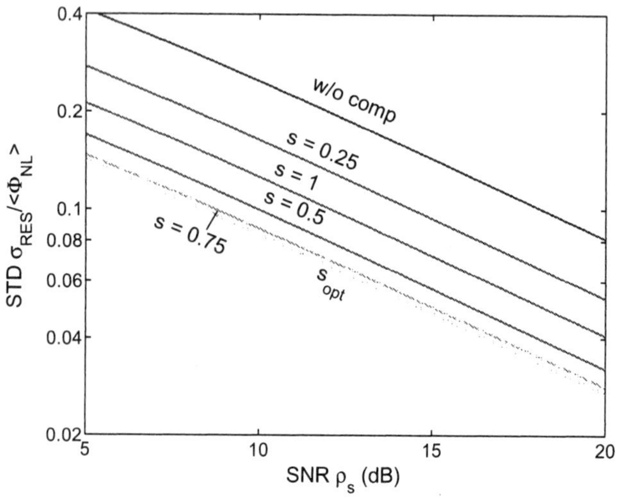

Figure 6.16. The ratio of STD to mean of nonlinear phase noise as a function of SNR ρ_s when the compensator located at various locations.
[Reprinted from Ho (2005a), copyright (2005), with permission from Elsevier]

From Fig. 6.16, the same as Sec. 6.2, a compensator preceding the receiver about halves the STD of the nonlinear phase noise. From Fig. 6.16, a compensator at the optimal location of $s = s_{\text{opt}}$ of Eq. (6.57) reduces the STD of nonlinear phase noise to about 1/3 of the case without compensation.

To illustrate the effects of compensator location to the performance of the nonlinear phase noise compensation scheme, Figure 6.17 shows the ratio of $\sigma_{\text{RES}}/\langle\Phi_{\text{NL}}\rangle$ as a function of compensator location for several values of SNR ρ_s. The residual nonlinear phase noise reduces when a compensator is used in the fiber link. For a location before about 2/3 of the fiber link, the STD of residual nonlinear phase noise decreases with the location of the compensator and reaches a value of about 1/3 of the value of STD without compensation when $s = 0$. For a location after about 2/3 of the fiber link, the STD of residual nonlinear phase noise increases with the location of the compensator and reaches a value of about 1/2 of the value of STD without compensation when the compensator precedes the receiver.

Mid-span compensation is commonly better than compensation preceding the receiver. The nonlinear phase noise of Eq. (5.32) is an integration of the noisy power of $|\sqrt{\rho_s} + b(t)|^2$ over distance of t. As an example, the mid-span intensity of $|\sqrt{\rho_s}+b(1/2)|^2$ has less noise than the received intensity of $|\sqrt{\rho_s} + b(1)|^2$. However, the nonlinear phase noise

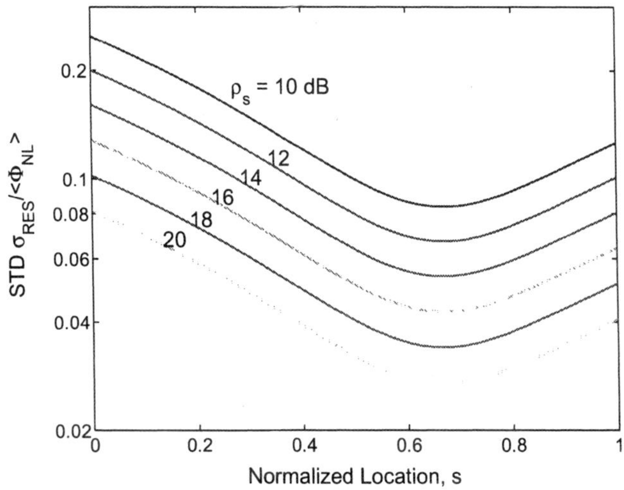

Figure 6.17. The ratio of STD to mean of nonlinear phase noise as a function of compensator location.
[Reprinted from Ho (2005a), copyright (2005), with permission from Elsevier]

after the compensator of $\int_{1/2}^{1} |\sqrt{\rho_s} + b(t)|^2 \mathrm{d}t$ can still be compensated using $|\sqrt{\rho_s}+b(1/2)|^2$. Because the intensity of $|\sqrt{\rho_s}+b(1/2)|^2$ is not suitable to compensate the nonlinear phase noise at the end of the fiber link, i.e., $\int_{0.8}^{1} |\sqrt{\rho_s} + b(t)|^2 \mathrm{d}t$, the optimal location needs to balance the usage of less noisy optical intensity and the ability to compensate the nonlinear phase noise at the end of the fiber link. The optimal location of about 2/3 is the balance between these two factors. The optimal compensation factor is $\alpha_{\text{opt}} \to 2/3$ at high SNR when $s = 2/3$. From the same length of fiber interval, the nonlinear phase noise of $\int_{0.8}^{1} |\sqrt{\rho_s} + b(t)|^2 \mathrm{d}t$ is significantly larger than $\int_{0.0}^{0.2} |\sqrt{\rho_s} + b(t)|^2 \mathrm{d}t$, the optimal compensator location is not at the middle of $s = 1/2$ but $s = 2/3$ toward the end of the fiber link.

It is also interesting to find the variance of nonlinear phase noise as a function of distance. The compensation factor of Eq. (6.55) minimizes the variance of residual nonlinear phase noise at the end of the fiber link. The compensator may increase the phase noise immediately at the output of the compensator. However, at the end of the fiber link, the residual nonlinear phase noise is definitely minimized. For a compensator located at s with a compensator factor of α, the variance of the residual nonlinear phase noise as a function of distance is equal to

$$\sigma_{\Phi}^2(t) + \alpha^2 \kappa(s,s) - 2\alpha\xi_t(s), \quad t \geq s, \tag{6.59}$$

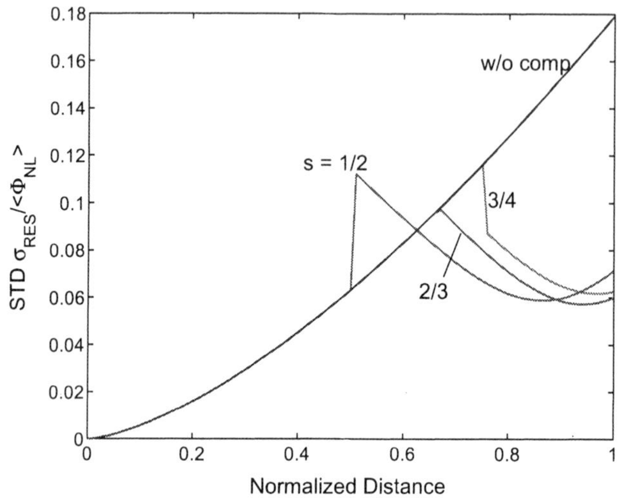

Figure 6.18. The ratio of STD to mean of nonlinear phase noise as a function of normalized distance when a compensator is located at $s = 1/2, 2/3$ and $3/4$. [Reprinted from Ho (2005a), copyright (2005), with permission from Elsevier]

where

$$\xi_t(s) = \int_0^t \kappa(s, \tau)\mathrm{d}\tau \tag{6.60}$$

is the correlation of the nonlinear phase noise at t with the intensity at s.

Figure 6.18 shows the ratio of $\sigma_{\mathrm{RES}}/\langle\Phi_{\mathrm{NL}}\rangle$ as a function of normalized distance when a compensator is located at $s = 1/2, 2/3$ and $3/4$. Without compensation, the STD of nonlinear phase noise as a function of distance is given by Eq. (6.59). Without compensation, the STD of nonlinear phase noise as a function of distance is given by

$$\sigma_\Phi^2(t) = \int_0^t \xi_t(s)\mathrm{d}s = \frac{2\rho_s}{3}t^3 + \frac{t^4}{6}. \tag{6.61}$$

When the amplifier noise is accumulated with distance, the nonlinear phase noise is increased faster at the end of the fiber link.

From Fig. 6.18, the STD of nonlinear phase noise is the largest at the compensator output for both $s = 1/2$ and $2/3$. When the compensator is located at $s = 1/2$, the compensator actually increases its output phase noise. When the compensator is located at $s = 3/4$, the compensator reduces its output phase noise. However, in all cases of Fig. 6.18, the compensator minimizes the nonlinear phase noise at the end of the fiber link.

4.2 Multiple Compensators

The system can further be improved when more than one compensator are used for mid-span compensation. If there are N_c compensators located at $s_1, s_2, \ldots, s_{N_c}$, the optimal compensation factors of $\alpha_1, \alpha_2, \ldots, \alpha_{N_c}$ minimize the variance of the residual nonlinear phase noise of

$$\Phi_\alpha = \Phi - \sum_{k=1}^{N_c} \alpha_k |\sqrt{\rho_s} + b(s_k)|^2. \tag{6.62}$$

The variance of the residual nonlinear phase noise is equal to

$$\sigma^2_{\Phi_\alpha} = \sigma^2_\Phi - 2\sum_{k=1}^{N_c} \alpha_k \xi(s_k) + \sum_{k=1}^{N_c}\sum_{l=1}^{N_c} \alpha_k \alpha_l \kappa(s_k, s_l). \tag{6.63}$$

where the second term is the covariance of the intensity at s_k of $|\sqrt{\rho_s} + b(s_k)|^2$ with the nonlinear phase noise of Eq. (5.32) and the third term is the covariance of the intensity at s_k and s_l.

The optimal compensation factors can be found by solving all equations of $\partial \sigma^2_{\Phi_\alpha} / \partial \alpha_k = 0$, or

$$\sum_{l=1}^{N_c} \alpha_l \kappa(s_k, s_l) = \xi(s_k), \qquad k = 1, \ldots, N_c. \tag{6.64}$$

In the case of $N_c = 1$, the solution is the same as that of Eq. (6.55). The N_c linear equations in Eq. (6.64) determine the N_c optimal compensation factors of $\alpha_k, k = 1, \ldots, N_c$. The optimal compensation factors are equal to

$$\vec{\alpha}_{\text{opt}} = \mathcal{K}^{-1} \vec{Z}, \tag{6.65}$$

where the covariance matrix has elements of $\mathcal{K}_{kl} = \kappa(s_k, s_l)$ and the correlation vector is $\vec{Z} = [\xi(s_1), \xi(s_2), \ldots, \xi(s_{N_c})]^T$. The solution of Eq. (6.65) is the same as that for most MMSE algorithms (Proakis, 2000, p. 626).

Figure 6.19 shows the ratio of $\sigma_{\text{RES}} / \langle \Phi_{\text{NL}} \rangle$ as a function of SNR ρ_s when multiple compensators are used. In Fig. 6.19(a), the N_c compensators are evenly located at $s_k = k/N_c, k = 1, 2, \ldots, N_c$, including one compensator preceding the receiver of $s_{N_c} = 1$. In Fig. 6.19(b), the N_c compensators are evenly located at $s_k = 2k/(2N_c + 1), k = 1, 2, \ldots N_c$. The compensator spacings in Fig. 6.19(a) and (b) are $1/N_c$ and $2/(2N_c + 1)$, respectively. The two curves in Fig. 6.16 without compensation and with one compensator of $s = 1$ are also shown in Fig. 6.19 for comparison.

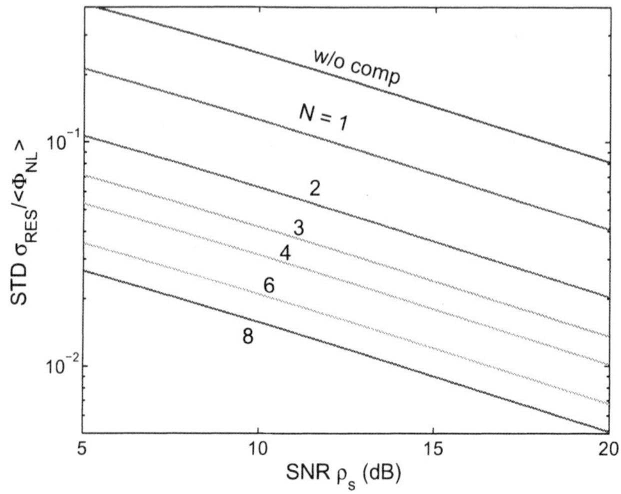

(a) $s_k = k/N_c, k = 1, \ldots, N_c$

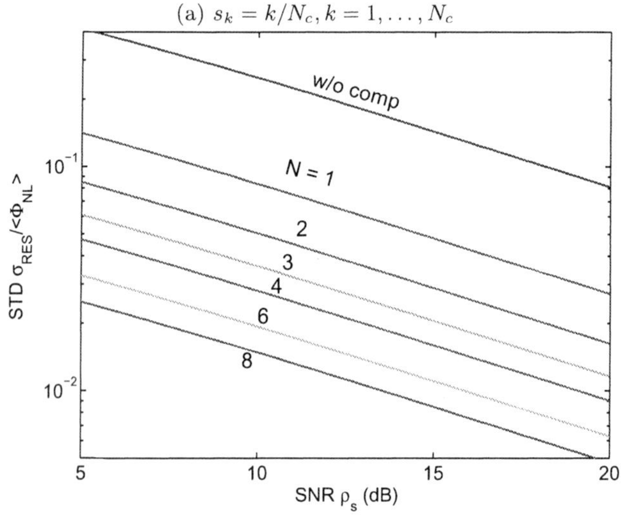

(b) $s_k = 2k/(2N_c + 1), k = 1, \ldots, N_c$

Figure 6.19. The ratio of STD to mean of nonlinear phase noise as a function of SNR ρ_s for systems with multiple compensators.
[Reprinted from Ho (2005a), copyright (2005), with permission from Elsevier]

From Fig. 6.19(a), the STD of nonlinear phase noise is reduced by the factor of about $2N_c$. From Fig. 6.19(b), the STD of nonlinear phase noise is reduced by the factor of about $2N_c + 1$. Because the intensity at $s_{N_c} = 1$ has larger noise, better performance is achieved when $s_{N_c} =$

$1 - 1/(2N_c + 1)$. In Fig. 6.17, the STD of the residual nonlinear phase noise is very close to its minimum value for a large range of locations from $s = 0.6$ to $s = 0.75$. Because the compensation scheme is optimized over a large range of locations, the locations of $s_k = 2k/(2N_c + 1)$ are very close to the optimal locations.

The optimal compensation factor of $\vec{\alpha}_{\text{opt}}$ can also be evaluated numerically. With the compensator location of $s_k = k/N_c$, the optimal compensation factors are $\alpha_{\text{opt},k} \rightarrow 1/N_c, k = 1, \ldots, N_c - 1$, and $\alpha_{\text{opt},N_c} \rightarrow 1/2N_c$. With the compensator locations of $s_k = 2k/(2N_c+1)$, the optimal compensation factors all approach the normalized values of $\alpha_{\text{opt},k} \rightarrow 2/(2N_c + 1), k = 1, \ldots, N_c$, that is the same as the compensator spacing. With the compensator locations of $s_k = 2k/(2N_c + 1)$, all compensators have approximately the same compensation factor.

From Fig. 6.19, when sufficient number of compensators is used, nonlinear phase noise can be completely eliminated. We may infer that if one compensator is used per fiber span in the system of Sec. 6.2, nonlinear phase noise can be completely eliminated.

When N_c compensators are used in optimal locations, the STD of nonlinear phase noise is reduced by a factor of about $2N_c+1$. If nonlinear phase noise is the dominant noise source, the transmission distance can be increased by a factor of $2N_c+1$. However, the amplifier noises modeled by the Wiener process of $b(t)$ also limits the transmission distance by adding noise into the signal for system with more fiber spans. With higher tolerance of nonlinear phase noise, the launched power to the system can be increased accordingly. Assumed the same amplifier noises, the system SNR is increased with the launched power. If the operation point is estimated when the variances of linear and nonlinear phase noise are equal, the overall transmission distance can be increased by a factor of $\sqrt{2N_c + 1}$.

Another method to reduce nonlinear phase noise is the usage of phase conjugation (McKinstrie et al., 2003). With a single compensator, the optimal location is also at 2/3 the fiber link and the STD is also reduced by a factor of 1/3. However, the techniques and methods for analysis are different. Instead of phase conjugator in McKinstrie et al. (2003), a phase modulator is used here instead. While the analysis of McKinstrie et al. (2003) is only for soliton, the method here is applicable to more general signals.

The inclusion of compensators in the fiber link requires additional amplification to compensate for the compensator loss. If the compensator is properly located, for example, in the mid-stage of an amplifier, compensator loss gives minimal increase to the amplifier noise figure and overall amplifier noise to the system.

In this section for mid-span compensation, with a single compensator, the optimal placement of the compensator is not preceding the receiver but at about 2/3 of the fiber link. With this optimal location, a single compensator reduces the STD of nonlinear phase noise by a factor of 3 instead of a factor of 2 for a compensator preceding the receiver. The optimal compensation factor is approximately equal to 2/3.

If N_c compensators are located at $2k/(2N_c + 1)$ of the fiber link, the STD of nonlinear phase noise is reduced by a factor of $2N_c + 1$. The optimal compensation factors for all compensators are all approximately equal to $2/(2N_c + 1)$.

The analysis of this section follows Ho (2005a).

5. Nonlinear Compensation

In previous sections, linear compensator is investigated in detail. From Sec. 6.2, in term of variance, the linear MMSE compensator performs very close to the optimal nonlinear MMSE compensator. Usually, in digital communications, most types of compensator or equalizer are designed based on the MMSE criterion for simplicity (Proakis, 2000, chs. 10, 11).

However, as from Sec. 6.3, it is obvious that the MMSE compensator does not necessary perform the same as MAP compensator that minimizes the error probability. The linear MAP compensator was found to perform up to 0.35 dB better than the linear MMSE compensator for binary phase-shift keying (PSK) signals.

In order to find the optimal nonlinear MAP compensator, this section first derives the joint distribution of the received amplitude and phase. The optimal nonlinear MAP compensator is then derived for PSK signal with nonlinear phase noise. The error probability of both nonlinear MAP and MMSE compensators is calculated using the joint distribution of received amplitude and phase.

As shown in Sec. 6.1, a nonlinear compensator can be implemented as a Yin-Yang detector with a spiral decision boundary or equivalently to add a nonlinear compensated phase to the received phase.

The MAP detector provides the ultimate performance that is possible for a PSK signal with nonlinear phase noise. Nonlinear compensator can be implemented by phase modulator with nonlinear driver circuits or electronic signal processing techniques. The optimal nonlinear detector derived here is for a compensator preceding or combined with a receiver, distributed nonlinear compensator in the middle of fiber link is not considered here.

5.1 Joint Distribution of the Received Amplitude and Phase

The received phase of Eq. (5.63) is confined to the range of $[-\pi, +\pi)$. The joint p.d.f. of received amplitude and phase $p_{R,\Phi_r|\theta_0}(r, \theta)$ can be modeled as a periodic function of θ with a period of 2π and expanded as a Fourier series as

$$p_{R,\Phi_r|\theta_0}(r, \theta) = \frac{p_R(r)}{2\pi} + \frac{1}{\pi}\sum_{m=1}^{+\infty} \Re\left\{C_m(r)e^{jm(\theta-\theta_0)}\right\}, \quad r \geq 0, \qquad (6.66)$$

where

$$p_R(r) = 2r \exp\left[-(r^2 + \rho_s)\right] I_0(2r\sqrt{\rho_s}) \qquad (6.67)$$

is the Rice p.d.f. of the received amplitude, $C_m(r)$ is the mth Fourier coefficient as a function of the received amplitude r, and $\Re\{\cdot\}$ denotes the real part of a complex number. The p.d.f. of Eq. (6.66) has been simplified using the relationship of $C_{-m}(r) = C_m^*(r)$. It is also obvious that the marginal p.d.f. of amplitude is the Rice distribution of $\int_{-\pi}^{+\pi} p_{R,\Phi_r|\theta_0}(r, \theta)\mathrm{d}\theta = p_R(r)$.

As the Fourier series of the p.d.f. of $p_{R,\Phi_r|\theta_0}(r, \theta)$, the Fourier coefficients of $C_m(r)$ are the Fourier transform of the p.d.f. of Eq. (6.66) with respect to θ at integer "angular frequency" of m. The Fourier transform of a p.d.f. is its characteristic function and $C_m(r)$ can come from the "partial" Fourier transform of the joint p.d.f. of nonlinear phase noise, received intensity, and the phase of amplifier noise.

Using $q_{\Phi,Y,\Theta_n}(\nu, y, m) = \mathcal{F}_\omega^{-1}\left\{\Psi_{\Phi,Y,\Theta_n}\right\}(\nu, y, m)$ to denotes the partial p.d.f. and characteristic function as the inverse Fourier transform of the characteristic function of $\Psi_{\Phi,Y,\Theta_n}(\nu, y, m)$ from Eq. (6.B.2), we has $q_{\Phi,R,\Theta_n}(\phi, r, m) = 2rq_{\Phi,Y,\Theta_n}(\phi, r^2, m)$ for the amplitude of R instead of intensity of Y. The Fourier coefficients of $C_m(r)$ are

$$C_m(r) = 2rq^*_{\Phi,Y,\Theta_n}\left(-m\frac{\langle\Phi_{\mathrm{NL}}\rangle}{\rho_s + 1/2}, r^2, m\right). \qquad (6.68)$$

Using Eq. (6.B.10) and $\theta_0 = 0$, we obtain (Gradshteyn and Ryzhik, 1980, §8.431)

$$q_{\Phi,Y,\Theta_n}(\nu, y, m) = \frac{\Psi_\Phi(\nu)}{2\sigma_\nu^2}\exp\left(-\frac{y + |\vec{\xi}_\nu|^2}{2\sigma_\nu^2}\right)I_m\left(\frac{\sqrt{y}|\vec{\xi}_\nu|}{\sigma_\nu^2}\right). \qquad (6.69)$$

Based on Eqs. (6.68) and (6.69), we obtain

$$C_m(r) = \frac{r\Psi_m}{s_m}\exp\left(-\frac{r^2 + \alpha_m^2}{2s_m}\right)I_m\left(\frac{\alpha_m r}{s_m}\right), \quad m \geq 1, \qquad (6.70)$$

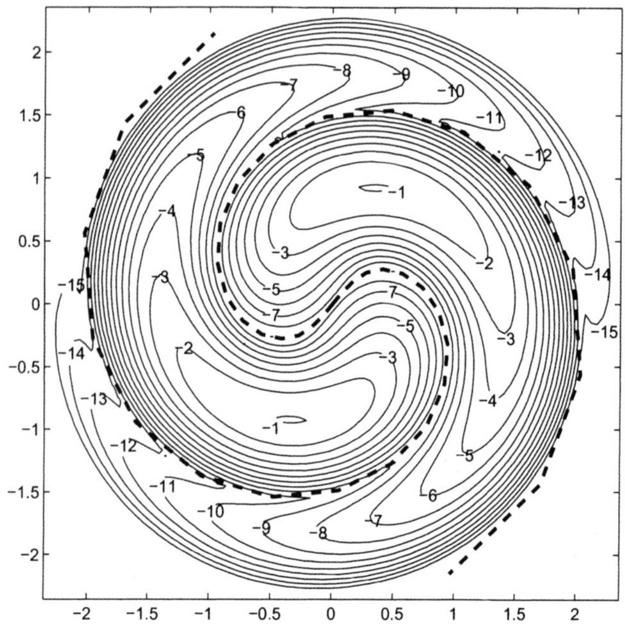

Figure 6.20. The distribution of received electric field for binary PSK signal. The contour lines are in logarithmic scale and both x and y axes are normalized to unity mean amplitude. The dashed line is the decision boundary. The SNR is $\rho_s = 18$ (12.6 dB) and the mean nonlinear phase shift is $\langle \Phi_{\mathrm{NL}} \rangle = 2$ rad.

with

$$\Psi_m = \Psi_\Phi \left(m \frac{\langle \Phi_{\mathrm{NL}} \rangle}{\rho_s + 1/2} \right), \tag{6.71}$$

$$\alpha_m = \sqrt{\rho_s} \sec \left[\left(jm \frac{\langle \Phi_{\mathrm{NL}} \rangle}{\rho_s + 1/2} \right)^{\frac{1}{2}} \right], \tag{6.72}$$

and

$$s_m = \frac{1}{2} \left(jm \frac{\langle \Phi_{\mathrm{NL}} \rangle}{\rho_s + 1/2} \right)^{-\frac{1}{2}} \tan \left[\left(jm \frac{\langle \Phi_{\mathrm{NL}} \rangle}{\rho_s + 1/2} \right)^{\frac{1}{2}} \right], \tag{6.73}$$

where $\Psi_\Phi(\nu)$ is the characteristic function of the normalized nonlinear phase noise of Eq. (5.48).

For a PSK signal with $\theta_0 \in \{0, \pi\}$, Figure 6.20 shows the distribution of the received electric field similar to Fig. 5.1(b). The contour lines of Fig. 6.20 are the logarithmic of the p.d.f. of $\frac{1}{2} \left[p_{R,\Phi_r|0}(r, \theta) + p_{R,\Phi_r|\pi}(r, \theta) \right]$ after the conversion to rectangular coordinate. The SNR of $\rho_s = 18$ (12.6

dB) is chosen for an error probability of 10^{-9} for binary PSK signal without nonlinear phase noise. The mean nonlinear phase shift of $\langle \Phi_{\mathrm{NL}} \rangle = 2$ rad is close to the optimal operating point for linearly compensated PSK signals from Fig. 6.10.

Similar to those of Fig. 5.1, the helix shape distribution of Fig. 6.20 clearly shows that the rotation due to nonlinear phase noise is correlated with the amplitude (the distance to the origin). The optimal MAP detector is derived here.

The derivation of joint p.d.f. here follows Huang and Ho (2003). Using method in quantum field theory, a p.d.f. similar to Eq. (6.66) was derived in Turitsyn et al. (2003). Similar p.d.f. was also derived in Mecozzi (1994a,b).

5.2 Optimal MAP Detector

For a PSK signal with $\theta_0 \in \{0, \pi\}$, using the Neyman-Peason lemma with unity likelihood ratio (McDonough and Whalen, 1995, §5.3), the optimal decision regions of the MAP detector are given by

$$\mathcal{R}_0 = \Big\{r, \theta \mid p_{R,\Phi_r|0}(r,\theta) \geq p_{R,\Phi_r|\pi}(r,\theta)\Big\}, \tag{6.74}$$

$$\mathcal{R}_1 = \Big\{r, \theta \mid p_{R,\Phi_r|\pi}(r,\theta) > p_{R,\Phi_r|0}(r,\theta)\Big\} \tag{6.75}$$

for the transmitted phases of $\theta_0 = 0$ and $\theta_0 = \pi$, respectively. The error probability is

$$\begin{aligned} p_e &= \frac{1}{2}\int_{\mathcal{R}_0} p_{R,\Phi_r|\pi}(r,\theta)\mathrm{d}r\mathrm{d}\theta + \frac{1}{2}\int_{\mathcal{R}_1} p_{R,\Phi_r|0}(r,\theta)\mathrm{d}r\mathrm{d}\theta \\ &= \int_{\mathcal{R}_0} p_{R,\Phi_r|\pi}(r,\theta)\mathrm{d}r\mathrm{d}\theta \\ &= \int_{\mathcal{R}_1} p_{R,\Phi_r|0}(r,\theta)\mathrm{d}r\mathrm{d}\theta. \end{aligned} \tag{6.76}$$

The decision regions of $\mathcal{R}_0$ and $\mathcal{R}_1$ have a boundary of $\theta_c(r) \pm \pi/2, r \geq 0$, where $\theta_c(r)$ is the center phase of the decision regions of $\mathcal{R}_0$. With the joint p.d.f. of Eq. (6.66), the error probability for MAP detector of Eq. (6.76) becomes

$$p_e = \frac{1}{2} - \frac{2}{\pi}\sum_{k=0}^{+\infty}\frac{(-1)^k}{2k+1}\int_0^\infty \Re\Big\{C_{2k+1}^*(r)e^{j(2k+1)\theta_c(r)}\Big\}\,\mathrm{d}r. \tag{6.77}$$

Using only the p.d.f. of $p_{R,\Phi_r|0}(r,\theta)$, because of the relationship of $p_{R,\Phi_r|\pi}(r,\theta) = p_{R,\Phi_r|0}(r,\theta-\pi)$, the center phase of $\theta_c(r)$ can be determined by the relationship of

$$p_{R,\Phi_r|0}\left(r, \theta_c(r) + \frac{\pi}{2}\right) = p_{R,\Phi_r|0}\left(r, \theta_c(r) - \frac{\pi}{2}\right). \tag{6.78}$$

Figure 6.20 also shows the decision boundary of $\theta_c(r) \pm \pi/2$ based on Eqs. (6.74) and (6.75) with center phase given by Eq. (6.78). The decision boundary is the "valley" between two peaks of the p.d.f. of $p_{R,\Phi_r|0}(r,\theta)$ and $p_{R,\Phi_r|\pi}(r,\theta)$, respectively.

The optimal MAP detector can be implemented based on the decision boundary of Fig. 6.20 of the Yin-Yang detector. The same MAP detector can also be implemented as a nonlinear compensator by adding an angle of $\theta_c(r)$ from Eq. (6.78) to the received phase. The Yin-Yang detector is equivalent to the nonlinear compensator.

In this section, the optimal MAP detector is derived based on $\theta_c(r)$ as a function of the received amplitude. Because the relationship between intensity and amplitude is a monotonic function of $Y = R^2, R \geq 0$, the nonlinear phase noise is considered to be compensated by the received intensity instead of the received amplitude.

The more popular differential phase-shift keying (DPSK) signals can also be compensated using the differential phase of $\theta_c(r_1) - \theta_c(r_2)$, where r_1 and r_2 are the received amplitudes in two consecutive symbols. While the compensator for DPSK signals can be constructed using the center phase of Eq. (6.78), the evaluation of the error probability of nonlinearly compensated DPSK signals is difficult.

5.3 Optimal MMSE Detector

The optimal MMSE compensator estimates the normalized nonlinear phase using the received intensity by the conditional mean of $E\{\Phi|R\} = E\{\Phi|Y\} = E\{\Phi|R^2\}$ from Appendix 6.A. The conditional mean is the Bayes estimator that minimizes the variance of the residual nonlinear phase noise without the constraint of linearity (McDonough and Whalen, 1995, §10.2). We call the estimation of $E\{\Phi|R\}$ an estimation based on received intensity although it is actually based on received amplitude. To find the conditional mean $E\{\Phi|R\}$, either the conditional p.d.f. of $p_{\Phi|R}(\theta)$ or the conditional characteristic function of $\Psi_{\Phi|R}(\nu)$, or equivalently $\Psi_{\Phi|Y}(\nu)$, is required. The conditional characteristic function of $\Psi_{\Phi|Y}(\nu)$ can be found using the relationship of

$$\Psi_{\Phi|Y}(\nu) = \frac{q_{\Phi,Y}(\nu, y)}{p_Y(y)}, \tag{6.79}$$

where $q_{\Phi,Y}(\nu, y)$ of Eq. (6.B.8) is the partial characteristic function of normalized nonlinear phase noise and p.d.f. of received intensity and

$p_Y(y)$ of Eq. (6.B.9) is the p.d.f. of the received intensity. Using both Eqs. (6.B.8) and (6.B.9), we obtain

$$\Psi_{\Phi|Y}(\nu) = \frac{\Psi_\Phi(\nu)\exp\left[-\dfrac{y+|\vec{\xi}_\nu|^2}{2\sigma_\nu^2}\right] I_0\left(\sqrt{y}\dfrac{|\vec{\xi}_\nu|}{\sigma_\nu^2}\right)}{2\sigma_\nu^2 \exp\left[-(y+\rho_s)\right] I_0\left(2\sqrt{y\rho_s}\right)}. \tag{6.80}$$

The optimal MMSE compensator is the conditional mean of the normalized nonlinear phase noise given the received intensity of $Y = R^2$, we obtain

$$\begin{aligned} E\{\Phi|R\} &= -j\,\frac{d}{d\nu}\Psi_{\Phi|Y}(\nu|y)\bigg|_{\nu=0,y=r^2} \\ &= \frac{1}{6}+\frac{1}{3}\rho_s+\frac{1}{3}r^2+\frac{\sqrt{\rho_s}r}{3}\frac{I_1(2r\sqrt{\rho_s})}{I_0(2r\sqrt{\rho_s})}. \end{aligned} \tag{6.81}$$

Other than the normalization, the MMSE compensator of Eq. (6.81) is similar to that of Eq. (6.A.15) for finite number of fiber spans from Appendix 6.A and can be derived by the method there. The optimal MMSE compensator Eq. (6.81) depends solely on the system SNR.

The normalized residual nonlinear phase noise is $\phi_e = \phi - E\{\Phi|R\}$, its conditional variance is

$$\begin{aligned} \sigma^2_{\Phi_e|R}(r) &= -\frac{d^2}{d\nu^2}\Psi_{\Phi|Y}(\nu|y)\bigg|_{\nu=0,y=r^2} - E\{\Phi|R\}^2 \\ &= \frac{1+4(\rho_s+r^2)}{90}+\frac{r^2\rho_s}{9}+\frac{\sqrt{\rho_s}r}{45}\frac{I_1(2r\sqrt{\rho_s})}{I_0(2r\sqrt{\rho_s})}-\frac{r^2\rho_s}{9}\frac{I_1^2(2r\sqrt{\rho_s})}{I_0^2(2r\sqrt{\rho_s})}, \end{aligned} \tag{6.82}$$

that also depends solely on SNR.

The variance of the normalized residual nonlinear phase noise can be numerically integrated as

$$\sigma^2_{\Phi_e} = \int_0^\infty \sigma^2_{\Phi_e|R}(r)p_R(r)\mathrm{d}r, \tag{6.83}$$

where $p_R(r)$ is the p.d.f. of the received amplitude of Eq. (6.67).

Similar to the error probability of Eq. (6.77), the error probability for MMSE detector is

$$p_e = \frac{1}{2}-\frac{2}{\pi}\sum_{k=0}^{+\infty}\frac{(-1)^k}{2k+1}\int_0^\infty \Re\left\{C^*_{2k+1}(r)\exp\left[\frac{j(2k+1)\langle\Phi_{\mathrm{NL}}\rangle}{\rho_s+1/2}E\{\Phi|R\}\right]\right\}\mathrm{d}r. \tag{6.84}$$

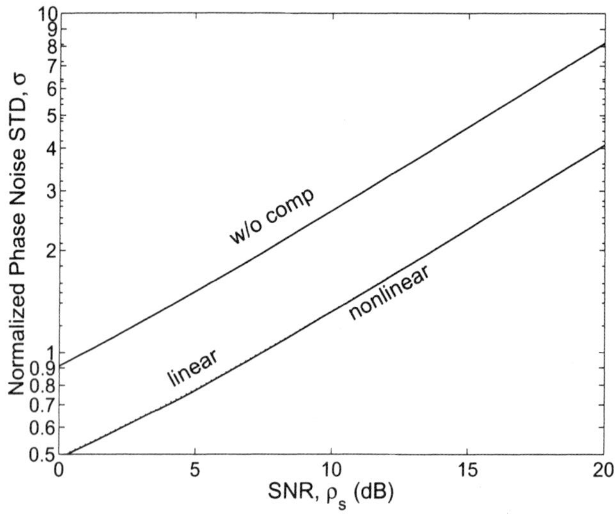

Figure 6.21. The STD of normalized nonlinear phase noise with and without compensation. The STD with linear and nonlinear compensator are showed as dotted- and solid lines, respectively.

The factor of $\langle\Phi_{\mathrm{NL}}\rangle/(\rho_s + 1/2)$ scales the normalized nonlinear phase noise Φ in Eq. (6.81) to the actual nonlinear phase noise Φ_{NL}.

Figure 6.21 plots the STD of the normalized nonlinear phase noise with and without compensation as a function of SNR ρ_s. The STD of residual nonlinear phase noise with linear and nonlinear compensator are shown as almost overlapped solid and dotted lines, respectively. The STD of nonlinear compensator is about 0.2% less than that of linear compensator. Figure 6.21 confirms the results of Fig. 6.4 that linear and nonlinear MMSE compensator performs the same in term of the variance of residual nonlinear phase noise. However, as shown later, the error probability of nonlinear MMSE compensator is far better that for linear MMSE compensator.

5.4 Numerical Results

Figure 6.22 plots the center phase of $\theta_c(r)$ from Eq. (6.78) as a function of the normalized received intensity of r^2/ρ_s. The system parameters of Fig. 6.22 are the same as that of Fig. 6.20. As discussed earlier, the center phase of Yin-Yang detector is the same as the compensated phase of a compensator. The center (or compensated) phase of the nonlinear MMSE compensator of Eq. (6.81) is also plotted in Fig. 6.22 for comparison. The phase of Eq. (6.81) is scaled by $\langle\Phi_{\mathrm{NL}}\rangle/(\rho_s+\frac{1}{2})$. The

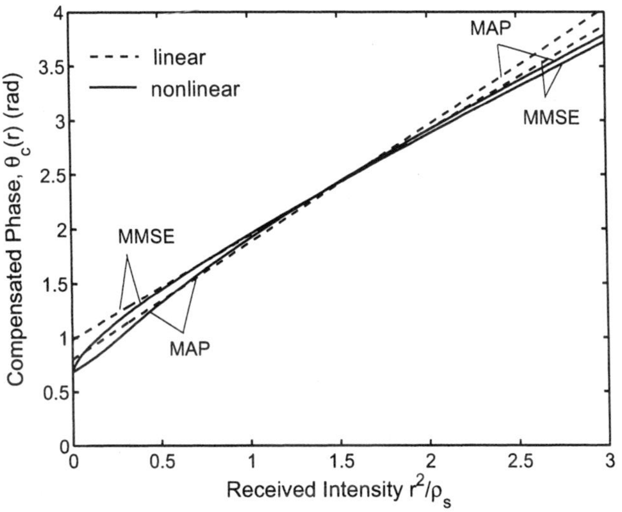

Figure 6.22. The compensated phase as a function of received intensity. System parameters are the same as that in Fig. 6.20.

compensated phases of the linear compensator designed by MMSE or MAP criteria are also plotted in Fig. 6.22 as dashed-lines for comparison. In Fig. 6.22, the received intensity is normalized with respect to the SNR of ρ_s.

From Fig. 6.22, nonlinear MMSE or MAP compensated curves are very close to the linear compensated curves, especially when the received intensity is near its mean value of about $r^2/\rho_s = 1$. When the STD of Fig. 6.21 is evaluated, the STD is mainly contributed from the region where the random variable is close to its mean value. Compared the linear and nonlinear compensated phases of Fig. 6.22, the STD of Fig. 6.21 for linear and nonlinear MMSE compensator should not have significant difference. From Fig. 6.21, the linear and nonlinear MAP compensated phases are also very close to each other. In the region closes to $r^2/\rho_s = 1$, because the nonlinear MAP compensated phase is closer to the linear MAP compensated phase than the nonlinear MMSE compensated phase, the linear MAP compensated phase has better performance than the nonlinear MMSE compensated phase.

Figure 6.23 shows the error probability given by both Eqs. (6.77) and (6.84) for optimal MAP and MMSE detectors, respectively, as a function of SNR ρ_s. Figure 6.23 also plots the error probability of a PSK signal without nonlinear phase noise of Eq. (3.78). From Fig. 6.23, the optimal MMSE detector does not minimize the error probability. The optimal

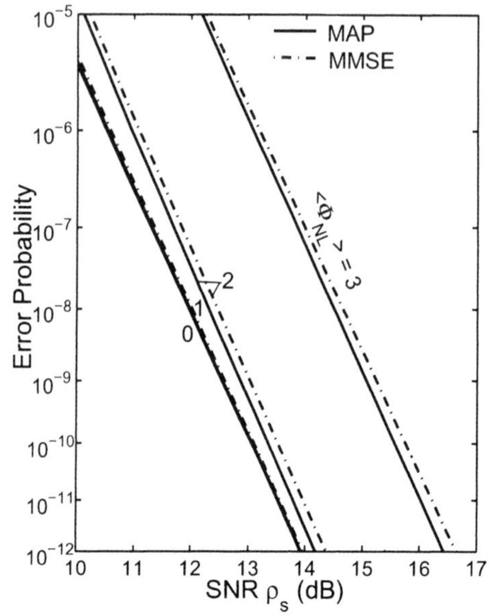

Figure 6.23. Error probability of a PSK signal with optimal MAP or MMSE detector.

compensated phase of Eq. (6.78) always gives a smaller error probability that the compensated phase of Eq. (6.81).

Figure 6.24 shows the SNR penalty of PSK signal for an error probability of 10^{-9} calculated by the MAP of Eq. (6.77) and MMSE of Eq. (6.84) error probability formulae. The SNR penalty with linear compensator is also plotted as dashed lines for comparison. The SNR penalty without compensation is also shown for comparison.

From Fig. 6.24, the nonlinear MMSE compensator performs up to 0.23 dB better than the linear MMSE compensator. Although Fig. 6.21 shows that linear and nonlinear MMSE compensators perform almost the same in term of the STD of residual nonlinear phase noise, the error probability has significant difference. The comparison between Figs. 6.21 and 6.24 shows that the variance does not correlate well with the error probability. From Fig. 6.24, the optimal linear MAP detector performs very close to the optimal nonlinear MAP detector. The optimal nonlinear MAP detector performs only up to 0.14 dB better than the linear MAP detector.

For binary PSK signal, Table 6.1 shows the mean nonlinear phase shift corresponding to a system with 1-dB SNR penalty and the optimal operating point. The optimal operating point is found by the condition

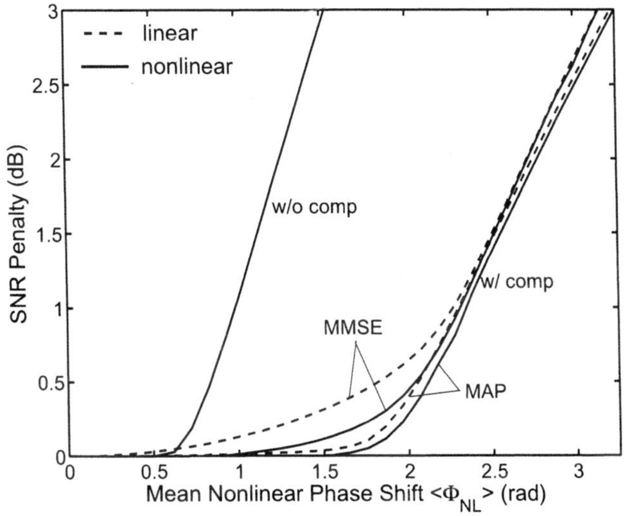

Figure 6.24. The SNR penalty of PSK signal as a function of mean nonlinear phase shift $\langle\Phi_{\mathrm{NL}}\rangle$.

Table 6.1. PSK Signals with Nonlinear Phase Noise Compensation.

		Mean Nonlinear Phase Shift $\langle\Phi_{\mathrm{NL}}\rangle$		Max. Diff. to Nonlin. MAP (dB)
		1-dB Penalty (rad)	Optimal Point (rad)	
Without Compensation		1.00	1.25	—-
MMSE	Linear	2.26	2.30	0.41
	Nonlinear	2.31	2.20	0.21
MAP	Linear	2.30	2.15	0.14
	Nonlinear	2.35	2.12	0.00

that the increase of SNR penalty is less than the increase of SNR that is proportional to the mean nonlinear phase shift. Because of the steepness of the slope, the optimal MAP detector actually gives smaller optimal operating point than other compensation schemes.

On the basis of the distribution of a received signal with nonlinear phase noise, the optimal MAP detector is derived for a phase-modulated signal to minimize the error probability. The error probability of nonlinear MAP and MMSE detector is also evaluated using the distribution of the received signal. Having the same variance of residual nonlinear phase noise, nonlinear MMSE compensator performs up to 0.23 dB better than linear MMSE compensator. The optimal nonlinear MAP compensator performs very close to the optimal linear MAP compensator with a difference less than 0.14 dB.

This section discussed the optimal nonlinear compensator for PSK systems with infinite number of fiber spans based on the model of Appendix 5.A and 6.B. For system with finite number of fiber span, similar nonlinear compensator can be derived using the model of Appendix 5.B.

6. Summary

The compensation of nonlinear phase noise is studied in detail. Based on the correlation of received intensity with nonlinear phase noise, the simplest compensator is linear compensator to use a scale version of the received intensity to reduce nonlinear phase noise. The linear compensator can be optimized using the MMSE criterion to minimize the variance of the residual nonlinear phase noise. The variance of the residual nonlinear phase noise is about a quarter of that of nonlinear phase noise, doubling the transmission distance if nonlinear phase noise is the major impairment.

While the MMSE criterion does not provide a minimum error probability, the linear MAP compensator optimized by numerical methods is able to well approximate the optimal nonlinear detector. In practice, the optimal detector can be implemented as a Yin-Yang detector or a compensator.

The optimal compensation scheme is distributed or mid-span compensator. Even with a single compensator, if located 2/3 from the beginning of the fiber link, the STD of residual nonlinear phase noise is reduced to 1/3 that without compensation, far better then a compensator co-located with the receiver.

Ideally, the nonlinear phase noise compensators should minimize the error probability of a phase-modulated system with nonlinear phase noise. The nonlinear MAP compensator is found numerically as the optimal detector to minimize the error probability. In term of variance, nonlinear MMSE compensator performs the same as linear MMSE compensator. Nonlinear MAP compensator performs up to about 0.21 dB better than nonlinear MMSE compensator but just 0.14 dB better than linear MAP compensator.

APPENDIX 6.A: Nonlinear MMSE Compensation

Using the received intensity of $P_N = r_r^2$, the optimal compensator is the MMSE estimator based on the conditional mean of

$$E\{\Phi_{\mathrm{NL}}|r_r\}, \tag{6.A.1}$$

where $E\{\cdot\}$ denotes expectation.

The variance of the difference of $\Phi_{\mathrm{NL}} - E\{\Phi_{\mathrm{NL}}|r_r\}$ is minimized using the nonlinear MMSE compensator of Eq. (6.A.1). The optimal compensator can be implemented using a phase modulator driven by the waveform of Eq. (6.A.1) or an electrical compensator to calculate $\vec{E}_r \exp[jE\{\Phi_{\mathrm{NL}}|r_r\}]$ using electronic signal processing. With the received electric field E_r, the detector can also be implemented as a Yin-Yang detector. If the received phase is demodulated as $\Phi_r = \arg(E_r)$, the compensator is a combiner of $\Phi_r + E\{\Phi_{\mathrm{NL}}|r_r\}$. There is no difference between $E\{\Phi_{\mathrm{NL}}|r_r^2\}$ and the expectation of Eq. (6.A.1) because the received amplitude is a positive number $r_r \geq 0$. The estimator of Eq. (6.A.1) is called a correction by received intensity although the expression of Eq. (6.A.1) is based on received amplitude.

Without loss of generality, assume that the transmitted electric field is $E_0 = A$, we need to calculate the estimation of term by term of

$$E\left\{|A + \vec{n}_1 + \cdots + \vec{n}_k|^2 \Big| |\vec{E}_0 + \vec{n}_1 + \cdots + \vec{n}_{N_A}|\right\}. \tag{6.A.2}$$

The estimation of Eq. (6.A.2) is a special case of the estimation of

$$\varphi(\sigma_{\zeta_1}, \sigma_{\zeta_2}) = E\left\{|\vec{E}_0 + \vec{\zeta}_1|^2 \Big| \rho_2\right\}, \tag{6.A.3}$$

for a real value A and two zero-mean complex circular Gaussian variables ζ_1 and ζ_2, where $\rho_2 = |E_0 + \zeta_1 + \zeta_2|$, $E\{\zeta_1\} = E\{\zeta_2\} = 0$, $E\{|\zeta_1|^2\} = 2\sigma_{\zeta_1}^2$, and $E\{|\zeta_2|^2\} = 2\sigma_{\zeta_2}^2$. The estimation of Eq. (6.A.3) is used to derive the MMSE estimator of Eq. (6.A.1) and the value of Eq. (6.A.2). The real value of A can be used to represent the amplitude of $|E_0|$ that is a constant for both PSK and DPSK signals.

Define $x_1 + jy_1 = A + \zeta_1$ and $x_2 + jy_2 = A + \zeta_1 + \zeta_2$, the joint p.d.f. of x_1, y_1, x_2, y_2 is

$$\begin{aligned} f_{xy}(x_1, y_1, x_2, y_2) = \frac{1}{(2\pi\sigma_{\zeta_1}\sigma_{\zeta_2})^2} \exp\left[-\frac{(x_1 - A)^2 + y_1^2}{2\sigma_{\zeta_1}^2}\right] \\ \times \exp\left[-\frac{(x_2 - x_1)^2 + (y_2 - y_1)^2}{2\sigma_{\zeta_2}^2}\right]. \end{aligned} \tag{6.A.4}$$

Using the joint p.d.f. of Eq. (6.A.4), changing the variable to $x_2 = \rho_2 \cos\theta_2$ and $y_2 = \rho_2 \sin\theta_2$, the marginal p.d.f. of ρ_2 is a Rice distribution of

$$\begin{aligned} f_{\rho_2}(\rho_2) &= \int_{-\infty}^{+\infty}\int_{-\infty}^{+\infty}\int_{-\pi}^{+\pi} f_{xy}(x_1, y_1, \rho_2\cos\theta_2, \rho_2\sin\theta_2)\, \rho_2 \mathrm{d}\theta_2 \mathrm{d}x_1 \mathrm{d}y_1 \\ &= \frac{\rho_2}{(\sigma_1^2 + \sigma_2^2)^{1/2}} \exp\left(-\frac{A^2 + \rho_2^2}{\sigma_{\zeta_1}^2 + \sigma_{\zeta_2}^2}\right) I_0\left(\frac{A\rho_2}{\sigma_{\zeta_1}^2 + \sigma_{\zeta_2}^2}\right). \end{aligned} \tag{6.A.5}$$

The estimation of Eq. (6.A.3) is

$$\varphi(\sigma_{\zeta_1}, \sigma_{\zeta_2}) = \int_{-\infty}^{+\infty}\int_{-\infty}^{+\infty} (x_1^2 + y_1^2) f_{xy|\rho_2}(x_1, y_1|\rho_2)\, \mathrm{d}x_1 \mathrm{d}y_1, \tag{6.A.6}$$

where the conditional p.d.f. is

$$f_{xy|\rho_2}(x_1, y_1|\rho_2) = \frac{1}{f_{\rho_2}(\rho_2)} \int_{-\pi}^{+\pi} f_{xy}(x_1, y_1, \rho_2 \cos\theta_2, \rho_2 \sin\theta_2)\, \rho_2 \mathrm{d}\theta_2. \tag{6.A.7}$$

The integration of Eq. (6.A.6) becomes

$$\frac{\rho_2}{f_{\rho_2}(\rho_2)} \int_{-\pi}^{+\pi} \int_{-\infty}^{+\infty} \int_{-\infty}^{+\infty} (x_1^2 + y_1^2) f_{xy}(x_1, y_1, \rho_2 \cos\theta_2, \rho_2 \sin\theta_2)\, \mathrm{d}x_1 \mathrm{d}y_1 \mathrm{d}\theta_2. \tag{6.A.8}$$

Integrated first over x_1 and x_2, we obtain

$$\frac{1}{2\pi(\sigma_{\zeta_1}^2 + \sigma_{\zeta_2}^2)^2 I_0\left(\frac{A\rho_2}{\sigma_{\zeta_1}^2 + \sigma_{\zeta_2}^2}\right)} \int_{-\pi}^{+\pi} e^{-A\rho_2 \cos\theta_2/(\sigma_{\zeta_1}^2 + \sigma_{\zeta_2}^2)} \Big[\sigma_{\zeta_1}^4 (\rho_2^2 + 2\sigma_{\zeta_2}^2)$$
$$+ \sigma_{\zeta_2}^4 (A^2 + 2\sigma_{\zeta_1}^2) + 2A\rho_2 \sigma_{\zeta_1}^2 \sigma_{\zeta_2}^2 \cos\theta_2\Big] \mathrm{d}\theta_2. \tag{6.A.9}$$

The integration of Eq. (6.A.6) is

$$\begin{aligned} \varphi(\sigma_{\zeta_1}, \sigma_{\zeta_2}) &= \frac{\sigma_{\zeta_1}^4 (\rho_2^2 + 2\sigma_{\zeta_2}^2) + \sigma_{\zeta_2}^4 (A^2 + 2\sigma_{\zeta_1}^2)}{(\sigma_{\zeta_1}^2 + \sigma_{\zeta_2}^2)^2} \\ &\quad + \frac{2A\rho_2 \sigma_{\zeta_1}^2 \sigma_{\zeta_2}^2}{(\sigma_{\zeta_1}^2 + \sigma_{\zeta_2}^2)^2} \frac{I_1\left(A\rho_2/(\sigma_{\zeta_1}^2 + \sigma_{\zeta_2}^2)\right)}{I_0\left(A\rho_2/(\sigma_{\zeta_1}^2 + \sigma_{\zeta_2}^2)\right)}. \end{aligned} \tag{6.A.10}$$

For the case of high SNR, the expectation of Eq. (6.A.3) can be approximated by

$$\begin{aligned} \varphi(\sigma_{\zeta_1}, \sigma_{\zeta_2}) &= E\Big\{|E_0 + \zeta_1|^2 \Big| |E_0 + \zeta_1 + zeta_2|^2\Big\} \\ &\approx E\Big\{A^2 + 2A(x_1 - A) \Big| A^2 + 2A(x_2 - A)\Big\}, \end{aligned} \tag{6.A.11}$$

Using the approximation of Eq. (6.A.11), the expectation of Eq. (6.A.3) becomes

$$\varphi(\sigma_{\zeta_1}, \sigma_{\zeta_2}) \approx \frac{\sigma_{\zeta_2}^2}{\sigma_{\zeta_1}^2 + \sigma_{\zeta_2}^2} A^2 + \frac{\sigma_{\zeta_1}^2}{\sigma_{\zeta_1}^2 + \sigma_{\zeta_2}^2} \rho_2^2. \tag{6.A.12}$$

Compared the estimation of Eq. (6.A.2) with (6.A.3), we have $\zeta_1 = n_1 + \cdots + n_k$ and $\zeta_2 = n_{k+1} + \cdots + n_{N_A}$ that are zero-mean complex circular Gaussian random variables with variance of $E\{|\zeta_1|^2\} = 2k\sigma_0^2$ and $E\{|\zeta_2|^2\} = 2(N_A - k)\sigma_0^2$. Using the function of $\varphi(\sigma_{\zeta_1}, \sigma_{\zeta_2})$ defines in Eq. (6.A.3), we obtain

$$E\left\{|A + n_1 + \cdots + n_k|^2 \Big| \rho_r\right\} = \varphi\left(\sqrt{k}\sigma_0, \sqrt{N_A - k}\sigma_0\right). \tag{6.A.13}$$

For the estimator of Eq. (6.A.1), we obtain

$$E\left\{\Phi_{\mathrm{NL}}|\rho_r\right\} = \gamma L_{\mathrm{eff}} \sum_{k=1}^{N_A} \varphi\left(\sqrt{k}\sigma_0, \sqrt{N_A - k}\sigma_0\right). \tag{6.A.14}$$

Substitute the function of $\varphi(\sigma_{\zeta_1}, \sigma_{\zeta_2})$ of Eq. (6.A.10) to Eq. (6.A.14), we get

$$\begin{aligned} E\left\{\Phi_{\mathrm{NL}}|\rho_r\right\} = \gamma L_{\mathrm{eff}} \Bigg\{ &\frac{(2N_A - 1)(N_A - 1)}{6N_A} A^2 + \frac{N_A^2 - 1}{3} \sigma_0^2 \\ &+ \frac{(2N_A + 1)(N_A + 1)}{6N_A} \rho_r^2 + \frac{(N_A^2 - 1)A\rho_r}{3N_A} \frac{I_1\left(\frac{A\rho_r}{N_A\sigma_0^2}\right)}{I_0\left(\frac{A\rho_r}{N_A\sigma_0^2}\right)} \Bigg\}. \end{aligned} \tag{6.A.15}$$

To derive Eq. (6.A.15), the relationships of Eqs. (6.18) to (6.20) are used.

Because $I_1(x)/I_0(x) \sim x/2$ for small $x < 0.5$, the phase estimator of Eq. (6.A.15) is a linear function of the received intensity of ρ_r^2 when the received signal is small. Because $I_1(x)/I_0(x) \sim 1$ for large $x > 10$, the phase estimator of Eq. (6.A.15) is a quadratic function of the received amplitude of ρ_r when the received signal is large. The interested systems usually have high SNR and the received signal has an amplitude around A, i.e., $\rho_r \sim A$, the phase estimator of Eq. (6.A.15) mostly functions as a nonlinear compensator.

For large number of fiber span of $N_A \gg 1$ and high SNR, the phase estimator of Eq. (6.A.15) is approximately

$$E\left\{\Phi_{\mathrm{NL}}|\rho_r\right\} \approx \frac{\langle\Phi_{\mathrm{NL}}\rangle}{3}\left[1 + \frac{\rho_r^2}{A^2} + \frac{\rho_r}{A}\frac{I_1\left(\rho_s A/\rho_r\right)}{I_0\left(\rho_s A/\rho_r\right)}\right], \tag{6.A.16}$$

where the mean nonlinear phase noise $\langle\Phi_{\mathrm{NL}}\rangle$ is given by Eq. (6.16). The estimation of Eq. (6.A.16) divides the estimation into three parts, the first term is one-third the mean nonlinear phase shift, and the second term proportional to received intensity, and third term is the nonlinear term.

While most of the above discussion is related to PSK systems, the optimal compensator for DPSK systems follows the same principle. For DPSK signals, the differential phase of $\Phi_r(t) - \Phi_r(t-T)$ is detected, where T is the symbol period. The optimal MMSE compensator should be

$$E\left\{\Phi_{\mathrm{NL}}(t)|\rho_r(t)\right\} - E\left\{\Phi_{\mathrm{NL}}(t-T)|\rho_r(t-T)\right\}, \tag{6.A.17}$$

similar to Eq. (6.A.15).

APPENDIX 6.B: Joint Characteristic Function

Here, the joint characteristic function of nonlinear phase noise, received intensity, and the phase of amplifier noise is derived. Corresponding to the Fourier coefficients, only the characteristic function at integer "frequency" of the phase of amplifier noise of Θ_n is derived. Using the partial p.d.f. and characteristic function of Eq. (5.A.10), with m as an non-negative integer, we obtain (Gradshteyn and Ryzhik, 1980, §8.431)

$$\Psi_{\Phi,Y,\Theta_n}(\nu,\omega,m) = \frac{\Psi_\Phi(\nu)}{2\sigma_\nu^2}\int_0^\infty \exp\left[-\frac{y+|\vec{\xi}_\nu|^2}{2\sigma_\nu^2}\right] I_m\left(\frac{\sqrt{y}|\vec{\xi}_\nu|}{\sigma_\nu^2}\right) e^{j\omega y}dy, \qquad m \geq 0, \tag{6.B.1}$$

and $\Psi_{\Phi,Y,\Theta_n}(\nu,\omega,-m) = \Psi_{\Phi,Y,\Theta_n}(\nu,\omega,m)$, where $I_m(\cdot)$ is the mth-order modified Bessel function of the first kind.

Using Gradshteyn and Ryzhik (1980, §6.614, §9.220), we get

$$\Psi_{\Phi,Y,\Theta_n}(\nu,\omega,m) = \Psi_\Phi(\nu)\exp\left[-\frac{|\vec{\xi}_\nu|^2}{2\sigma_\nu^2}\right]\frac{\Gamma(m/2+1)}{m!} \times \frac{\sqrt{2}\sigma_\nu\gamma_{\nu,\omega}^{(m+1)/2}}{(1-2j\omega\sigma_\nu^2)^{1/2}|\vec{\xi}_\nu|}\,{}_1F_1\left(\frac{m}{2}+1; m+1; \gamma_{\nu,\omega}\right), \qquad m \geq 0, \tag{6.B.2}$$

where

$$\begin{aligned}\gamma_{\nu,\omega} &= \frac{1}{1-2j\omega\sigma_\nu^2}\frac{|\vec{\xi}_\nu|^2}{2\sigma_\nu^2} \\ &= \frac{2j\nu}{\left(\sqrt{j\nu}-j\omega\tan\sqrt{j\nu}\right)\sin\left(2\sqrt{j\nu}\right)}\rho_s,\end{aligned} \tag{6.B.3}$$

$\Gamma(\cdot)$ is the Gamma function, and ${}_1F_1(a;b;\cdot)$ is the confluent hypergeometric function of the first kind.

Using γ_ν defined by Eq. (5.A.12), we can rewrite Eq. (6.B.2) as

$$\Psi_{\Phi,Y,\Theta_n}(\nu,\omega,m) = \Psi_\Phi(\nu)\frac{e^{-\gamma_\nu}\Gamma(\frac{1}{2}m+1)}{m!}\frac{\gamma_{\nu,\omega}^{\frac{m}{2}+1}}{\gamma_\nu}{}_1F_1\left(\frac{m}{2}+1;m+1;\gamma_{\nu,\omega}\right). \tag{6.B.4}$$

With $\omega = 0$ in Eq. (6.B.4), using Gradshteyn and Ryzhik (1980, §9.212), the joint Fourier coefficients of nonlinear phase noise and the phase of amplifier noise are

$$\begin{aligned}\Psi_{\Phi,\Theta_n}(\nu,m) &= \Psi_\Phi(\nu)\frac{e^{-\gamma_\nu}\gamma_\nu^{\frac{m}{2}}}{m!}\Gamma\left(\frac{m}{2}+1\right){}_1F_1\left(\frac{m}{2}+1;m+1;\gamma_\nu\right) \\ &= \frac{\Psi_\Phi(\nu)\gamma_\nu^{\frac{m}{2}}}{m!}\Gamma\left(\frac{m}{2}+1\right){}_1F_1\left(\frac{m}{2};m+1;-\gamma_\nu\right),\end{aligned} \tag{6.B.5}$$

the same as Eq. (5.62). We may also derive Eq. (5.62) using the method here.

If $m = 0$ in Eq. (6.B.4), using the relationship of ${}_1F_1(1;1;z) = e^z$ (Gradshteyn and Ryzhik, 1980, §9.215), the joint characteristic function of nonlinear phase noise and the received intensity is

$$\begin{aligned}\Psi_{\Phi,Y}(\nu,\omega) &= \Psi_\Phi(\nu)\frac{\gamma_{\nu,\omega}}{\gamma_\nu}e^{-\gamma_{\nu,0}+\gamma_{\nu,\omega}} \\ &= \frac{\Psi_\Phi(\nu)}{1-2j\omega\sigma_\nu^2}\exp\left[\frac{j\omega|\vec{\xi}_\nu|^2}{1-2j\omega\sigma_\nu^2}\right],\end{aligned} \tag{6.B.6}$$

or

$$\Psi_{\Phi,Y}(\nu,\omega) = \frac{1}{\cos\sqrt{j\nu}-j\omega\frac{\sin\sqrt{j\nu}}{\sqrt{j\nu}}}\exp\left[\rho_s\sqrt{j\nu}\tan\sqrt{j\nu}+\frac{j\omega\rho_s}{\cos^2\sqrt{j\nu}-j\omega\frac{\sin(2\sqrt{j\nu})}{2\sqrt{j\nu}}}\right]. \tag{6.B.7}$$

The characteristic function of Eq. (6.B.6) is the same as the characteristic function of noncentral χ^2 distribution with two degrees of freedom. Using the joint characteristic function of Eq. (6.B.6), take the inverse Fourier transforms, we obtain

$$\mathcal{F}_\omega^{-1}\left\{\Psi_{\Phi,Y}(\nu,\omega)\right\} = \frac{\Psi_\Phi(\nu)}{2\sigma_\nu^2}\exp\left[-\frac{y+|\vec{\xi}_\nu|^2}{2\sigma_\nu^2}\right]I_0\left(\frac{\sqrt{y}|\vec{\xi}_\nu|}{\sigma_\nu^2}\right), \tag{6.B.8}$$

similar to a noncentral χ^2 distribution with two degrees of freedom. With $\nu = 0$, the p.d.f. of received intensity is

$$p_Y(y) = \exp[-(y+\rho_s)]I_0(2\sqrt{y\rho_s}). \tag{6.B.9}$$

To simplify the characteristic function of Eq. (6.B.4) using the modified Bessel functions, from Gradshteyn and Ryzhik (1980, §9.212, §9.238) and similar to Jain

(1974), Jain and Blachman (1973) and Blachman (1981, 1988), we get

$$\Psi_{\Phi,Y,\Theta_n}(\nu,\omega,m) = \Psi_{\Phi}(\nu)\frac{\sqrt{\pi}\gamma_{\nu,\omega}^{3/2}}{2\gamma_{\nu}} \exp\left(-\gamma_{\nu} + \frac{\gamma_{\nu,\omega}}{2}\right)\left[I_{\frac{m-1}{2}}\left(\frac{\gamma_{\nu,\omega}}{2}\right) + I_{\frac{m+1}{2}}\left(\frac{\gamma_{\nu,\omega}}{2}\right)\right]. \tag{6.B.10}$$

In this chapter, the joint characteristic function of Eq. (6.B.10) is used to calculate the error probability of phase-modulated signal with linearly compensated nonlinear phase noise. The received phase is the linear combination of Φ, Y, and Θ_n.

Chapter 7

INTRACHANNEL PULSE-TO-PULSE INTERACTION

When optical pulses are propagated in an optical fiber, the pulses are broadened by fiber chromatic dispersion as shown in Sec. 4.4. When the broadened optical pulses overlap with each other, phase modulation is induced from adjacent pulses due to fiber Kerr effect. Those phase modulation from adjacent pulses gives both intrachannel cross-phase modulation (IXPM) and intrachannel four-wave-mixing (IFWM). For systems using on-off keying, IXPM and IFWM induce timing and amplitude jitter to existing optical pulses, respectively. IFWM also gives ghost pulses in the time slots without optical pulses.

For phase-modulated signal, in term of intensity, as shown in Fig. 2.17, the launched signal from the transmitter is a constant pulse train. Unlike on-off keying, IXPM from differential phase-shift keying (DPSK) signal induces both constant timing jitter and phase modulation to all DPSK pulses. The ghost pulses from IFWM are equivalently noise to other optical pulses. This chapter focuses on the analysis of IFWM, mostly for return-to-zero (RZ)-DPSK signal.

Intrachannel pulse-to-pulse interaction is significant only for system with pulse overlap. For fiber link with small chromatic dispersion and limited pulse broadening, the dominant effect should be nonlinear phase noise in Chapter 5. Nonlinear phase noise cannot be avoided but can be compensated as shown in Chapter 6. Other than general methods to reduce or compensate for fiber nonlinearities (Brener et al., 2000, Paré et al., 1996, Pepper and Yariv, 1980), there is no known method to compensate for IFWM but IFWM can be reduced using better dispersion management, modified signal format, and other techniques.

A broadened optical pulse has a peak amplitude inversely proportional to the pulse width. Nonlinear phase noise is reduced with the increase of fiber chromatic dispersion. However, the phase noise from the beating of signal with noise also spreads to adjacent pulse, causing IXPM-induced nonlinear phase noise. The combined nonlinear phase noise of dispersive transmission system is larger than the impact of IFWM-induced ghost pulses.

1. Pulse Overlap in Dispersive Fiber

Fiber dispersion broadens the optical pulses and limits the transmission distance of an optical signal. Chromatic dispersion was studied in Sec. 4.4 based on computer simulation. While the impact of chromatic dispersion to practical signal must be studied using computer simulation, analytical results are available for Gaussian pulse in a dispersive fiber.

For simplicity, assume that the pulse launched into the optical fiber has a low pass representation of

$$u_0(0, t) = A_0 \exp\left[-\frac{1}{2}\left(\frac{t}{T_0}\right)^2\right], \tag{7.1}$$

as a Gaussian pulse with a peak amplitude of A_0, where T_0 is the $1/e$ width of the pulse intensity, and $1.66T_0$ is full-width-half-maximum (FWHM) pulse width. Based on the simple model of Eq. (4.46) in Sec. 4.4, the pulse spectrum at the distance of z is

$$B(z, \omega) = A_0\sqrt{2\pi}T_0 \exp\left[-\frac{\omega^2 T_0^2 - j\beta_2\omega^2}{2}\right], \tag{7.2}$$

with a pulse shape of

$$u_0(z, t) = \frac{A_0 T_0}{(T_0^2 - j\beta_2 z)^{1/2}} \exp\left[-\frac{t^2}{2(T_0^2 - j\beta_2 z)}\right], \tag{7.3}$$

where β_2 is the group velocity dispersion coefficient. After a distance of z, the pulse of Eq. (7.1) broadens to an $1/e$ pulse width of

$$\tau(z) = \sqrt{T_0^2 + \beta_2^2 z^2/T_0^2}. \tag{7.4}$$

A Gaussian pulse is broadened with distance from Eq. (7.4), for short pulse with small initial pulse width of T_0, the pulse width increases linearly with distance as $\tau(z) \approx |\beta_2|z/T_0$. For 40-Gb/s system with $T = 25$ ps and using RZ pulses with duty-cycle of 50% of $T_0 = 7.53$ ps. In typical standard single-mode fiber, $\beta_2 = -21.7$ ps^2/km, corresponding to $D = 17$ ps/km/nm at the wavelength of 1.55 μm. The pulse width is doubled just after a short distance of about 4.5 km, far shorter than the effective nonlinear length of $L_{\text{eff}} \approx 1/\alpha = 21.7$ km. When $z = L_{\text{eff}}$, the pulse width is already $8.5T_0$ in this particular case. For low-power linear transmission, dispersion compensation can be used at the end of the fiber to return the original launched pulse. When optical amplifiers are used periodically to boost up the power, nonlinear pulse-to-pulse interaction is very strong in the system due to pulse overlap.

Pulse overlap is reduced for system with small fiber dispersion. For example, if fiber dispersion coefficient is $D = 3.5$ ps/km/nm, typically for nonzero dispersion-shifted fiber (NZDSF), the pulse width is doubled after a fiber distance of 22.3 km. For low speed system with long pulse width, for example, 10-Gb/s system with $T_0 = 30.1$ ps, the pulse width is doubled after a fiber distance of 71.4 km for $D = 17$ ps/km/nm. Pulse-to-pulse interaction is a particular problem mostly for 40 Gb/s systems, especially for system using standard single-mode fiber with large dispersion.

The above analysis for Gaussian pulse is largely based on Marcuse (1980, 1981), and Agrawal (2002). We assume a zero-chirp Gaussian pulse for simplicity. Both Marcuse (1981) and Agrawal (2002) derive the pulse broadening for chirp Gaussian pulse. For non-Gaussian pulses or fiber with large third-order dispersion, there is no analytical solution for the pulse shape but the pulse width (Marcuse, 1980, 1981, Miyagi and Nishida, 1979).

Without pulse broadening, there is no pulse-to-pulse interaction. However, pulse broadening is not necessary bad for pulse-to-pulse interaction. From Eq. (7.3), the amplitude of the pulse decreases with pulse width. After sufficient pulse broadening, the effects of pulse-to-pulse interaction decrease accordingly.

For illustration purpose, for two optical pulses of

$$E_1(t) = A_0 e^{-t^2/2\tau^2}, \qquad E_2(t) = A_0 e^{-(t-T)^2/2\tau^2}, \tag{7.5}$$

with the same $1/e$ width of τ and separated by the time of T, the action of cross-phase modulation (XPM) from $E_2(t)$ to $E_1(t)$ can be described by $dE_1(t)/dz = j\gamma E_1(t)|E_2(t)|^2$. With a phase shift of $\gamma|E_2(t)|^2$, the instantaneous phase shift is $d\delta\phi_1(t)/dz = \gamma|E_2(t)|^2$ with a mean phase shift of

$$\frac{d\Delta\phi_1}{dz} = \frac{\int \frac{d\delta\phi_1(t)}{dz}|E_1(t)|^2 \mathrm{d}t}{\int |E_1(t)|^2 \mathrm{d}t}, \tag{7.6}$$

or

$$\frac{d\Delta\phi_1}{dz} = \frac{\gamma P_0}{\sqrt{2\pi}}\frac{T}{\tau}\exp\left(-\frac{T^2}{2\tau^2}\right), \tag{7.7}$$

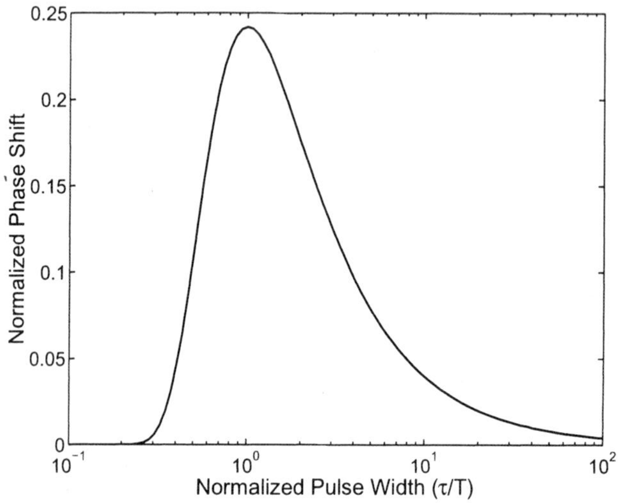

Figure 7.1. The normalized phase shift function as a function of τ/T

where the optical power of the pulse train is equal to

$$P_0 = \sqrt{\pi} A_0^2 \tau / T. \tag{7.8}$$

Figure 7.1 shows the function of $\frac{1}{\sqrt{2\pi}}(\tau/T)e^{-T^2/2\tau^2}$ from Eq. (7.7) as a function of τ/T. With no pulse overlap when $\tau \ll T$, there is also no phase shift as from Fig. 7.1. The phase shift is the largest at $\tau = T$ with moderated pulse overlap but decreases with the occurrence of further pulse overlap when $\tau \gg T$. Figure 7.1 shows two regions with small pulse-to-pulse interaction effect when τ is either much smaller or larger than the pulse separation of T.

In practical communication systems, a train of pulse is transmitted to carry information. Pulse broadening reduces the effect of each interacting component as from Fig. 7.1. However, two far away pulses with large time separation may also overlap with each other and gives more phase shift. Systems with large dispersion and fast pulse broadening may but not necessary provide better performance.

Intrachannel interaction in very dispersive fiber is first observed by Shake et al. (1998) and Essiambre et al. (1999). The reduction of intrachannel pulse-to-pulse interaction by pulse broadening was first proposed in Mamyshev and Mamysheva (1999). The derivation here follows the method of Mamyshev and Mamysheva (1999). As a comparison, in soliton-to-soliton interaction, as shown in Gordon (1983), distortion increases monotonically with pulse overlap.

2. Intrachannel Four-Wave Mixing

The combination of fiber Kerr effect, fiber loss and dispersion can be described by the nonlinear Schrödinger equation of

$$\frac{\partial A}{\partial z} + \frac{j\beta_2}{2}\frac{\partial^2 A}{\partial z^2} + \frac{\alpha}{2}A = j\gamma|A|^2 A, \tag{7.9}$$

where $A(z,t)$ is the low-pass representation of the optical signal. At the location of z, assume that there are pulses of

$$u_k(z,t) = \frac{A_k T_0}{(T_0^2 - j\beta_2 z)^{1/2}} \exp\left[-\frac{(t-kT)^2}{2(T_0^2 - j\beta_2 z)}\right], \tag{7.10}$$

where $A_k = \{0, A_0\}$ for on-off keying and $A_k = \pm A_0$ for either phase-shift keying (PSK) or DPSK signal. The overall optical signal is $A(z,t) = e^{-\alpha z/2}\sum_k u_k(z,t)$.

The nonlinear force of $j\gamma|A|^2 A$ for nonlinear Schrödinger equation is $F_0(z,t) = j\gamma u_l n_m u_n^* e^{-3\alpha z/2}$ when the pulses of u_l, u_m, and u_n are considered in $A(z,t) = u_l + u_m + u_n + \cdots$. Self-phase modulation (SPM) is induced to the pulse of u_l if $m = n = l$. SPM from $m = n = l$ is called intrachannel SPM (ISPM) in this chapter. Intrachannel cross-phase modulation (IXPM) is induced to the pulse of u_l when $m = n \neq l$ and the pulse of u_m when $l = n \neq m$. As shown later, intrachannel four-wave mixing (IFWM) is induced to the pulse of u_k if $k = l + m - n$, $l \neq n$, and $m \neq n$.

The nonlinear force term of $F_0(z,t)$ is equal to

$$F_0(z,t) = j\gamma A(z)\exp\left[-a(z)t^2 + b(z)Tt - c(z)T^2\right] \tag{7.11}$$

where

$$A(z) = \frac{A_l A_m A_n^* T_0^3 e^{-3\alpha z/2}}{\sqrt{|\xi|^2\xi}}, \qquad a(z) = \frac{1}{2}\left(\frac{1}{\xi^*} + \frac{2}{\xi}\right),$$

$$b(z) = \frac{n}{\xi^*} + \frac{l+m}{\xi}, \qquad c(z) = \frac{1}{2}\left(\frac{n^2}{\xi^*} + \frac{l^2+m^2}{\xi}\right) \tag{7.12}$$

with $\xi = T_0^2 - j\beta_2 z$. The Fourier transform of $F_0(z,t)$ is

$$\hat{F}_0(z,\omega) = j\gamma A(z)\sqrt{\frac{\pi}{a(z)}}\exp\left\{-c(z)T^2 - \frac{[\omega + jb(z)T]^2}{4a(z)}\right\}. \tag{7.13}$$

With large pulse overlap when $|\beta_2|z \gg T_0$, $\xi \sim -j\beta_2 z$ is close to a pure imaginary number, we obtain $b(z) \sim (l+m-n)/\xi$ and $a(z) \sim 1/2\xi$. Ignored some constant factors independent of ω, we obtain

$$\hat{F}_0(z,\omega) \sim \exp\left[-\frac{j}{2}\beta_2 z\omega^2 - j(l+m-n)\omega T\right]. \tag{7.14}$$

In time domain, the nonlinear force of $F_0(z,t)$ is centered at the time of $(l+m-n)T$, similar to four-wave mixing effects, the pulse width is equal to $\sqrt{|\beta_2|z/2}$.

The nonlinear force of $F_0(z,t)$ propagates for the remaining distance of $L-z$ to reach the destination of L. From the location of z to L, there is a fiber loss of $e^{-\alpha(L-z)/2}$, an optical amplifier with gain of $e^{\alpha L/2}$, and a dispersion compensator for an overall dispersion of $\beta_2 L$ for the whole fiber span, all located at the destination of L. At $z = L$, the nonlinear force of $\hat{F}_0(z,\omega)$ gives an electric field with Fourier transform of

$$\hat{F}_0(z,\omega)e^{-j\beta_2 z\omega^2/2}e^{\alpha z/2}. \tag{7.15}$$

For the whole fiber span from 0 to L, the overall nonlinear force in time domain is accumulated to

$$\Delta u_{l,m,n}(L,t) = \int_0^L \mathcal{F}_\omega^{-1}\left\{\hat{F}_0(z,\omega)e^{-j\beta_2 z\omega^2/2}\right\} e^{\alpha z/2}\mathrm{d}z, \tag{7.16}$$

where $\mathcal{F}_\omega^{-1}$ denotes inverse Fourier transform with respect to ω. We obtain

$$\begin{aligned}\mathcal{F}_\omega^{-1}\left\{\hat{F}_0(z,\omega)e^{-j\beta_2 z\omega^2/2}\right\} &= \frac{j\gamma A(z)}{\sqrt{1+2j\beta_2 z a(z)}} \\ &\times \exp\left\{-\frac{a(z)t^2 - b(z)Tt - jb^2(z)\beta_2 z/2}{1+2j\beta_2 z a(z)} - c(z)T^2\right\}. \end{aligned} \tag{7.17}$$

The above electric field can be simplified for $k = l+m-n$ in the general case. With a constant time shift, when $l+m-n = 0$ for the ghost pulses that affect $u_0(z,t)$ at $t = 0$, after some simplifications, we obtain

$$
\begin{aligned}
\Delta u_{l,m,l+m}(L,t) &= j\gamma A_l A_m A_{l+m}^* \exp\left(-\frac{t^2}{6T_0^2}\right) \\
&\times \int_0^L \frac{e^{-\alpha z}}{\sqrt{1+2j\beta_2 z/T_0^2+3(\beta_2 z/T_0^2)^2}} \\
&\times \exp\left\{-\frac{3(2t/3-lT)(2t/3-mT)}{T_0^2+3j\beta_2 z}\right. \\
&\left. -\frac{(m-l)^2T^2}{T_0^2\left[1+2j\beta_2 z/T_0^2+3(\beta_2 z/T_0^2)^2\right]}\right\}.
\end{aligned} \tag{7.18}
$$

Unlike on-off keying, both PSK and DPSK signals are a constant pulse train as shown in Fig. 2.17. IXPM is induced to u_0 if $l = n, m = 0$ or $m = n, l = 0$, i.e., either $l = 0$ or $m = 0$. Assume that $l = 0$, the factor outside of the integration in Eq. (7.18) is $j\gamma A_0|A_m|^2$, i.e., a phase modulation proportional to $\gamma|A_m|^2$. While A_m induced phase modulation to u_0, A_{m+1} induces phase modulation to u_1. Because $|A_m|^2$ and $|A_{m+1}|^2$ has the same intensity of $|A_0|^2$ for PSK and DPSK signals, the phase difference between adjacent pulses u_0 and u_1 does not change due to IXPM. However, for on-off keying, there is 1/4 probability that $|A_m|^2 = |A_0|^2$ but $|A_{m+1}|^2 = 0$. In this particular case, the phase modulation changes the frequency of the pulse of u_0 but not u_1. Combined with chromatic dispersion of the fiber, the frequency modulation of on-off keying signal is translated to timing jitter. For phase-modulated signal, the timing jitter for all pulses is the same, i.e., no timing jitter variation. The timing jitter for on-off keying signal was considered in Ablowitz and Hirooka (2001, 2002), Kumar et al. (2002), Mårtensson et al. (2001), and Mecozzi et al. (2000a,b, 2001). IXPM does not degrade the DPSK signals. Similarly, ISPM also does not degrade DPSK signals.

The pulse-to-pulse interaction is small if optical pulses do not overlap. Significant interaction occurs only when $|\beta_2|z/T_0^2 \gg 1$. In this limit and for lossless fiber with $\alpha = 0$, we obtain

$$
\Delta u_{l,m,l+m}(L,t) = j\gamma A_l A_m A_{l+m}^* e^{-t^2/6T_0^2} \frac{T_0^2}{\sqrt{3}|\beta_2|} E_1\left(-j\frac{lmT^2}{\beta_2 L}\right), \tag{7.19}
$$

where $E_1(x) = \int_1^{+\infty} t^{-1}e^{-xt}dt$ is the exponential integral (Gradshteyn and Ryzhik, 1980, §8.21). From Eq. (7.19), the pulse induced from the nonlinear force is Gaussian shape with a pulse width of $\sqrt{3}T_0$. From the channel power of Eq. (7.8), the pulse amplitude of Eq. (7.19) to A_0 is

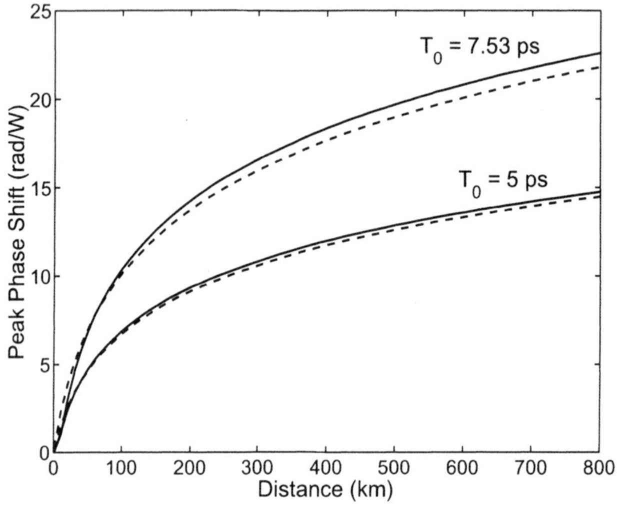

Figure 7.2. The peak phase shift due to IFWM as a function of distance for lossless fiber. The solid lines are the exact integration of Eq. (7.18) and the dashed lines are the approximation of Eq. (7.19).

proportional to $P_0 T_0$, or

$$\frac{\Delta u_{l,m,l+m}(L,0)}{A_0} \propto P_0 T_0. \tag{7.20}$$

The amplitude variation is proportional to the pulse width. With shorter pulse, the pulse width increases very fast due to fiber chromatic dispersion. The ghost pulses from IFWM reduce accordingly. The usage of short optical pulse in highly dispersive transmission reduces IFWM induced ghost pulses.

Figure 7.2 shows the peak phase shift of $2\Delta u_{+1,-1,0}/A_0$ due to the two degenerated IFWM from $l = +1, m = -1$ and $l = -1, m = +1$. The peak phase shift is expressed in the unit of rad/W for a pulse width of $T_0 = 7.53$ and 5 ps, and the bit interval of $T = 25$ ps for 40-Gb/s system. The fiber nonlinear coefficient is $\gamma = 1.24$/km/W. The group-velocity dispersion coefficient is $\beta_2 = -21.7$ ps^2. For lossless fiber, Figure 7.2 shows that the approximation of Eq. (7.19) is very close to the integration of Eq. (7.18). From Fig. 7.2, the amount of IFWM also increases with the initial pulse width of T_0.

3. Impact to DPSK Signals

Although IXPM does not affect DPSK signals and is excluded here right now, IFWM gives ghost pulses exactly located at the signal pulse. In the simplest model, the ghost pulse can be modeled as Gaussian noise although numerical simulation shows a distribution having significant different with Gaussian distribution. This section first studies the statistical properties of IFWM. The error probability of DPSK signals with IFWM is evaluated based on both empirical Gaussian approximation and a semi-analytical model.

3.1 Statistics of Intrachannel Four-Wave Mixing

When many pulses in the fiber are interacted with each other through IFWM, the overall IFWM induced ghost pulses have a peak amplitude of

$$\begin{aligned}\Delta u_0 &= \sum_{l,m,l\neq 0,m\neq 0} \Delta u_{l,m,l+m}(L,0), \\ \Delta u_1 &= \sum_{l,m,l\neq 1,m\neq 1} \Delta u_{l,m,l+m-1}(L,0)\end{aligned} \tag{7.21}$$

co-located with u_0 and u_1 at $t = 0$ and $t = T$, respectively. If N_A identical fiber spans with a length of L are used repeatedly one after another, IFWM adds coherently one span after another. The overall IFWM ghost pulse has a peak amplitude N_A times that of Eq. (7.21), or equivalently speaking, induced by a single-span fiber link with a launched power of $N_A P_0$. In practice, the repetition of identical fiber span gives the largest accumulation of fiber nonlinearities, representing the worst-case system design. For a fiber link with arbitrary configuration, each ghost pulse of $\Delta u_{l,m,l+m}$ can be calculated using N_A different integrations of Eq. (7.18).

Figure 7.3(a) shows the distribution of the peak normalized electric field of $\Delta u_0/A_0$. Figure 7.3(b) shows the distribution of the peak relative phase shift of $\Im\{\Delta u_0\}/A_0$ versus $\Im\{\Delta u_1\}/A_1$ between two consecutive time intervals, where $\Im\{\cdot\}$ denotes the imaginary part of a complex number. Figure 7.3(c) shows the distribution of $\Re\{\Delta u_0\}/A_0$ versus $\Re\{\Delta u_1\}/A_1$ between two consecutive time intervals, where $\Re\{\cdot\}$ denotes the real part of a complex number. Figures 7.3 are obtained for 100-km long fiber span with a normalized launched power of $\langle\Phi_{\rm NL}\rangle = N_A \gamma L_{\rm eff} P_0 = 1$ rad, 100% dispersion compensation at the end of each fiber link. The fiber link has an attenuation coefficient of $\alpha = 0.2$ dB/km. The pulse width is $T_0 = 7.53$ ps, corresponding to a FWHM pulse width of 12.5 ps, or 50% of the bit interval of $T = 25$ ps for 40-Gb/s DPSK signals. The distributions of Figs. 7.3 are shown as a gray scale inten-

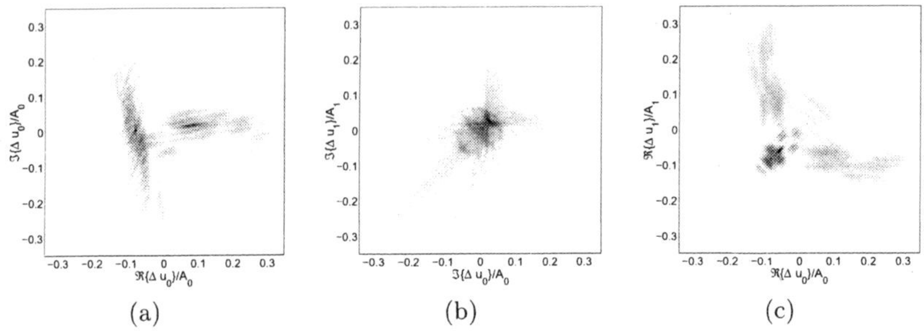

Figure 7.3. Distribution of IFWM ghost pulses. (a) The distribution of the complex electric field of $\Delta u_0/A_0$. (b) The distribution of the relative phase of $\Im\{\Delta u_0\}/A_0$ versus $\Im\{\Delta u_1\}/A_1$ for two consecutive symbols. (C) The distribution of the real parts of the electric field of $\Re\{\Delta u_0\}$ versus $\Re\{\Delta u_1\}/A_1$ for two consecutive symbols.

sity. The simulation of Figs. 7.3 uses 20-bit DPSK signal with about 1 million different combinations. Figure 7.3(a) is for ghost pulse at the center bit and Fig. 7.3(b) is the imaginary parts of the ghost pulses of the two center bits. If the ghost pulses have peak amplitude significantly smaller than the pulse amplitude of $|A_0|$, Figure 7.3(b) is approximately the relative phase shift between two consecutive pulses, i.e., the x-axis is $\Im\{\Delta u_0\}/A_0$ and the y-axis is $\Im\{\Delta u_1\}/A_1$. The simulation of Figs. 7.3 ignores all the contribution from ISPM and IXPM that do not degrade DPSK signals. When identical fiber span is repeated one after another, IFWM induced ghost pulses add coherently one after another. Figures 7.3 are valid for single- and multi-span fiber link with mean nonlinear phase shift of $\langle \Phi_{\mathrm{NL}} \rangle = 1$ rad.

From Figs. 7.3, the IFWM induced ghost pulse does not have a regular distribution. While the imaginary part in Fig. 7.3(a) has larger peak amplitude, the real part of the ghost pulse is still very significant and has twice the variance of the imaginary part. Although $\Delta u_0/A_0$ is zero mean, the distribution of Fig. 7.3(a) trends toward positive real part but spreads more or less equally in both positive and negative imaginary part.

The IFWM ghost pulses induce phase changes of $\Im\{\Delta u_0\}/A_0$ and $\Im\{\Delta u_1\}/A_1$ that are correlated with each other. The correlation coefficient of that in Fig. 7.3(b) is about 0.58. The positive correlation reduces the impact of IFWM to the error probability of DPSK signal. As adjacent pulses trend to have similar phase shift, the phase difference is reduced accordingly.

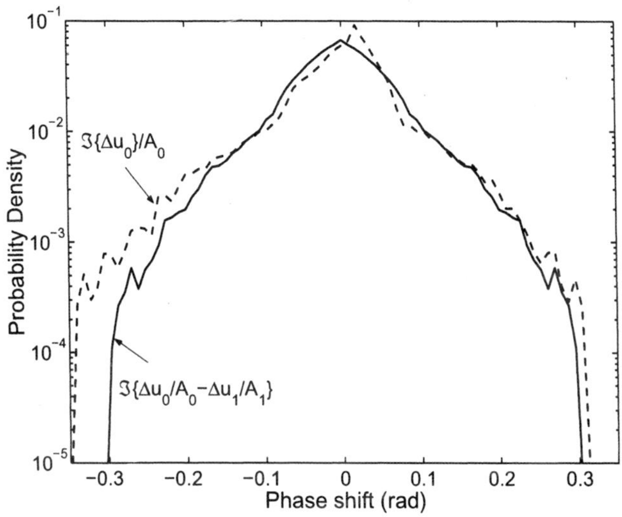

Figure 7.4. The phase distribution of IFWM.

The real part correlation of Fig. 7.3(c) is different with the correlation of Fig. 7.3(b) for imaginary part. If $\Re\{\Delta u_0\}/A_0$ is very large and positive, $\Re\{\Delta u_1\}/A_1$ is usually very small. The correlation coefficient is negative and equal to about -0.55. When the real part of $\Re\{\Delta u_0\}/A_0$ mostly has large positive values but $\Re\{\Delta u_1\}/A_1$ has small value, the error probability reduces as the two constellation points of two consecutive symbols are further apart. The correlation property of Fig. 7.3(c) also reduces the error probability for DPSK signal.

When A_k all change sign to $-A_k$, $\Delta u_0/A_0$ remains the same. The ghost pulses of Δu_0 is symmetrical with the origin and equal to that in Fig. 7.3(a) and its rotation by 180° for negative A_0. When A_k is changed to $(-1)^k A_k$ with sign change in only odd position, Δu_0 remains the same but Δu_1 changes sign. However, $\Im\{\Delta u_0\}/A_0$ and $\Im\{\Delta u_1\}/A_1$ remains the same. The distribution of $\Im\{\Delta u_0\}$ versus $\Im\{\Delta u_1\}$ is the 90°, 180°, and 270° rotations of Fig. 7.3(b).

Figure 7.4 shows the phase distribution due to IFWM, including the phase shift of $\Im\{\Delta u_0\}/A_0$ and the phase difference of $\Im\{\Delta u_0\}/A_0 - \Im\{\Delta u_1\}/A_1$. Figure 7.4 is shown for the same parameters of Figs. 7.3. Numerical calculation shows the variance of the IFWM-induced phase difference of $\Im\{\Delta u_0\}/A_0 - \Im\{\Delta u_1\}/A_1$ is about the same as the variance of the phase of $\Im\{\Delta u_0\}/A_0$, showing that the phase shift at two consecutive symbols are correlated with each other.

The distribution of Fig. 7.3(b) is basically the same as that in Wei and Liu (2003, Fig. 2). The distribution of Fig. 7.4 is also similar to that in Wei and Liu (2003, Fig. 3). For Figs. 7.3, we consider a fiber with loss instead of the lossless case of Wei and Liu (2003). For lossy fiber with span-by-span dispersion compensation, ghost pulses from IFWM are not reduced by 50% precompensation of chromatic dispersion. IFWM can be reduced by dispersion management, usually using symmetric dispersion compensation (Mecozzi et al., 2001, Striegler and Schmauss, 2004). However, the compensation scheme of Mecozzi et al. (2001) does not use dispersion management along the fiber, similar to the recent system demonstration of Vaa et al. (2004) with lump dispersion compensation at the beginning and end of the fiber link.

3.2 Error Probability for DPSK Signals

The error probability of DPSK signal with IFWM ghost pulse is difficult to find analytically. From Figs. 7.3, the distribution of the IFWM induced ghost pulses is not Gaussian distributed. Of course, similar to the simulated error probability of Figs. 5.10 and 5.14, the impact of IFWM to DPSK signal can be studied based on numerical simulation. However, with the distribution of Fig. 7.3 evaluated numerically, the error probability of DPSK signal with IFWM can be calculated semi-analytically.

Similar to the approaches of Secs. 3.4.2 and 4.2, ignored the constant factor of interferometer loss and photodiode responsivity, the photocurrent is

$$i(t) = |E(t) + \Delta u_1 + E(t-T) + \Delta u_0|^2 - |E(t) + \Delta u_1 - E(t-T) - \Delta u_0|^2, \tag{7.22}$$

where Δu_0 and Δu_1 are ghost pulses due to IFWM, located at two consecutive time intervals of $t = 0$ and $t = T$, respectively. Assumed for simplicity that the transmitted phases at $t = 0$ and $t = T$ are identical and, without loss of generality, $E_s(t) = E_s(t-T) = A_0 > 0$. With $E(t) = A_0 + n(t)$, we obtain

$$i(t) = |2A_0+\Delta u_1+\Delta u_0+n(t)+n(t-T)|^2-|\Delta u_1-\Delta u_0+n(t)-n(t-T)|^2. \tag{7.23}$$

A decision error occurs if $i(t) < 0$. Given Δu_0 and Δu_1, the two terms of Eq. (7.23) are independent of each other and have a noncentral chi-square (χ^2) distribution. Each term of Eq. (7.23) has the same noise variance of $4\sigma_n^2$ where $E\{|n(t)|^2\} = 2\sigma_n^2$. The noncentralities of two terms of Eq. (7.23) are $|2A_0+\Delta u_1+\Delta u_0|^2$ and $|\Delta u_1-\Delta u_0|^2$, respectively.

From Appendix 3.A, the probability of $i(t) < 0$ is equal to

$$p_e(\Delta u_0, \Delta u_1) = Q(a, b) - \frac{1}{2} e^{-(a^2+b^2)/2} I_0(ab), \tag{7.24}$$

where

$$a^2 = \frac{\rho_s}{2} \left| 2 + \frac{\Delta u_1}{A_0} + \frac{\Delta u_0}{A_0} \right|^2, \tag{7.25}$$

$$b^2 = \frac{\rho_s}{2} \left| \frac{\Delta u_1}{A_0} - \frac{\Delta u_0}{A_0} \right|^2, \tag{7.26}$$

where $\rho_s = A_0^2/2\sigma_n^2$ is the signal-to-noise ratio (SNR) without taking into account the ghost pulses.

The error probability is equal to the expectation of

$$p_e = E\left\{p_e(\Delta u_0, \Delta u_1)\right\} \tag{7.27}$$

that can be evaluated numerically based on the distribution of, for example, Figs. 7.3. From the distribution of Figs. 7.3(b) and (c), when $|\Delta u_0|$ is very large, $|\Delta u_1|$ is also very large too. When $\Delta u_0 + \Delta u_1$ has its peak values, either positive or negative, $\Delta u_0 - \Delta u_1$ is usually very small. Similarly, when $\Delta u_0 - \Delta u_1$ has its peak values, $\Delta u_0 + \Delta u_1$ is usually very small. The error probability of Eq. (7.27) only considers the case when $A_0 = A_1$. As discussed above, the distribution for $A_0 = -A_1$ of the ghost pulses for Figs. 7.3(b) and (c) are the same as that for $A_0 = A_1$. If the transmitted symbols of two consecutive symbols have a phase difference of 180°, the error probability is the same as that given by Eq. (7.27) for $A_0 = A_1$ without phase difference.

For the real parts of the ghost pulses of $\Delta u_0/A_0$ and $\Delta u_1/A_1$, the correlation coefficient is negative, giving a variance of $\Re\{\Delta u_0\}/A_0 - \Re\{\Delta u_1\}/A_0$ larger than the variance of $\Re\{\Delta u_0\}/A_0 + \Re\{\Delta u_1\}/A_0$. The negative correlation supposes to increase the error probability. However, from Fig. 7.3(c), when $\Re\{\Delta u_0\}/A_0$ has large positive values, $\Re\{\Delta u_1\}/A_1$ has small value, the error probability reduces as the two constellation points of two consecutive symbols are further apart. The combined effect of negative correlation but special distribution should reduce the effect of IFWM ghost pulses to error probability.

Figure 7.5 shows the error probability as a function of SNR ρ_s for DPSK signal with IFWM induced ghost pulses. The semi-analytical formula of Eq. (7.27) with (7.26) is used to calculate the error probability based on IFWM ghost pulses distribution of Figs. 7.3 from a 16-bit permutation of a binary sequence. The number of bits is reduced for Fig. 7.5 as large number of bits increases the computation time but

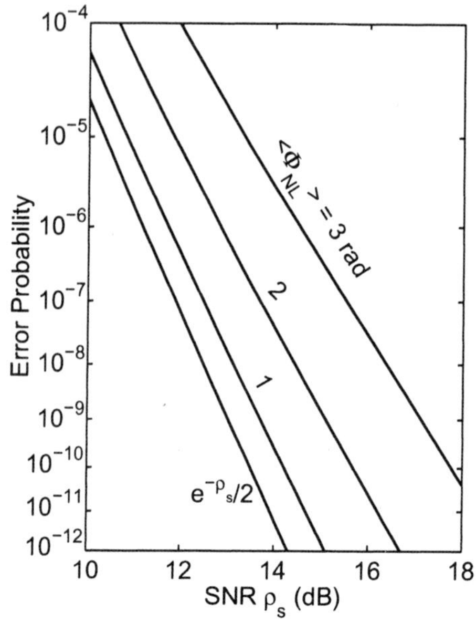

Figure 7.5. The error probability as a function of SNR for DPSK signal with IFWM induced ghost pulses.

not the accuracy of the curves in Fig. 7.5. The system configuration is the same as that for Figs. 7.3. The mean nonlinear phase shift is $\langle\Phi_{\rm NL}\rangle = N_A\gamma L_{\rm eff}P_0$. Assume that identical fiber span is repeated one after another, Figure 7.5 is valid for single and multiple systems with the same $\langle\Phi_{\rm NL}\rangle$. Of course, the repeat of the same configuration span after span is the worst-case with the largest accumulation of distortion due to fiber nonlinearities.

Figure 7.6 shows the SNR penalty as a function of mean nonlinear phase shift $\langle\Phi_{\rm NL}\rangle$. In additional to the SNR penalty corresponding to Fig. 7.5, Figure 7.6 also shows the SNR for the case when the initial pulse width is $T_0 = 5$ ps, for a duty cycle of about 1/3. Followed the estimation of Fig. 7.2, a system with short initial pulse width has less IFWM and smaller IFWM induced SNR penalty.

If the IFWM induced ghost pulses are assumed as Gaussian distributed electric field, the noise is increased to $n(t) + \Delta u_0$ at $t = 0$. The SNR including the ghost pulse has a SNR ratio of $A_0^2/E\{|n(t) + \Delta u_0|^2\}$. However, this definition of SNR ignores the correlation between Δu_0 and Δu_1.

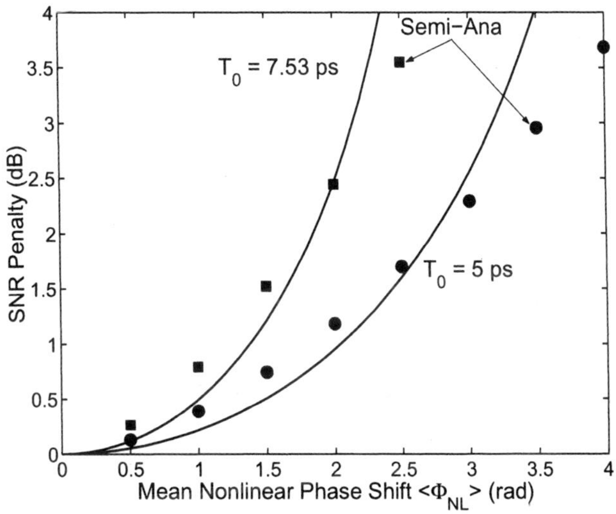

Figure 7.6. The SNR penalty as a function of mean nonlinear phase shift of $\langle \Phi_{\text{NL}} \rangle$. [Adapted from Ho (2005c)]

The variance of the phase of $A_0 + n(t)$ is approximately equal to $1/2\rho_s$ from Eq. (4.A.15). Similar to that for the Q-factor of Eq. (5.81), the differential phase has a phase variance of $1/\rho_s$. For IFWM induced ghost pulses, the corresponding differential phase variance is

$$\sigma^2_{\Delta\phi} = E\left\{ [\Im\{\Delta u_0\}/A_0 - \Im\{\Delta u_1\}/A_1]^2 \right\}. \tag{7.28}$$

With this additional differential phase variance, the SNR penalty can be estimated as

$$-10 \cdot \log_{10}\left(1 - 20\sigma^2_{\Delta\phi}\right), \tag{7.29}$$

where 20 (13 dB) is the required SNR to give an error probability of 10^{-9} for DPSK signal as from Fig. 3.13.

Figure 7.6 also shows the SNR penalty calculated by Eq. (7.29). For a SNR penalty less than 2 dB, the SNR penalty of Eq. (7.29) underestimates the SNR penalty by up to 0.25 dB. For SNR penalty larger than 2 dB, the SNR penalty of Eq. (7.29) overestimates the SNR penalty.

Note that the usage of the approximation of Eq. (7.29) does not greatly simplify the analysis of DPSK signal with IFWM induced ghost pulses. To calculate the variance of $\sigma^2_{\Delta\phi}$ of Eq. (7.28), IFWM terms of Δu_0 induce by many different combinations of bit sequence must be evaluated. The semi-analytical error probability just requires one further step to find the error probability of Eq. (7.26) for each combination of Δu_0 and Δu_1. The average error probability of Eq. (7.27) becomes the error probability for the system. The variance of Eq. (7.29) in different mean nonlinear phase shift of $\langle \Phi_{\rm NL} \rangle$ increases with $\langle \Phi_{\rm NL} \rangle^2$. The whole curves of Fig. 7.6 requires, to certain extend, one evaluation of the variance of $\sigma^2_{\Delta\phi}$. However, using the same set of $\Delta u_0/A_0$ and $\Delta u_1/A_1$, different point of the semi-analytical results in Fig. 7.6 requires the evaluation of a new set of error probabilities of Eq. (7.26) to find the average of Eq. (7.27).

The above analysis and numerical results always used the peak amplitude of the ghost pulses and the signal pulses. The pulse width of the ghost pulses is ignored for simplicity. However, the IFWM terms of Eq. (7.18) does not have the same pulse shape as the signal but in fact approximately broadens to $\sqrt{3}$ times wider than the signal. As the power is proportional to the pulse width, the ghosts pulse has $\sqrt{3}$ times larger power than the signal. In the worst case, the corresponding curves of Fig. 7.6 must be scaled by a factor of $3^{1/4} = 1.32$ to take that into account. However, the discrepancy due to pulse width depends on the bandwidth of the optical and electric filters in the receiver.

If optical matched filter is used preceding the polarized direct-detection DPSK receiver and the electric filter at the receiver has a wide bandwidth that does not distort the signal, assume the ghost pulse is $\sqrt{3}$ times wider than the signal pulse, the IFWM amplitude is increased by a factor of $\sqrt{3/2} = 1.22$, less than the ratio of $3^{1/4}$ due to power increase.

If both the optical and electric filter has a very wide bandwidth, a very bad case that allows too much noise to the receiver, the peak amplitude directly transfers to the receiver. IFWM does not increase from the peak amplitude for Eq. (7.18). If the optical filter has a wide bandwidth but the electrical filter is a Bessel filter having a bandwidth 0.75 time the data rate, IFWM increases by a factor of 1.19 relative to the signal pulse.

In practical system design, Figure 7.6 must be modified to take into account the design of both receiver and transmitter. Also note that the results of this section assume an optical matched filter but ignored the pulse width of ghost pulses.

IFWM and IXPM were first observed by Shake et al. (1998), and Essiambre et al. (1999) and first analyzed by Mamyshev and Mamysheva (1999). The analysis here follows similar method in Ablowitz and Hirooka (2002), Mecozzi et al. (2000b), Striegler and Schmauss (2004), and Wei and Liu (2003). The formula of Eq. (7.18) was first derived in Mecozzi et al. (2001) but modified by Essiambre et al. (2002) and Chowdhury and Essiambre (2004).

IFWM can be suppressed using symmetric dispersion compensation (Mecozzi et al., 2001, Striegler and Schmauss, 2004, Wei and Plant,

2004) and alternating polarization (Liu et al., 2004c, Xie et al., 2004). As a general method, optical phase conjugation can also reduce IFWM (Brener et al., 2000, Chowdhury and Essiambre, 2004, Chowdhury et al., 2005, Pepper and Yariv, 1980).

While phase modulation is another method to suppress IFWM for on-off keying (Alic and Fainman, 2004, Appathurai et al., 2004, Cheng and Conradi, 2002, Forzati et al., 2002, Gill et al., 2003, Liu et al., 2002a), DPSK signal always has phase modulation with higher tolerance to IFWM than on-off keying signal. For on-off keying signal, all terms of $A_l A_m A^*_{l+m}$ are positive and sum together. With phase modulation, the factors of $A_l A_m A^*_{l+m}$ may be positive and negative. Of course, if the terms of $A_l A_m A^*_{l+m}$ for different l and m are independent of each other, the variance of the ghost pulse remains the same as the case without phase modulation. However, because of the dependence between terms for different l and m, the overall IFWM reduces by phase modulation. Phase modulation is intrinsically used in DPSK signals and IFWM can thus be reduced accordingly.

IXPM also dose not degrade PSK signal as the phase-locked loop (PLL) tracks out the constant phase. Like usual PSK signal receiver, if the demodulator of PSK signal does not take into account the correlation between two adjacent ghost pulses. The SNR penalty for PSK signal is larger than that for DPSK signal and performs worst than DPSK signal in high IFWM. Using a method similar to Eq. (7.29), the SNR penalty for PSK signal can be estimated as

$$-10 \cdot \log_{10}\left(1 - 36E\left\{[\Im\{\Delta u_0/A_0\}]^2\right\}\right). \tag{7.30}$$

When the phase variance and the differential phase variance is approximately the same from Fig. 7.4, the SNR penalty for PSK signal may be larger than that for DPSK signals. Of course, practical PSK receiver may be designed to anticipate the correlate phase due to IFWM. In practice, the phase error for PSK signal must be analyzed based on Eq. (4.26) for a phase-locked loop with closed-loop transfer function of $H(s)$. The correlation of IFWM ghost pulses must be included in $S_{\phi_n}(f)$ for Eq. (4.26). Currently, the power spectral density of the IFWM ghost pulses is not known.

4. Nonlinear Phase Noise Versus Intrachannel Four-Wave-Mixing

The previous Chapters 5 and 6 all consider non-return-to-zero (NRZ) signals without pulse distortion when the optical signal is propagated through the optical fiber. In Sec. 5.1, the mean nonlinear phase shift for an optical pulse is given by Eq. (5.8) for NRZ or continuous-wave signal.

However, SPM induced nonlinear phase noise for short pulse requires special treatment other than the method in Chapter 5, especially must be consistent with previous section for a fair comparison with IFWM.

The mean nonlinear phase shift of $\langle \Phi_{\mathrm{NL}} \rangle = \gamma N_A P_0 L_{\mathrm{eff}}$ does not change with signal format of RZ-OOK or RZ-DPSK. ISPM phase noise, or SPM-induced nonlinear phase noise, is given by the beating of the optical pulse itself with amplifier noises. For system without chromatic dispersion, for the same average power, the peak amplitude of the signal is inversely proportional to the pulse width. In additional to giving more nonlinear phase noise to the signal, the usage of RZ pulse gives time-dependence to the nonlinear phase noise. The variance of nonlinear phase noise is a function of time and the peak variance of nonlinear phase noise increases with the decrease of pulse width.

For system with chromatic dispersion, from Eq. (7.3), the optical pulse broadens with transmission distance. The pulse is usually broadened faster for shorter pulse. Nonlinear phase noise decreases with chromatic dispersion, mostly due to pulse broadening. With pulse overlap, the adjacent pulses beat with the amplifier noise located with the center pulse. IXPM phase noise, or IXPM-induced nonlinear phase noise, increases with pulse broadening. As shown later, the combined ISPM and IXPM phase noises are comparable to the nonlinear phase noise of NRZ signal without pulse broadening.

To be consistent with the method to derive Eq. (7.18), we consider the nonlinear force of the interaction of signal with noise. For signal with amplifier noise, the SPM-induced nonlinear force including amplifier noises is equal to

$$j\gamma[u_0(z,t) + n(z,t)]|u_0(z,t) + n(z,t)|^2. \tag{7.31}$$

The nonlinear force for ISPM is $j\gamma u_0|u_0|^2$ from the signal alone, the same as that for Eq. (7.18) with $l = m = 0$. The nonlinear force associated with nonlinear phase noise has two different terms of

$$2j\gamma|u_0(z,t)|^2 n(z,t), \quad \text{and} \quad j\gamma u_0^2(z,t) n^*(z,t), \tag{7.32}$$

when all quadratic or higher-order terms of the noise are ignored at high SNR. The IXPM term of $2j\gamma|u_m(z,t)|^2 n(z,t)$ also gives IXPM phase noise to the pulse of $u_0(z,t)$ at $t = 0$. For system with large SNR, the IFWM related noise and signal beating of $2j\gamma u_l(z,t) u_m^*(z,t) n(z,t)$ with $l \neq m$ is much smaller than the IFWM induced ghost pulses from the same effects. While IFWM phase noise is ignored, the ISPM phase noise is considered but its result is also applicable to IXPM phase noise, almost without modification. As shown in last section, pulse-to-pulse interaction due to both ISPM and IXPM gives the same phase shift to all pulses and does not degrade a DPSK signal. While the pulse-to-pulse interaction due to IFWM gives ghost pulses, both ISPM and IXPM phase noises are from the interaction of pulses with amplifier noises. Due to the difference of amplifier noise located at adjacent pulses, ISPM and IXPM phase noises are not the same at adjacent pulses.

For the nonlinear force from $2j\gamma|u_0(z,t)|^2 n(z,t)$, the overall nonlinear force is equal to

$$\Delta u_n(t) = 2j\gamma \int_0^L \left[|u_0(z,t)|^2 n(z,t)\right] \otimes h_{-z}(t) e^{-\alpha z} \mathrm{d}z, \tag{7.33}$$

where $\otimes$ denotes convolution. The impulse response of $h_{-z}(t)$ provides dispersion compensation for $h_z(t)$, where $h_z(t)$ is an impulse response from fiber chromatic dispersion, the corresponding transfer function is $H_z(\omega) = \exp(j\beta_2 z\omega^2/2)$.

Due to fiber chromatic dispersion, from Eq. (7.3), the pulse of $|u_0(z,t)|^2$ has an $1/e$ width of $\tau(z)/\sqrt{2}$ and a Fourier transform of

$$V_0(z,\omega) = \sqrt{\pi}|A_0|^2 T_0 \exp\left[-\tau^2(z)\omega^2/4\right], \tag{7.34}$$

where $\tau(z)$ is the pulse width as a function of distance given by Eq. (7.4). The optical amplifier noise of $n(z,t)$ is also changed due to fiber dispersion. Although fiber dispersion does not change the spectrum of the signal, the time dependence of the signal is changed by chromatic dispersion. At the input of the fiber, $E\left\{n(0,t+\tau)n^*(0,t)\right\} = 2\sigma_n^2\delta(\tau)$ as additive white Gaussian noise, where σ_n^2 is the noise variance per dimension. With fiber dispersion, $n(z,t) = n(0,t) \otimes h_z(t)$ and $E\left\{n(z,t+\tau)n(z,t)\right\} = 2\sigma_n^2\delta(\tau)$, but

$$\int_{-\infty}^{+\infty} E\left\{n(z_1,t+\tau)n^*(z_2,t)\right\} e^{-j\omega\tau} \mathrm{d}\tau = 2\sigma_n^2 H_{z_1}(\omega) H_{z_2}^*(\omega) = 2\sigma_n^2 e^{j\beta_2(z_1-z_2)\omega^2/2}. \tag{7.35}$$

The temporal profile of $\Delta u_n(t)$ can be represented by the variance of $\Delta u_n(t)$ as a function of time, or

$$\begin{aligned}\sigma_{\Delta u_n}^2(t) &= E\left\{|\Delta u_n(t)|^2\right\} \\ &= 4\gamma^2 \int_0^L \int_0^L \int_{-\infty}^{+\infty} \int_{-\infty}^{+\infty} |u_0(z_2,\tau_2)|^2 |u_0(z_1,\tau_1)|^2 \\ &\quad \times E\{n(z_1,\tau_1)n^*(z_2,\tau_2)\} h_{-z_2}^*(t-\tau_2) h_{-z_1}(t-\tau_1) \\ &\quad \times e^{-\alpha z_1 - \alpha z_2} \mathrm{d}\tau_1 \mathrm{d}\tau_2 \mathrm{d}z_1 \mathrm{d}z_2.\end{aligned} \tag{7.36}$$

Replacing $E\{n(z_1,\tau_1)n^*(z_2,\tau_2)\}$, $h_{-z_1}(t)$, and $h_{-z_2}(t)$ with their corresponding Fourier transforms, we obtain

$$\sigma^2_{\Delta u_n}(t) = \frac{\gamma^2\sigma_n^2}{\pi^3}\int_0^L\int_0^L\int_{-\infty}^{+\infty}\int_{-\infty}^{+\infty}\int_{-\infty}^{+\infty} V_0(z_1,\omega_1-\omega)V_0(z_2,\omega_2-\omega) \times e^{j\beta_2(z_1-z_2)\omega^2/2}H^*_{-z_2}(\omega_2)H_{-z_1}(\omega_1) \times e^{j(\omega_1-\omega_2)t-\alpha(z_1+z_2)}\mathrm{d}\omega_1\mathrm{d}\omega_2\mathrm{d}\omega\mathrm{d}z_1\mathrm{d}z_2. \quad (7.37)$$

Note that both $V_0(z,\omega)$ and $H_z(\omega)$ are "Gaussian" shape and the integrations over z_1 and z_2 are complex conjugate of each other, the time-depending variance is equal to

$$\sigma^2_{\Delta u_n}(t) = \frac{4\gamma^2\sigma_n^2T_0^2A_0^4}{\pi}\int_{-\infty}^{+\infty}\left|\int_0^L\frac{\exp\left(-\frac{t^2+j\tau^2(z)\omega t+\beta_2^2z^2\omega^2}{\tau^2(z)-2j\beta_2 z}-\alpha z\right)}{\sqrt{\tau^2(z)-2j\beta_2 z}}\mathrm{d}z\right|^2\mathrm{d}\omega. \quad (7.38)$$

Similarly, the variance profile corresponding to $\gamma u_0^2(z,t)n^*(z,t)$ can be calculated by replacing $\tau^2(z)$ by $T_0^2 - j\beta_2 z$ and the spectral density for $n^*(z,t)$ is the complex conjugate of that for $n(z,t)$. We obtain

$$\sigma^2_{\Delta u_n'}(t) = \frac{\gamma^2\sigma_n^2T_0^2A_0^4}{\pi}\int_{-\infty}^{+\infty}\left|\int_0^L\frac{\exp\left[-\frac{(t-\beta_2 z\omega)(t-jT_0^2\omega)}{T_0^2+j\beta_2 z}-\alpha z\right]}{\sqrt{T_0^4+\beta_2^2z^2}}\mathrm{d}z\right|^2\mathrm{d}\omega. \quad (7.39)$$

Figure 7.7 shows the temporal profile, both the standard deviation (STD) of $\sigma_{\Delta u_n}(t)$ and $\sigma_{\Delta u_n'}(t)$, of the nonlinear force of both $\Delta u_n(t)$ and $\Delta u_n'(t)$ for typical fiber dispersion coefficients of $D = 17$ and 3.5 ps/km/nm. The initial launched pulse has an $1/e$ width of $T_0 = 5$ ps. Figure 7.7 shows that the nonlinear force of $\Delta u_n(t)$ due to the beating of $|u_0(z,t)|^2$ with $n(z,t)$ is far larger than the nonlinear force of $\Delta u_n'(t)$ due to the beating of $u^2(z,t)$ with $n^*(z,t)$. In term of power, the variance of $\sigma^2_{\Delta u_n'(t)}$ is less than 1% of $\sigma^2_{\Delta u_n(t)}$. The noise term of $\Delta u_n(t)$ also has more spreading over time than $\Delta u_n'(t)$. The spreading of $\Delta u_n(t)$ is more obvious for a high dispersion coefficient of $D = 17$ ps/km/nm. From Fig. 7.7, the contribution of $\Delta u_n'(t)$ to the nonlinear phase noise of the system can be ignored.

The temporal spreading of $\sigma_{\Delta u_n}(t)$ shows that ISPM phase noise spreads to adjacent symbols. Unlike Eq. (7.18) with an $1/e$ pulse width of about $\sqrt{3}T_0$, the profile of Fig. 7.7 has a pulse profile narrower than $\sqrt{3}T_0$ near the peak but a slow decreasing tail. For 40-Gb/s signal with $T = 25$ ps, the STD of $\sigma_{\Delta u_n}(t)$ at ± 25 ps is still very large, indicating that IXPM phase noise may be significant to the system with large

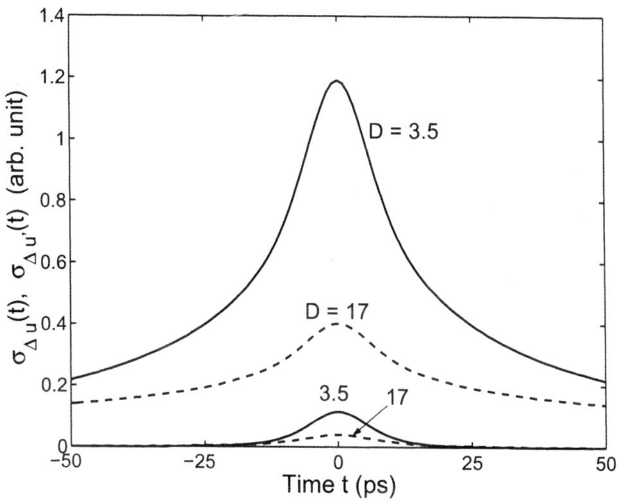

Figure 7.7. The temporal distribution of nonlinear force due to the beating of signal with noise. The upper two curves are $\sigma_{\Delta u_n}(t)$ and the lower two curves are $\sigma_{\Delta u'_n}(t)$.

pulse overlap. For the case with $D = 3.5$ ps/km/nm, $\sigma_{\Delta u_n}(t)$ at ± 25 ps is relatively smaller than that for the case with $D = 17$ ps/km/nm.

The temporal profile of Fig. 7.7 does not provide a direct answer to whether the spreading of $\sigma_{\Delta u_n}(t)$ can be ignored. Later part of this section will either include or exclude the IXPM phase noise. The temporal profile of Fig. 7.7 cannot be used directly to estimate the dependence between the nonlinear phase at $t = 0$ and, for example, $t = T$. As a trivial example for the signal of a system without chromatic dispersion and pulse distortion, the nonlinear force is always proportional to $|u_0(0,t)|^2 n(0,t)$. As white noise, the noises of $n(0,t)$ at $t = 0$ and $t = T$ are independent of each other. In this trivial case, the profile corresponding to Fig. 7.7 is proportional to $|u_0(0,t)|^2$.

If the nonlinear force of $\Delta u_n(t)$ is passing through an optical filter with an impulse response of $h_o(t)$, the filter output at the time of mT is

$$\begin{aligned}\zeta_{0,m} &= \int_{-\infty}^{+\infty} h_o(mT - t)\Delta u_n(t)\mathrm{d}t \\ &= 2j\gamma \int_{-\infty}^{+\infty}\int_0^L h_o(mT - t)\left\{\left[|u_0(z,t)|^2 n(z,t)\right] \otimes h_{-z}(t)\right\} e^{-\alpha z}\mathrm{d}z\mathrm{d}t.\end{aligned} \tag{7.40}$$

The ISPM phase noise from $\zeta_{0,0}$ is the noise generated by the beating of $|u_0(z,t)|^2$ with $n(z,t)$ and affects the DPSK pulse at $t = 0$. The IXPM phase noise from $\zeta_{0,1}$ is from the beating of $|u_0(z,t)|^2$ with $n(z,t)$ and affect the DPSK pulse at $t = T$. Due to IXPM, the DPSK pulse at $t = 0$ also affects by the beating of $|u_1(z,t)|^2$ (the pulse at $t = T$) with $n(z,t)$ to give the IXPM phase noise of $\zeta_{1,0}$. Other than the temporal location, $\zeta_{1,0}$ is statistically the same as $\zeta_{0,-1}$. In general, $\zeta_{k,m}$ is statistically the same as $\zeta_{0,m-k}$. The term of $\zeta_{0,0}$ is ISPM phase noise and other terms of $\zeta_{l,m}$ with $l \neq m$ are IXPM phase noises.

For DPSK signal, the same as Fig. 7.3(b) to consider the differential phase, the differential nonlinear phase noise from both ISPM and IXPM phase noises is approximately equal to

$$\delta\phi_n = \frac{1}{A_0}\Im\left\{\sum_m \zeta_{m,0}\right\} - \frac{1}{A_1}\Im\left\{\sum_m \zeta_{m,1}\right\}. \tag{7.41}$$

For the optical pulses of $u_0(0,t)$ and $u_1(0,t)$, the optical filter of $h_o(t)$ is assumed to give an output of A_0 and A_1 at $t = 0$ and $t = T$, respectively. If the optical filter of $h_o(t)$ has a Gaussian impulse response and an $1/e$ pulse width of t_o, the impulse response is

$$h_o(t) = \frac{1}{\sqrt{2\pi}} \frac{\sqrt{t_o^2 + T_0^2}}{t_o T_0} \exp\left(-\frac{t^2}{2t_o^2}\right), \tag{7.42}$$

and

$$H_o(\omega) = \sqrt{1 + \frac{t_o^2}{T_0^2}} \exp\left(-\frac{1}{2}t_o^2\omega^2\right), \tag{7.43}$$

For simplicity, $A_0 = A_1$ is assumed for the same transmitted symbols in consecutive time intervals. As a circular complex random variables, the random variables of $\zeta_{0,m}$ has the property that $\Re\{\zeta_{0,m}\}$ and $\Im\{\zeta_{0,m}\}$ are independent and identically distributed. The variance of $\delta\phi_n$ is

$$\sigma_{\delta\phi_n}^2 = \frac{1}{A_0^2}\left[\sum_{m_1}\sum_{m_2} E\left\{\zeta_{m_1,0}\zeta_{m_2,0}^*\right\} - \sum_{m_1}\sum_{m_2} E\left\{\zeta_{m_1,0}\zeta_{m_2,1}^*\right\}\right]. \tag{7.44}$$

Defined a function of $f_m(\omega)$ as

$$f_m(\omega) = 2\gamma|A_0|^2(T_0^2 + t_o^2)^{\frac{1}{2}} \int_0^L \frac{\exp\left\{-\frac{1}{2}t_o^2\omega^2 + \frac{\left[(t_o^2 - j\beta_2 z)\omega + jmT\right]^2}{\tau(z)^2 - 2j\beta_2 z + 2t_o^2}\right\}}{\sqrt{\tau(z)^2 - 2j\beta_2 z + 2t_o^2}} e^{-\alpha z}\mathrm{d}z, \tag{7.45}$$

where t_o is the $1/e$-width for the impulse response of $h_o(t)$, assumed that it is Gaussian pulse shape, we obtain

$$E\left\{\zeta_{m_1,0}\zeta_{m_2,0}^*\right\} = \frac{\sigma_n^2}{\pi}\int_{-\infty}^{+\infty} f_{m_1}(\omega) f_{m_2}^*(\omega)\mathrm{d}\omega \tag{7.46}$$

and

$$E\left\{\zeta_{m1,0}\zeta_{m2,1}^*\right\} = \frac{\sigma_n^2}{\pi}\int_{-\infty}^{+\infty} f_{m1}(\omega) f_{m2-1}^*(\omega) e^{j\omega T} \mathrm{d}\omega. \tag{7.47}$$

Numerical results show that the correlation of $E\left\{\zeta_{m1,0}\zeta_{m2,1}^*\right\}$ is far smaller than $E\left\{\zeta_{m1,0}\zeta_{m2,0}^*\right\}$, i.e., the cross-correlation between the nonlinear phase noise from adjacent symbols is small. The variance of $E\left\{\zeta_{0,0}\zeta_{0,0}^*\right\}$ is the largest and the variances of $E\left\{\zeta_{m,0}\zeta_{m,0}^*\right\}$, $m \neq 0$, are also very large, i.e., IXPM phase noise is very large.

For a system with N_A fiber spans, the energy per pulse is $\sqrt{\pi}T_0|A_0|^2$ with a SNR of $\rho_s = \sqrt{\pi}T_0|A_0|^2/(2N_A\sigma_n^2)$. Unlike the variance of $\sigma_{\Delta\phi}^2$ from Eq. (7.28) that has equal contribution from all fiber spans, amplifier noise is the smallest in the first span and the largest in the last span. The nonlinear phase noise is accumulated span after span with a variance of $\sigma_{\Delta\phi_n}^2 \approx N_A^3\sigma_{\delta\phi_n}^2/3$ similar to that in of Eq. (6.14), corresponding to $2\sigma_{\Phi_{\mathrm{NL}}}^2$ for system without pulse broadening with the factor of 2 for differential phase. With a mean nonlinear phase shift of $\langle\Phi_{\mathrm{NL}}\rangle = N_A\gamma L_{\mathrm{eff}}P_0 = N_A\gamma\sqrt{\pi}L_{\mathrm{eff}}|A_0|^2T_0/T$, the variance of nonlinear phase noise is proportional to $\langle\Phi_{\mathrm{NL}}\rangle^2/\rho_s$, similar to that in Eq. (6.14).

For a system with an optical matched filter of $t_o = T_0$, for the case without dispersion of $\beta_2 = 0$ and without signal distortion along the fiber, we obtain the variance of nonlinear phase noise of

$$\sigma_{\delta\phi_n}^2 = \frac{8\sigma_n^2\gamma^2|A_0|^2L_{\mathrm{eff}}^2}{\sqrt{3\pi}T_0} \tag{7.48}$$

by using only the ISPM phase noises of $\zeta_{0,0}$ and $\zeta_{1,1}$. The overall phase noise has a variance of

$$\sigma_{\Delta\phi_n}^2 \approx \frac{4T^2\langle\Phi_{\mathrm{NL}}\rangle^2}{3\sqrt{3\pi}T_0^2\rho_s}. \tag{7.49}$$

Figure 7.8 shows the STD of $\sigma_{\Delta\phi}$ of Eq. (7.28) from IFWM and $\sigma_{\Delta\phi_n}$ of Eq. (7.44) from nonlinear phase noise, both as a function of the fiber dispersion coefficient of the optical fiber. The system has a mean nonlinear phase shift of $\langle\Phi_{\mathrm{NL}}\rangle = 1$ rad and SNR of $\rho_s = 20$ (13 dB). Figure 7.8 is for a Gaussian signal with $T_0 = 5$ ps and 40-Gb/s DPSK signal with $T = 25$ ps. The group velocity dispersion coefficient is $\beta_2 = -21.7$ ps^2/km and the fiber loss coefficient is $\alpha = 0.2$ dB/km. In Fig. 7.8, for $T_0 = 5$ ps, the STD of $\sigma_{\Delta\phi}$ from IFWM scales up by the factor of $\sqrt{3/2} = 1.22$ to take into account the $\sqrt{3}$ times increase in the width of IFWM ghost pulses. For $T_0 = 7.53$ ps, the STD of $\sigma_{\Delta\phi}$ from IFWM scales up by a factor of 1.36 with an extra factor to take into account

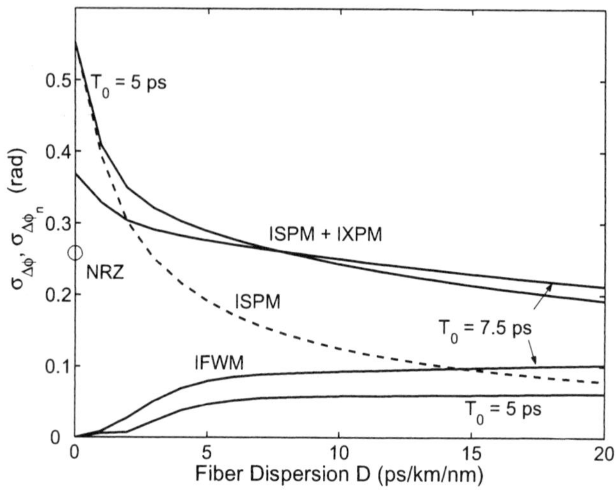

Figure 7.8. The phase noise STD of $\sigma_{\Delta\phi}$ and $\sigma_{\Delta\phi_n}$ due to IFWM and nonlinear phase noise. Also included is the dashed line for ISPM phase noise from $\zeta_{0,0}$ and $\zeta_{1,1}$ alone.

the overlap of adjacent IFWM ghost pulses. The FWHM of the IFWM ghost pulse is 21.6 ps, very close to the bit interval of $T = 25$ ps. Using only the peak amplitude, there is no scaling in Figs. 7.5 and 7.6.

The IFWM-induced ghost pulses give a phase noise variance increase with fiber dispersion. From Fig. 7.1 and Eq. (7.19), IFWM ghost pulse decreases with fiber dispersion. However, both Fig. 7.1 and Eq. (7.19) are for a single IFWM term. With large fiber chromatic dispersion and significant pulse overlap, more terms induce ghost pulses and the overall contribution from IFWM increases slowly with fiber dispersion as shown in Fig. 7.8. From Fig. 7.8, the increase of number of terms and the reduce of the amplitude of each term balance each other at large dispersion. For $D > 7$ ps/km/nm, the contribution from IFWM increases very slow with the increase of fiber dispersion.

The contribution from nonlinear phase noise of $\sigma_{\Delta\phi_n}$ is far larger than that from IFWM. Even with large fiber dispersion, $\sigma_{\Delta\phi_n}$ from nonlinear phase noise is two to three times larger than $\sigma_{\Delta\phi}$ from IFWM. In additional to the overall $\sigma_{\Delta\phi_n}$, Figure 7.8 also shows ISPM phase noise with contribution from $\zeta_{0,0}$ and $\zeta_{1,1}$ alone for the case of $T_0 = 5$ ps. With fiber dispersion, the contribution from IXPM phase noise must be included as from Fig. 7.8. In term of power or variance, IXPM phase noise is larger than ISPM phase noise for $D > 5$ ps/km/nm.

Also shown in Fig. 7.8 is $\sqrt{2}\sigma_{\Phi_{\mathrm{NL}}}$ for the differential phase of NRZ signal from Eq. (6.14). Without dispersion of $D = 0$ (or $\beta_2 = 0$), the RZ signal has a STD of $\sigma_{\Delta\phi_n}$ about twice or one and half time larger than that of NRZ signal when $T_0 = 5$ and 7.53 ps, respectively. In the range of $D > 5$ ps/km/nm, the STD of $\sigma_{\Delta\phi_n}$ is about that of NRZ signal. At large dispersion, the nonlinear phase noise from shorter pulse of $T_0 = 5$ ps is smaller than that form larger pulse of $T_0 = 7.53$ ps. At small fiber dispersion, nonlinear phase noise decreases rapidly with fiber dispersion. At large fiber dispersion, the decreasing of nonlinear phase noise slows down and reaches a level comparable to that of NRZ signal at $D = 0$.

At large fiber dispersion with significant pulse broadening, the optical intensity of the signal is similar to a continuous-wave signal with rapid variation. The amplifier noise is beating with a continuous-wave signal and gives a nonlinear phase noise similar to that of NRZ signal without distortion. Figure 7.8 shows that this first-order interpretation is loosely correct.

Following the estimation of Eq. (7.29) for the SNR penalty of DPSK signal with IFWM, with both nonlinear phase noise and IFWM, the required SNR of ρ_s can be calculated from

$$\frac{1}{20} = \frac{1}{\rho_s} + k_1^2 \langle\Phi_{\mathrm{NL}}\rangle^2 + k_2^2 \frac{20\langle\Phi_{\mathrm{NL}}\rangle^2}{\rho_s}, \qquad (7.50)$$

where $k_1 = 0.1$ and $k_2 = 0.22$ when $D = 17$ ps/km/nm as from Fig. 7.8. The SNR penalty is

$$-10\log_{10}\left(1 - 20k_1^2\langle\Phi_{\mathrm{NL}}\rangle^2\right) + 10\log_{10}\left(1 + 20k_2^2\langle\Phi_{\mathrm{NL}}\rangle^2\right), \qquad (7.51)$$

to show that SNR penalty is additive with the first term the same as Eq. (7.29) and the second term is from nonlinear phase noise alone. Unfortunately, the above formula of Eq. (7.51) overestimates the SNR penalty for nonlinear phase noise as compared with the Nicholson model of Eq. (5.82). Note that the Nicholson model assumes Gaussian distribution of nonlinear phase noise. Although both ISPM and IXPM phase noise has a phase noise STD two to three times larger than that for IFWM ghost pulses induced phase shift, the SNR penalty from nonlinear phase noise is typically 50% to 60% larger than that from IFWM ghost pulses.

The SNR penalty of Eq. (7.51) gives at least a theoretical background for the assumption that SNR penalty is additive. In principle, the variance of nonlinear phase noise is equal to the last term of Eq. (7.50). The Nicholson model of Eq. (5.82), or similarly that of Eq. (4.40), may be used to calculate the SNR penalty for nonlinear phase noise. The semi-analytical model of Eq. (7.27) or the empirical model of Eq. (7.29) may be used to estimate the SNR penalty for IFWM. The summation of the respective SNR penalty is that for a system with the combined effects of nonlinear phase noise and IFWM.

Figure 7.8 shows that nonlinear phase noise has far larger impact to the systems than IFWM induced ghost pulse. The reduction of nonlinear phase noise with dispersion is first studied in Green et al. (2003) based on the method of Cartaxo et al. (1999). McKinstrie and Xie (2002) also considers phase jitter in dispersion-managed soliton systems. However, both McKinstrie and Xie (2002) and Green et al. (2003) are for lossless system. Green et al. (2003) is for continuous-wave signal and McKinstrie and Xie (2002) ignores IXPM phase noise.

In Green et al. (2003) for continuous-wave signal, the influence of fiber nonlinearity on the fiber transfer function is taken into account based on the model of Cartaxo et al. (1999) instead of Wang and Petermann (1992) and Hui et al. (1997).

5. Summary

Intrachannel pulse-to-pulse interaction degrades the performance of DPSK signals but smaller than the impact of nonlinear phase noise. Unlike on-off keying signal, IXPM does not degrade the performance of DPSK signal. The IFWM induced ghost pulses between two consecutive symbols are also correlated with each other. Although the distribution of IFWM ghost pulses is not Gaussian, for performance evaluation purpose with SNR penalty less than 2 dB, the phase of IFWM induced ghost pulses can be approximated as Gaussian distributed.

In term of phase shift, ISPM and IXPM are both far larger than IFWM. While both ISPM and IXPM from signal beating do not affect DPSK signals, ISPM and IXPM phase noise from the beating of signal with noise induce nonlinear phase noise. Even for system with large chromatic dispersion and high SNR, ISPM and IXPM phase noise is far larger than IFWM ghost pulse.

Chapter 8

WAVELENGTH-DIVISION-MULTIPLEXED DPSK SIGNALS

The low-loss window of typical optical fiber has a wavelength bandwidth of about 140 nm from 1.48 to 1.62 μm, translated to a frequency bandwidth of about 17 THz, or 240,000 times the initial bidirectional 70 MHz bandwidth of GSM cellular phone around 900 MHz. This wide bandwidth can be divided into different channel slots, each carrying independent data channel as shown earlier in Fig. 1.5. Many channels can be multiplexed in the fiber and the Erbium-doped fiber amplifiers (EDFA) can amplifier all wavelength-division-multiplexed (WDM) channels together. Typical EDFA also have a wide bandwidth in both the C- and L-bands. For on-off keying or direct-detection differential phase-shift keying (DPSK) signals, the number of channels and channel separation are limited only by the available optical filter technology.

In additional to a simple point-to-point link of Fig. 1.5, reconfigurable optical add/drop multiplexer (ROADM) or wavelength router can be used in dynamic WDM networks. Both ROADM and wavelength router use WDM as an underlying technology to simplify the management and provisioning of an optical network.

When many WDM channels co-exist in the same optical fiber, they interact with each other due to fiber nonlinearities. Nonlinear phase noise induced by other channels degrades phase-modulated optical signals. While previous chapters consider single channel systems, multichannel WDM systems are considered in this chapter.

This chapter first discusses some essential WDM components to build an optical network. Homodyne crosstalk may arise due to insufficient crosstalk rejection of a WDM demultiplexer. Afterward, nonlinear phase noise due to cross-phase modulation (XPM) is studied for DPSK signals. Depending on fiber chromatic dispersion and its management, XPM-induced nonlinear phase noise may be larger than self-phase modulation (SPM) induced nonlinear phase noise of Chapter 5.

1. WDM Based Optical Networking

The point-to-point WDM link of Fig. 1.5 is the simplest method to utilize the enormous bandwidth in an optical fiber. The optical bandwidth is subdivided into many channels and each channel transmits its signal individually. In the fiber link of Fig. 1.5, all channels begin in the same source and transmit to the same destination with no signal add-drop in between.

Figure 8.1(a) shows a fiber ring based on WDM technology with ROADM in the ring. The same ROADM may locate in the fiber link of Fig. 1.5. Figure 8.1(a) assumes an unidirectional ring for simplicity. A bidirectional ring can be constructed using two unidirectional rings of Fig. 8.1 for each direction. In Fig. 8.1(a), optical amplifiers are usually located before and after each node, and may be in the middle of the fiber link between nodes. Not shown in Fig. 8.1(a), optical amplifiers are required and are implicitly assumed for loss compensation.

Two popular ROADM architectures are shown in Figs. 8.1(b) and (c), respectively. Figure 8.1(b) is a demultiplexing, switching, and multiplexing ROADM. All WDM channels are demultiplexed into their respective ports using a WDM demultiplexer. Using a switching network, each individual channel is either drop to the node, add back to the fiber after processing, or just directly passing through the node. In the architecture of Fig. 8.1(b), each WDM channel is processed individually. There are some variations of Fig. 8.1(b), for example, certain channels are always passing through the node or group of the channels are dropped/added/passed the node together as a band of channels. As all WDM channels are processed in the ROADM, the architecture of Fig. 8.1(b) is usually very costly and has high fiber-to-fiber loss.

Figure 8.1(c) is a broadcast and select architecture. The WDM signal is splitted into two paths. The dropped channels are selected using tunable filter and added back to the ring using another set of tunable transmitters. A tunable wavelength blocker is used between the add and drop ports to block the dropped channels. In practice, the wavelength blocker and tunable filter may be built together as a single optical component. Usually, the broadcast and select architecture of Fig. 8.1(c) has both low cost and loss. The architecture of Fig. 8.1(c) can also broadcast a WDM channel to many nodes simultaneously. However, the architecture of Fig. 8.1(b) can process all channels together but the architecture of Fig. 8.1(c) may not scale well with the increase of channel number.

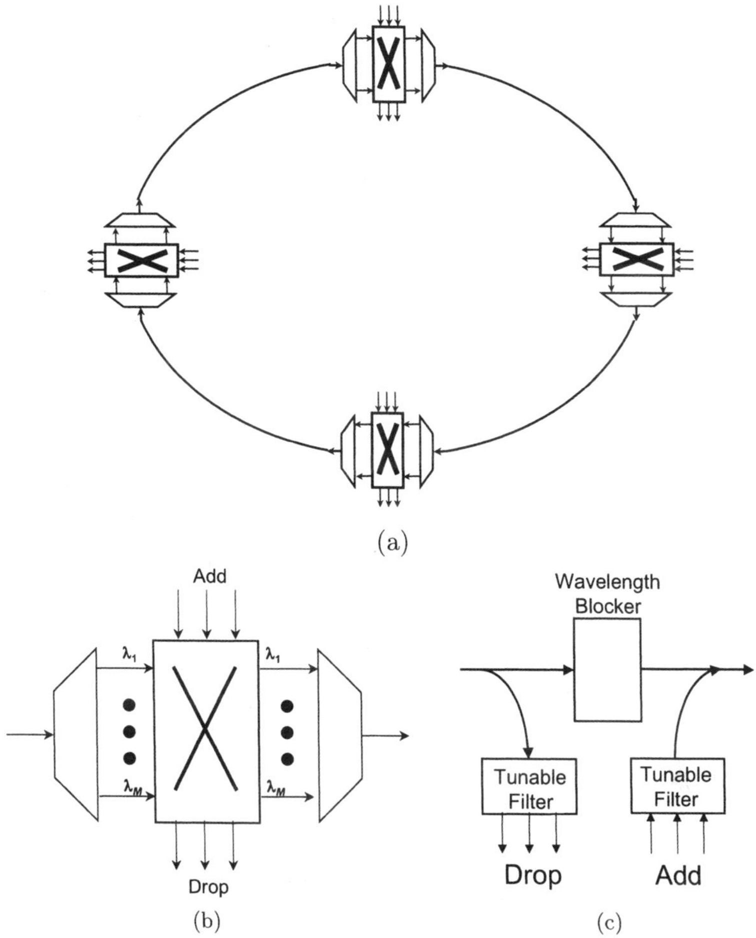

Figure 8.1. (a) A WDM ring with ROADM. Two ROADM designs based on (b) demultiplexing/switching/multiplexing and (c) broadcasting and selection.

In the ring network of Fig. 8.1(a) or long fiber link of Fig. 1.5, the ROADM of Figs. 8.1(b) and (c) can be used. The ROADM of both Figs. 8.1(b) and (c) has only one input and one output fiber. For a mesh network in Fig. 8.2(a), a wavelength router of Fig. 8.2(b) is required that have multiple input and output fibers. A fully functional, path-protection bidirectional WDM ring may also require 2 by 2 ROADM or wavelength router. Ideally, the wavelength router of Fig. 8.2(b) can pick

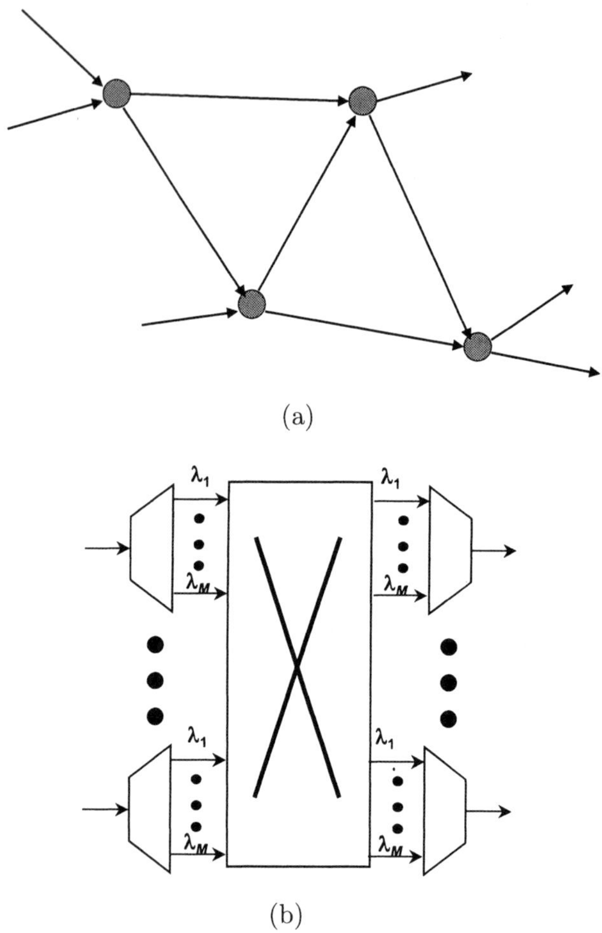

Figure 8.2. (a) A WDM based mesh network. (b) The required wavelength router in each node.

a wavelength channel from one of the input ports and switch it to one of the output ports, with or without wavelength conversion.

WDM multiplexer and demultiplexer are usually based on similar technology. For large number of channels, WDM multiplexer and demultiplexer are most based on arrayed-waveguide grating (AWG) (Dragone, 1991, 2005, Hibino, 2002, Smit and van Dam, 1996, Takahashi et al., 1990, Uetsuka, 2004). With very long history, dielectric multilayer thin-film filter is also used in WDM systems, mostly for systems with small number of channels (Domash et al., 2004, Janssen, 2001, Kajikawa and Kataoka, 1997, MacLeod, 2001, Minowa and Fujii, 1983).

EDFA intrinsically has a gain change with wavelength (Desurvire, 1994). For WDM applications, EDFA must be designed for flat gain for all WDM channels (Becker et al., 1999). Raman amplifier also has a gain changing with wavelength separation between the channel and Raman pump (Stolen, 1980). For WDM applications, multiple pumps are used to provide equal Raman gain for all WDM channels (Bromage, 2004, Islam, 2002). Additional optical filters may be used to flatten the gain of EDFA or Raman amplifier (Janssen, 2001, Lee and Lai, 2002, Vengsarkar et al., 1996).

Most of the large-scale optical switches are based on micro-electro-mechanical systems (MEMS) (Dadap et al., 2003, Lin et al., 2000, Neilson et al., 2004). Fiber Bragg grating may be another good choice (Chen and Lee, 1998, Moon and Kikuchi, 2003).

In optical fiber, signal can be propagated in both directions and bidirectional optical add/drop multiplexer was also proposed (Ho and Liaw, 1998b, Ho et al., 2000b, Kim et al., 1998, Tran et al., 2003).

2. Crosstalk Issues

Crosstalk arises due to insufficient crosstalk rejection of the WDM demultiplexer of Figs. 1.5, 8.1(b), and 8.2(b). For homodyne and heterodyne receivers in which the received signal beats with local oscillator (LO) laser, crosstalk from adjacent channels is eliminated by the low-pass or band-pass filter in the homodyne or heterodyne receivers, respectively. In direct-detection receiver, crosstalk from adjacent channels usually degrades the signal.

Homodyne crosstalk has the same wavelength as the signal that may arise due to many sources. In the broadcast and select ROADM of Fig. 8.1(c), for example, insufficient blocking by the wavelength blocker gives small amount of leakage directly from the input to output port. Having the same wavelength as the added channel, the leakage becomes homodyne crosstalk that propagates with the signal. In the wavelength switch of Fig. 8.2(c), homodyne crosstalk may be induced by signal with the same wavelength from other input ports due to component crosstalk inside the ROADM.

2.1 Linear Crosstalk

Linear crosstalk has a wavelength different with the signal channel. At the output of both homodyne and heterodyne receivers, linear crosstalk gives an interference with a frequency comparable to the channel separation. With a low-pass or band-pass filter at a homodyne and heterodyne receiver, linear crosstalk can be filtered out. The above analysis is only valid if balanced receiver is used or the signal has constant intensity. For single-branch receiver for amplitude-shift keying (ASK) signal, the intensity of linear crosstalk of $|x_t(t)|^2$ distorts the signal. With single-branch homodyne or heterodyne receiver, the crosstalk of $|x_t(t)|^2$ is much smaller than the receiver signal amplitude of $\sqrt{P_r P_{\mathrm{LO}}}$ with the assumption of small linear crosstalk of $P_r \gg |x_t(t)|^2$ and large LO power of P_{LO}.

Linear crosstalk is an issue for direct-detection receiver with insufficient crosstalk rejection. For direct-detection on-off keying signal, the signal of $|x_t(t)|^2$ gives an eye-closure penalty of $1-r_{\mathrm{xt}}$ to the eye-diagram, where r_{xt} is the crosstalk ratio. A crosstalk ratio of $r_{\mathrm{xt}} = 6.8$ dB gives an eye-closure penalty of 1 dB. However, experimental measurement found that the crosstalk ratio should be less than -15 dB (Ho and Liaw, 1998a).

For DPSK signals, using the approaches of Secs. 3.3.3 and 3.4.2, similar to Eq. (3.102) and assumed the usage of balanced receiver, with linear crosstalk, we obtain

$$r_d(t) = \Re\left\{\tilde{r}(t)\tilde{r}^*(t-T)\right\} + \Re\left\{x_t(t)x_t^*(t-T)e^{j\omega_{\mathrm{xt}}T}\right\}, \tag{8.1}$$

where ω_{xt} is the angular frequency of the crosstalk signal. With different wavelength, the crosstalk term of

$$\Re\left\{x_t(t)x_t^*(t-T)e^{j\omega_{\mathrm{xt}}T}\right\} \propto \cos\left[\omega_{\mathrm{xt}}T + \phi_{\mathrm{xt}}(t) - \phi_{\mathrm{xt}}(t-T)\right], \tag{8.2}$$

where $\phi_{\mathrm{xt}}(\cdot)$ is the phase of the crosstalk channel. If the crosstalk channel is DPSK signal with $\phi_{\mathrm{xt}}(t) - \phi_{\mathrm{xt}}(t-T) = 0, \pi$, whether the crosstalk channel affects the signal depends on the phase difference of $\omega_{\mathrm{xt}}T$. If $\omega_{\mathrm{xt}}T = k\pi + \pi/2$, where k is an integer, the crosstalk disappear. In the worst case of $\omega_{\mathrm{xt}}T = k\pi$, the received signal is equal to

$$r_d(t) = \Re\left\{\tilde{r}(t)\tilde{r}^*(t-T)\right\} \pm r_{\mathrm{xt}}A^2. \tag{8.3}$$

Following the scheme to calculate the error probability of envelope detection of orthogonal signal of Eq. (3.99) in Sec. 3.3.2, the error probability with linear crosstalk is approximately equal to

$$\begin{aligned} p_e &\approx \int_0^\infty p_1(r_1) \int_{r_1 - r_{\mathrm{xt}}A^2}^\infty p_2(r_2)\mathrm{d}r_2\mathrm{d}r_1 \\ &= \frac{1}{2}\exp\left[-\rho_s(1-r_{\mathrm{xt}})\right] \end{aligned} \tag{8.4}$$

with a SNR penalty of $(1-r_{\mathrm{xt}})^{-1}$, where the integration is lower to $r_1 - r_{\mathrm{xt}}A^2$ due to crosstalk.

With good WDM demultiplexer design, linear crosstalk is normally not a critical issue for a WDM optical network. Linear crosstalk for direct-detection on-off keying signal was analyzed by Hamdy and Humblet (1993), Humblet and Hamdy (1990), and Hill and Payne (1985). For on-off keying signal, linear crosstalk can be reduced using interference cancellation techniques (Ho and Kahn, 1996, Ho and Lin, 1996, Minardi and Ingram, 1995).

2.2 Gaussian Model for Homodyne Crosstalk

In optical networks, optical signal add-drop is performed by wavelength router or ROADM. A fundamental difficulty for the wavelength router is homodyne crosstalk from neighboring input ports or upstream nodes, causing severe degradation in system performance. Homodyne crosstalk has identical or very close wavelength to that of the signal, is difficult to eliminate by filtering, beats with the desire signal, and gives a new kind of noise at the receiver.

For the worst-case assumption that the crosstalk interferer has the same polarization as the signal, the received electric field corresponding to Eq. (3.2) is equal to

$$E_r(t) = [A_s(t)e^{j\phi_s(t)} + n_x(t) + x_t(t)]e^{j\omega_c t}\mathbf{x} + n_y(t)e^{j\omega_c t}\mathbf{y}, \tag{8.5}$$

where $x_t(t)$ is the electric field due to crosstalk with the same polarization as the signal.

If the crosstalk of $x_t(t)$ is assumed to be Gaussian distributed, the same model as the amplifier noise of $n_x(t)$, the SNR with crosstalk becomes

$$\rho_s = \frac{P_r}{S_{n_s}B_d + r_{\text{xt}}P_r}, \tag{8.6}$$

where $r_{\text{xt}} = E\{|x_t(t)|^2\}/P_r$ is the crosstalk ratio. The penalty due to homodyne crosstalk is equal to

$$\delta_p = \frac{1}{1 - r_{\text{xt}}\rho_{s0}}, \tag{8.7}$$

where ρ_{s0} is the required SNR for the signal. The above analysis is a conservative estimation for the SNR penalty. Typically, the distribution of the crosstalk of $x_t(t)$ is not Gaussian distributed and Gaussian approximation overestimates the penalty. In the worst-case assumption of Eq. (8.5), there is no crosstalk in the orthogonal polarization together with $n_y(t)$. For heterodyne receiver, the noise of $n_y(t)$ from orthogonal polarization does not contribute to the overall system noise and the model here is valid as the worst-case assumption. For direct-detection or polarization-diversity receiver, the penalty of Eq. (8.7) overestimates the penalty and provides a conservative criterion for system design. If the crosstalk is assumed to be evenly Gaussian distributed in both polarizations of **x** and **y**, the Gaussian approximation is also correct. The crosstalk ratio is also reduced by 3 dB if the crosstalk evenly distributes between the two polarizations.

When there are many interferers, the crosstalk of $x_t(t)$ is the sum of contributions from those interferers. Using the central-limit theorem, the summation of many more or less the same random variables becomes a Gaussian-distributed random variable. The Gaussian model is valid for more than 10 interferers (Ho, 1999a, Takahashi et al., 1996). If the number of interferers approaches infinity, the Gaussian model is an exact model (Kamalakis and Sphicopoulos, 2003). The Gaussian approximation represents the worst-case assumption and serving well for conservative system design.

In the Gaussian approximation, the crosstalk of $x_t(t)$ is assumed as Gaussian distributed. The decision random variable, of course, is not Gaussian distributed as shown in Secs. 3.3.3 and 3.4.2.

Homodyne crosstalk was first considered in Goldstein et al. (1994) but multiple reflections induced intensity noise is mathematically the same (Gimlett and Cheung, 1989, Tur and Goldstein, 1991). Even for a single interferer, previous analysis of homodyne crosstalk was largely based on Gaussian approximation (Gimlett and Cheung, 1989, Goldstein et al., 1994, Li and Tong, 1996, Takahashi et al., 1996). However, the intensity noise due to homodyne crosstalk from a single interferer has bounded amplitude (Gimlett and Cheung, 1989, Tur and Goldstein, 1991). Though valid for many interferers, there were reports and evidences to show that the Gaussian model is incorrect for single interferer (Dods and Tucker, 2001, Gimlett and Cheung, 1989, Ho, 1998, 1999a, Ho et al., 1998a, Kamalakis et al., 2003, Monroy and Tangdiongga, 1998, Tur and Goldstein, 1991). Typical optical network may be dominated by a single interferer from neighboring ROADM or wavelength router.

2.3 Single Interferer in Synchronous Receivers

If the system is dominated by only a single interferer, non-Gaussian model may better estimate the system performance. Furthermore, a non-Gaussian model can be used to verify the condition of the validity of the Gaussian approximation.

With a single interferer, for ASK and phase-shift keying (PSK) signal with homodyne crosstalk, the output of a synchronous receiver is

$$r_d(t) = A + \sqrt{r_{\text{xt}}} A \cos(\phi) + n_1(t), \tag{8.8}$$

where ϕ is the relative phase between the signal and crosstalk channel. In practice, the phase of ϕ is a time-dependent random variable that changes with laser phase noise, fiber temperature variation, and other external disturbances. This relative phase of ϕ is modeled as uniformly distributed random variable from $-\pi$ to $+\pi$. As all factors that change ϕ vary slowly with time, we assume that ϕ is a constant within one or two bit intervals. For PSK signal, this random variables of ϕ also includes the transmitted phase of the crosstalk channel.

For PSK signal, with the decision variable of Eq. (8.8), the error probability as a function of ϕ is

$$p_e(\phi) = \frac{1}{2}\mathrm{erfc}\left[\sqrt{\rho_s}(1 + \sqrt{r_{\mathrm{xt}}}\cos\phi)\right]. \tag{8.9}$$

and

$$p_e = \frac{1}{4\pi}\int_{-\pi}^{+\pi} \mathrm{erfc}\left[\sqrt{\rho_s}(1 + \sqrt{r_{\mathrm{xt}}}\cos\phi)\right] \mathrm{d}\phi. \tag{8.10}$$

Using the model of Ho (1998), the random variable of $r_{\mathrm{xt}}A\cos\phi + n_1(t)$ has a characteristic function of

$$J_0\left(\sqrt{r_{\mathrm{rx}}}A\nu\right) e^{-\sigma_n^2\omega^2/2}, \tag{8.11}$$

where $J_0(\cdot)$ is the zeroth-order Bessel function of the first kind. When $J_0\left(\sqrt{r_{\mathrm{rx}}}A\nu\right)$ is expanded as a power series of ν, after an inverse Fourier transform, the noise probability density function (p.d.f.) is a series expansion of

$$p_r(r) = \frac{1}{\sqrt{2\pi}\sigma_n}\sum_{k=0}^{\infty}\frac{(r_{\mathrm{xt}}A^2/2\sigma^2)^k}{2^k(k!)^2}H_{2k}\left(\frac{r}{\sigma_n}\right)e^{-r^2/2\sigma_n^2}, \tag{8.12}$$

where $H_m(x)$ is the Hermite polynomial defined by

$$H_m(x) = (-1)^m e^{x^2/2}\frac{d^m}{dx^m}e^{-x^2/2}. \tag{8.13}$$

For PSK signal with synchronous receiver degrades by homodyne crosstalk, the error probability is a series expansion of

$$p_e = \frac{1}{2}\mathrm{erfc}\sqrt{\rho_s} + \frac{e^{-\rho_s}}{\sqrt{2\pi}}\sum_{k=1}^{\infty}\frac{r_{\mathrm{xt}}^k\rho_s^k}{2^k(k!)^2}H_{2k-1}\left(\sqrt{2\rho_s}\right). \tag{8.14}$$

Numerical results show that the error probability from Eqs. (8.10) and (8.14) is identical.

For ASK signal, there is 1/2 probability that there is no homodyne crosstalk when the crosstalk channel transmits zero intensity and another

1/2 probability with homodyne crosstalk. The error probability using series expansion is similar to Eq. (8.14) as

$$p_e = \frac{1}{2}\mathrm{erfc}\sqrt{\frac{\rho_s}{2}} + \frac{e^{-\rho_s/2}}{2\sqrt{2\pi}} \sum_{k=1}^{\infty} \frac{r_{\mathrm{xt}}^k \rho_s^k}{(k!)^2} H_{2k-1}\left(\sqrt{\rho_s}\right). \tag{8.15}$$

The series expansion here is largely based on Ho (1998), Rice (1948), and McDonough and Whalen (1995).

2.4 Single Interferer for DPSK Signals

For DPSK signal and the worst-case analysis, we assume that homodyne crosstalk has the same polarization as the signal. The amplifier noise from the orthogonal polarization is first ignored for simplicity but included afterward. The homodyne crosstalk is also assumed to be from a DPSK signal. With homodyne crosstalk, the received signal is the first term of Eq. (8.5) of

$$E_r(t) = E_s(t) + x_t(t) + n(t), \tag{8.16}$$

where $E_s(t)$ is the transmitted signal, $x_t(t)$ is the homodyne crosstalk, and $n(t)$ is the amplifier noise, all in the same polarization. The signal-to-amplifier-noise ratio is defined as $\rho_{n0} = E\{|E_s(t)|^2\}/E\{|n(t)|^2\}$.

When the DPSK signal is directly detected using an asymmetric interferometer and a balanced receiver of Figs. 1.4(c) or 3.12, ignored those constant factors of interferometer loss and photodiode responsivity, similar to Sec. 3.4.2, the received photocurrent after the balanced receiver is

$$i(t) = |E_r(t) + E_r(t-T)|^2 - |E_r(t) - E_r(t-T)|^2. \tag{8.17}$$

Defined $r_1^2 = |E_r(t) + E_r(t-T)|^2$ and $r_2^2 = |E_r(t) - E_r(t-T)|^2$ and assume that the consecutive transmitted phases at $t - T$ and t are the same, the error probability is equal to

$$p_e = \Pr\left\{r_1^2 < r_2^2\right\}. \tag{8.18}$$

The error probability when $E_s(t)$ and $E_s(t-T)$ have a π phase difference is the same as that given by Eq. (8.18). Without loss of generality, $E_s(t) = E_s(t-T) = A$ is assumed with A as a positive amplitude. The crosstalk is

$$x_t(t) = \sqrt{r_{\mathrm{xt}}} A e^{j\phi} e^{j\theta_x(t)}, \tag{8.19}$$

where $\theta_x(t)$ is the transmitted phase of the crosstalk channel, the phase ϕ is the relative phase difference between the signal and crosstalk channel with the same wavelength. The crosstalk phases of $\theta_x(t) = \theta_x(t-T) = 0$, or $\theta_x(t) = 0$ and $\theta_x(t-T) = \pi$ have equal probability. In the short time from $t - T$ to t, the phase of ϕ is assumed the same as all factors that change ϕ are slowly varying with time. The initial phase of $\theta_x(t)$ may also be a random phase uniformly distributed from $-\pi$ to $+\pi$ but is combined together to the relative phase of ϕ.

With 1/2 probability for $\theta_x(t) = \theta_x(t-T) = 0$, we obtain

$$r_1^2 = |2A + 2A\sqrt{r_{\text{xt}}}e^{j\phi} + n(t) + n(t-T)|^2, \tag{8.20}$$

$$r_2^2 = |n(t) - n(t-T)|^2. \tag{8.21}$$

Because the two noises of $n(t) \pm n(t-T)$ are also independent of each other, without crosstalk of $r_{\text{xt}} = 0$, the error probability is that for DPSK signal of $p_{e1} = \frac{1}{2}e^{-\rho_{n0}}$ from Eq. (3.105). The probability of $\Pr\left\{r_1^2 < r_2^2\right\}$ as a function of the phase of ϕ is the error probability of DPSK signal with a SNR of $\rho_{n0}|1 + \sqrt{r_{\text{xt}}}e^{j\phi}|^2$, or

$$\begin{aligned} p_{e1}(\phi) &= \frac{1}{2}\exp\left(-\rho_{n0}\left|1 + \sqrt{r_{\text{xt}}}e^{j\phi}\right|^2\right) \\ &= \frac{1}{2}\exp\left[-\rho_{n0}\left(1 + r_{\text{xt}} + 2\sqrt{r_{\text{xt}}}\cos\phi\right)\right]. \end{aligned} \tag{8.22}$$

With an uniformly distributed relative phase of ϕ, we obtain

$$\begin{aligned} p_{e1} &= \frac{1}{2\pi}\int_{-\pi}^{+\pi} p_{e1}(\phi)\mathrm{d}\phi \\ &= \frac{1}{2}\exp\left[-\rho_{n0}\left(1 + r_{\text{xt}}\right)\right] I_0(2\sqrt{r_{\text{xt}}}\rho_{n0}), \end{aligned} \tag{8.23}$$

where $I_m(\cdot)$ is the mth-order modified Bessel function of the first kind.

With 1/2 probability for $\theta_x(t) = 0$ and $\theta_x(t-T) = \pi$, we obtain

$$r_1^2 = |2A + n(t) + n(t-T)|^2, \tag{8.24}$$

$$r_2^2 = |2A\sqrt{r_{\text{xt}}}e^{j\phi} + n(t) - n(t-T)|^2. \tag{8.25}$$

In this case, the random variables of r_1^2 and r_2^2 are independent of each other. From Appendix 3.A, the probability of $\Pr\left\{r_1^2 < r_2^2\right\}$ is independent of ϕ and equals to

$$p_{e2} = Q\left(\sqrt{2r_{\text{xt}}\rho_{n0}}, \sqrt{2\rho_{n0}}\right) - \frac{1}{2}e^{-\rho_{n0}(1+r_{\text{xt}})} I_0(2\sqrt{r_{\text{xt}}}\rho_{n0}), \tag{8.26}$$

where $Q(\cdot,\cdot)$ is Marcum Q function of Appendix 3.A.

Including amplifier noise only from the same polarization as the signal, the overall error probability is

$$p_e = \frac{1}{2}(p_{e1} + p_{e2}) = \frac{1}{2}Q\left(\sqrt{2r_{\text{xt}}\rho_{n0}}, \sqrt{2\rho_{n0}}\right). \tag{8.27}$$

The error probability of Eq. (8.27) is applicable to heterodyne DPSK receiver or polarized direct-detection DPSK receiver. However, direct-detection DPSK receiver usually does not have a polarizer to filter-out amplifier noise from orthogonal polarization.

With amplifier noise from orthogonal polarization, from the error probability of Eq. (3.163) without crosstalk, the error probability increases from $\frac{1}{2}e^{-\rho_{n0}}$ to $\frac{1}{2}e^{-\rho_{n0}}\left(1 + \frac{1}{4}\rho_{n0}\right)$. The error probability corresponding to Eq. (8.23) becomes

$$p_{e1} = \frac{1}{2}\exp\left[-\rho_{n0}\left(1 + r_{\text{xt}}\right)\right]\left\{I_0(2\sqrt{r_{\text{xt}}}\rho_{n0}) + \frac{\rho_{n0}}{4}\left[1 + r_{\text{xt}} + 2\sqrt{r_{\text{xt}}}I_1(2\sqrt{r_{\text{xt}}}\rho_{n0})\right]\right\}. \quad (8.28)$$

From Eq. (4.11) and Appendix 3.A, the error probability corresponding to Eq. (8.26) becomes

$$p_{e2} = Q\left(\sqrt{2r_{\text{xt}}\rho_{n0}}, \sqrt{2\rho_{n0}}\right) - \frac{1}{2}e^{-\rho_{n0}(1+r_{\text{xt}})}I_0(2\sqrt{r_{\text{xt}}}\rho_{n0}) + \frac{r_{\text{xt}}^{-1/2} - r_{\text{xt}}^{1/2}}{8}e^{-\rho_{n0}(1+r_{\text{xt}})}I_1(2\sqrt{r_{\text{xt}}}\rho_{n0}). \quad (8.29)$$

Including amplifier noise from orthogonal polarization, the error probability corresponding to Eq. (8.27) is

$$p_e = \frac{1}{2}Q\left(\sqrt{2r_{\text{xt}}\rho_{n0}}, \sqrt{2\rho_{n0}}\right) + \frac{\exp\left[-\rho_{n0}(1 + r_{\text{xt}})\right]}{16} \times \left\{\rho_{n0}(1 + r_{\text{xt}}) + \left[(2\rho_{n0} - 1)r_{\text{xt}}^{1/2} + r_{\text{xt}}^{-1/2}\right]I_1(2\sqrt{r_{\text{xt}}}\rho_{n0})\right\}. \quad (8.30)$$

Figure 8.3 shows the SNR penalty for both PSK and DPSK signals as a function of crosstalk ratio for an error probability of 10^{-9}. For DPSK signal, the error probability is calculated using Eqs. (8.27) and (8.30) when the amplifier noises from only the signal polarization and both polarizations are included, respectively. When amplifier noise from orthogonal polarization is included, the SNR penalty of DPSK signal for about 0.4 dB at low crosstalk level is from the amplifier noise but not crosstalk. The SNR penalty for PSK signal of -0.45 dB at low crosstalk level is also from the improvement of PSK to DPSK signal. In additional to the single interferer, Figure 8.3 also shows the SNR penalty with Gaussian distributed crosstalk for large number of interferers.

Figure 8.3 shows that for 1-dB SNR penalty and DPSK signal, the crosstalk ratio must be less than -16.5 and -17.8 dB for single and dual-

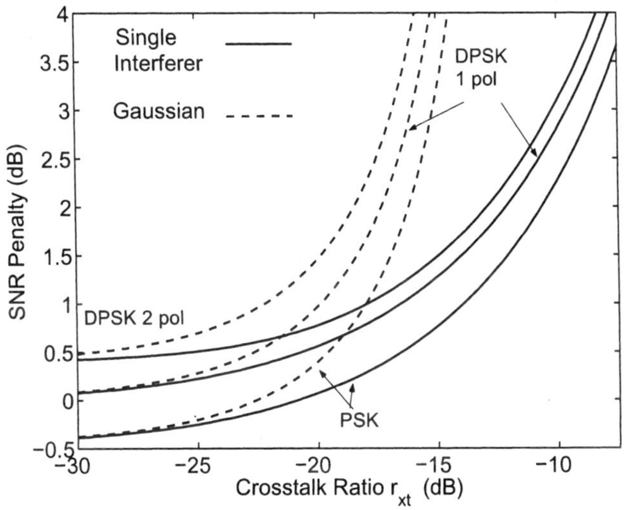

Figure 8.3. The SNR penalty as a function of crosstalk ratio for phase-modulated signals.

polarization amplifier noises, respectively. For 0.5-dB SNR penalty, using the single-polarization model, the crosstalk ratio must be less than -20.5 dB.

From Fig. 8.3, Gaussian model for homodyne crosstalk overestimates the SNR penalty for system with a single interferer. When Fig. 8.3 is compared with the experimental results in Liu et al. (2004d, Fig. 3), Figure 8.3 is almost the same (within ±0.2 dB) as the experimental measurement for the crosstalk ratio from -20 to -15 dB for a SNR penalty of about 0.5 to 1.3 dB. At high crosstalk ratio of -12 dB, Figure 8.3 has a penalty of 2.1 dB versus 2.5 dB in Liu et al. (2004d, Fig. 3), still a very close match. The difference may be attributed to measurement uncertainty. The crosstalk ratio for a SNR penalty of around and less than 1 dB is important for system design.

2.5 Single Interferer for On-Off Keying Signals

For direct-detection on-off keying signals limited by amplifier noise and homodyne crosstalk, if there is homodyne crosstalk, the received signal of on state depends on the relative phase of the signal and crosstalk channels. Similar to Eq. (3.125), the received signal at the on state is

$$r_{\mathrm{on}}(t) = [A(1+\sqrt{r_{\mathrm{xt}}}\cos\phi) + n_{x1}(t)]^2 + [\sqrt{r_{\mathrm{xt}}}A\sin\phi + n_{x2}(t)]^2 + n_{y1}^2(t) + n_{y2}^2(t), \qquad (8.31)$$

with p.d.f. as a function of ϕ similar to that of Eq. (3.126) of

$$p_1(y) = \frac{1}{2\sigma_n^2}\sqrt{\frac{y}{s^2}}e^{-(s^2+y)/2\sigma_n^2} I_1\left(\sqrt{y}\frac{s}{\sigma_n^2}\right), \qquad y \geq 0 \qquad (8.32)$$

where $s^2 = A^2\left(1 + r_{\mathrm{xt}} + 2\sqrt{r_{\mathrm{xt}}}\cos\phi\right)$ is the noncentrality parameter as a function of ϕ.

The signal of the off state is

$$r_{\mathrm{off}}(t) = [\sqrt{r_{\mathrm{xt}}}A\cos\phi + n_{x1}(t)]^2 + [\sqrt{r_{\mathrm{xt}}}A\sin\phi + n_{x2}(t)]^2 + n_{y1}^2(t) + n_{y2}^2(t). \qquad (8.33)$$

with p.d.f. independent of ϕ as

$$p_1(y) = \frac{1}{2\sigma_n^2}\sqrt{\frac{y}{r_{\mathrm{xt}}A^2}}e^{-(r_{\mathrm{xt}}A^2+y)/2\sigma_n^2} I_1\left(\frac{\sqrt{r_{\mathrm{xt}}y}A}{\sigma_n^2}\right), \qquad y \geq 0 \quad (8.34)$$

With a threshold of y_{th}, the error probability as a function of the phase of ϕ is

$$p_{e1}(r_{\mathrm{xt}}, \phi) = \frac{1}{2}\left[1 - Q_2\left(\frac{A\left(1 + r_{\mathrm{xt}} + 2\sqrt{r_{\mathrm{xt}}}\cos\phi\right)^{1/2}}{\sigma_n}, \frac{\sqrt{y_{\mathrm{th}}}}{\sigma_n}\right)\right] + \frac{1}{2}Q_2\left(\frac{\sqrt{r_{\mathrm{xt}}}A}{\sigma_n}, \frac{\sqrt{y_{\mathrm{th}}}}{\sigma_n}\right). \qquad (8.35)$$

where $Q_2(\cdot,\cdot)$ the second-order Marcum Q function given by Eq. (3.130). For the ASK signal, the error probability with homodyne crosstalk is

$$p_e = \frac{1}{4\pi}\int_{-\pi}^{\pi} p_{e1}(r_{\mathrm{xt}}, \phi)\mathrm{d}\phi + \frac{1}{2}p_{e1}(0, 0), \qquad (8.36)$$

where $p_{e1}(0,0)$ is the error probability when the interferer is at off state. Note that $p_{e1}(0,0)$ is the same as the error probability of Eq. (3.132).

Figure 8.4 shows the SNR penalty as a function of crosstalk ratio for ASK signals using direct-detection or synchronous receivers. The error probability of direct-detection ASK signal is calculated using Eq. (8.36) to include the amplifier noise from both parallel and orthogonal polarizations to the signal. The threshold of y_{th} is equal to either $A^2/4$ as that for Eq. (3.132) or optimized using numerical algorithm. The required

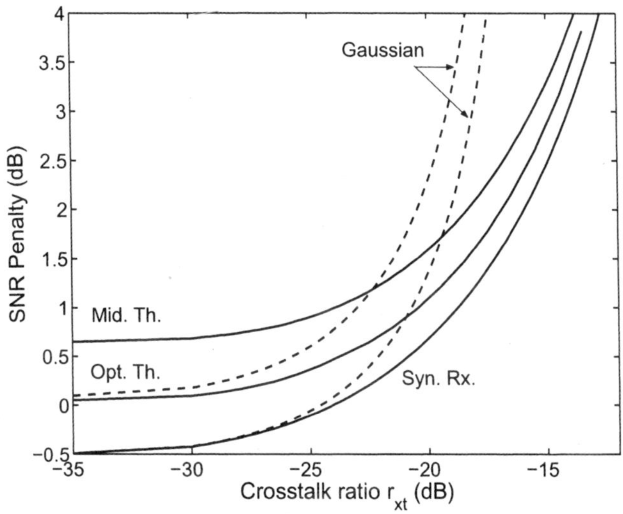

Figure 8.4. The SNR penalty as a function of crosstalk ratio for ASK signals.

SNR for an error probability of 10^{-9} is 41.1 (16.1 dB) from Fig. 3.11 with optimized threshold. In additional to direct-detection ASK signal, Figure 8.4 also includes the SNR penalty calculated using Eq. (8.15) for synchronous detection of ASK signal. Note that the SNR penalty and gain at low crosstalk level cannot attribute to crosstalk but due to the difference in receiver sensitivity without crosstalk.

For a 1-dB SNR penalty, the crosstalk ratio for ASK signal must be less than -20.5 dB, about 4 dB less than DPSK signal from Fig. 8.3. For a 0.5-dB SNR penalty, the crosstalk ratio must be less than -23.6 dB, about 3.1 dB less than the corresponding crosstalk ratio for DPSK signal. Because of this comparison, Liu et al. (2004d) concludes that DPSK signal can tolerate higher homodyne crosstalk ratio than ASK or on-off keying signal.

Figure 8.4 also shows that the Gaussian model overestimates the SNR penalty for single interferer. For small number of interferers, non-Gaussian model must be used for system analysis. Like Fig. 3.11, the decision threshold of the receiver must be optimized to account for the non-Gaussian characteristic of the homodyne crosstalk and amplifier noise.

Non-Gaussian model for single and small number of interferers had been developed by series expansion (Dods and Tucker, 2002, Ho, 1998, 1999a, Ho et al., 1998a), saddle-point approximation (Danielsen et al., 1998, Kamalakis et al., 2003, Monroy and Tangdiongga, 1998), numerical integration or simulation (Cornwell and Andonovic, 1996, Dods and Tucker, 2001, Jiang and Roudas, 2001). Other than Jiang and Roudas (2001), Kamalakis and Sphicopoulos (2003), Kamalakis et al. (2003), and Ho (1998), most of those works ignored the crosstalk-to-crosstalk beating term. The analysis here for direct-detection receiver takes into account the crosstalk-to-crosstalk beating.

This section considers only a homogeneous WDM system in which all channels have the same modulation format. The effect of crosstalk in a heterogeneous WDM system can be analyzed similarly.

3. Cross-Phase Modulation Induced Nonlinear Phase Noise

In this section, the error probability is derived analytically for DPSK signals contaminated by both self- and cross-phase modulation (SPM and XPM) induced nonlinear phase noise. SPM-induced nonlinear phase noise, often called Gordon-Mollenauer effect (Gordon and Mollenauer, 1990), was discussed in details in Chapter 5 as a non-Gaussian distributed noise that can be compensated using the received intensity as from Chapter 6. XPM-induced nonlinear phase noise is Gaussian distributed phase noise, similar to laser phase noise in Sec. 4.3. When fiber dispersion is compensated perfectly in each fiber span, XPM-induced nonlinear phase is summed coherently span after span and is the dominant nonlinear phase noise for typical WDM DPSK signals. For systems without or with XPM-suppressed dispersion compensation, SPM-induced nonlinear phase noise is usually the dominant nonlinear phase noise. With longer walk-off length, for the same mean nonlinear phase shift, 10-Gb/s systems are more sensitive to XPM-induced nonlinear phase noise than 40-Gb/s systems.

3.1 Variance of Nonlinear Phase Noise

Both Eqs. (5.80) and (6.7) gives the variance of SPM-induced nonlinear phase noise, the ratio of the variance of nonlinear phase noise induced by XPM to that by SPM is derived here. The variance is first derived for the two-channel pump-probe model and later extended to multi-span multichannel WDM systems.

Pump-probe model

For the simplest model of having two WDM channels, the overall nonlinear phase shift to the first channel is equal to

$$\Phi_{\mathrm{NL}} = \gamma \int_0^L \left[|E_1(z)|^2 + 2|E_2(z)|^2 \right] \mathrm{d}z, \tag{8.37}$$

where E_1 and E_2 are the electric field of the first and second channels, respectively. In Eq. (8.37), the first term of the right-hand-size is from SPM and has been considered in Chapter 5 and the second term is from XPM. If both the first and second channel propagates in the same speed in the fiber, the contribution from XPM is the same as that from SPM. With channel walk-off, the XPM term is an average over an interval of time.

Based on the pump-probe model, the phase modulation of channel 1 (probe) induced by channel 2 (pump) is

$$\phi_{1,\mathrm{XPM}}(L,t) = 2\gamma \int_0^L P_2(0, t + d_{12}z) e^{-\alpha z} \mathrm{d}z, \tag{8.38}$$

where $P_2(z,t)$ is the power of channel 2 as a function of position z and time t, γ is the fiber nonlinear coefficient given by Eq. (5.2), α is fiber attenuation coefficient, L is the fiber length, $d_{12} \approx D\Delta\lambda$ is the relative walk-off between two channels with wavelength separation of $\Delta\lambda$, and D is the dispersion coefficient of the fiber chromatic dispersion. The phase of $\phi_{1,\mathrm{XPM}}(L,t)$ assumes that the waveform of $P_2(z,t) = P(0, t - z/c_2)$ without distortion along the fiber, where c_2 is the speed of light at channel 2. While waveform distortion is ignored, the walk-off effect is included by the parameter of d_{12}. Some numerical results from Ho (1999b) showed that the waveform distortion is a second-order effect.

By taking the Fourier transform of the autocorrelation function, when the power spectral density of $P_2(0,t)$ is $\Phi_{P_2}(f)$, the power spectral density of $\phi_{1,\mathrm{XPM}}(L,t)$ is

$$\Phi_{\phi_1}(f) = \Phi_{P_2}(f)|H_{12}(f)|^2, \tag{8.39}$$

where $H_{12}(f) = 2\gamma \int_0^L e^{-\alpha z + j2\pi f d_{12} z} \mathrm{d}z$ or

$$H_{12}(f) = 2\gamma \frac{1 - e^{-\alpha L + j2\pi f d_{12} L}}{\alpha - j2\pi f d_{12}}. \tag{8.40}$$

At the DPSK receiver, after the asymmetric Mach-Zehnder interferometer, the differential nonlinear phase noise of $\Delta\phi_{1,\mathrm{XPM}}(L,t) = \phi_{1,\mathrm{XPM}}(L,t) - \phi_{1,\mathrm{XPM}}(L, t-T)$ adds to the differential phase of the signal, where T is the symbol interval. The power spectral density of $\Delta\phi_{1,\mathrm{XPM}}(L,t)$ is

$$\Phi_{\Delta\phi_1}(f) = 4\Phi_{P_2}(f)|H_{12}(f)|^2 \sin^2(\pi f T). \tag{8.41}$$

When the pump (channel 2) has amplifier noises, $P_2(0,t) = |E_2 + N_2|^2$ where E_2 and N_2 are the electric fields from both signal and noise, respectively. In the power of $P_2(0,t) = |E_2|^2 + E_2 \cdot N_2^* + E_2^* \cdot N_2 + |N_2|^2$, the dc-term of $|E_2|^2$ gives no nonlinear phase noise but a constant phase shift, the signal-noise beating of $E_2 \cdot N_2^* + E_2^* \cdot N_2$ gives a noise spectral density of $2|E_2|^2 S_{\mathrm{sp}}$, the noise-noise beating of $|N_2|^2$ gives a noise spectral density of $2S_{\mathrm{sp}}^2 \Delta\nu_{\mathrm{opt}}$, where S_{sp} is the spectral density of the amplifier noise and $\Delta\nu_{\mathrm{opt}}$ is the optical bandwidth of the amplifier noise. The optical SNR over an optical bandwidth of $\Delta\nu_{\mathrm{opt}}$ is $|E_2|^2/(2S_{\mathrm{sp}}\Delta\nu_{\mathrm{opt}})$. For a launched power of P_0 and a single optical amplifier with a noise variance of $S_{\mathrm{sp},1}$, we get $\Phi_{P_2}(f) = 2P_0 S_{\mathrm{sp},1} + 2S_{\mathrm{sp},1}^2 \Delta\nu_{\mathrm{opt}} \approx 2\sigma_n^2 P_0$ as a constant over frequency.

For a non-return-to-zero (NRZ) DPSK signal, $|E_2|^2$ is a dc-term and can be ignored. For the more popular return-to-zero (RZ)-DPSK signal, $|E_2|^2$ is a periodic function with a period of T and its power spectral density is tones at the frequencies of k/T, where k is integer. The differential transfer function of $4\sin^2(\pi f T)$ has notches at k/T and cancels all nonlinear phase shift due to $|E_2|^2$ even for RZ signal. For RZ-DPSK signal with pulse broadening due to fiber dispersion, if the dispersion is assumed to be a linear effect, for system without pulse overlapping, the differential transfer function can also completely eliminate XPM-induced nonlinear phase shift from $|E_2|^2$. Of course, this assumption is valid with no pulse distortion in the fiber with the relationship $P_2(z,t) = P_2(0, t - z/c_2)$. With pulse broadening such that two pulses overlap after a short fiber distance, later part of this chapter considers the nonlinear phase shift due to the signal of $|E_2|^2$.

The phase variance as a function of frequency separation is

$$\sigma_{\mathrm{XPM},0}^2(\Delta\lambda) = 4\Phi_{P_2}(f) \int_{-1/T}^{1/T} |H_{12}(f)|^2 \sin^2(\pi f T)\,\mathrm{d}f, \tag{8.42}$$

where the integration is reduced from $\pm\infty$ to $\pm 1/T$ by taking into account only the phase noise over a bandwidth confined within the bit-rate. Please note that $\Phi_{P_2}(f)$ is a constant independent of frequency. The dependence of the variance of Eq. (8.42) on the wavelength separation of $\Delta\lambda$ is originated from the dependence of $H_{12}(f)$ of Eq. (8.40) on $\Delta\lambda$.

SPM-induced nonlinear phase noise, using Eq. (8.42), is equal to

$$\sigma_{\mathrm{SPM}}^2 = \frac{1}{4}\sigma_{\mathrm{XPM},0}^2(0) = \frac{8\sigma_n^2\gamma^2 P_0 L_{\mathrm{eff}}^2}{T}, \tag{8.43}$$

where $L_{\mathrm{eff}} = (1 - e^{-\alpha L})/\alpha$ is the effective nonlinear length per span. The factor of $1/4$ is because the phase shift induced by XPM is twice larger than that induced by SPM for the same intensity. For a long fiber span with $L \gg 1/\alpha$ and large walk-off coefficient of d_{12}, using Gradshteyn and Ryzhik (1980, §3.824), we get

$$\begin{aligned}\frac{\sigma^2_{\mathrm{XPM},0}(\Delta\lambda)}{\sigma^2_{\mathrm{SPM}}} &= 8\alpha^2 T \int_0^{1/T} \frac{\sin^2(\pi f T)}{\alpha^2 + (2\pi f d_{12})^2} \mathrm{d}f \\ &\approx 8\alpha^2 T \int_0^{\infty} \frac{\sin^2(\pi f T)}{\alpha^2 + (2\pi f d_{12})^2} \mathrm{d}f \\ &= \alpha L_W (1 - e^{-\alpha L_W}), \end{aligned} \tag{8.44}$$

where $L_W = T/|d_{12}|$ is the walk-off length that is the distance for two aligned pump and probe pulses of length T becomes completely walk-off. Like that for stimulated Raman scattering of Ho (2000), even without approximation, the variances of XPM-induced nonlinear phase noise depends on the walk-off length of L_W only.

The pump-probe model is based on Chiang et al. (1994, 1996), Ho et al. (1999), and Leibrich et al. (2002). The definition of walk-off length follows Ho (2000). Kim (2003) first measured the effect of XPM-induced nonlinear phase noise to a WDM system.

Like most of other models (Cartaxo, 1999, Hui et al., 1999), we assume that the waveform does not have distortion over distance. Chromatic dispersion broadens a signal over distance as an intrachannel effect. Another aspect of chromatic dispersion, pulse walk-off between channels is also due to different propagating speed of two adjacent channels. As the channel separation is always larger than the signal spectrum, channel walk-off always has larger effect than pulse broadening. The assumption of waveform without distortion over distance is correct when considered together with walk-off effect.

Multi-span WDM systems

For a $(2M+1)$-channel WDM system, for the worst case of the center channel, in the first span, the nonlinear phase noise variance per span is equal to

$$\sigma^2_{\mathrm{XPM},1} = 2\sum_{k=1}^{M} \sigma^2_{\mathrm{XPM},0}(k\Delta\lambda), \tag{8.45}$$

where, using the same symbol as the above pump-probe model, $\Delta\lambda$ is the wavelength separation between adjacent channels. For a short walk-off length of L_W per channel separation of $\Delta\lambda$, a wavelength separation of $k\Delta\lambda$ gives a XPM-induced nonlinear phase noise variance of about a factor of $1/k^2$ smaller than that for a wavelength separation of $\Delta\lambda$. For large number of channels, using the relationship of $\sum_{k=1}^{\infty} \frac{1}{k^2} = \frac{\pi^2}{6}$

(Gradshteyn and Ryzhik, 1980, §0.233), we get

$$\lim_{M\to\infty} \frac{\sigma^2_{\mathrm{XPM},1}}{\sigma^2_{\mathrm{SPM}}} \approx \frac{\pi^2}{3}\alpha L_W(1 - e^{-\alpha L_W}). \tag{8.46}$$

For a system with N_A fiber spans, the variance of the overall XPM-induced nonlinear phase noise depends on the method of dispersion compensation. Similar to Eq. (8.41), the power spectrum density of XPM-induced nonlinear phase noise after K fiber spans is

$$\Phi_{K,\Delta\phi_1}(f) = 4\Phi_{P_2}(f)|H_{12}(f)|^2 \sin^2(\pi f T)\left|\frac{\sin[K\pi f(1-\kappa)d_{12}L]}{\sin[\pi f(1-\kappa)d_{12}L]}\right|^2, \tag{8.47}$$

where κ is the fraction of dispersion compensation per span, i.e., $\kappa = 1$ and $\kappa = 0$ for perfect and without dispersion compensation, respectively. The spectrum density of Eq. (8.47) assumes K cascaded identical fiber spans with the same configuration.

In the worst case of perfect dispersion compensation with $\kappa = 1$ such that all channels are well aligned when launch to each fiber span, the variance of Eq. (8.42) is increased by a factor of K^2 after K fiber spans. With perfect dispersion compensation, the variance of the overall XPM-induced nonlinear phase noise for an N_A-span fiber system is

$$\sigma^2_{\mathrm{XPM,max}} = N_A^2\sigma^2_{\mathrm{XPM},1} + (N_A - 1)^2\sigma^2_{\mathrm{XPM},2} + \cdots + \sigma^2_{\mathrm{XPM},N_A}, \tag{8.48}$$

where the first term of $\sigma^2_{\mathrm{XPM},1}$ is the nonlinear phase noise induced from the amplifier noise from the first span, the second term is the nonlinear phase noise induced from the amplifier noise from the second span, and so on. Because of perfect dispersion compensation, the same noise from the first span is perfectly (or coherently) added N_A times into the overall nonlinear phase noise. The noise from the second span $\sigma^2_{\mathrm{XPM},2}$ adds into the overall nonlinear phase noise $N_A - 1$ times. With the assumption that all spans have the same configuration, $\sigma^2_{\mathrm{XPM},1} = \sigma^2_{\mathrm{XPM},2} = \cdots = \sigma^2_{\mathrm{XPM},N}$. The maximum nonlinear phase noise of Eq. (8.48) is

$$\sigma^2_{\mathrm{XPM,max}} = \frac{1}{6}N_A(N_A + 1)(2N_A + 1)\sigma^2_{\mathrm{XPM},1}. \tag{8.49}$$

For SPM-induced nonlinear phase noise, the amplifier noise from the first span is also perfectly aligned in all fiber spans afterward. The relationship between single- and multi-span phase noise variance is the same as that of Eq. (8.49). With perfect dispersion compensation, the ratio of the variance of XPM- to SPM-induced nonlinear phase noise is independent of the number of fiber spans in WDM systems. The ratio of Eq. (8.46) is the ratio of XPM- to SPM-induced nonlinear phase noise in the worst-case of perfect dispersion compensation.

When dispersion compensation is conducted channel-by-channel with XPM-suppression (Bellotti and Bigo, 2000, Bellotti et al., 2000), the factor of κ is the offset of pump and probe at the output of a fiber span with respect to the fiber walk-off. If $|\kappa| \gg 1$ such that the nonlinear phase noise induced to every fiber span is independent of each other, we obtain

$$\sigma^2_{\mathrm{XPM,ind}} = N_A \sigma^2_{\mathrm{XPM},1} + (N_A - 1)\sigma^2_{\mathrm{XPM},2} + \cdots + \sigma^2_{\mathrm{XPM},N_A}, \tag{8.50}$$

or

$$\sigma^2_{\mathrm{XPM,ind}} = \frac{1}{2} N_A (N_A + 1) \sigma^2_{\mathrm{XPM},1}, \tag{8.51}$$

and the ratio of

$$\frac{\sigma^2_{\mathrm{XPM,max}}}{\sigma^2_{\mathrm{XPM,ind}}} = \frac{1}{3}(2N_A + 1). \tag{8.52}$$

In Eq. (8.50), the amplifier noise from the first span induces nonlinear phase noise in each fiber span. However, the nonlinear phase noises induced to the first and second span are from the amplifier noise in completely non-overlapped time interval and independent of each other. In an N_A-span system, the overall nonlinear phase noise is generated from $\frac{1}{2}N_A(N_A+1)$ "segments" of independent amplifier noise. When the fiber link is very long with $L \gg L_W$ with many walk-off intervals, with non-prefect dispersion compensation of $\kappa \neq 1$, the variance of Eq. (8.50) is also valid.

The last term of Eq. (8.47) is from multi-span effect of

$$\frac{\sin[K\pi f(1-\kappa)d_{12}L]}{\sin[\pi f(1-\kappa)d_{12}L]}, \tag{8.53}$$

that has peak values of K or $-K$ when $f(1-\kappa)d_{12}L = m$, where m is a positive or negative integer. The last term of Eq. (8.41) due to differential operation of $\sin^2(\pi f T)$ has notches at $f = m/T$ with m as an integer. If the notches of $\sin^2(\pi f T)$ match to the peaks of Eq. (8.53), XPM-induced nonlinear phase noise is approximately minimized. With short walk-off length of $L_W < L$, the dispersion compensation factor of

$$\kappa = 1 - \frac{L_W}{L} \tag{8.54}$$

approximately gives minimum XPM-induced nonlinear phase noise. Similar to resonance effect in Nelson et al. (1999), certain combinations of

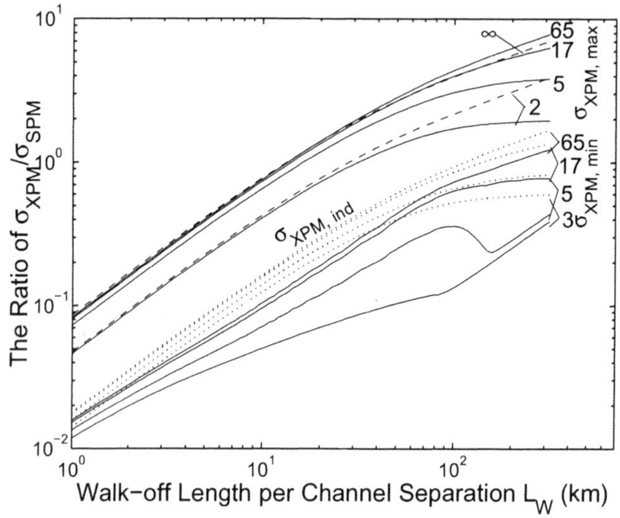

Figure 8.5. The ratio of $\sigma_{\text{XPM}}/\sigma_{\text{SPM}}$ as a function of walk-off length per channel separation. The dashed curves are the approximation of Eqs. (8.44) and (8.46). The ratio for the worst-case of $\sigma_{\text{XPM,max}}$ is independent of number of spans. The ratios for the typical cases of $\sigma_{\text{XPM,ind}}$ and the minimum case of $\sigma_{\text{XPM,min}}$ are for $N_A = 32$ fiber spans and show as dotted and solid lines, respectively. The label of each curve is the number of WDM channels. [Adapted from Ho (2004b)]

walk-off length and dispersion compensation factor minimize the variance of XPM-induced nonlinear phase noise. Numerical results show that the compensation factor of Eq. (8.54) approximately minimizes the variance of XPM-induced nonlinear phase noise.

Figure 8.5 shows the ratio of the standard deviation (STD) of XPM- to SPM-induced nonlinear phase noise as a function of the walk-off length L_W per channel separation $\Delta\lambda$. The ratio of $\sigma_{\text{XPM,min}}/\sigma_{\text{SPM}}$ shown as dotted lines is calculated numerically and uses the dispersion compensation factor of Eq. (8.54) for $L_W < L$ and $\kappa = 0$ for $L_W \geq L$. The fiber attenuation coefficient is $\alpha = 0.20$ dB/km. The fiber span length is $L = 80$ km. For the pump-probe model (labeled as "2" WDM channels), the approximation of Eq. (8.44) is valid for L_W less than 50 km.

For multi-channel WDM systems, in both the worst and independent cases, the approximation of Eq. (8.46) is valid for L_W less than 100 km and number of channels larger than 17. The approximation of Eq. (8.46) can be used to model typical WDM systems. For example, typical 10-Gb/s 50-GHz spacing system using nonzero dispersion-shifted fiber (NZDSF) with $D = 4$ ps/km/nm has a walk-off length of 62.5 km.

Typical 40-Gb/s systems have a walk-off length of 7.8 km (100 GHz spacing and $D = 4$ ps/km/nm).

When the walk-off effect is weak with long walk-off length ($L_W > 100$ km), from the pump-probe model, the nonlinear phase noise STD due to XPM approaches twice that due to SPM. With long walk-off length, nonlinear phase noise induced by XPM is much larger than that induced by SPM.

For a multi-span system, from Eq. (6.14), the variance of the SPM-induced nonlinear phase noise is equal to

$$\sigma^2_{\mathrm{SPM}} \approx \frac{4 <\Phi_{\mathrm{NL}}>^2}{3\rho_s}, \tag{8.55}$$

where ρ_s is the SNR defined for optical matched filter and a single polarization. The variance of Eq. (8.55) is twice that of Eq. (6.14) for differential operation. Combined Eq. (8.55) with the ratio in Fig. 8.5, the variance of XPM-induced nonlinear phase noise σ^2_{XPM} can be calculated.

The analysis here follows that in Chiang et al. (1996) for multi-span systems. The exact variance of Eq. (8.55) is given in Ho and Kahn (2004a), Gordon and Mollenauer (1990), or Eq. (6.7).

3.2 Error Probability for DPSK Signals

For performance analysis, we need to find the error probability of DPSK signal with SPM- and XPM-induced nonlinear phase noise. The error probability is important to characterize a system.

Assume that nonlinear phase noise and phase of amplifier noise are independent of each other, from Eq. (5.78), with only SPM-induced nonlinear phase noise, the error probability of a DPSK signal is

$$p_{e,\mathrm{SPM}} = \frac{1}{2} - \frac{\rho_s e^{-\rho_s}}{2} \sum_{k=0}^{\infty} \frac{(-1)^k}{2k+1} \left[I_k\!\left(\frac{\rho_s}{2}\right) + I_{k+1}\!\left(\frac{\rho_s}{2}\right) \right]^2 \left|\Psi_{\Phi_{\mathrm{NL}}}(2k+1)\right|^2, \tag{8.56}$$

where $I_k(\cdot)$ is a kth-order modified Bessel function of the first kind, $\Psi_{\Phi_{\mathrm{NL}}}(\nu)$ is the characteristic function of SPM-induced nonlinear phase noise given by Eq. (5.22). Simulation by error counting confirms the validity of the error probability of Eq. (8.56) in Fig. 5.14.

When independent phase noises from different sources are summed together, the coefficients of the Fourier series of the overall p.d.f. are the product of the corresponding Fourier coefficients of each individual component. The error probability with Gaussian distributed phase noise follows both Eqs. (4.40) and (5.82). If the XPM-induced nonlinear phase noise is Gaussian distributed, the error probability of the DPSK signal

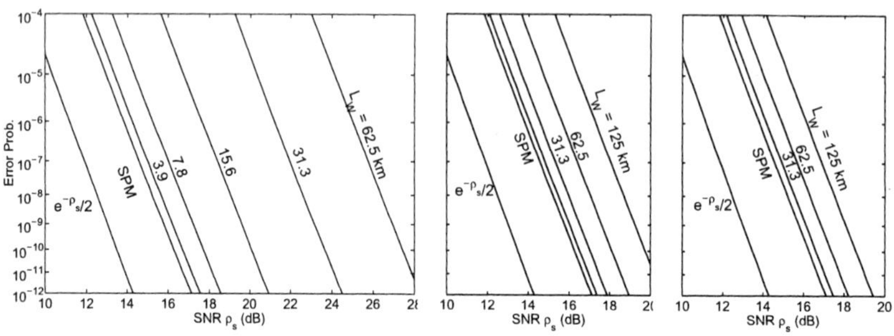

(a) $\sigma^2_{\text{XPM,max}}$ (b) $\sigma^2_{\text{XPM,ind}}$ (c) $\sigma^2_{\text{XPM,min}}$

Figure 8.6. Error probability as a function of SNR for WDM DPSK systems with various dispersion compensation schemes, corresponding to (a) maximum, (b) independent, and (c) minimum variance of XPM-induced nonlinear phase noise. [Adapted from Ho (2004b)]

is

$$p_{e,\text{XPM}} = \frac{1}{2} - \frac{\rho_s e^{-\rho_s}}{2} \sum_{k=0}^{\infty} \frac{(-1)^k}{2k+1} \left[I_k\!\left(\frac{\rho_s}{2}\right) + I_{k+1}\!\left(\frac{\rho_s}{2}\right) \right]^2 \times \left| \Psi_{\Phi_{\text{NL}}}[(2k+1)j] \right|^2 \exp\left[-\frac{2k+1}{2} \sigma^2_{\text{XPM}} \right]. \tag{8.57}$$

Because XPM-induced nonlinear phase noise is generated by the interaction of many bits or WDM channels, the Gaussian approximation is valid from the central limit theorem (Papoulis, 1984). If the walk-off length of L_W is small, the nonlinear phase noise is induced by at least about $2 \times L_{\text{eff}}/L_W$ independent bits from two adjacent channels. If the walk-off length is large, many adjacent channels induce more or less the same amount of nonlinear phase noise. In both cases, the central limit theorem leads to Gaussian distribution. Similar argument is also given in Forghieri et al. (1995, 1997) and Ho (2000) for stimulated Raman scattering induced signal distortion.

Figures 8.6 show the error probability of DPSK signal as a function of SNR ρ_s. The error probability is calculated by Eqs. (8.56) and (8.57) for system with and without XPM-induced nonlinear phase noise, respectively. The system in Figs. 8.6 has $N_A = 32$ identical fiber spans, mean nonlinear phase shift of $<\Phi_{\text{NL}}>= 1$ rad, and 65 WDM channels. The mean nonlinear phase shift of 1 rad is the optimal operation point from Sec. 5.3.3. Figures 8.6 also plot the error probability of $\exp(-\rho_s)/2$ without nonlinear phase noise of Eq. (3.105).

The walk-off length of Figs. 8.6 forms a geometric series and corresponds to typical 10- and 40-Gb/s systems in NZDSF and standard single-mode fiber with dispersion coefficients of about $D = 4$ and 16 ps/km/nm, respectively. For example, the walk-off length $L_W = 7.8$ km is that of 10-Gb/s 100-GHz ($\Delta\lambda = 0.8$ nm) system in standard single-mode fiber and 40-Gb/s 100-GHz system in NZDSF.

Figures 8.6(a), (b), and (c) are calculated using the XPM-induced nonlinear phase noise variances of $\sigma^2_{\mathrm{XPM,max}}$, $\sigma^2_{\mathrm{XPM,ind}}$, and $\sigma^2_{\mathrm{XPM,min}}$ in Eq. (8.57), respectively, corresponding to the worst, independent, and best case dispersion compensation, respectively. Comparing Figs. 8.6(b) and (c) with Fig. 8.6(a), DPSK system requires the reduction of XPM-induced nonlinear phase noise using, for example, XPM suppressor or optimal dispersion compensation factor of Eq. (8.54). With perfect dispersion compensation, XPM-induced cross-phase modulation is negligible if the walk-off length is $L_W < 3.9$ km. Without dispersion compensation or with XPM suppressor, XPM-induced cross-phase modulation is negligible if $L_W < 15.8$ km.

Figures 8.7 show the SNR penalty for an error probability of 10^{-9} as a function of mean nonlinear phase shift of $\langle \Phi_{\mathrm{NL}} \rangle$ for the same system as Figs. 8.6. Figures 8.7(a), (b), and (c) are calculated using the XPM-induced nonlinear phase noise variances of $\sigma^2_{\mathrm{XPM,max}}$, $\sigma^2_{\mathrm{XPM,ind}}$, and $\sigma^2_{\mathrm{XPM,min}}$ in Eq. (8.57), respectively, corresponding to the worst, independent, and best case dispersion compensation, respectively.

For the systems in Fig. 8.7(a), with perfect dispersion compensation such that XPM-induced nonlinear phase noise adds coherently, XPM-induced nonlinear phase noise gives the same SNR penalty as SPM-induced nonlinear phase noise when the walk-off length is about $L_W = 7.8$ km for $\langle \Phi_{\mathrm{NL}} \rangle$ less than 1 rad. For a 10-Gb/s WDM system in standard single mode fiber, the channel spacing must be larger than or equal to 100 GHz (or 0.8 nm). For a 10-Gb/s WDM system in NZDSF, the channel spacing must be larger than or equal to 400 GHz (or 3.2 nm). For a 10-Gb/s WDM system with typical channel spacing of 50 or 100 GHz, with perfect dispersion compensation, the effect of XPM-induced nonlinear phase noise is larger than that induced by SPM. For a 40-Gb/s WDM system in NZDSF, the channel spacing must be larger than or equal to 100 GHz. With channel spacing larger than 100 GHz, typical 40-Gb/s WDM systems have a SNR penalty from XPM-induced nonlinear phase noise about that from SPM.

For the systems in Figs. 8.7(b) and (c), with dispersion compensation factor of Eq. (8.54) or with XPM suppressor, XPM-induced nonlinear phase noise gives the same SNR penalty as SPM-induced nonlinear phase noise when the walk-off length is about $L_W = 62.5$ km. For a 10-Gb/s

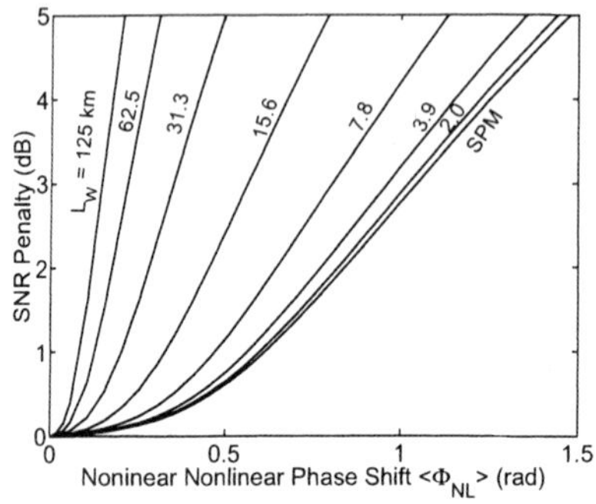

(a) $\sigma^2_{\text{XPM,max}}$

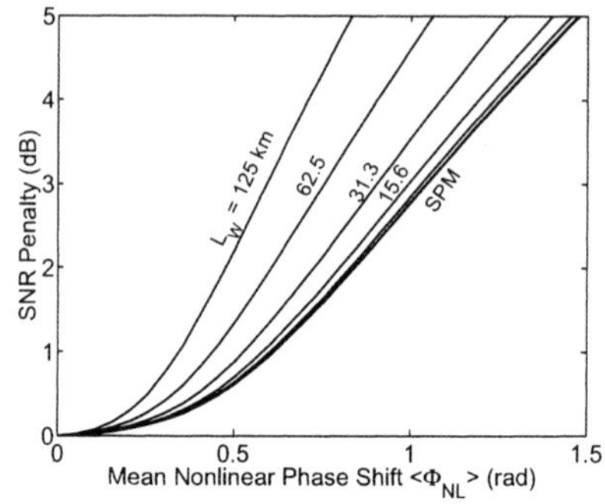

(b) $\sigma^2_{\text{XPM,ind}}$

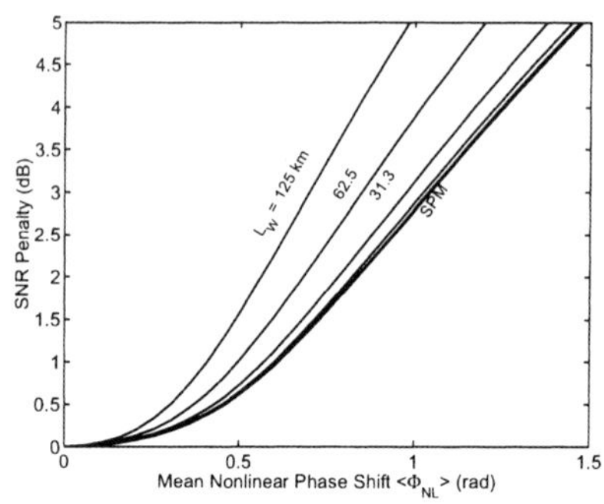

(c) $\sigma^2_{\text{XPM,min}}$

Figure 8.7. SNR penalty as a function of mean nonlinear phase shift $\langle \Phi_{\text{NL}} \rangle$ for WDM DPSK systems with various dispersion compensation schemes, corresponding to (a) maximum, (b) independent, and (c) minimum variance of XPM-induced nonlinear phase noise. [From Ho (2004b), © 2005 IEEE]

WDM system in standard single mode fiber, the channel spacing must be larger than or equal to 12.5 GHz (or 0.1 nm). For a 10-Gb/s WDM system in NZDSF, the channel spacing must be larger than or equal to 50 GHz (or 0.4 nm). For a 10-Gb/s WDM system with typical channel spacing of 50 or 100 GHz, the effect of XPM-induced nonlinear phase noise is much smaller than that induced by SPM. Typical 40-Gb/s WDM systems have a SNR penalty from XPM-induced nonlinear phase noise far less than that from SPM.

For the same channel spacing and fiber type, 10-Gb/s systems have a walk-off length four times that of 40-Gb/s systems. From Figs. 8.7, with the same mean nonlinear phase shift, the nonlinear phase noise induced SNR penalty of 10-Gb/s systems is smaller than that for the corresponding 40-Gb/s systems. However, 40-Gb/s systems have four times the bandwidth and require four times the power of the corresponding 10-Gb/s systems for the same SNR ρ_s defined in Eq. (3.36) and Table 3.1. Because the mean nonlinear phase shift is proportional to the launched power, for the same SNR and system configuration, the mean nonlinear phase shift of 40-Gb/s systems is four times larger than that for 10-Gb/s systems.

When the dependence between nonlinear phase noise and amplifier noise is taking into account, we derive the exact error probability of DPSK signals with nonlinear phase noise in Sec. 5.3.3. With the distributed assumption, the difference between exact and approximated error probability is less than 0.23 dB in term of SNR penalty from Fig. 5.13. With the discrete model, the difference between exact and approximate error probability is smaller than 0.23 dB from Fig. 5.17. From both Figs. 5.13 and 5.17, the model of Eq. (8.56) is accurate for DPSK signal to within 0.23 dB.

SPM-induced nonlinear phase noise correlates with the received intensity and can be compensated using the correlation properties as shown in Chapter 6. Other than using simultaneous multi-channel detection, XPM-induced nonlinear phase noise cannot be compensated.

Kim (2003) first measured the impact of XPM-induced nonlinear phase noise in a WDM system. This section is based on Ho (2004b) that first analyzed the DPSK signal with XPM-induced nonlinear phase noise. As shown before, the formulas of Eqs. (8.56) and (8.57) are similar to that with noisy phase reference (Jain, 1974), laser phase noise (Nicholson, 1984), phase error (Blachman, 1981), or laser phase noise together with phase error (Jacobsen and Garrett, 1987). The terms within the summation of Eq. (8.57) is the product of that due to SPM-induced nonlinear phase noise of Sec. 5.3.3 or Ho (2003b) and laser phase noise of Sec. 4.3 or Nicholson (1984).

Ideally, exact error probability for DPSK system with SPM-induced nonlinear phase noise is given by Eq. (5.83) based on the model of Appendix 5.B. This section uses the approximation of Eq. (8.56) from (5.78) for simplicity. The difference is very small as from Fig. 5.16, especially for system with small number of fiber spans. Both Figs. 8.6 and 8.7 use walk-off length of L_W to form a geometric series and suitable for the dispersion coefficient of $D = 4$ and 16 ps/km/nm. These two values of dispersion coefficient are within that for standard single-mode fiber and NZDSF as from Sec. 4.4.

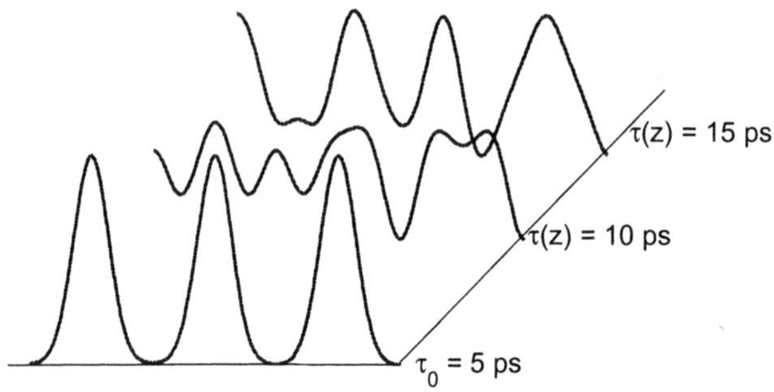

Figure 8.8. The intensity of optical signal with overlap pulse.

4. Cross-Phase Modulation from Overlapped Pulses

Similar to that induces intrachannel four-wave-mixing (IFWM) in Chapter 7, the optical pulses are broadened and overlap with each other when propagates in fiber with large chromatic dispersion. In single-channel DPSK system, pulse overlap gives ghost pulses that are typically smaller than the impact of nonlinear phase noise. The XPM-induced phase noise due to overlap pulses from another channel is also expected to be small. Here, we develop model to calculate the XPM from overlap pulses, mostly using the notations and methods of Chapter 7.

Assume that the systems launch Gaussian pulse with an $1/e$-pulse width of T_0. The pulse along fiber is $u_k(z,t)$ of Eq. (7.10) at $t = kT$. Figure 8.8 plots the intensity profile of the pulse train for $T_0 = 5$ ps and $T = 25$ ps with $A_k = A_0, k \neq 1$ and $A_1 = -A_0$. The bit-interval of $T = 25$ ps corresponds to 40-Gb/s DPSK signal. Figure 2.18 shows the intensity profiles of $\tau(0) = T_0 = 5$ ps and the cases with double and triple pulse width of $\tau(z) = 10, 15$ ps after certain distance of z. A pulse width of $\tau(z) = 15$ ps gives a FWHM pulse width the same as T. Other than the initial intensity profile with $T_0 = 5$ ps, pulse overlap gives non-periodic intensity profile of $|\sum_k u_k(z,t)|^2$ for the specific position of z. Figure 8.8 clearly shows that the intensity profiles with $\tau(z) = 10$ and 15 ps are not periodic intensity like the case of at $z = 0$ with $\tau(0) = T_0 = 5$ ps.

With $A(z,t) = \sum_k u_k(z,t)$, also using the pump-probe model, the overall amount of XPM from the signal is the second term of Eq. (8.37) of

$$\phi_{\mathrm{XPM}}(t) = 2\gamma \int_0^L |A(z, t - d_{12}z)|^2 \otimes h_{\phi,z}(t) e^{-\alpha z} \mathrm{d}z, \qquad (8.58)$$

where $\otimes$ denotes convolution, $h_{\phi,z}(t)$ is the impulse response corresponding the frequency response of $\cos(\beta_2 z\omega^2/2)$. The impulse response of $h_{\phi,z}(t)$ is the phase-to-phase transfer characteristic due to chromatic dispersion (Wang and Petermann, 1992). To be consistent with Eq. (8.38) for nonlinear phase noise, the approach here is not that same as that in Sec. 7.4 that finds the nonlinear force consistent with the model of intra-channel four-wave-mixing. As shown in Fig. 2.18 with pulse overlapping, the pulse train of $|A(z, t - d_{12}z)|^2$ is not a periodic function any more. The nonlinear phase shift of Eq. (8.58) is the same as that of Eq. (8.38) but including the effect of chromatic dispersion.

The XPM-induced phase noise is

$$\phi_{\mathrm{XPM}}(t) = 2\gamma \sum_{k=-\infty}^{+\infty} \sum_{m=-\infty}^{+\infty} b_k b_{k+m} \int_0^L [u_0(z, t - d_{12}z - kT) \times u_0^*(z, t - d_{12}z - kT - mT)] \otimes h_{\phi,z}(t) e^{-\alpha z} \mathrm{d}z, \quad (8.59)$$

or

$$\phi_{\mathrm{XPM}}(t) = \sum_{k=-\infty}^{+\infty} \sum_{m=-\infty}^{+\infty} b_k b_{k+m} q_m(t - kT). \quad (8.60)$$

where

$$q_m(t) = 2\gamma \int_0^L [u_0(z, t - d_{12}z) u_0^*(z, t - d_{12}z - mT)] \otimes h_{\phi,z}(t) e^{-\alpha z} \mathrm{d}z, \quad (8.61)$$

and $b_m = A_m/A_0 = \pm 1$ is the phase modulation of each optical Gaussian pulse.

While the summation of m in Eq. (8.60) is from $-\infty$ to $+\infty$, there is a small number of m such that $q_m(t) \neq 0$, i.e., there is small number of pulse-overlap terms. The phase noise of Eq. (8.60) is similar to a digital modulated signal with $b_k b_{k+m}$ as the digital data and $q_m(t - kT)$ as the pulse shape. Similar to typical method to find the spectral density of digital signals (Proakis, 2000), the average autocorrelation function of $\phi_{\mathrm{XPM}}(t)$ is

$$\overline{R}_\phi(\tau) = \frac{1}{T} \sum_{k=-\infty}^{+\infty} \sum_{l=-\infty}^{+\infty} \sum_{m=-\infty}^{+\infty} \sum_{n=-\infty}^{+\infty} E\{b_k b_l b_{k+m} b_{l+n}\} \times \int_0^T q_m(t - kT) q_n^*(t + \tau - lT) \mathrm{d}t, \quad (8.62)$$

or

$$\overline{R}_\Phi(\tau) = \frac{1}{T}\sum_{m=-\infty}^{+\infty}\int_{-\infty}^{+\infty} q_0(t)q_0^*(t+\tau-mT)\mathrm{d}t$$
$$+\frac{1}{T}\sum_{m=-\infty,m\neq 0}^{+\infty}\int_{-\infty}^{+\infty}\left[q_m(t)q_m^*(t+\tau)+q_m(t)q_{-m}^*(t+\tau-mT)\right]\mathrm{d}t, \tag{8.63}$$

because $E\{b_k b_{k+l} b_{k+m} b_{k+n}\} = 1$ only if $m = 0$ and $l = n$, $l = 0$ and $m = n$, $n = 0$ and $l = m$ when two pairs of bits are equal among those four bits.

The power spectral density of $\phi_{\mathrm{XPM}}(t)$ is

$$\Phi_\phi(\omega) = \frac{1}{T}\sum_{m=-\infty,m\neq 0}^{+\infty}\left[|Q_m(\omega)|^2 + Q_m(\omega)Q_{-m}^*(\omega)e^{jm\omega T}\right]$$
$$+\frac{2\pi}{T^2}\sum_{m=-\infty}^{+\infty}\left|Q_0\left(\frac{2\pi m}{T}\right)\right|^2\delta\left(\omega-\frac{2\pi m}{T}\right), \tag{8.64}$$

or

$$\Phi_\phi(\omega) = \frac{1}{T}\sum_{m=1}^{+\infty}\left[|Q_m(\omega)|^2 + |Q_{-m}(\omega)|^2 + 2\Re\left\{Q_m(\omega)Q_{-m}^*(\omega)e^{jm\omega T}\right\}\right]$$
$$+\frac{2\pi}{T^2}\sum_{m=-\infty}^{+\infty}\left|Q_0\left(\frac{2\pi m}{T}\right)\right|^2\delta\left(\omega-\frac{2\pi m}{T}\right), \tag{8.65}$$

where $\Re\{\cdot\}$ denotes the real part of a complex number and $Q_m(\omega)$ is the Fourier transform of $q_m(t)$. Without pulse overlap, $Q_m(\omega) = 0, m \neq 0$ and $\Phi_\phi(\omega)$ are tones at the frequency of m/T with m as an integer.

The spectrum of $Q_m(\omega)$ comes almost directly from Eq. (8.61). The signal of $u_0(z,t)u_0^*(z,t-mT)$ is a Gaussian pulse by itself with an $1/e$-pulse width of $\frac{T_0}{\sqrt{2}}\sqrt{1+\beta_2^2 z^2/T_0^4}$ and Fourier transform of

$$A_0^2 T_0\sqrt{\pi}\exp\left[-j\frac{mT}{2}\omega - \frac{m^2T^2}{4T_0^2} - \frac{mT\beta_2 z\omega}{2T_0^2} - \frac{T_0^2\omega^2}{4}\left(1+\frac{\beta_2^2 z^2}{T_0^4}\right)\right]. \tag{8.66}$$

We obtain

$$Q_m(\omega) = 2\gamma P_0 T\exp\left(-j\frac{mT}{2}\omega - \frac{m^2T^2}{4T_0^2} - \frac{T_0^2\omega^2}{4}\right)$$
$$\times\int_0^L e^{-\alpha z - jd_{12}\omega z - \frac{mT\beta_2 z\omega}{2T_0^2} - \frac{\omega^2\beta_2^2 z^2}{4T_0^2}}\cos\left(\frac{1}{2}\beta_2 z\omega^2\right)\mathrm{d}z. \tag{8.67}$$

or

$$\begin{aligned} Q_m(\omega) &= \gamma P_0 T \exp\left(-j\frac{mT}{2}\omega - \frac{m^2T^2}{4T_0^2} - \frac{T_0^2\omega^2}{4}\right) \\ &\quad \times \eta^{-1}\left\{e^{\kappa_+^2}\left[\operatorname{erfc}(\kappa_+) - \operatorname{erfc}\left(\kappa_+ + \frac{\eta L}{2}\right)\right]\right. \\ &\quad \left. + e^{\kappa_-^2}\left[\operatorname{erfc}(\kappa_-) - \operatorname{erfc}\left(\kappa_- + \frac{\eta L}{2}\right)\right]\right\}, \end{aligned} \tag{8.68}$$

where erfc(·) is the complementary error function, $\eta = |\beta_2\omega|/T_0$, and

$$\kappa_\pm = \eta^{-1}\left(\alpha + jd_{12}\omega + \frac{mT\beta_2\omega}{2T_0^2} \pm \frac{j}{2}\beta_2\omega^2\right). \tag{8.69}$$

With the same amplitude response, the spectra of $Q_m(\omega)$ and $Q_{-m}(\omega)$ are related to each other by the relationship of

$$Q_{-m}(\omega) = Q_m(-\omega)\exp\left(-jmT\omega\right). \tag{8.70}$$

For DPSK signal, the variance of the differential phase of $\phi_{\text{XPM}}(t) - \phi_{\text{XPM}}(t-T)$ due to pulse overlap is equal to

$$\begin{aligned} \sigma^2_{\Delta\phi_{\text{XPM}}} &= \frac{1}{2\pi}\int_{-2\pi/T}^{+2\pi/T} \Phi_\phi(\omega)\left|1 - e^{-j\omega T}\right|^2 \mathrm{d}\omega, \\ &= 8\int_0^{1/T} \Phi_\phi(2\pi f)\sin^2(\pi f T)\,\mathrm{d}f, \end{aligned} \tag{8.71}$$

where the integration is between $\pm 2\pi/T$ to include only the phase noise within the data bandwidth, also the first lobe of $\left|1 - e^{-j\omega T}\right|^2$.

Figure 8.9 shows the spectral density of $\Phi_\phi(\omega)$ for a fiber link with dispersion coefficient of $D = 17$ and 3.5 ps/km/nm. The channel separation is 100 GHz, corresponding to a wavelength separation of $\Delta\lambda = 0.8$ nm at the wavelength around 1.55 μm. The fiber loss coefficient is $\alpha = 0.2$ dB/km and the fiber length is 100 km. The launched pulse has an initial $1/e$ pulse width of $T_0 = 5$ ps. The DPSK signal has a data rate of 40-Gb/s with a bit interval of $T = 25$ ps.

Figure 8.9 shows a peak near the frequency separation of 100 GHz independent of fiber dispersion, may be due to the interaction of channel walk-off and modulation instability. The peaks are far smaller than that at low frequency and should not affect the performance of the system.

Figure 8.9 shows large variation at the spectral density at low frequency. Peaks and notches appear in the spectral density at certain frequencies. Because the spectra of $Q_m(\omega)$ have peaks at different frequencies, the combination of many terms of $Q_m(\omega)$ in Eq. (8.65) gives the resonance effect with many peaks.

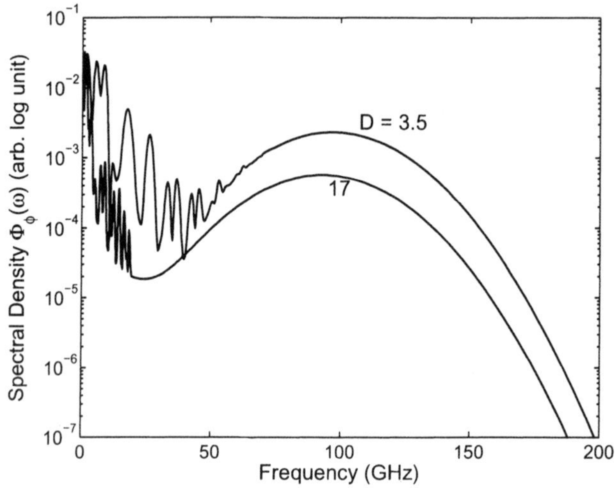

Figure 8.9. The spectral density of $\Phi_\phi(\omega)$ for system with dispersion coefficients of $D = 17$ and 3.5 ps/km/nm and channel separation of $\Delta\lambda = 0.8$ nm.

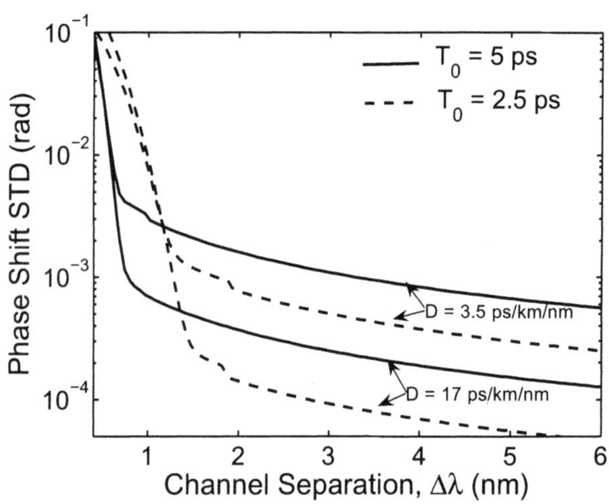

Figure 8.10. The pulse-overlap/XPM induced phase STD of $\sigma_{\Delta_{\rm XPM}}$ as a function of channel separation of $\Delta\lambda$.

Figure 8.10 shows the phase STD of $\sigma_{\Delta_{\rm XPM}}$ induced by pulse-overlap and XPM as a function of channel separation of $\Delta\lambda$ normalized for a mean nonlinear phase shift of $\langle\Phi_{\rm NL}\rangle = 1$ rad. Evaluated numerically, the large $\sigma_{\Delta_{\rm XPM}}$ at very small channel separation is due to spectral overlap between two adjacent channels. At small channel spacing, the channel-to-channel crosstalk is the dominant impairment. At the tail of Fig. 8.10, the STD decreases with channel separation and largely proportional to $\Delta\lambda^{-1}$ or the walk-off length L_W, i.e., the same as Eq. (8.44). Similar to intrachannel four-wave-mixing, the effect of pulse overlap also decreases with the initial pulse width of T_0.

For the system considered in Fig. 8.9 with $T_0 = 5$ ps, increase linearly with mean nonlinear phase shift of $\langle\Phi_{\mathrm{NL}}\rangle$, the STD of the nonlinear phase noise from overlap pulses is $\sigma_{\Delta\phi_{\mathrm{XPM}}} = 0.02\,\langle\Phi_{\mathrm{NL}}\rangle$ for $D = 3.5$ ps/km/nm and $\sigma_{\Delta\phi_{\mathrm{XPM}}} = 4.5 \times 10^{-3}\,\langle\Phi_{\mathrm{NL}}\rangle$ for $D = 17$ ps/km/nm, respectively. Comparing with the nonlinear phase noise STD from Fig. 7.8, the STD from overlap pulses is about 10% of the STD from Fig. 7.8, equivalent to 1% of power in term of variance.

The STD from pulse overlap may also compare with XPM-induced nonlinear phase noise of last section. Using the ratio of Eq. (8.44) and Fig. 8.5, the ratio the STD from XPM- to SPM-induced nonlinear phase noise is 0.37 and 0.083 for $D = 3.5$ and 17 ps/km/nm, respectively. The STD from XPM-induced phase noise due to overlap pulses is about four times less than that from XPM-induced nonlinear phase noise.

So far in this section, the variance of $\sigma^2_{\Delta\phi_{\mathrm{XPM}}}$ is calculated for a single-span two-channel WDM system. The results can extend to multi-channel WDM systems easily. For multi-span WDM systems, the accumulation of pulse-overlap induced nonlinear phase shift also depends on dispersion compensation. If the system has perfect dispersion compensation, the phase shift from overlapped pulse adds coherently span after span and the overall phase noise STD increases linearly with the number of fiber spans of N_A. If the system has dispersion compensation to suppress XPM induced phase noise, the overall phase noise STD increases with the square-root of the number of fiber spans of $\sqrt{N_A}$. Since pulse-overlap induced XPM-phase noise does not induce very large penalty to the system, its impact to DPSK signal will not discuss in further detail.

This section just considers all first-order effects due to the interaction of fiber chromatic dispersion and XPM-induced phase noise. As second-order effect, phase modulation (PM) converts to amplitude-modulation (AM) noise by fiber dispersion. Combined with XPM, AM noise gives nonlinear phase noise to other channels (Norimatsu and Iwashita, 1993). When phase modulation converts to amplitude-modulation, only high frequency AM noise is induced by a transfer function of $\sin(\beta_2 z\omega^2/2)$ (Wang and Petermann, 1992). With the low-pass characteristic of $H_{12}(f)$ of Eq. (8.40), the combined effects of AM-PM conversion with XPM should be small.

For RZ-DPSK signal with pulse broadening, the XPM-induced nonlinear phase noise is similar to Fig. 8.5 for NRZ signal without pulse broadening in the fiber. Detail calculation is complicated without providing further accuracy.

5. Summary

In WDM DPSK signal, homodyne crosstalk with the same wavelength as the signal is studied carefully in this chapter. With large number of interferers, homodyne crosstalk can be approximated as Gaussian distributed that overestimates the SNR penalty and serves well for conservative system design. For single interferer, a closed-form formula is provided to find the error probability with homodyne crosstalk based on Marcum Q function or Hermite polynomial.

Closed-form formulae are derived for the error probability of WDM DPSK signals contaminated by both SPM- and XPM-induced nonlinear phase noise. The error probability is derived based on the assumption that the phase of amplifier noise is independent of nonlinear phase noise. While SPM-induced nonlinear phase noise is not Gaussian-distributed from Chapter 5, XPM-induced nonlinear phase noise is assumed to be Gaussian-distributed when either the walk-off length is small or number of WDM channels is large.

When fiber chromatic dispersion is compensated perfectly in each fiber span, XPM-induced nonlinear phase is summed coherently span after span and is the dominant nonlinear phase noise for typical multi-span WDM DPSK systems. For system without dispersion compensation or with XPM suppressor, the dominant nonlinear phase noise is typically induced by SPM. In general, with longer walk-off length, 10-Gb/s systems are more likely to be dominated by XPM-induced nonlinear phase noise than 40-Gb/s systems.

Broadened by fiber chromatic dispersion, two optical pulses overlap with each other and give non-periodic intensity profile. Combined with XPM, the non-periodic intensity gives phase noise and degrades the performance of DPSK signal. Detail analysis shows that XPM-induced phase noise from overlap pulses is smaller than the corresponding XPM-induced nonlinear phase noise from the beating of the signal with amplifier noises.

Chapter 9

MULTILEVEL SIGNALING

In order to increase the spectral efficiency to more than 1 b/s/Hz, without polarization-division multiplexing (PDM), multilevel signaling must be used to send more than two different choices of waveform per symbol. From Sec. 1.4 and Fig. 1.6, one of the major advantages of phase-modulated or coherent optical communication systems is the improvement of spectral efficiency with minimal degradation to receiver sensitivity. From the comparison of on-off keying (OOK) with quadrature-amplitude modulation (QAM) in Table 1.1, the energy per bit of 64-QAM signal is about the same as 4-OOK signal but with three times the spectral efficiency. In this chapter, we will first discuss the methods to generate M-ary phase-shift-keying (PSK) and QAM signals with special emphasize in the popular differential quadrature phase-shift keying (DQPSK) signals. The performance of M-ary PSK and QAM signals are analyzed when synchronous receiver is used with a phase-locked loop (PLL).

This chapter also analyzes the popular direct-detection DQPSK signal from Table 1.3. The direct-detection DQPSK receiver uses two interferometers to demodulate the two bits within the same symbol. The method in previous chapters to analyze differential phase-shift keying (DPSK) signal can be generalized to DQPSK signal. The performance of DQPSK signal with interferometer phase error, laser phase noise, and nonlinear phase noise are also calculated, mostly based on the series expansion of Appendix 4.A.

Although there are some initial experiments (Miyazaki and Kubota, 2004) and proposals (Han et al., 2004, Ohm, 2004), DQPSK signal remains the only choice for multilevel signal. Together with polarization-division multiplexing (PDM), the spectral efficiency of DQPSK signals is more than 2 b/s/Hz (Cho et al., 2004a,b).

1. Generation of Multilevel Signals

From the low-pass representation of the signal of $A_s(t)e^{j\phi_s(t)}$, there are many methods to generate multilevel signal. In an M-ary PSK signal, the phase is divided into M evenly levels for M different waveforms. In an M-ary QAM signal, both the amplitude of $A_s(t)$ and the phase of $\phi_s(t)$ are used for M different waveforms. While there are other variations to use M different frequencies or only a real number of both positive and negative $A_s(t)$, those methods usually do not provide better spectral efficiency than M-ary QAM signals. For superior spectral efficiency, we focus on both M-ary PSK and QAM signals.

An M-ary PSK signal has a constant amplitude of $A_s(t) = A$ and M evenly spacing phases of $\theta_k = \pi(2k-1)/M$, $k = 1, 2, \ldots, M$. The M different signal waveforms are

$$\begin{aligned} s_k(t) &= \Re\left\{Ae^{j\pi(2k-1)/M}e^{j\omega_c t}\right\}, \\ &= A\cos\left[\frac{\pi(2k-1)}{M}\right]\cos\omega_c t - A\sin\left[\frac{\pi(2k-1)}{M}\right]\sin\omega_c t, \\ &\qquad k = 1, 2, \ldots, M. \end{aligned} \tag{9.1}$$

The Euclidean distance between the k and lth symbols is equal to

$$\sqrt{2}A\left|\sin\left(\frac{(k-l)\pi}{M}\right)\right| \tag{9.2}$$

with a minimum Euclidean distance of $\sqrt{2}A\sin\pi/M$.

Both quadrature phase-shift keying (QPSK) and DQPSK systems use the waveforms of Eq. (9.1) with $M = 4$ to transmit four possible different phases in a single time interval. Currently, most spectral efficiency systems use DQPSK signals as from Table 1.3. Figure 9.1 shows the constellation of (D)QPSK signals with normalized constellation points of

$$\pm 1 \pm j. \tag{9.3}$$

Like the same transmitted signal for binary PSK and DPSK signals, QPSK and DQPSK systems transmit the same signal. QPSK signal uses synchronous receiver based on the quadrature receiver of Fig. 3.4. DQPSK signal uses an asynchronous receiver similar to the DPSK receiver of Figs. 1.4(b) and (c). DQPSK signal requires a precoder to eliminate error propagating in the receiver but QPSK signal typically

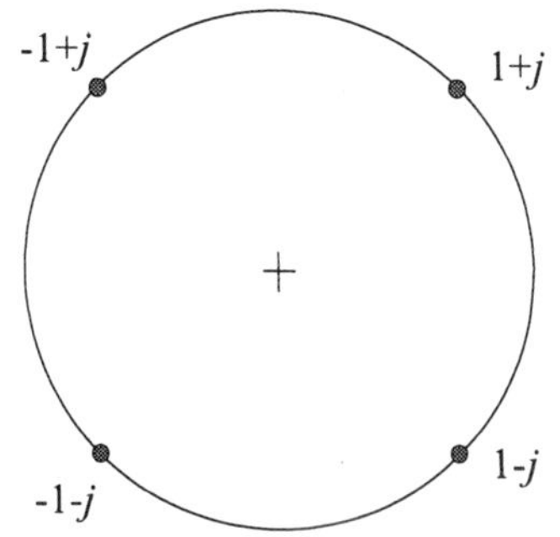

Figure 9.1. The constellation of (D)QPSK signals

does not require a precoder. The design of DQPSK precoder is considered later.

QAM signal provides the highest spectral efficiency among all types of modulation scheme. In an M-ary QAM systems, if $\sqrt{M}$ is an integer, the M different signal waveforms are

$$\begin{aligned} s_k(t) &= A\Re\left\{(a_k + jb_k)e^{j\omega_c t}\right\} \\ &= a_k A\cos\omega_c t - b_k A \sin\omega_c t, \\ & \quad a_k, b_k = -\sqrt{M}+1, -\sqrt{M}+3, \ldots, -1, +1, \ldots, \sqrt{M}-1. \end{aligned} \tag{9.4}$$

Figure 1.6 shows a 64-QAM constellation in a square grid. For simplicity, we assume that all 64 points are used and transmit with equal probability. In general, QAM constellation may locate in irregular location for optimal performance (Foschini et al., 1974, Gockenbach and Kearsley, 1999). M-ary PSK signal can be considered as a special case of QAM signal. Different constellation point of a QAM signal also does not necessary to transmit in the same probability. The system may achieve better performance with unequal probability for different constellation point by signal shaping (Calderbank and Ozarow, 1990, Laroia et al., 1994).

This section discusses two methods that generate QAM signal based on special (usually integrated) modulator or conventional dual-drive Mach-Zehnder modulator of Fig. 2.13. The generation of QAM signal is used as an example.

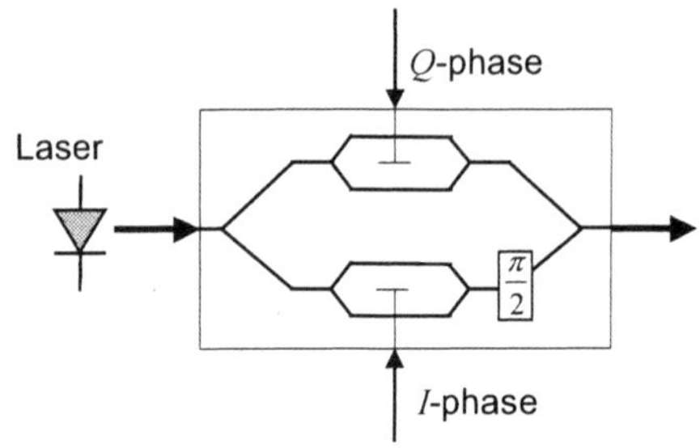

Figure 9.2. QAM transmitter based on two Mach-Zehnder modulators in an interferometer. [From Ho and Cuei (2005), © 2005 IEEE]

1.1 Conventional Quadrature Signal Generator

Figure 9.2 shows a conventional QAM transmitter based on two Mach-Zehnder modulators within an interferometer. A QAM signal can be represented as

$$s(t) = a(t)\cos\omega_c t + b(t)\sin\omega_c t, \tag{9.5}$$

where $a(t)$ and $b(t)$ are the two independent quadrature signals and ω_c is the angular frequency of the optical carrier. Compared Eq. (9.4) and (9.5), $a(t)$ and $b(t)$ are one of the $a_k A$ and $-b_k A$, $m = 1, 2, \ldots, M$ that change from time to time to encode digital data. The transmitter of Fig. 9.2 is the direct generation of the quadrature signals of Eq. (9.5). For the 64-QAM rectangular constellation of Fig. 1.6, $a(t)$ and $b(t)$ are both eight-level signals.

The two Mach-Zehnder modulators of Fig. 9.2 generate the signal for the two quadrature components with a phase difference of $\pi/2$, i.e., cosine and sine. If the upper branch of Fig. 9.2 has a carrier of $\cos\omega_c t$, with 90° phase difference, the lower branch has a carrier of $-\sin\omega_c t$. The baseband complex representation of the output of the transmitter of Eq. (9.5) is the complex number of $a(t) + b(t)e^{-j\pi/2} = a(t) - jb(t)$. The two Mach-Zehnder modulators of Fig. 9.2 should have zero chirp. From the transfer characteristic of Eq. (2.51), both modulators should operate with a peak-to-peak drive voltage of $2V_\pi$ such that both $a(t)$ and $b(t)$ can be both positive and negative values with minimum signal loss. The output electric field is thus

$$E_o = \Re\{[a(t) - jb(t)]e^{j\omega_c t}\} = a(t)\cos(\omega_c t) + b(t)\sin(\omega_c t), \tag{9.6}$$

the same as Eq. (9.5), where $\Re\{\cdot\}$ denotes the real part of a complex number. The signal of $a(t)$ and $b(t)$ of Eqs. (9.5) and (9.6) are from the two independent drive signals of the two Mach-Zehnder modulators of

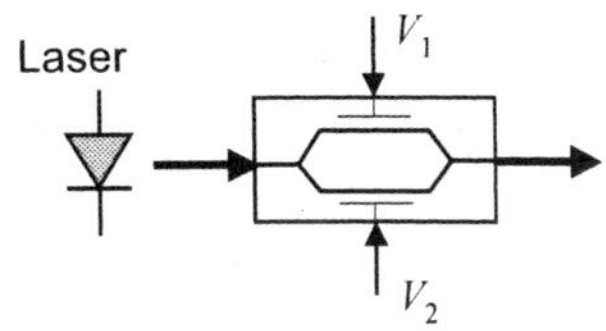

Figure 9.3. Basic structure of a dual-drive Mach-Zehnder modulator with two independent phase modulators in a Mach-Zehnder interferometer.

$V_1(t)$ and $V_2(t)$ in Fig. 9.2. Due to the nonlinear transfer characteristic of Eq. (2.51), the driving voltages of $V_1(t)$ and $V_2(t)$ are not necessary evenly spacing for evenly spacing signal of Eq. (9.5).

With two Mach-Zehnder modulators and an interferometer, the special modulator of Fig. 9.2 is difficult to fabricate. If the transmitter of Fig. 9.2 is implemented using discrete components of two Mach-Zehnder modulators within an interferometer, the transmitter is costly with many components. The transmitter in Fig. 9.2 also requires two bias controls for the Mach-Zehnder modulators and a phase control of the $\pi/2$ phase shifter. Most DQPSK demonstrations of Griffin and Carter (2002), Griffin et al. (2003), Zhu et al. (2004b), and Cho et al. (2003, 2004a,b) use an integrated transmitter the same as that shown in Fig. 9.2. The conventional QAM transmitter may be fabricated using electro-optic effect in semiconductor (Griffin et al., 2002, 2003, 2004) or $LiNbO_3$ (Cho et al., 2004b). As shown later, QAM signal can also be generated using a single dual-drive Mach-Zehnder modulator of Fig. 2.13.

1.2 Generation of QAM Signal Using a Single Dual-Drive Modulator

A QAM signal can also be generated using a conventional dual-drive Mach-Zehnder modulator. Figure 9.3 is the basic structure of a dual-drive Mach-Zehnder modulator, the same as that of Fig. 2.13. The dual-drive Mach-Zehnder modulator consists of two phase modulators that can be operated independently.

As shown in Sec. 2.3.2, Mach-Zehnder modulator can be made using various materials and $LiNbO_3$ is the most popular material. Almost all commercial long-haul dense wavelength-division-multiplexed (WDM) systems use $LiNbO_3$ Mach-Zehnder modulator. In the dual-drive structure of Fig. 9.3, the modulator chirp is adjustable (Gnauck et al., 1991, Ho and Kahn, 2004c). Currently, dual-drive 40 Gb/s modulator with a V_π of less than 2 V has been fabricated (Sugiyama et al., 2002).

For simplicity, assume an operation in steady-state for the modulator. Similar to Eq. (2.51) but the two paths of the dual-drive Mach-Zehnder modulator have independent drive voltages of V_1 and V_2, with an input electric field of E_i, the output electric field is,

$$E_o = \frac{E_i}{2}\left[\exp\left(j\pi\frac{V_1}{V_\pi}\right) + \exp\left(j\pi\frac{V_2}{V_\pi}\right)\right], \tag{9.7}$$

where V_π is the voltage to provide a 180° phase shift of each phase modulator. In the most trivial case, the Mach-Zehnder modulator is operated as a phase modulation if $V_1 = V_2$. The input and output relationship of Eq. (9.7) is rewritten in the following normalized form of

$$E_o = \frac{r_{\max}}{2}\left(e^{j\phi_1} - e^{j\phi_2}\right), \tag{9.8}$$

where $\phi_1 = \pi V_1/V_\pi$ and $\phi_2 = \pi V_2/V_\pi + \pi$. The output electric field of E_o is the difference of two vectors in the circle having a radius of $r_{\max}/2$. The modulator of Eq. (9.8) is biased at the minimum transmission point or the null point and the maximum output electric field has an amplitude of $r_{\max}$ when $V_1 = V_2$ or ϕ_1 and ϕ_2 are antipodal phases. The representation of Eq. (9.8) gives a simple geometric representation of the operation of a dual-drive Mach-Zehnder modulator with two independent phase modulators.

Assumed an M-ary signal constellation that can be represented as complex numbers of

$$s_k = r_k e^{j\theta_k}, \quad r_k > 0, \quad 0 \le \theta_k < 2\pi, \qquad k = 0, \ldots, M-1, \tag{9.9}$$

with maximum amplitude of

$$r_{\max} = \max\{r_0, r_1, \ldots, r_{M-1}\}. \tag{9.10}$$

With two phases of

$$\phi_{k1} = \theta_k + \cos^{-1}(r_k/r_{\max}) \tag{9.11}$$

$$\phi_{k2} = \theta_k - \cos^{-1}(r_k/r_{\max}) + \pi, \tag{9.12}$$

we obtain

$$s_k = \frac{r_{\max}}{2}\left(e^{j\phi_{k1}} - e^{j\phi_{k2}}\right). \tag{9.13}$$

Figure 9.4 shows the procedure to find the two phases of ϕ_{k1} and ϕ_{k2} in the circle with radius of $r_{\max}/2$ for the constellation point of s_k. The real number of r_k is equal to the sum of two conjugate symmetric complex numbers of $\frac{1}{2}r_k \pm jy_k$ in the circle with radius of $\frac{1}{2}r_{\max}$, i.e., $\frac{1}{4}r_k^2 + y_k^2 =$

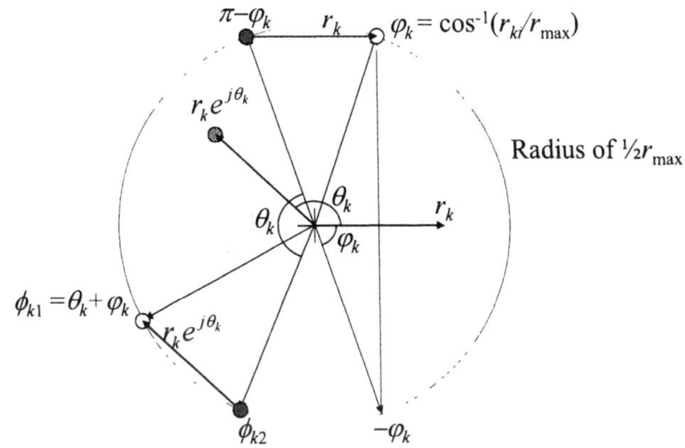

Figure 9.4. The procedure to find ϕ_{k1} and ϕ_{k2} for $s_k = r_k e^{j\theta_k}$. [Adapted from Ho and Cuei (2005)]

$\frac{1}{4}r^2_{\max}$. With $\varphi_k = \cos^{-1}(r_k/r_{\max})$, we obtain $\frac{1}{2}r_k \pm jy_k = \frac{1}{2}r_{\max}e^{\pm j\varphi_k}$. Figure 9.4 represents the two complex numbers of $\frac{1}{2}r_{\max}e^{\pm j\varphi_k}$ as two vectors in the phase angle of $\pm\varphi_k$ with $r_k = \frac{1}{2}r_{\max}e^{j\varphi_k} + \frac{1}{2}r_{\max}e^{-j\varphi_k}$. Alternatively, we may rewrite the summation as the difference of

$$r_k = \frac{1}{2}r_{\max}e^{j\varphi_k} - \frac{1}{2}r_{\max}e^{j(\pi-\varphi_k)}. \tag{9.14}$$

Figure 9.4 also shows the difference of Eq. (9.14) to find r_k. The signal of $s_k = r_k e^{j\theta_k}$ is a rotation of θ_k from r_k. If $e^{j\theta_k}$ is multiply to both sides of Eq. (9.14), we obtain the expression of Eq. (9.13) with ϕ_{k1} and ϕ_{k2} given by Eq. (9.11) and Eq. (9.12), respectively. In Fig. 9.4, the three vectors of r_k, $\frac{1}{2}r_{\max}e^{j\varphi_k}$, and $\frac{1}{2}r_{\max}e^{j(\pi-\varphi_k)}$ are rotated by an angle of θ_k to obtain $r_k e^{j\theta_k}$, $\frac{1}{2}r_{\max}e^{j\phi_{k1}}$, and $\frac{1}{2}r_{\max}e^{j\phi_{k2}}$, respectively.

All constellation points of Eq. (9.9) can be generated based on two phase modulators having the phases of Eqs. (9.11) and (9.12), respectively. The decomposition of a QAM signal or arbitrary quadrature signals into two phase-modulated signals was originated from Cox (1974) and Cox and Leck (1975). These types of linear amplification using nonlinear components (LINC) transmitter are very popular in wireless communications (Casadevall and Valdovinos, 1995, Shi and Sundstrom, 2000, Zhang et al., 2001). The two phases of Eqs. (9.11) and (9.12) drive the two phase modulators of Fig. 9.3 of the dual-drive Mach-Zehnder modulator.

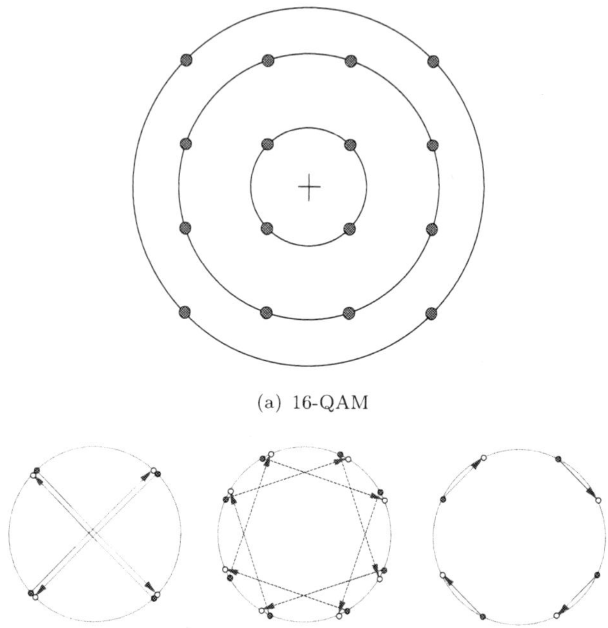

(a) 16-QAM

(b) Big QPSK (c) 8PSK (d) Small QPSK

Figure 9.5. (a) 16-QAM constellation and its separation into two QPSK and an 8-PSK signals, (b), (c), (d) are the two phases of ϕ_1 (empty circles) and ϕ_2 (solid circles) to generate the 16-QAM signal. [Adapted from Ho and Cuei (2005)]

1.3 Generation of 16-QAM Signal

Figure 9.5(a) shows a 16-QAM constellation in a square grid. If the 16-QAM signal is generated by the conventional transmitter of Fig. 9.3(a), both $a(t)$ and $b(t)$ must be a four-level signal corresponding to the x- and y-axes of the constellation of Fig. 9.5(a). In the steady state for equal spacing, we get $a(t) = \pm 1, \pm 1/3$ and $b(t) = \pm 1, \pm 1/3$. If the push-pull modulator of Eq. (2.53) is used, the four amplitude levels of the drive signal for equal spacing $a(t)$ and $b(t)$ are $\pm V_\pi$ and $\pm 0.216 V_\pi$ after proper biasing.

Figures 9.5(b)-(d) show the phases of Eqs. (9.11) and (9.12) for the dual-drive Mach-Zehnder modulator to generate the constellation. Instead of representing all 16 constellation points in the same figure, Figure 9.5(a) separates the 16-QAM signal into two QPSK and one 8-PSK signals. Figures 9.5(b)-(d) are the corresponding ϕ_{k1} and ϕ_{k2} of all 16 points according to the separation of Fig. 9.5(a). The empty and solid circles correspond to ϕ_{k1} and ϕ_{k2}, respectively. Figure 9.5(b) gives the QPSK signal of Fig. 9.5(a) with maximum amplitude of r_{max}. For illustration purpose, the two closest points of Fig. 9.5(b) are the same but draw differently for the phases of two different signals. Figure 9.5(c) generates the 8-PSK signal of Fig. 9.5(a). Figure 9.5(d) gives the QPSK signal of Fig. 9.5(a) with the smallest amplitude.

From Figs. 9.5, the generation of a 16-QAM signal using a dual-drive Mach-Zehnder modulator requires the usage of a 16-level drive voltage and thus very complicated drive circuits. In the transmitter of Fig. 9.2, the two drive signals are two independent four-level drive signals corresponding to the x and y axis of Fig. 9.5(a). Compared with the conventional transmitter of Fig. 9.2, the simplification of the optical components using a single dual-drive Mach-Zehnder modulator creates a high complexity in the electronic driving circuits. With the allowance of higher modulator loss, as shown later, the number of drive levels can be reduced.

2. Transmitter of (D)QPSK Signals

DQPSK signal is by far the most popular multilevel signal for phase-modulated optical communications. The transmitter of (D)QPSK signal is considered here in more detail.

To generate the (D)QPSK signal using the conventional transmitter of Fig. 9.2, $V_1(t), V_2(t) = \pm V_\pi$ to provide the real and imaginary parts of Eq. (9.3) of ± 1, respectively. With the phase shift of $\pi/2$, the output of the transmitter becomes that of Eq. (9.3) with $s(t) = \pm \cos(\omega_c t) \pm \sin(\omega_c t)$.

A (D)QPSK signal can also be generated by the cascade of two phase modulators both driven by binary signal. One modulator provides a phase modulation of 0 and π similar to DPSK signal and may used a zero-chirp amplitude modulator of Eq. (2.53). Another modulator provides a phase modulation of $\pm\pi/2$ using a phase modulator. The amplitude ripples from the driving signal of the phase modulator transfers to the phase ripple. However, the amplitude ripples for the amplitude modulator does not transfer to to the phase ripple.

Alternatively, the (D)QPSK signal of Eq. (9.3) can also generate using a phase modulator driven by a four-level signal to give the phases of $\pm\pi/4$ and $\pm 3\pi/4$. However, the four-level signal is difficult to operate.

When the dual-drive modulator transmitter of Fig. 9.3 is used to generate (D)QPSK signals with a constellation of Fig. 9.1. Similar to that of Fig. 9.5(b), Fig. 9.6(a) is the trivial case of operating the dual-drive Mach-Zehnder modulator as a phase modulator when ϕ_1 and ϕ_2 are antipodal phases. The four phases of Fig. 9.1 are generated by a four-

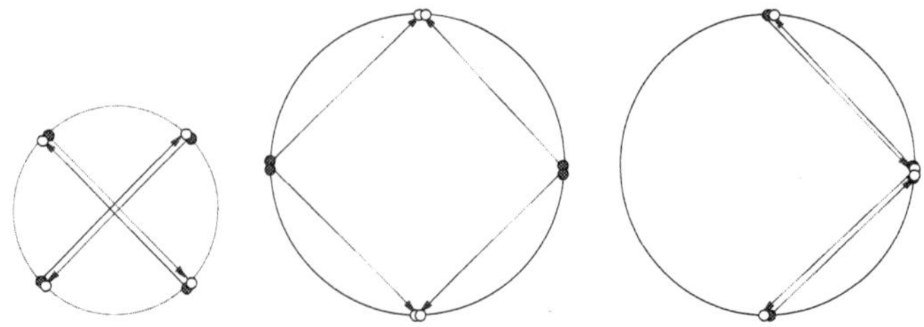

Figure 9.6. The generation of (D)QPSK signal using dual-drive modulator with (a) four-, (b) two-, and (c) three-level drive signal. [Adapted from Ho and Cuei (2005)]

level drive signal. The usage of a dual-drive Mach-Zehnder modulator as a phase modulator may not be an interesting application.

A (D)QPSK signal may be generated with higher loss but smaller number of levels if the four constellation points are generated with phases of ϕ_1 or ϕ_2 the same as one another. With only two values of ϕ_1 and ϕ_2, Figure 9.6(b) drives the dual-drive Mach-Zehnder modulator in Fig. 9.3 with a two-level drive signal. Figure 9.6(c) has three different values of ϕ_1 and ϕ_2 and requires a three-level drive signal. In Figs. 9.6(b) and (c), all phases of ϕ_1 and ϕ_2 are located in circle with the same diameter, representing the phase modulator with the same input signals. In Figs. 9.6, the vectors representing the output electric fields all have the same magnitude. The diameter of the circle of Fig. 9.6(a) is about $1/\sqrt{2}$ that of the circles of Figs. 9.6(b) and (c), requiring 1/2 the power at the input of the phase modulator to generate an output electric field of the same length. Alternatively, the schemes using two- and three-level drive signal have 3-dB extra loss more than that of Fig. 9.6(a).

The two level transmitter is for particular interested for it simplicity. Comparing Eq. (9.7) with (9.3), if the dual-drive modulator is driven with a binary signal such that $\exp(j\pi V_1/V_\pi) = \pm 1$ and $\exp(j\pi V_2/V_\pi) = \pm j$, all constellation points of Eq. (9.3) can be generated. The drive voltages are $V_1 = 0, V_\pi$ and $V_2 = \pm V_\pi/2$. The dual-drive modulator requires just a single bias control to ensure that the means of V_1 and V_2 have a difference of $V_\pi/2$. The peak-to-peak drive voltages of V_1 and V_2 are identical and equal to V_π.

In the two phase modulators of Fig. 9.3, the peak-to-peak drive voltages are proportional to the maximum phase difference of ϕ_1 or ϕ_2, respectively. The maximum phase difference of the 4-drive level of Fig. 9.6(a) is $3\pi/2$ and that of the two- and three-drive level of Figs. 9.6(b)

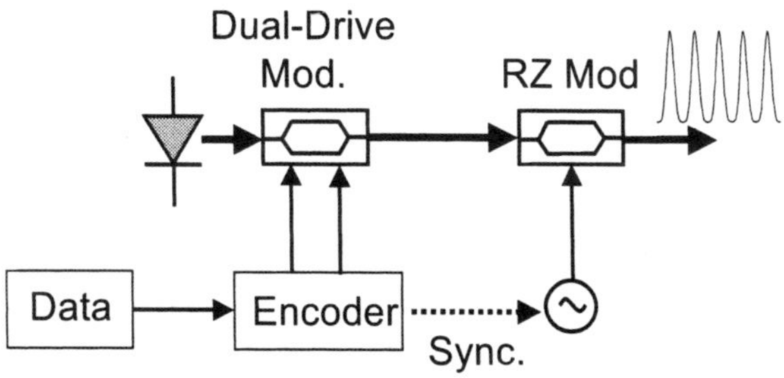

Figure 9.7. A RZ-(D)QPSK transmitter with a dual-drive Mach-Zehnder modulator followed by a RZ modulator. [From Ho and Cuei (2005), © 2005 IEEE]

and (c) is π. Both the peak-to-peak drive voltage and the number of drive levels are reduced in Figs. 9.6(b) and (c).

Figures 9.6 show the operation of the (D)QPSK transmitter in steady state when both ϕ_1 and ϕ_2 are in the correct phase angles. However, when the phases are in transition from one to another between consecutive transmitted symbols, the dynamic of the transmitter requires careful studies.

Similar to the conventional transmitter of Fig. 9.2, both the two- and three-level transmitters of Figs. 9.6(b) and (c) do not have constant intensity between consecutive symbols. Similar to Fig. 2.17 for return-to-zero (RZ)-DPSK, Figure 9.7 shows a RZ-(D)QPSK transmitter with the assumption of the usage of a dual-drive Mach-Zehnder modulator followed by a RZ modulator. In practice, the RZ modulator can either precede or follow the (D)QPSK generator. Without the RZ modulator, the transmitter of Fig. 9.7 gives non-return-to-zero (NRZ) (D)QPSK signal. The conventional transmitter of Fig. 9.2 can also be used in Fig. 9.7 to generate the (D)QPSK signals. The RZ modulator should be operated in the interval when the (D)QPSK signal generator is in steady state. Regardless of the transmitter types to generate the (D)QPSK signal, after the RZ modulator, the optical intensity is ideally a constant pulse train without ripple between symbols. However, the output signal of the dual-drive Mach-Zehnder modulator may have either overshoot and undershoot ripples.

Figures 9.8 show the eye-diagram of the drive signal and the optical intensity between the RZ modulator and the (D)QPSK transmitter. Figure 9.8(a) is the eye-diagram when the conventional transmitter of Fig. 9.2 is used with two two-level drive signals having a peak-to-peak drive voltage of $2V_\pi$. Using the dual-drive transmitter of Fig. 9.3, the

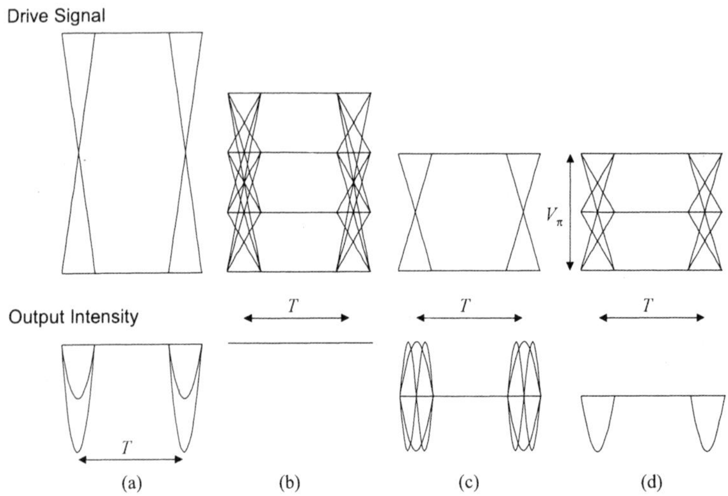

Figure 9.8. The eye-diagram of the drive signal and output intensity between two modulators of the transmitter of Fig. 9.7 when (a) the conventional transmitter of Fig. 9.2 and the dual-drive transmitter of Fig. 9.3 with (b) four-, (c) two-, and (d) three-level drive signals are used. [From Ho and Cuei (2005), © 2005 IEEE]

peak-to-peak drive voltage is reduced from $1.5V_\pi$ for four-level signal to V_π for two- and three-level drive signals.

The output intensity of the conventional transmitter has optical intensity ripples between consecutive symbols. With two or three levels of drive signal, the output intensity of the dual-drive Mach-Zehnder modulator also has ripples between consecutive symbols. The simplest two-level scheme of Figs. 9.6(b) and 9.8(c) has overshoot ripples doubling the output intensity. The ripples of the three-level signal of Fig. 9.8(d) with dual-drive transmitter are similar to that of Fig. 9.8(a) with conventional transmitter. For NRZ signal without RZ modulator, overshoot ripples are equivalent to short optical pulses that are particularly detrimental and potentially give high signal distortion due to fiber nonlinearities. Without the RZ modulator, two-level drive signals of Fig. 9.6(b) cannot be used for NRZ signals due to the overshoot ripples.

While Figs. 9.8 show the eye-diagram of the optical intensity, Figures 9.9 show the trace of the output electric field with both the real and imaginary parts. The four signal constellation points of Fig. 9.1 are shown in Fig. 9.9 as solid circles. The electric field of Fig. 9.9 other than the four signal constellation points occurs during the transition between consecutive (D)QPSK symbols. Other than Fig. 9.9(b) with all

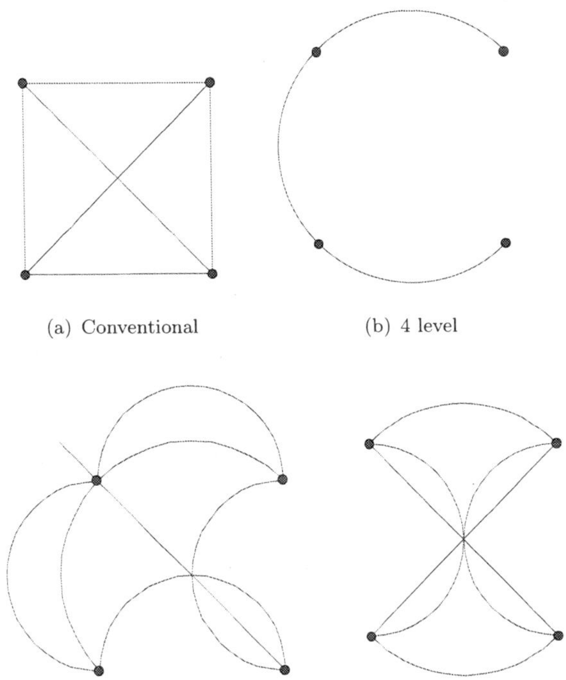

(a) Conventional (b) 4 level

(c) 2 level (d) 3 level

Figure 9.9. The electric field locus of a (D)QPSK signal generated using (a) conventional transmitter of Fig. 9.2, and dual-drive transmitter of Fig. 9.3 with (b) four-, (c) two-, and (d) three-level drive signals. [From Ho and Cuei (2005), © 2005 IEEE]

electric field in the same circle with equal distance to the origin (constant intensity), other traces show that the electric field has a wide range of variations. Those variations of electric field are equivalent to frequency chirp (Gnauck et al., 1991, Koyama and Oga, 1988).

In the conventional transmitter of Fig. 9.9(a), the electric field may pass through the origin and has an intensity equal to zero as shown in the optical intensity of Fig. 9.8(a). The electric field trace of the two-level signal of Figs. 9.6(b) and 9.8(c) shows a peak electric field about $\sqrt{2}$ of the signal points, corresponding to optical intensity ripples twice the steady-state intensity. The three-level signal of Fig. 9.9(d) has the same behavior as Fig. 9.9(a). Of course, the three-level signal of Fig. 9.9(d) has transition with constant intensity but all transitions of the conventional signal of Fig. 9.9(a) have ripples.

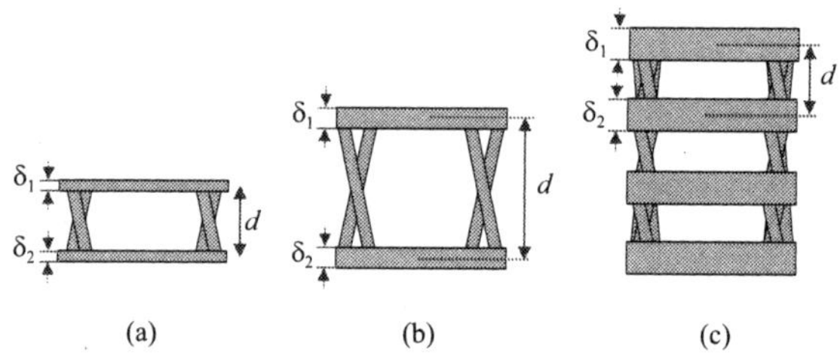

Figure 9.10. The increases of eye-spreading for a multi-level signal. [From Ho and Cuei (2005), © 2005 IEEE]

Figure 9.8 also shows the eye-diagram of the electrical drive signal to the dual-drive modulator with a peak-to-peak voltage of V_π. Similar to the conventional DQPSK transmitter of Fig. 9.2, for NRZ signals without the RZ modulator, the transmitter of Fig. 9.3 has intensity ripples between consecutive symbols. Together with the overshoot intensity ripples, phase jitter occurs between consecutive symbols. With an RZ modulator followed or preceded the DQPSK generator, both the intensity ripple and phase jitter can be eliminated. The RZ modulator has to be operated within the region when the DQPSK generator is in the steady state in the middle of its eye-diagram. In Fig. 9.8, the RZ modulator gives an RZ pulse having a duty-cycle of 50%. The drive signals of the dual-drive modulator must have a short rise- and fall-time to ensure that the both the overshoot ripple and phase jitter are outside the pulses generated by the RZ modulator.

The ripples of the drive signal transfer to the optical signal. With conventional transmitter of Fig. 9.2, no amplitude ripple of the drive signal transfers to phase ripple. Even when the drive signal has large ripple, the intensity ripple of the transmitted signal is compressed by the nonlinear transfer function of the modulator.

To generate the (D)QPSK signal of Figs. 9.1 using a phase modulator, the ripples from the drive signal increases by the transmitter. Figure 9.10 illustrates the increase of eye-spreading when two two-level signals are summed to a four-level signal. If the eye-spreading is defined as $\Delta_e = (\delta_1 + \delta_2)/d$, where δ_1 and δ_2 are the ripple (or spreading) in the upper and lower level, and d is the high of the eye-diagram, the eye-closure is $1 - \frac{1}{2}\Delta_e$ and the eye-penalty is equal to $10 \cdot \log_{10}\left(1 - \frac{1}{2}\Delta_e\right)$. From Fig. 9.10, when two two-level signals of Figs. 9.10(a) and (b) are summed to the four-level signal of Fig. 9.10(c), the eye-spreading is increased by a factor of three. When phase modulator is used to generate the (D)QPSK signal, the eye-spreading of the phase increases accordingly.

Table 9.1. The Transfer of Amplitude Ripples from Drive Signal to the Receiver for DQPSK Signals.

Schemes	Two-Level	Multi-Level	RZ Pulses	Received Signal
Four-Level	10%	30%	1.3%	47%
Two-Level	10%	Nil	31%	33%
Three-Level	Nil	10%	15%	16%
Conventional	10%	Nil	0.7%	0.6%

Table 9.1 shows the eye-spreading of Δ_e for all RZ-DQPSK transmitters discussed in this section and the conventional transmitter by assuming an "initial" eye-spreading of 10% for convenience. In the four-level transmitter, the initial 10% eye-spreading is increased to 30% as illustrate in Fig. 9.10. For QPSK signals without the differential operation, the spreading of the received signal is also equal to 30%. When the dual-drive modulator functions as a phase modulator, the RZ pulses at the output of the transmitter do not have eye-spreading but the DQPSK receivers have an eye-spreading of 47% due to phase distortion. The two-level transmitter provides a received signal spreading of about 33%, less than that of four-level transmitter. The internal operation of the two-level transmitter may be similar to the illustration of Fig. 9.10. For three-level transmitter, the eye-spreading is increased by a factor of 1.5. In conventional transmitter, ripples of the drive signal do not transfer to phase ripple but only small amplitude ripple. Table 9.1 shows that a drive signal ripple of 10% gives a received signal ripples less than 1%.

For the DQPSK transmitter using dual-drive modulator, the eye-spreading of the received signal increases almost linearly with the eye-spreading of the drive signal. For an eye-penalty less than 1 dB at the received signal, 41% eye-spreading is allowed, translating to a maximum eye-spreading of the two-level signal of 8.7% and 12.8% for four- and two-level transmitters, respectively. For an eye-penalty less than 0.5 dB, 22% eye-spreading is allowed, translating to a maximum eye-spreading of the two-level signal of 5.7% and 6.9% for four- and two-level transmitters, respectively.

As long as the electrical drive signal has very good quality and RZ modulator is used, the RZ-(D)QPSK transmitter using dual-drive modulator can generate mathematically the same signal as conventional transmitter. RZ-(D)QPSK signal broadens the bandwidth of the signal inversely proportional to the duty-cycle of the RZ pulses. Ripples or eye-spreading transferred from the electrical drive signal are the major degradation using a phase modulator or dual-drive modulator to generate (D)QPSK signals.

Most DQPSK demonstrations used an integrated transmitter in Griffin and Carter (2002), Griffin et al. (2003), Zhu et al. (2004b), and Cho et al. (2003, 2004a,b). DQPSK signals were also generated by the cascade of practically two phase modulators driven by two binary sequences in Kim and Essiambre (2003), Tokle et al. (2004), Wree et al. (2003a), and Yoshikane and Morita (2004a,b, 2005).

3. Synchronous Detection of Multilevel Signals

Synchronous receiver with PLL to track the received phase gives the best receiver sensitivity. With large number of signal levels, synchronous receiver is usually 3-dB better than the corresponding asynchronous receiver. The multilevel quadrature signals may be M-ary PSK or QAM signals. Multilevel frequency-shift keying (FSK) signals do not provide good spectral efficiency and are not discussed here.

3.1 M-ary PSK Signal

The synchronous quadrature receiver of Fig. 3.4 can be used to detect the M-ary PSK signal of Eq. (9.1) where the M possible phases of $\theta_k = \pi(2k-1)/M, k = 1, 2, \ldots, M$ are used to carry the information. Assume that the receiver is limited by amplifier noise, when the phase of θ_k is the transmitted, the phase distribution is $p_{\Theta_n}(\theta - \theta_k)$, where $p_{\Theta_n}(\theta)$ is from Eq. (4.A.7) in Appendix 4.A. When Gray code is used, the bit error probability is equal to

$$p_e = \frac{1}{\log_2 M}\left[1 - \int_{-\pi/M}^{+\pi/M} p_{\Theta_n}(\theta)\mathrm{d}\theta\right]. \tag{9.15}$$

Using the series expansion of Eq. (4.A.12), the bit error probability is equal to

$$\begin{aligned} p_e &= \frac{1}{\log_2 M}\left\{1 - \frac{1}{M} - \frac{\sqrt{\rho_s}e^{-\rho_s/2}}{\sqrt{\pi}}\right. \\ &\quad \left.\times \sum_{m=1}^{\infty} \frac{\sin(m\pi/M)}{m}\left[I_{\frac{m-1}{2}}\left(\frac{\rho_s}{2}\right) + I_{\frac{m+1}{2}}\left(\frac{\rho_s}{2}\right)\right]\right\}. \end{aligned} \tag{9.16}$$

When $M = 2$ for binary PSK signal, the series of Eq. (9.16) is the same as that from Eq. (4.A.16), or $\frac{1}{2}\mathrm{erfc}(\sqrt{\rho_s})$ of Eq. (3.78).

The series summation of Eq. (9.16) is difficult to evaluate from Appendix 4.A. The minimum Euclidean distance between the two closest

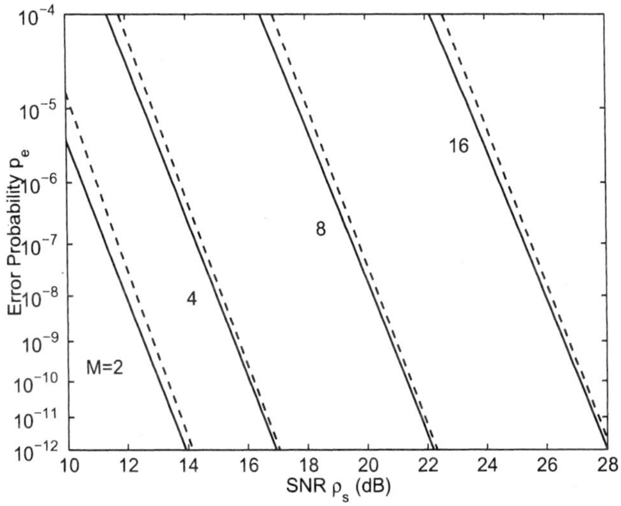

Figure 9.11. The bit-error probability of M-ary PSK signal. The solid lines is exact results from the series of Eq. (9.16) and the dashed lines are the approximation from Eq. (9.17).

signal points is $\sqrt{2}A \sin \pi/M$ and the error probability is approximately equal to

$$p_e \approx \frac{2}{\log_2 M} \text{erfc}\left(\sqrt{\rho_s} \sin \frac{\pi}{M}\right). \tag{9.17}$$

Figure 9.11 shows the bit-error probability calculated using the exact series summation of Eq. (9.16) and the approximation of Eq. (9.17). For $M \geq 8$, the approximation of Eq. (9.17) is very accurate. For binary PSK, the exact error probability is given by Eq. (3.78). For QPSK signals, the constellation is the same as Fig. 9.1 with the combination of two sets of antipodal phases. The minimum distance is reduced by $\sqrt{2}$ as compared with binary PSK. The correct probability is when both quadratures are correct of $(1 - p_e)^2$, where p_e the error probability of binary PSK of Eq. (3.78) with $\sqrt{2}$ less power. The exact symbol error probability for QPSK is $1 - (1 - p_e)^2$, and the bit error probability is

$$\frac{1}{4} \text{erfc} \sqrt{\frac{\rho_s}{2}} \left[1 - \frac{1}{4} \text{erfc} \sqrt{\frac{\rho_s}{2}}\right]. \tag{9.18}$$

QPSK signal is about 3-dB worse than binary PSK signal but half the bandwidth and half the noise. QPSK and binary PSK signals require the same SNR per bit.

The error probability of Eq. (9.16) was first derived by Prabhu (1969).

3.2 Quadrature Amplitude Modulation

The M-ary QAM signal of Eq. (9.4) can also also be detected using the synchronous quadrature receiver of Fig. 3.4. The average energy per symbol of the QAM signal of Eq. (9.4) is

$$E_{\mathrm{av}} = \frac{2A^2}{M} \sum_{k=1}^{\sqrt{M}} \left(-\sqrt{M} + 2k - 1 \right)^2 = \frac{2A^2}{3}(M-1). \tag{9.19}$$

For each dimension, the symbol error probability is

$$p_x = p_y = \left(1 - \frac{1}{\sqrt{M}}\right) \mathrm{erfc}\sqrt{\frac{A^2}{2\sigma_n^2}}. \tag{9.20}$$

The symbol error probability for QAM signal is when both quadrature components are correct of $1-(1-p_x)(1-p_y)$. The bit error probability is

$$\begin{aligned} p_e &= \frac{2}{\log_2 M}\left(1 - \frac{1}{\sqrt{M}}\right) \mathrm{erfc}\sqrt{\frac{3\rho_s}{2(M-1)}} \\ &\quad \times \left[1 - \frac{1}{2}\left(1 - \frac{1}{\sqrt{M}}\right) \mathrm{erfc}\sqrt{\frac{3\rho_s}{2(M-1)}}\right] \end{aligned} \tag{9.21}$$

For $M = 4$, the error probability is the same as that of Eq. (9.18) for QPSK signal.

In high SNR, we obtain

$$p_e \approx \frac{2}{\log_2 M}\left(1 - \frac{1}{\sqrt{M}}\right) \mathrm{erfc}\sqrt{\frac{3\rho_s}{2(M-1)}}. \tag{9.22}$$

Figure 9.12 shows the bit-error probability of M-ary QAM signals for $M = 4, 8, 16, 32$, and 64. The 4-QAM signal is the same as QPSK signal of Eq. (9.18). The required SNR of the signal increases with the number of levels in the constellation. For $M = 8$ and 32, the error probability of Eq. (9.21) is modified for the non-square constellation.

QPSK signal was demonstrated using different types of synchronous receiver (Derr, 1990, Kahn et al., 1992, Norimatsu et al., 1992). QPSK signal was analyzed by Barry and Kahn (1992) and Yamazaki and Emura (1990) in detail. An intradyne receiver was proposed by Derr (1992) using signal processing for demodulation.

M-ary FSK was also analyzed by Jeromin and Chan (1986) and demonstrated in Alexander et al. (1990). Although M-ary FSK signals

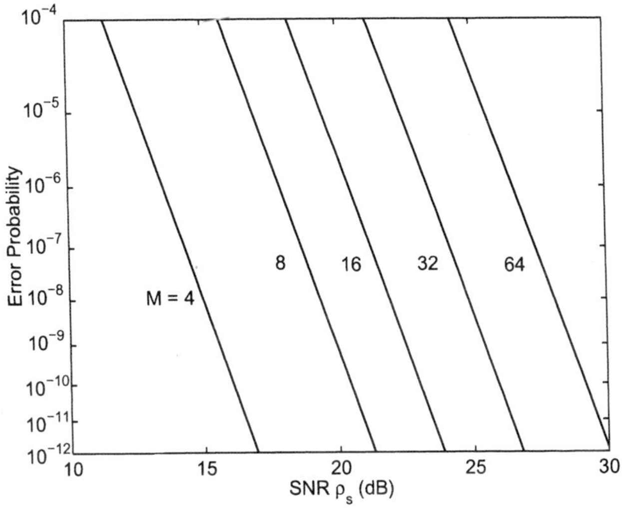

Figure 9.12. The bit-error probability of M-ary QAM signal.

do not have good spectral efficiency, using more bandwidth to obtain better performance, M-ary FSK signals can achieve very small SNR per bit.

Both M-ary PSK and QAM signals for large M have very small tolerance to laser and nonlinear phase noise. Using the 16-QAM signal of Fig. 9.5(a) as an example, laser phase noise spreads out the phase of all constellation point equally. Separated by an angle of $\pi/4$, the middle 8-PSK signal is affected the most by the laser phase noise. If the system is dominated by laser phase noise, the optimal signal constellation is four concentric circles of QPSK signal or two concentric circles of 8-PSK signal with equal angles between two adjacent constellation points.

For system with nonlinear phase noise, the amount of nonlinear phase noise depends on the distance of the signal to the origin. The big QPSK in 16-QAM signal of Fig. 9.5(a) has nine times larger nonlinear phase noise than the small QPSK signal. For system with large nonlinear phase noise, the optimal signal constellation may have QPSK signal at larger amplitude but 8-PSK signal at smaller amplitude, differed with conventional signal design.

Currently, QAM signal with more than four levels is rarely used for optical communications. Of course, subcarrier multiplexing may use high-level QAM signals for cable modem or digital video applications (Phillips and Darcie, 1997, Way, 1998). High-speed subcarrier multi-

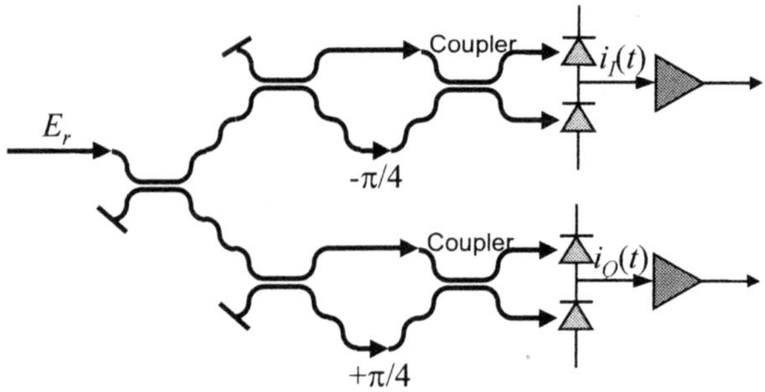

Figure 9.13. Direct-Detection Receiver for DQPSK Signals.

plexing may also use high-level and high-speed QAM signals (Chen and Way, 2004, Hui et al., 2002, Urick et al., 2004). Equivalently an 8-QAM signal, DQPSK signal can also use together with amplitude-shift keying (ASK) signal (Hayase et al., 2004, Miyazaki and Kubota, 2004).

4. Direct-Detection of DQPSK Signal

DPSK signal is demodulated using a direct-detection receiver based on the interferometric receiver of Fig. 1.4(c). DQPSK receiver uses two asymmetric Mach-Zehnder interferometers for the differential phase of each quadrature component. Direct-detection DQPSK receiver is the most popular receiver for multilevel phase-modulated optical communications from as Table 1.3.

4.1 Receiver Structure and Ideal Performance

Figure 9.13 shows the direct-detection receiver for DQPSK signals. The received electric field of E_r is splitted into two paths of $\pm E_r/\sqrt{2}$ and passing through two asymmetric interferometers with phase difference of $\pi/2$. The two interferometers also have a path difference of the symbol time of T. With the received signal of $E_r = Ae^{j\phi_s(t)} + n(t)$, assuming ideal 3-dB coupler and balanced receiver, the photocurrent of upper branch is

$$i_I(t) = \frac{A^2}{2}\cos[\phi_s(t) - \phi_s(t-T) + \pi/4] + \text{noise terms.} \tag{9.23}$$

the photocurrent of the lower branch is

$$i_Q(t) = \frac{A^2}{2}\cos[\phi_s(t) - \phi_s(t-T) - \pi/4] + \text{noise terms.} \tag{9.24}$$

Without noise, a phase difference of $\phi_s(t) - \phi_s(t-T) = 0°$ gives an output of $i_I(t) = i_Q(t) = A^2/2$. A phase difference of 90° gives an output of $i_I(t) = A^2/2$ and $i_Q(t) = -A^2/2$. A phase difference of $\phi_s(t) - \phi_s(t-T) = 180°$ gives an output of $i_I(t) = i_Q(t) = -A^2/2$. A phase difference of $-90°$ gives an output of $i_I(t) = -A^2/2$ and $i_Q(t) = A^2/2$. The signs of both $i_I(t)$ and $i_Q(t)$ can map to the detected bits of the data.

In the simplest method, the error probability of DQPSK signal can be analyzed using the series expansion of Eq. (4.A.18). When the transmitted phases of consecutive symbols are the same, error occurs when the received differential phase is outside $\pm\pi/4$ or

$$\begin{aligned} p_e &= \frac{1}{2}\left(1 - \int_{-\pi/4}^{+\pi/4} p_{\Delta\Theta_n}(\theta)\mathrm{d}\theta\right) \\ &= \frac{3}{8} - \frac{\rho_s e^{-\rho_s}}{4}\sum_{m=1}^{\infty}\frac{\sin(m\pi/4)}{m}\left[I_{\frac{m-1}{2}}\left(\frac{\rho_s}{2}\right) + I_{\frac{m+1}{2}}\left(\frac{\rho_s}{2}\right)\right]^2. \end{aligned} \tag{9.25}$$

Alternatively, a DQPSK signal is the same as the correlated binary signals of Sec. 3.3.6 with correlation coefficient of $\rho = 1/\sqrt{2}$. The error probability is that of Eq. (3.119)

$$p_e = Q(a,b) - \frac{1}{2}e^{-(a^2+b^2)/2}I_0(ab) \tag{9.26}$$

with

$$a = \sqrt{\rho_s\left(1 - \sqrt{1/2}\right)}, \qquad b = \sqrt{\rho_s\left(1 + \sqrt{1/2}\right)}. \tag{9.27}$$

Equivalently speaking, the error probability of Eq. (9.26) is the same as that of DPSK signal with phase error of Eq. (4.10) when the phase error is $\theta_e = 45°$.

The error probability of Eq. (9.26) ignores amplifier noise from orthogonal polarization by assuming a polarized receiver. If the amplifier noise from orthogonal polarization is included, the error probability is the same as that of Eq. (4.11) for DPSK signal with the parameters a and b given by Eq. (9.27).

Figure 9.14 shows the error probability of DQPSK signal as a function of SNR ρ_s. The error probability with and without amplifier noise from orthogonal polarization is almost the same. The error probability of binary PSK and QPSK signals of Eqs. (3.78) and (9.18) is also shown for comparison. The error probability for DPSK signal of Eq. (3.105) is also shown in Fig. 9.14.

For an error probability of 10^{-9}, DQPSK signal requires a SNR of ρ_s = 17.9 dB, 4.9 dB larger than the requirement of DPSK signal

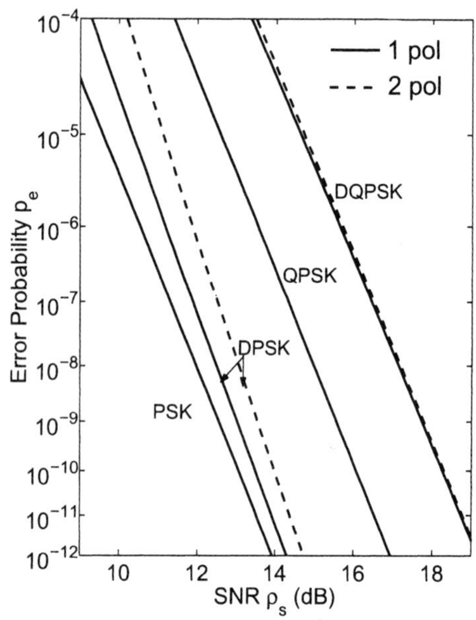

Figure 9.14. The error probability of binary and quaternary PSK and DPSK signals.

for a heterodyne receiver, or 4.5 dB larger than an unpolarized direct-detection receiver. In term of SNR per bit, DQPSK signal requires 1.9 dB larger than DPSK signal with heterodyne receiver. DQPSK signal is 2.2 dB worse than QPSK signal.

For M-ary DPSK signal, the signal is demodulated according to the phase difference of $\Delta\Theta_n = \Theta_n(t) - \Theta_n(t-T)$. For $M \geq 8$, both Θ_n and $\Theta_n(t-T)$ can assume to be both Gaussian distributed. The phase variance of $\Delta\Theta_n$ is twice that of $\Theta_n(t)$ and $\Theta_n(t-T)$. The error probability is the same as that of Eq. (9.17) but with twice the noise. For $M \geq 8$, the bit-error probability for M-ary DPSK signal is approximately given by

$$p_e \approx \frac{2}{\log_2 M} \mathrm{erfc}\left(\sqrt{\frac{\rho_s}{2}} \sin\frac{\pi}{M}\right). \tag{9.28}$$

Recently, there are proposals to use 8-level DPSK signal to improve the spectral efficiency (Han et al., 2004, Ohm, 2004).

4.2 Impairment to DQPSK Signals

Similar to DPSK signal as analyzed in Chapter 4, DQPSK signal is degraded by both interferometer phase error and laser phase noise.

Similar to DPSK signal of Chapter 5, DQPSK signal is also degraded by nonlinear phase noise.

Interferometer Phase Error

The performance of DPSK signal with interferometer phase error is discussed in Sec. 4.2.1 based on an equivalent correlation coefficient. Without going into details, with phase error, the bit-error probability for DQPSK signal is

$$\begin{aligned} p_e &= \frac{1}{2}\Bigg\{ Q_1(a_+, b_+) - \frac{1}{2}e^{-(a_+^2+b_+^2)/2} I_0(a_+ b_+) \\ &\qquad + Q_1(a_-, b_-) - \frac{1}{2}e^{-(a_-^2+b_-^2)/2} I_0(a_- b_-) \Bigg\}, \\ & a_\pm = \sqrt{\rho_s \left[1 - \cos\left(\frac{\pi}{4} \pm \theta_e\right)\right]}, \\ & b_\pm = \sqrt{\rho_s \left[1 + \cos\left(\frac{\pi}{4} \pm \theta_e\right)\right]}, \end{aligned} \qquad (9.29)$$

where θ_e is the phase error of the asymmetric interferometer.

From the two terms of Eq. (9.29), due to the phase error, the two adjacent points of the signal constellation of the DQPSK signal have different correlation coefficients. The correlated angles to the adjacent points are increased (corresponding to a_+ and b_+) or decreased (corresponding to a_- and b_-) by the phase error of θ_e from $\pi/4$. In practice, the two asymmetric Mach-Zehnder interferometers of Fig. 9.13 may have two phase errors. The error probability is the average of the error probability given by Eq. (9.29) when the two phase errors are used in Eq. (9.29).

Figure 9.15 shows the SNR penalty as a function of phase error for DQPSK signals for polarized DQPSK receiver. The SNR penalty from Fig. 4.3 for DPSK signal with phase error is also shown for comparison. The SNR penalty is calculated for a BER of 10^{-9}, corresponding to a required SNR of 13 and 17.9 dB for DPSK and DQPSK signals, respectively.

The curve of SNR penalty of a DQPSK signal in Fig. 9.15 has insignificant difference with the corresponding curves in Kim and Winzer (2003, Fig. 3), Bosco and Poggiolini (2003, Fig. 2), Bosco and Poggiolini (2004b, Fig. 2), and Winzer and Kim (2003, Fig.5) (required the adjusting of x-axis).

For the same SNR penalty, the DQPSK signal is about 2.7 times more sensitive to the phase error than the DPSK signal, more or less the same ratio as the experimental and simulated results in Kim and

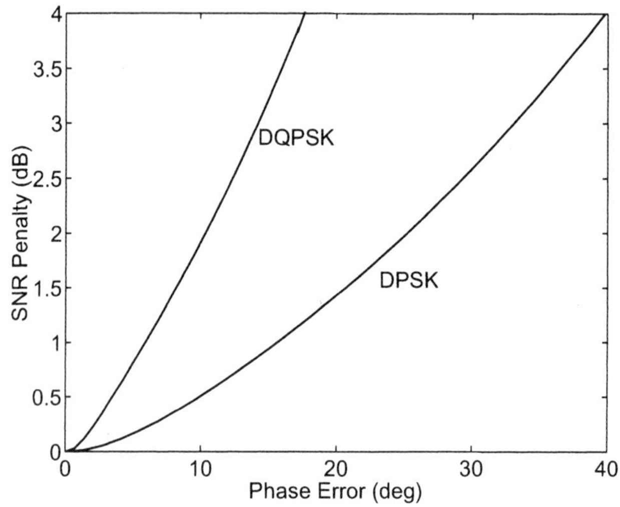

Figure 9.15. SNR penalty as a function of interferometer phase error for DQPSK signals. [From Ho (2004a), © 2005 IEEE]

Winzer (2003). The phase error for a SNR penalty of 1 dB is about 6° (or 1.7% of 360°) for a DQPSK signal, the same as the ratio of mismatched frequency to the symbol rate of Kim and Winzer (2003, Fig. 3) by simulation. The measurement of Kim and Winzer (2003) shows a larger penalty than that of Fig. 9.15 due to non-ideal signal source (Winzer and Kim, 2003). The analysis for interferometer phase error to DQPSK signal here follows that in Ho (2004a).

Nonlinear and Laser Phase Noise

DQPSK signal is affected by both nonlinear and laser phase noise. As shown in Fig. 5.15, the impact of nonlinear phase noise to DPSK signal does not have the same model as that of laser phase noise. However, unlike DPSK signal, the impact of nonlinear and laser phase noise to DQPSK can use the same model and determine by the variance of the phase noise.

Similar to the error probability of Eqs. (9.16) and (5.77), with Gray code, the bit-error probability for DQPSK signal is always equal to

$$
\begin{aligned}
p_e &= \frac{1}{2}\left[1 - \int_{-\pi/4}^{+\pi/4} p_{\Delta\Phi}(\theta)\mathrm{d}\theta\right] \\
&= \frac{3}{8} - \frac{1}{\pi}\sum_{m=1}^{\infty} \frac{1}{m}\sin\left(\frac{m\pi}{4}\right)\left|\Psi_{\Phi_r}(m)\right|^2, \qquad (9.30)
\end{aligned}
$$

where the factor of 1/2 arises because a symbol error induces one bit error in the two-bit symbol. A symbol error occurs if the differential phase is outside the angles of $\pm\pi/4$. The summation of Eq. (9.26) includes all contribution from terms other than those in which m are integer multiple of 4 with $\sin(m\pi/4) = 0$. The corresponding error probability for DPSK signals in Eq. (5.77) includes only all odd number terms.

With the exact model as from Eq. (5.70), ignored some constant phases, the coefficients for Eq. (9.30) are

$$
\Psi_{\Phi_r}(m) = \frac{\sqrt{\pi r_m}}{2} e^{-r_m/2}\Psi_\Phi\left[\frac{m\langle\Phi_{\mathrm{NL}}\rangle}{\rho_s + 1/2}\right]\left[I_{\frac{m-1}{2}}\left(\frac{r_m}{2}\right) + I_{\frac{m+1}{2}}\left(\frac{r_m}{2}\right)\right], \quad m \geq 1, \qquad (9.31)
$$

with parameters similar to Eq. (5.71) of

$$
r_m = \frac{2\left[\dfrac{jm\langle\Phi_{\mathrm{NL}}\rangle}{\rho_s + 1/2}\right]^{\frac{1}{2}}}{\sin\left\{2\left[\dfrac{jm\langle\Phi_{\mathrm{NL}}\rangle}{\rho_s + 1/2}\right]^{\frac{1}{2}}\right\}}\rho_s, \qquad (9.32)
$$

where $\Psi_\Phi(\nu)$ is the characteristic function of normalized nonlinear phase noise depending solely on SNR ρ_s given by Eq. (5.48). The parameters of r_m are the "angular frequency" depending SNR.

When the nonlinear phase noise is approximated as Gaussian distributed, similar to both Eqs. (4.40) and (5.82), the coefficients become

$$
\left|\Psi_{\Phi_r}(m)\right|^2 \approx \frac{\pi\rho_s}{4}\exp\left(-\rho_s - m^2\sigma^2_{\Phi_{\mathrm{NL}}}\right)\left[I_{\frac{m-1}{2}}\left(\frac{\rho_s}{2}\right) + I_{\frac{m+1}{2}}\left(\frac{\rho_s}{2}\right)\right]^2, \qquad (9.33)
$$

where $\sigma^2_{\Phi_{\mathrm{NL}}}$ is the variance of nonlinear phase noise given by Eq. (5.80).

Figure 9.16 plots the SNR penalty of DQPSK signals as a function of mean nonlinear phase shift of $\langle\Phi_{\mathrm{NL}}\rangle$. The SNR penalty is calculated for a bit-error probability of 10^{-9}. Without nonlinear phase noise, the required SNR for an error probability of 10^{-9} is 17.9 dB from Fig. 9.14.

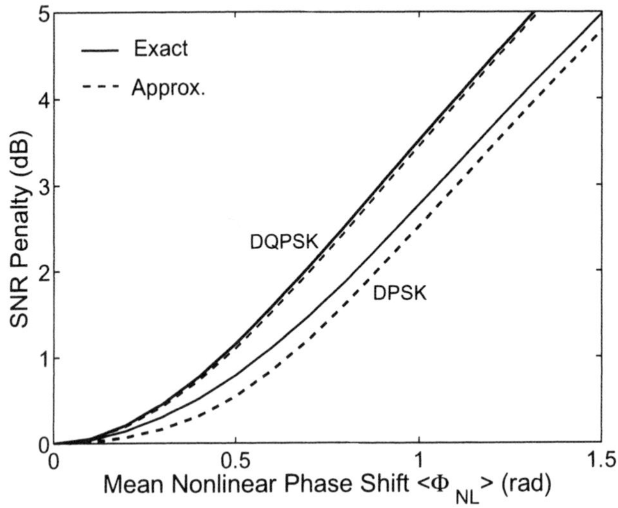

Figure 9.16. The SNR penalty as a function of mean nonlinear phase shift. The penalty is calculated by exact model or approximating the nonlinear phase noise as Gaussian distributed.

For comparison purpose, the SNR penalty of DPSK signal from Fig. 5.13 is also shown in Fig. 9.16. For the same mean nonlinear phase shift, the performance of DPSK signal is 0.5 to 0.8 dB better than DQPSK signal.

For a SNR penalty of less than 1 dB, the mean nonlinear phase shift must be less than 0.50 rad. The performance of a system increases with the SNR that is also proportional to the launched power of the signal. As shown earlier by Eq. (6.16), the mean nonlinear phase shift is proportional to the launched power of the system. From Fig. 9.16, the SNR penalty increases with the mean nonlinear phase shift. The optimal operating point can be found by the condition that the increase of SNR penalty is smaller than the SNR improvement. From Fig. 9.16, the optimal operating point is for a mean nonlinear phase shift of $\langle \Phi_{\mathrm{NL}} \rangle =$ 0.89 rad. The optimal operating point is about 0.1 rad less than 0.97 rad for DPSK signal from Table 5.1, corresponding to a difference of 0.75 dB.

In term of mean nonlinear phase shift, DQPSK and DPSK systems have more or less the same tolerance to nonlinear phase noise. The mean nonlinear phase shift is proportional the product of the number of fiber spans and the launched power per span by Eq. (6.16). Alternatively

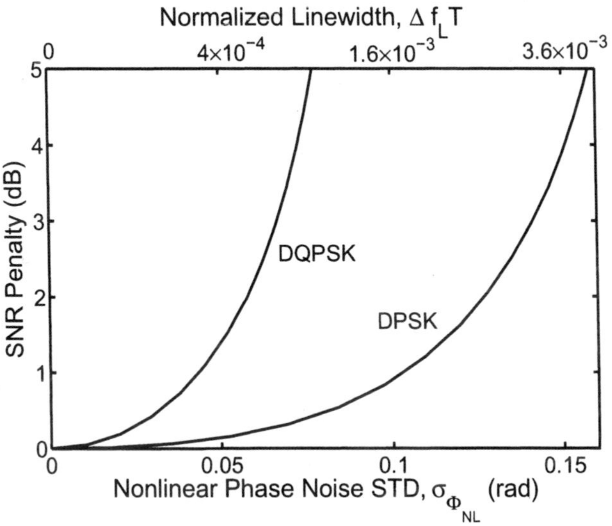

Figure 9.17. The SNR penalty as a function of the STD of nonlinear phase noise and normalized laser linewidth.

speaking, DQPSK and DPSK systems can tolerate the same amount of self-phase modulation.

The claim that DQPSK and DPSK systems can have the same amount of nonlinear phase noise is counter-intuitive. In fact, the variance of nonlinear phase noise decreases with SNR. From Fig. 9.14, comparing with the 13.0 dB SNR requirement for DPSK signal, DQPSK signals with an SNR requirement of 17.9 dB have a variance of nonlinear phase noise 4.9 dB less than that for DPSK signal for the same mean nonlinear phase shift. Due to larger SNR requirement, DQPSK signals can tolerate a mean nonlinear phase shift close to DPSK signals although the constellation points are closer than that for DPSK signal.

Based on the Gaussian approximation of nonlinear phase noise, Figure 9.17 shows the SNR penalty of DQPSK and DPSK signals as a function of the standard deviation (STD) of nonlinear phase noise. DQPSK signal can tolerate a nonlinear phase noise STD about half of that for DPSK signal. At the optimal operation point, DQPSK signal can tolerate a nonlinear phase noise STD about 53% that for DPSK signal.

Requiring about 4.9 dB larger SNR but a bandwidth half of that of DPSK signals, for the same spectral density of amplifier noise, DQPSK signals require about 1.9 dB larger launched power for the same SNR.

Table 9.2. The Logic Operations of the DQPSK Precoder.

Data $a_k b_k$	Phase Diff. $\Delta\theta_k$	Output Logic p_k	q_k
00	180°	$\overline{p}_{k-1}$	$\overline{q}_{k-1}$
01	90°	q_{k-1}	$\overline{p}_{k-1}$
10	270°	$\overline{q}_{k-1}$	p_{k-1}
11	0°	p_{k-1}	q_{k-1}

Precoder operations:

$$
\begin{aligned}
p_k &= \overline{a}_k \overline{b}_k \overline{p}_{k-1} + \overline{a}_k b_k q_{k-1} + a_k \overline{b}_k \overline{q}_{k-1} + a_k b_k p_{k-1}. \\
q_k &= \overline{a}_k \overline{b}_k \overline{q}_{k-1} + \overline{a}_k b_k \overline{p}_{k-1} + a_k \overline{b}_k p_{k-1} + a_k b_k q_{k-1}.
\end{aligned}
$$

Assume that nonlinear phase noise and amplifier noise are the dominant noise, for the same mean nonlinear phase shift, DQPSK systems have about 55% the reach of DPSK signal. Current experiment of DQPSK signal (Tokle et al., 2004) reaches about 50% the distance of similar DPSK experiments (Cai et al., 2003b, 2004).

Both Gaussian distributed, the SNR penalty of Fig. 9.17 is the same as that for DQPSK signal with laser phase noise. Figure 9.17 also shows the SNR penalty as a function of the normalized laser linewidth of $\Delta f_L T$. The phase error due to laser phase noise has an variance equal to $\sigma_{\phi_e}^2 = 2\pi\Delta f_L T$ from Eq. (4.38), where T is the symbol interval of the DQPSK signal. The SNR penalty for DQPSK signal due to laser phase noise is that of Fig. 9.17 with a variance of phase noise calculated differently. For 1-dB SNR penalty, the σ_{ϕ_e} must be less than about 0.04, corresponding to a normalized linewidth of $\Delta f_L T$ about 2.5×10^{-4}. With longer symbol time, the required linewidth for 10 and 40-Gb/s DQPSK system is 1.3 and 5.1 MHz, respectively.

The impact of laser phase noise to DQPSK signal was analyzed by Gene et al. (2004) by computer simulation and by Savory and Hadjifotiou (2004) using Marcum Q function. The linewidth requirement for QPSK signal was also analyzed by Barry and Kahn (1992).

The effect of chromatic dispersion to DQPSK signal was also studied by Gene et al. (2004) and Griffin et al. (2003).

4.3 DQPSK Precoder

DQPSK signal is demodulated by two interferometers of Fig. 9.13 corresponding to the phase difference of the DQPSK signals of Eq. (9.3) in consecutive symbols.

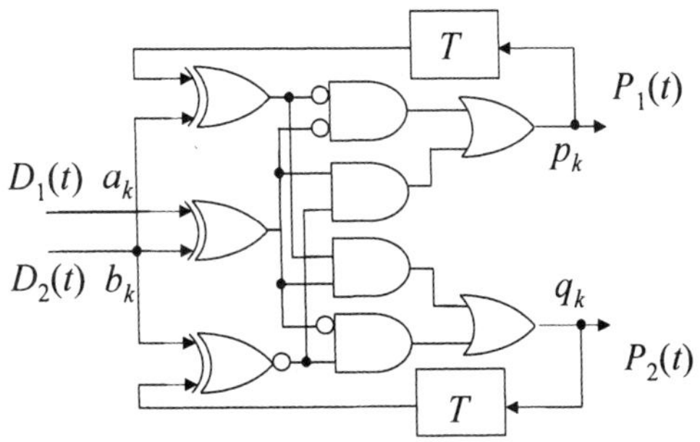

Figure 9.18. The precoder for DQPSK signal.

The logical operation of the precoder is derived in Table 9.2 and given by Fig. 9.18. At the receiver of Fig. 9.13, the output of $i_I(t)$ corresponds to $2b_k - 1$ and the output of $i_Q(t)$ corresponds to $2a_k - 1$. The input data of $D_1(t)$ and $D_2(t)$ are represented by the Boolean variables of a_k and b_k in Table 9.2. The output of the precoder of $P_1(t)$ and $P_2(t)$ are represented by the Boolean variables of p_k and q_k. In the receiver using two interferometers, the phase differences of $\Delta\theta_k = 0°, 90°, 180°, 270°$ are decoded to two-bit output of 11, 01, 00, and 10, respectively. The input data of $a_k b_k$ should map to the corresponding phase differences as shown in Table 9.2.

In the interface between the precoder and the driver amplifier of Fig. 9.3, an "1" logic is mapped to high voltage. The binary signal of $P_1(t) = 0, 1$ or Boolean variable of p_k is mapped to the drive voltage of $V_1(t)$ based on $V_1(t) = V_\pi P_1(t)$ or $V_1(t) = p_k V_\pi$. The binary signal of $P_2(t) = 0, 1$ or Boolean variable of q_k is mapped to the drive voltage of $V_2(t)$ based on $V_2(t) = V_\pi \left[P_2(t) - \frac{1}{2}\right]$ or $V_2(t) = \left[q_k - \frac{1}{2}\right] V_\pi$.

In Table 9.2, with input data of $a_k = b_k = 1$ with a phase difference of $\Delta\theta_k = 0°$, there is no phase difference or no need to change p_{k-1} and q_{k-1} to p_k and q_k. With $a_k = b_k = 1$, we obtain $p_k = p_{k-1}$ and $q_k = q_{k-1}$. When the input data are $a_k = b_k = 0$ with a phase difference of $\Delta\theta_k = 180°$, both p_k and q_k must be the opposite of p_{k-1} and q_{k-1} for the 180° phase shift. With $a_k = b_k = 0$, we obtain $p_k = \overline{p}_{k-1}$ and $q_k = \overline{q}_{k-1}$. When $a_k = 0$ and $b_k = 1$ with $\Delta\theta = 90°$ phase difference, the 90° rotation changes $p_{k-1}q_{k-1} \to p_k q_k$ according to the cyclic mapping of $00 \to 01 \to 11 \to 10 \to 00$ or the logic of $p_k = q_{k-1}$ and $q_k = \overline{q}_{k-1}$. The operation of $a_k = 1$ and $b_k = 0$ with $\Delta\theta = 270°$ can be derived similarity. The overall logic operations of the precoder are also shown in Table 9.2. To minimize the number of logic gates, the precoder can be simplified to

$$\begin{aligned} p_k &= \overline{a_k \oplus b_k} \cdot \overline{b_k \oplus p_{k-1}} + (a_k \oplus b_k) \cdot \overline{b_k \oplus q_{k-1}}, \\ q_k &= \overline{a_k \oplus b_k} \cdot \overline{b_k \oplus q_{k-1}} + (a_k \oplus b_k) \cdot (b_k \oplus p_{k-1}), \end{aligned} \quad (9.34)$$

where $\oplus$ denotes exclusive-OR operation. The precoder of Eq. (9.34) is also shown in Fig. 9.18.

5. Direct-Detection of Multilevel On-Off Keying Signals

On-off keying signals may also have more than two levels. A four-level on-off keying is shown in Fig. 1.6 as an example. Including the noise from orthogonal polarization and assuming an optical matched filter preceding the receiver, the received signal is the same as that of Eq. (3.125)

$$r_k(t) = [A_k + n_{x1}(t)]^2 + n_{x2}^2(t) + n_{y1}^2(t) + n_{y2}^2(t), \quad k = 1, \ldots, M, \quad (9.35)$$

for M-ary on-off keying signal, where A_k is the amplitude for the kth signal waveform. The M waveforms are assumed to be transmitted with equal probability but signal shaping may provide better performance (Shiu and Kahn, 1999). Also the same as that in Eq. (3.126), the probability density function (p.d.f.) of the output signal is a noncentral χ^2 distribution with four degrees of freedom of

$$p_k(y) = \frac{1}{2\sigma_n^2}\sqrt{\frac{y}{A_k^2}}\, e^{-(A_k^2+y)/2\sigma_n^2} I_1\left(\sqrt{y}\frac{A_k}{\sigma_n^2}\right), \qquad y \geq 0 \qquad (9.36)$$

with a mean and variance of

$$m_{yk} = A_k^2 + 4\sigma_n^2, \qquad (9.37)$$

$$\sigma_{yk}^2 = 4\sigma_n^2 A_k^2 + 8\sigma_n^4. \qquad (9.38)$$

Figure 9.19 shows the conditional p.d.f. of Eq. (9.36) for equal and unequal spacing signal. The variance of σ_{yk}^2 increases with A_k^2 and the upper two levels have the largest noise variance. If the on-off keying signal has equal spacing in m_{yk}, from Fig. 9.19(a), the system is limited by the upper two levels. The system should be designed for equal error probability between levels, as shown in Fig. 9.19(b), with unequal spacing in m_{yk}. While the exact analysis is difficult, a Gaussian approximation using Q factor similar to that of Eq. (3.140) is very simple. The analysis can further be simplified if the variance of Eq. (9.38) is approximated $\sigma_k^2 \approx 4\sigma_n^2 A_k^2$ for $k > 1$. With the approximation, the Q

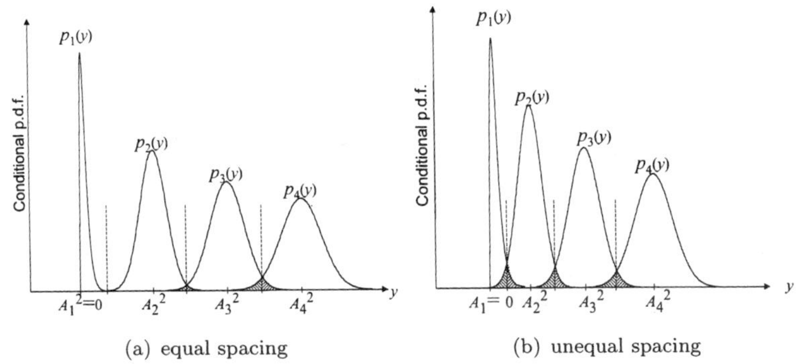

Figure 9.19. The conditional p.d.f. of Eq. (9.36) for (a) equal and (b) unequal spacing signal.

factor between kth and $(k-1)$th levels is equal to

$$Q_{k-1} = \frac{A_k^2 - A_{k-1}^2}{2\sigma_n A_k + 2\sigma_n A_{k-1}} = \frac{A_k - A_{k-1}}{2\sigma_n}. \tag{9.39}$$

To equalize the Q factor, the electric field of A_k should be equally spacing but the output of y or m_{my} should be quadratic spacing.

Figure 9.20 shows the error probability of two- and four-level on-off keying signal. The error probability of two-level on-off keying is calculated using the Q-factor of Eq. (3.149) with the approximation of Eq. (3.140), the same as the corresponding curve in Fig. 3.11. Multilevel on-off keying gives a very large SNR penalty as from Fig. 9.20.

Multilevel on-off keying was proposed for a long time (Muoi and Hullett, 1975). Multilevel on-off keying was used mostly to improve spectral efficiency (Cimini Jr. and Foschini, 1993, Hatami-Hanza et al., 1997) or extend the dispersion-limited transmission distance (Walklin and Conradi, 1999). The quadratic level can further be optimized (Ho and Kahn, 2004b, Muoi and Hullett, 1975, Rebola and Cartaxo, 2000).

Other than multilevel on-off keying, only one axis of QAM signal can be used, called pulse-amplitude modulation (PAM) in digital communications (Proakis, 2000). Four-level PAM signal was used in Ohm and Speidel (2003) and Hansryd et al. (2004) together with differential phase detection.

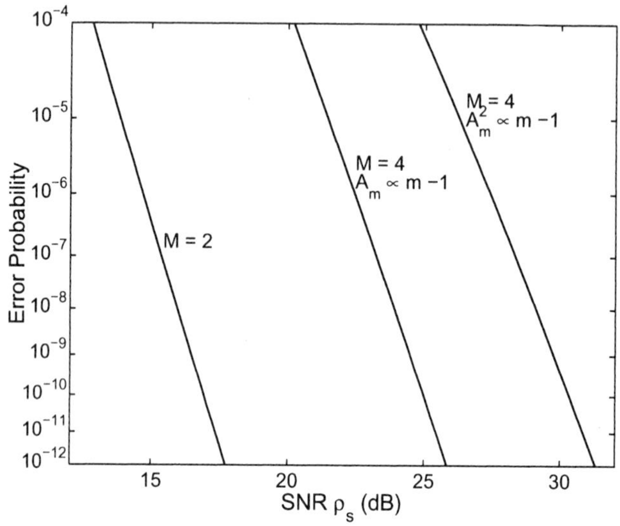

Figure 9.20. The error probability for two- and four-level on-off keying signal.

6. Comparison of Multilevel Signals

From Eq. (9.17), the penalty for M-ary PSK to binary PSK signal is approximately equal to

$$\delta_p = \left(\sin\frac{\pi}{M}\right)^{-2} \tag{9.40}$$

in term of SNR of ρ_s. In term of SNR per bit, the penalty of SNR per bit is

$$\delta_p = \frac{1}{\log_2 M \sin^2 \pi/M}. \tag{9.41}$$

From Eq. (9.21), the penalty for M-ary QAM to binary PSK signal is approximately equal to

$$\delta_p = \frac{2}{3}(M-1), \quad M \geq 4 \tag{9.42}$$

in SNR or

$$\delta_p = \frac{2}{3\log_2 M}(M-1), \quad M \geq 4 \tag{9.43}$$

in SNR per bit.

Assume equally spacing in electric field and quadratic spacing in optical intensity, in SNR per bit, the penalty due from two to M-ary on-off

Table 9.3. SNR Penalty for Multilevel Signals.

Level	SNR Penalty (dB)				SNR Penalty per Bit(dB)			
M	PSK	QAM	OOK	DPSK	PSK	QAM	OOK	DPSK
2	0.0	-	3.5	0.45	0.0	-	3.5	.45
4	3.0	3.0	11.9	5.2	0.0	0.0	8.9	2.2
8	8.3	7.4	18.5	11.3	3.5	2.6	13.7	6.5
16	14.2	10.0	-	17.2	8.2	4.0	-	11.2
32	20.2	13.0	-	-	13.2	6.0	-	-
64	-	16.2	-	-	-	8.5	-	-

keying (OOK) can be estimated by

$$\delta_p = \frac{2}{M \log_2 M} \sum_{m=1}^{M} (m-1)^2 = \frac{(2M-1)(M-1)}{3 \log_2 M}. \tag{9.44}$$

Please also note the 3-dB penalty from PSK to binary OOK.

Table 9.3 shows the SNR penalty for multilevel signal compared to binary PSK signal for an error probability of 10^{-9}. With the superior receiver sensitivity provided by coherent detection, 64-QAM is even 0.4 dB better than four-level on-off keying, similar to the assessment of Fig. 1.6. Table 9.3 also shows that QPSK signal has the same SNR per bit as the optimal PSK signal.

With an interferometric based direct-detection receiver, DQPSK signal doubles the spectral efficiency compared with binary on-off keying signals with about 1.3 dB of sensitivity improvement.

In summary, this chapter studies multilevel phase-modulated optical communications. QAM signals can improve the spectral efficiency without a large penalty in receiver sensitivity.

Multilevel signal is used for system with high SNR to improve spectral efficiency. For system with low SNR, binary signal is the optimal modulation format.

Chapter 10

PHASE-MODULATED SOLITON SIGNALS

Other than Sec. 7.4, previous chapters mainly focus on the nonlinear phase noise for non-return-to-zero (NRZ) pulses, in certain sense, a continuous-wave signal. From Table 1.2, virtually all differential phase-shift keying (DPSK) experiments use return-to-zero (RZ) pulses. The nonlinear phase noise of previous chapters is induced by interaction of amplitude noise with fiber Kerr effect. While Sec. 7.4 gives the variance of nonlinear phase noise for RZ pulse, the probability density function (p.d.f.) is implicitly assumed as Gaussian. With well-developed perturbation theory, soliton can provide analytical results on the p.d.f. of nonlinear phase noise. To certain extend, a soliton DPSK system may be a good approximation to phase modulated dispersion-managed soliton or RZ signal (McKinstrie and Xie, 2002, Nakazawa et al., 2000, Smith et al., 1997, Suzuki et al., 1995, Takushima et al., 2002).

The phase jitter of soliton due to amplifier noise, like Gordon-Haus timing jitter (Gordon and Haus, 1986), is usually assumed to be Gaussian distributed (Blow et al., 1992, Hanna et al., 2000, 2001, Iannone et al., 1998). When the phase jitter of soliton was studied, the phase jitter variance was given or measured and the statistics of soliton phase is not discussed (Blow et al., 1992, Hanna et al., 1999, 2000, Leclerc and Desurvire, 1998, McKinstrie and Xie, 2002). With well-developed perturbation theory (Georges, 1995, Iannone et al., 1998, Kaup, 1990, Kivshar and Malomed, 1989), the distribution of the soliton phase jitter can be derived analytically.

The error probability of DPSK soliton signal was calculated in Shum et al. (1997) using the methods of Shum and Ghafouri-Shiraz (1996) and Humblet and Azizoğlu (1991) without taking into account the effect of phase jitter. If the phase jitter is Gaussian distributed, the system can be analyzed by the formulae of both Eqs. (4.40) and (5.82) similar to laser phase noise. The phase jitter may be indeed Gaussian distributed in certain regimes around the center of the distribution (Hanna et al., 2000, Holzlohner et al., 2002), especially if the p.d.f. is plotted in linear scale. The tail probability less than, for example, 10^{-9}, is certainly not Gaussian distributed. As optical communication systems are aimed for very low error probability, a careful study of the statistics of the soliton phase is necessary to characterize the performance of the system.

After the characteristic function of soliton phase jitter is derived analytically, we can also find the error probability of a DPSK soliton system. The method is similar to Appendix 4.A using the Fourier expansion of the p.d.f. Similar to Sec. 5.3, different models for the nonlinear phase noise can be used to derive the error probability.

1. Soliton Perturbation

Solitary pulse was first proposed by Hasegawa and Tappert (1973) to transmit information without distortion and was first observed by Mollenauer et al. (1980) in optical fiber. The simplest implementation is on-off keying soliton system to encode information with and without the presence of a soliton. The information can also be encoded in the phase of the soliton like a phase-shift keying (PSK) signal or the phase difference like a DPSK signal. With amplifier noise, the soliton is distorted. However, for small amount of amplifier noise, the soliton pulse shape can be maintained along the fiber. The signal propagation in fiber is governed by the nonlinear Schrödinger equation of Eq. (7.9). The soliton arising from Eq. (7.9) is analyzed in detail in Iannone et al. (1998) and not shown here. The changes of soliton parameters can be analyzed based on the well-known soliton perturbation theory (Georges, 1995, Iannone et al., 1998, Kaup, 1990, Kivshar and Malomed, 1989).

Without going into detail, from the first-order perturbation theory, with amplifier noise, the soliton parameters evolve according to the following normalized equations

$$\frac{dA}{d\zeta} = \Im\left\{\int \mathrm{d}\tau f_A n(\zeta,\tau)\right\} \tag{10.1}$$

$$\frac{d\Omega}{d\zeta} = \Re\left\{\int \mathrm{d}\tau f_\Omega n(\zeta,\tau)\right\} \tag{10.2}$$

$$\frac{dT}{d\zeta} = -\Omega + \Im\left\{\int \mathrm{d}\tau f_T n(\zeta,\tau)\right\} \tag{10.3}$$

$$\frac{d\phi}{d\zeta} = \frac{1}{2}\left(A^2 - \Omega^2\right) + T\frac{d\Omega}{d\zeta} + \Re\left\{\int \mathrm{d}\tau f_\phi n(\zeta,\tau)\right\} \tag{10.4}$$

where $\Re\{\cdot\}$ and $\Im\{\cdot\}$ denote the real and imaginary parts of a complex number, respectively, $n(\zeta, \tau)$ is the amplifier noise with the correlation of

$$E\{n(\zeta_1, \tau_1)n(\zeta_2, \tau_2)\} = \sigma_n^2 \delta(\zeta_1 - \zeta_2)\delta(\tau_1 - \tau_2), \tag{10.5}$$

$A(\zeta)$, $\Omega(\zeta)$, $T(\zeta)$, and $\phi(\zeta)$ are the amplitude, frequency, timing, and phase parameters of the perturbed soliton of

$$q_0(\tau, \zeta) = A(\zeta)\mathrm{sech}\left\{A(\zeta)[\tau - T(\zeta)]\right\} \exp\left[-j\Omega(\zeta)\tau + j\phi(\zeta)\right] \tag{10.6}$$

with initial values of $A(0) = A$ and $\Omega(0) = \phi(0) = T(0) = 0$.

Functions related to soliton parameters are

$$
\begin{aligned}
f_A &= q_0^*, &(10.7)\\
f_\Omega &= \tanh[A(\tau - T)]q_0^*, &(10.8)\\
f_T &= \frac{\tau - T}{A} q_0^*, &(10.9)\\
f_\phi &= -\frac{1}{A}\left\{1 - A(\tau - T)\tanh[A(\tau - T)]\right\} q_0^*. &(10.10)
\end{aligned}
$$

The soliton perturbation theory was independently developed by two groups of Karpman and Maslov (1977) and Kaup (1976, 1990), Kaup and Newell (1978) and comprehensively reviewed by Kivshar and Malomed (1989). The book of Iannone et al. (1998) gave a good derivation and explanation of Eqs. (10.1) to (10.4). The above equations of Eqs. (10.1) to (10.4) use the approaches of Georges (1995) but the notation of Iannone et al. (1998).

From both Eqs. (10.1) and (10.2), we get

$$
\begin{aligned}
A(\zeta) &= A + w_A(\zeta), &(10.11)\\
\Omega(\zeta) &= w_\Omega(\zeta), &(10.12)
\end{aligned}
$$

where w_A and w_Ω are two independent zero-mean Wiener process with autocorrelation functions of

$$
\begin{aligned}
E\{w_A(\zeta_1)w_A(\zeta_2)\} &= \sigma_A^2 \min(\zeta_1, \zeta_2), &(10.13)\\
E\{w_\Omega(\zeta_1)w_\Omega(\zeta_2)\} &= \sigma_\Omega^2 \min(\zeta_1, \zeta_2), &(10.14)
\end{aligned}
$$

where $\sigma_A^2 = A\sigma_n^2$ and $\sigma_\Omega^2 = A\sigma_n^2/3$. Defined for the amplitude, the signal-to-noise ratio (SNR) as a function of distance is

$$\rho_s = \frac{A^2}{2\sigma_A^2 \zeta} = \frac{A}{2\sigma_n^2 \zeta}. \tag{10.15}$$

Using Eqs. (10.3) and (10.12), the timing jitter is

$$T(\zeta) = -\int_0^\zeta w_\Omega(\zeta_1)\mathrm{d}\zeta_1 + w_T(\zeta), \tag{10.16}$$

where w_T is a zero-mean Wiener process with autocorrelation function of

$$E\{w_T(\zeta_1)w_T(\zeta_2)\} = \sigma_T^2 \min(\zeta_1, \zeta_2) \tag{10.17}$$

with

$$\sigma_T^2 = \frac{\pi^2}{12}\frac{\sigma_n^2}{A}. \tag{10.18}$$

The timing jitter of Eq. (10.16) is the Gordon-Haus timing jitter (Gordon and Haus, 1986). The integration of the Wiener process of $w_\Omega(\zeta)$ over distance gives a timing jitter variance increase cubically with distance ζ.

Using Eqs. (10.3), (10.11), and (10.16), the phase jitter is

$$\phi(\zeta) = \frac{1}{2}\int_0^\zeta [A + w_A(\zeta_1)]^2 \mathrm{d}\zeta_1 - \frac{1}{2}\int_0^\zeta w_\Omega^2(\zeta_1)\mathrm{d}\zeta_1 + \int_0^\zeta \left(-\int_0^{\zeta_1} w_\Omega(\zeta_2)\mathrm{d}\zeta_2 + w_T(\zeta_1)\right)\mathrm{d}w_\Omega(\zeta_1) + w_\phi(\zeta), \tag{10.19}$$

where w_ϕ is a zero-mean Wiener process with autocorrelation function of

$$E\{w_\phi(\zeta_1)w_\phi(\zeta_2)\} = \sigma_\phi^2 \min(\zeta_1, \zeta_2) \tag{10.20}$$

with

$$\sigma_\phi^2 = \frac{\sigma_n^2}{3A}\left(1 + \frac{\pi^2}{12}\right). \tag{10.21}$$

The Wiener processes of w_A, w_Ω, w_T, and w_ϕ are independent of each other. The amplitude [Eq. (10.11)], frequency [Eq. (10.12)], and timing [Eq. (10.16)] jitters are all Gaussian distributed. From Eq. (10.19), it is obvious that the phase jitter is not Gaussian distributed. If Eq. (10.4) is linearized or all higher-order terms of Eq. (10.19) are ignored, other than a constant phase shift, the phase jitter is approximately equal to

$$\phi(\zeta) \approx A\int_0^\zeta w_A(\zeta_1)\mathrm{d}\zeta_1 + w_\phi(\zeta). \tag{10.22}$$

As a linear combination of the Gaussian process of both $w_A(\zeta)$ and $w_\phi(\zeta)$, the approximated phase jitter of Eq. (10.22) is obviously Gaussian distributed.

The approximated phase jitter of Eq. (10.22) is almost the same as the timing jitter of Eq. (10.16). The statistical properties of soliton phase jitter are approximately the same as that for timing jitter.

2. Statistics of Soliton Phase Jitters

In the phase jitter of Eq. (10.19), there are three independent contributions from amplitude jitter (the first term), frequency and timing jitter (the second and third terms), and the projection of amplifier noise to phase jitter w_ϕ. In this section, the characteristic functions of each individual component are derived and the overall characteristic function of phase jitter is the product of the characteristic functions of each independent contribution.

2.1 Amplitude-Induced Nonlinear Phase Noise

The first term of Eq. (10.19) is the nonlinear phase noise of

$$\phi_{\mathrm{GM}}(\zeta) = \frac{1}{2}\int_0^\zeta [A + w_A(\zeta_1)]^2 \mathrm{d}\zeta_1, \tag{10.23}$$

induced by the interaction of fiber Kerr effect and amplifier noise, affecting phase-modulated signals as shown in Chapters 5 and 6. This self-phase modulation induced nonlinear phase noise is often referred as Gordon-Mollenauer effect.

The phase jitter of Eq. (10.23) is the same as the corresponding expression for the normalized nonlinear phase noise of Φ in Eq. (5.32). Comparing Eq. (10.23) with the non-soliton case of Eq. (5.32), the mean and standard deviation (STD) of the Gordon-Mollenauer phase noise of soliton are about half of that of non-soliton case with the same amplitude A as the signal level for NRZ or continuous-wave signals.

The characteristic function of Gordon-Mollenauer nonlinear phase noise is given by Eq. (5.48). For the specific case of Eq. (10.23), we obtain

$$\Psi_{\phi_{\mathrm{GM}}(\zeta)}(\nu) = \sec^{\frac{1}{2}}\left(\zeta\sigma_A\sqrt{j\nu}\right)\exp\left[\frac{A^2}{2\sigma_A}\sqrt{j\nu}\tan\left(\zeta\sigma_A\sqrt{j\nu}\right)\right]. \tag{10.24}$$

The above characteristic function Eq. (10.24) can also be derived from Eq. (10.A.7) of Appendix 10.A.

The mean and variance of the phase jitter of Eq. (10.23) are

$$\langle\phi_{\mathrm{GM}}(\zeta)\rangle = -j\,\frac{d}{d\nu}\Psi_{\phi_{\mathrm{GM}}(\zeta)}(\nu)\bigg|_{\nu=0} = \frac{1}{2}A^2\zeta + \frac{1}{4}\sigma_A^2\zeta^2, \tag{10.25}$$

and

$$\begin{aligned}\sigma^2_{\phi_{\mathrm{GM}}(\zeta)} &= -\frac{d^2}{d\nu^2}\Psi_{\phi_{\mathrm{GM}}(\zeta)}(\nu)\bigg|_{\nu=0} - <\phi_{\mathrm{GM}}(\zeta)>^2 \\ &= \frac{1}{3}A^2\sigma_A^2\zeta^3 + \frac{1}{12}\sigma_A^4\zeta^4,\end{aligned} \tag{10.26}$$

respectively. The first term of Eq. (10.26) increases with ζ^3, conforming to that of Gordon and Mollenauer (1990) for the cubical increase of nonlinear phase noise with distance. Given a large fixed SNR of $A^2/(2\sigma_A^2\zeta)$ from Eq. (10.15), the second term of Eq. (10.26) is much smaller than the first term and also increases with ζ^3. Note that the first term of the mean of Eq. (10.25) is also larger than the second term for large SNR.

The characteristic function of Eq. (10.24) depends on two parameters: $A^2\zeta/2$ and the SNR of Eq. (10.15). As shown later, the mean nonlinear phase shift is approximately equal to $A^2\zeta/2$. Given a fixed mean nonlinear phase shift of $A^2\zeta/2$, the shape of the distribution depends only on the SNR, the same as the normalized nonlinear phase noise of Eq. (5.48).

2.2 Frequency and Timing Effect

The frequency and timing jitter contributes to phase jitter by

$$\phi_{\Omega,T}(\zeta) = -\frac{1}{2}\int_0^\zeta w_\Omega^2(\zeta_1)\mathrm{d}\zeta_1 - \int_0^\zeta \int_0^{\zeta_1} w_\Omega(\zeta_2)\mathrm{d}\zeta_2 \mathrm{d}w_\Omega(\zeta_1) + \int_0^\zeta w_T(\zeta_1)\mathrm{d}w_\Omega(\zeta_1) \tag{10.27}$$

as the second and third terms of Eq. (10.19).

By changing the order of integration for the second term of Eq. (10.27), we obtain

$$\phi_{\Omega,T}(\zeta) = \frac{1}{2}\int_0^\zeta w_\Omega^2(\zeta_1)\mathrm{d}\zeta_1 + \int_0^\zeta w_T(\zeta_1)\mathrm{d}w_\Omega(\zeta_1) - w_\Omega(\zeta)\int_0^\zeta w_\Omega(\zeta_1)\mathrm{d}\zeta_1. \tag{10.28}$$

From Eq. (10.A.12) of Appendix 10.A, the characteristic function of $\phi_{\Omega,T}(\zeta)$ is

$$\Psi_{\phi_{\Omega,T}(\zeta)}(\nu) = \Psi_{\varphi_1,\varphi_2,\varphi_3}\left(\frac{\nu}{2}, \nu, -\nu\right). \tag{10.29}$$

The mean and variance of the phase jitter of Eq. (10.27) are

$$\langle \phi_{\Omega,T}(\zeta) \rangle = -j\,\frac{d}{d\nu}\Psi_{\phi_{\Omega,T}(\zeta)}(\nu)\bigg|_{\nu=0} = -\frac{1}{4}\sigma_\Omega^2\zeta^2, \tag{10.30}$$

and

$$\begin{aligned}\sigma^2_{\phi_{\Omega,T}(\zeta)} &= -\frac{d^2}{d\nu^2}\Psi_{\phi_{\Omega,T}(\zeta)}(\nu)\bigg|_{\nu=0} - <\phi_{\Omega,T}(\zeta)>^2 \\ &= \frac{1}{2}\sigma_\Omega^2\sigma_T^2\zeta^2 + \frac{1}{4}\sigma_\Omega^4\zeta^4,\end{aligned} \tag{10.31}$$

respectively.

Comparing the means of Eqs. (10.25) and (10.30), in terms of absolute value, the mean nonlinear phase shift due to Gordon-Mollenauer effect is much larger than that due to frequency and timing effect. Comparing the variances of Eqs. (10.26) and (10.31), the variance of nonlinear phase noise due to Gordon-Mollenauer effect is also much larger than that due to frequency and timing effect.

Unlike the Gordon-Mollenauer effect, the characteristic function of Eq. (10.29), from Appendix 10.A, is not determined only on the SNR and the mean nonlinear phase shift.

2.3 Linear Phase Noise

The last term of Eq. (10.19) gives the linear phase noise of

$$\phi_{\mathrm{LN}}(\zeta) = w_\phi(\zeta) \tag{10.32}$$

with a characteristic function of

$$\Psi_{\phi_{\mathrm{LN}}(\zeta)}(\nu) = \exp\left(-\frac{1}{2}\sigma_\phi^2\zeta\nu^2\right) \tag{10.33}$$

the same as a Gaussian-distributed characteristic function with variance of $\sigma_\phi^2\zeta$. From the characteristic function of Eq. (10.33), the linear phase noise depends solely on the SNR.

The characteristic function of the overall phase jitter $\phi(\zeta)$ is the multiplication of the characteristic functions of Eqs. (10.24), (10.29), and (10.33).

Although the actual mean nonlinear phase shift for the soliton is

$$\langle\phi(\zeta)\rangle = \langle\phi_{\Omega,T}(\zeta)\rangle + \langle\phi_{\mathrm{GM}}(\zeta)\rangle = \frac{1}{2}A^2\zeta + \frac{1}{6}\sigma_A^2\zeta^2, \tag{10.34}$$

we mostly call $A^2\zeta/2$ the mean nonlinear phase shift as a good approximation in high SNR.

2.4 Numerical Results

The p.d.f. of a random variance is the inverse Fourier transform of the corresponding characteristic function. Figures 10.1 show the evolution of the distribution of the phase jitter [Eq. (10.19)] with distance. The system parameters are $A = 1$ and $\sigma_n^2 = 0.05$. Those parameters are chosen for typical distribution of the phase jitter.

Figures 10.1(a), (b), (c) are the distribution of Gordon-Mollenauer nonlinear phase noise of Eq. (10.24), frequency and timing nonlinear phase noise of Eq. (10.29), and the linear phase noise of Eq. (10.33), respectively, as components of the overall phase jitter of Eq. (10.19). Figure 10.1(d) is the distribution of the overall phase jitter of Eq. (10.19).

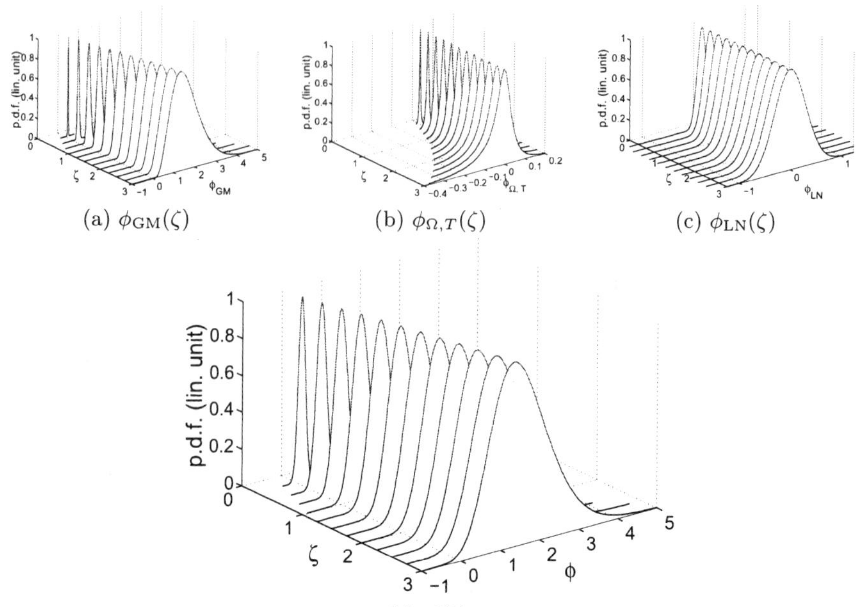

(a) $\phi_{\text{GM}}(\zeta)$ (b) $\phi_{\Omega,T}(\zeta)$ (c) $\phi_{\text{LN}}(\zeta)$

(d) $\phi(\zeta)$

Figure 10.1. The distributions of soliton phase jitter for difference distance for $A = 1$, $\sigma_n^2 = 0.05$. The distributions are normalized for a unity peak. The x-axis is not in the same scale. [From Ho (2004e)]

The p.d.f. in Figs. 10.1 are normalized to an unity peak value for illustration purpose. The x-axis of individual figure of Figs. 10.1 does not have the same scale. From Figs. 10.1, the nonlinear phase noises from Gordon-Mollenauer effect and frequency and timing effect are obvious not Gaussian distributed. With small mean and variance, the nonlinear phase noise from frequency and timing effect has a very long tail.

Figures 10.2 plot the p.d.f. of Figs. 10.1 in logarithmic scale for the cases of $\zeta = 1, 2$. The Gaussian approximation is also plotted in Figs. 10.2 for the overall phase jitter $\phi(\zeta)$. In both cases of $\zeta = 1, 2$, the Gaussian approximation is not close to the exact p.d.f. in the tails. However, if the p.d.f. are plotted in linear scale, Gaussian approximation may be very close to the actual distribution, especially for large phase jitter. The p.d.f. in Figs. 10.2 are not normalized to an unity peak.

From both Figs. 10.1 and 10.2, the nonlinear phase noises of ϕ_{GM} and $\phi_{\Omega,T}$ are not symmetrical with respect to their corresponding means. While ϕ_{GM} spreads further to positive phase, $\phi_{\Omega,T}$ spreads further to negative phase. Plotted in the same scale, the nonlinear phase noise of

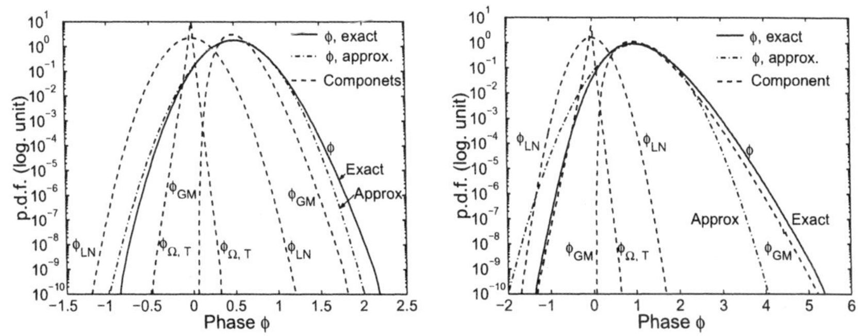

(a) $\zeta = 1$ (b) $\zeta = 2$

Figure 10.2. The distributions of soliton phase jitter for two distances of (a) $\zeta = 1$ and (b) $\zeta = 2$. [Adapted from Ho (2004e)]

ϕ_{GM} due to Gordon-Mollenauer effect is much larger than the nonlinear phase noise of $\phi_{\Omega,T}$ due to frequency and timing effect.

The p.d.f.'s in Figs. 10.1 cannot cover all possible cases. While both the Gordon-Mollenauer and linear phase noises depend on the mean nonlinear phase shift $A^2\zeta/2$ and SNR, the nonlinear phase noise induced by frequency and timing effect does not have a simple scaled relationship.

For a mean nonlinear phase shift of $\frac{1}{2}A^2\zeta = 1$ rad, Figures 10.3 plot the distribution of the overall phase jitter [Eq. (10.19)] for a SNR of 10 for $\zeta = 1, 10$. After a scale factor, the distributions of both Gordon-Mollenauer and linear phase noise are the same as that in Figs. 10.2. In additional to the overall phase jitter, Figures 10.3 also plot the distribution of the nonlinear phase noise from frequency and timing effect of $\phi_{\Omega,T}$.

For a fixed mean nonlinear phase shift and SNR, from Fig. 10.3, the nonlinear phase noise from frequency and timing effect of $\phi_{\Omega,T}(\zeta)$ has less effect to the overall phase jitter for long distance than short distance systems. Figures 10.1 are plotted for short distance of $\zeta \leq 3$ to show the contribution of frequency and timing jitter to nonlinear phase noise. The effect of $\phi_{\Omega,T}(\zeta)$ is smaller for large SNR of 20 than small SNR of 10. The main contribution to the overall phase jitter is always the Gordon-Mollenauer effect and the linear phase noise.

3. Error Probability of Soliton DPSK Signals

In previous chapter, the soliton perturbation equations of Eqs. (10.1) to (10.4) are all expressed in normalized form. Before discussing the evaluation of the error probability, the characteristic functions of Eqs. (10.24), (10.29), and (10.33) must be presented in a correct parametric format

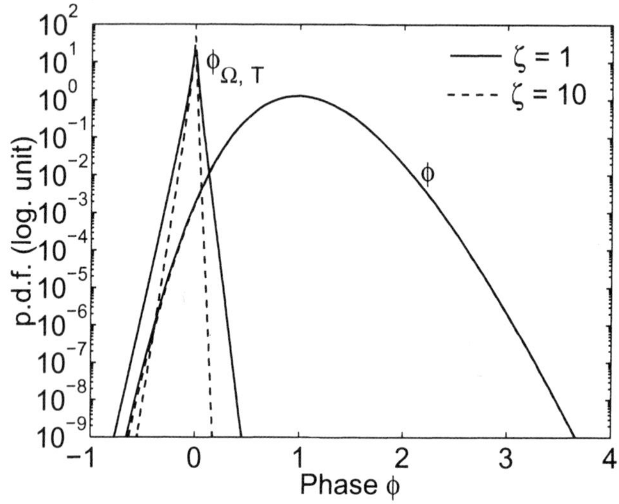

Figure 10.3. The distributions of soliton phase jitter for SNR of 10. [Adapted from Ho (2004e)]

without normalization. For a mean nonlinear phase shift of $\langle\Phi_{\rm NL}\rangle$, the overall phase jitter of Eq. (10.19) does not have the simple scale relationship like Eq. (5.50). Of course, the nonlinear phase noise from Gordon-Mollenauer effect of Eq. (10.23) still has the same scale relationship similar to Eq. (5.50). Using Eq. (10.25), the scale relationship similar to Eq. (5.50) is modified to

$$\frac{2\langle\Phi_{\rm NL}\rangle}{A^2\zeta + \frac{1}{2}\sigma_A^2\zeta^2}. \tag{10.35}$$

The term of $\sigma_A^2\zeta^2$ can be ignored in practice for system with high SNR.

The error probability is evaluated for the overall phase jitter of

$$\Phi_r = \frac{2\langle\Phi_{\rm NL}\rangle}{A^2\zeta + \frac{1}{2}\sigma_A^2\zeta^2}\left[\phi_{\rm GM}(\zeta) + \phi_{\Omega,T}(\zeta)\right] + w_\phi(\zeta) \tag{10.36}$$

for different SNR of ρ_s. There are many approaches to study the error probability of DPSK signal, depending mostly on the model of $w_\phi(\zeta)$. If the linear phase noise of $w_\phi(\zeta)$ is assumed to be Gaussian distributed from the perturbation method of previous section, the received phase has a characteristic function of

$$\Psi_{\Phi_r}(\nu) \approx \Psi_{\phi_{\mathrm{GM}}(\zeta)}\left(\frac{2\,\langle\Phi_{\mathrm{NL}}\rangle}{A^2\zeta+\frac{1}{2}\sigma_A^2\zeta^2}\nu\right) \times \Psi_{\phi_{\Omega,T}(\zeta)}\left(\frac{2 <\Phi_{\mathrm{NL}}>}{A^2\zeta+\frac{1}{2}\sigma_A^2\zeta^2}\nu\right)\Psi_{\phi_{\mathrm{LN}}(\zeta)}(\nu). \tag{10.37}$$

Based Eq. (10.37), similar to Eq. (5.77), the error probability of DPSK signal is approximately equal to

$$p_e = \frac{1}{2} - \frac{2}{\pi}\sum_{k=1}^{\infty}\frac{(-1)^k}{2k+1}|\Psi_{\Phi_r}(2k+1)|^2 \tag{10.38}$$

with $\Psi_{\Phi_r}(\cdot)$ given by Eq. (10.37). Note that the error probability of Eq. (10.38) is an exact error probability provided that the characteristic function of $\Psi_{\Phi_r}(\cdot)$ is exact expression. The characteristic function of Eq. (10.37) is an approximation for the phase jitter of the soliton.

However, as from Fig. 4.A.2, the distribution of the linear phase of $w_\phi(\zeta)$ is certainly non-Gaussain. If the nonlinear phase noise is assumed to be independent of the linear phase noise, from Chapter 5 and ignored a constant phase, the coefficient in the error probability of Eq. (10.38) is

$$\Psi_{\Phi_r}(2k+1) \approx \Psi_{\phi_{\mathrm{GM}}(\zeta)}(\nu_k)\,\Psi_{\phi_{\Omega,T}(\zeta)}(\nu_k) \times \frac{\sqrt{\pi\rho_s}}{2}e^{-\rho_s/2}\left[I_k\left(\frac{\rho_s}{2}\right)+I_{k+1}\left(\frac{\rho_s}{2}\right)\right], \tag{10.39}$$

where

$$\nu_k = \frac{2(2k+1) <\Phi_{\mathrm{NL}}>}{A^2\zeta+\frac{1}{2}\sigma_A^2\zeta^2}. \tag{10.40}$$

As shown in Chapter 5, the nonlinear phase noise and the phase of amplifier noise have very weak dependence between each other. However, the dependent model cannot directly apply to the received phase of Eq. (10.36). In the Gordon-Mollenauer effect from the first-order perturbation of Eq. (10.23), only the in-phase amplifier noise from the same quadrature as the signal contributes to nonlinear phase noise. While the model has sufficient accuracy for the independent model, the same model cannot be used to derive the joint characteristic function of the received electric field and Gordon-Mollenauer effect induced nonlinear phase noise.

When the amplifier noise from the orthogonal quadrature component of the signal is taken into account, Gordon-Mollenauer effect gives a nonlinear phase noise of

$$\phi_{\mathrm{GMD}}(\zeta) = \frac{1}{2}\int_0^{\zeta} [A + w_A(z)]^2 dz + \frac{1}{2}\int_0^{\zeta} w_I^2(z) dz, \tag{10.41}$$

where $w_I^2(z)$ is the amplifier noise in the direction orthogonal to the signal with the same statistical properties as $w_A(z)$. When the dependence between linear and nonlinear phase noise is taken into account, more accurate error performance of soliton DPSK system can be found.

By similar procedure as Chapter 5 and from Eq. (5.56), the characteristic function of real amplitude and complex noise induced nonlinear phase noise is

$$\Psi_{\phi_{\mathrm{GMD}}}(\nu) = \sec\left(\zeta\sigma_A\sqrt{j\nu}\right) \exp\left[\frac{A^2}{2\sigma_A}\sqrt{j\nu}\tan\left(\zeta\sigma_A\sqrt{j\nu}\right)\right]. \tag{10.42}$$

Compared to Eqs. (10.24) and (10.42), the impact of additional zero mean noise from orthogonal polarization is to multiply $\sec^{1/2}(\zeta\sigma_A\sqrt{j\nu})$ with the characteristic function of single polarization noise of Eq. (10.24). The mean nonlinear phase shift becomes

$$\langle\phi_{\mathrm{GMD}}(\zeta)\rangle = \frac{1}{2}A^2\zeta + \frac{1}{2}\sigma_A^2\zeta^2, \tag{10.43}$$

which is slightly larger than Eq. (10.25).

Similar to the joint statistics of Eq. (5.62), the joint characteristic function is

$$\Psi_{\Phi_{\mathrm{NLD}},\Theta_n}(\nu, m) = \frac{\Psi_{\mathrm{NLD}}(\nu)\sqrt{\pi\gamma_\nu}}{2} e^{-\gamma_\nu/2}\left[I_{\frac{m-1}{2}}\left(\frac{\gamma_\nu}{2}\right) + I_{\frac{m+1}{2}}\left(\frac{\gamma_\nu}{2}\right)\right] \tag{10.44}$$

for $m \geq 0$, where $\Psi_{\mathrm{NLD}}(\nu)$ is the product of $\Psi_{\phi_{\mathrm{GM}(\zeta)}}(\nu)$ and $\Psi_{\phi_{\Omega,T(\zeta)}}(\nu)$ and

$$\gamma_\nu = \frac{2\sigma_A\zeta\sqrt{j\nu}}{\sin(2\sigma_A\zeta\sqrt{j\nu})}\rho_s. \tag{10.45}$$

With the dependent model, the characteristic function of the received phase is equal to

$$\Psi_r(2k+1) = \Psi_{\Phi_{\mathrm{NLD}},\Theta_n}\left(\frac{2(2k+1)\langle\Phi_{\mathrm{NL}}\rangle}{A^2\zeta + \sigma_A^2\zeta^2}, 2k+1\right). \tag{10.46}$$

In the model of this chapter, all statistics properties and parameters can be fully represented by SNR, mean nonlinear phase shift, and propagation distance ζ. With fixed distance, for example $\zeta = 2$, variance of both linear and nonlinear phase noise is reduced with increased SNR.

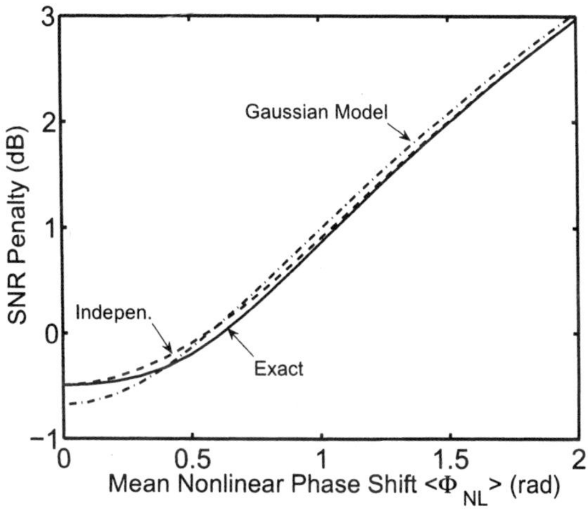

Figure 10.4. The SNR penalty of soliton DPSK signal as a function of mean nonlinear phase shift $\langle \Phi_{\mathrm{NL}} \rangle$ with $\zeta = 2$ and fixed an error probability of 10^{-9}.

The variance of nonlinear phase noise increases with mean nonlinear phase shift. The variance of linear phase noise is independent of nonlinear phase shift. The ratio of linear and nonlinear phase noise is unity in the mean nonlinear phase shift of 0.96 rad, corresponding to the 3 dB SNR difference between 0 and 1 rad.

Figure 10.4 shows the SNR penalty of soliton DPSK signal for an error probability of 10^{-9} versus mean nonlinear phase shift based on the error probability formulae derived based on the Gaussain linear phase noise, and the independent and dependent models. The SNR penalty of both independent and dependent models gives very similar amount of SNR penalty when mean nonlinear phase shift is larger than 1 rad. The independent model always underestimates the SNR penalty but the Gaussian model overestimates the SNR penalty for a mean nonlinear phase shift larger than 0.7 rad. For the Gaussian, independent, and dependent models, the mean nonlinear phase shift must be less than 0.57, 0.63, and 0.56 rad for a SNR penalty less than 1 dB, respectively. The SNR penalty is negligible for both dependent and independent models when the mean nonlinear shift is less than 0.4 rad. For 3 dB penalty due to nonlinear phase jitter that doubled SNR requirement, Figure 10.4 indicates that the mean nonlinear phase shifts are 1.01, 1.06, and 1.05 rad for the Gaussian, independent, and dependent models, respectively.

The largest SNR penalty difference between independent and dependent models is 0.21 dB for a mean nonlinear phase shift of about 0.5 rad.

For the nonlinear phase shift between 0.25 and 0.95 rad, the difference between independent and dependent models is larger than 0.1 dB. The largest SNR penalty difference between the dependent and the Gaussian models is 0.18 dB for a mean nonlinear phase shift of 1.3 rad. For nonlinear phase noise with zero mean nonlinear phase shift, the required SNR for Gaussian model is 0.37 dB smaller than that for the dependent model.

Considering the error probability of 10^{-9}, compared to the dependent model, the Gaussian model underestimates the required SNR about 0.32 dB with zero mean nonlinear phase shift and overestimates the require SNR about 0.15 dB with large mean nonlinear phase shift. The reason of underestimating the SNR at zero mean nonlinear phase shift without nonlinear phase noise is the difference between Q-factor approximation for Gaussian model and the theoretical limit from Fig. 5.15.

The SNR requirement for dependent model to achieve error performance of 10^{-9}, according to Fig. 10.4 with mean nonlinear phase shifts of 0, 0.5, 1, and 1.5 rad are 13.0, 13.8, 15.8, and 18.0 dB, respectively. For system performance requirements with an error probability down to 10^{-12}, the SNR requirements for corresponding mean nonlinear phase shifts are 14.4, 15.2, 17.2, and 19.5 dB. To improve error performance for 3 orders from 10^{-9} to 10^{-12}, 1.5 dB larger SNR is required.

As shown in previous section, the time-frequency effect is very small compared with the amplitude jitter induced phase jitter. Further numerical results show that the SNR penalty of the system with time-frequency effect is slightly larger than the system without time-frequency effect of about 0.01 dB for high error probability (10^{-4}) region and no observable difference for low error probability at the region of 10^{-9} to 10^{-12}.

4. Further Remarks and Summary

The phase jitter of Eq. (10.19) is derived based on the first-order perturbation theory of Eqs. (10.1) to (10.4). Followed Ho (2004e), the non-Gaussian distribution is induced by the higher-order terms of Eq. (10.19) or the nonlinear terms of Eq. (10.4). Second- and higher-order soliton perturbation (Haus et al., 1997, Kaup, 1991) may give further non-Gaussian characteristic to the phase jitter. Currently, there is no comparison between contributions of the higher-order terms of Eq. (10.4) and higher-order soliton perturbation.

Like almost all other literatures (Blow et al., 1992, Georges, 1995, Gordon and Haus, 1986, Gordon and Mollenauer, 1990, Iannone et al., 1998, Kaup, 1990, Kivshar and Malomed, 1989, Leclerc and Desurvire, 1998, McKinstrie and Xie, 2002), the impact of amplitude jitter to the noise variances of σ_A^2, σ_Ω^2, σ_T^2, and σ_ϕ^2 is ignored. The noise variances of σ_A^2,

σ_Ω^2, σ_T^2, and σ_ϕ^2 are assumed independent of distance. If the amplitude noise variance is $\sigma_A^2 = A(\zeta)\sigma_n^2$ with dependence on the instantaneous amplitude jitter, amplitude, frequency, and timing jitters are all non-Gaussian distributed (Ho, 2003c). As an example, amplitude jitter is noncentral χ^2 distributed (Ho, 2003c, Moore et al., 2003). However, the statistics of phase jitter [Eq. (10.19)] does not have a simple analytical solution when the noise variance depends on amplitude jitter. With a high SNR, the amplitude jitter is alway smaller than the amplitude $A(0) = A$. Even in high SNR, the phase jitter is non-Gaussian based on Eq. (10.19).

The statistics of the phase jitter of fiber soliton are analyzed based on the first-order perturbation theory. The time-frequency effect is assumed to be independent of other contributions to the nonlinear phase noise. The dominant effect for the overall nonlinear phase noise is the Gordon-Mollenauer effect that is induced by the interaction of amplifier noise and fiber Kerr effects. In the perturbation theory, the amplifier noise is mapped to the phase jitter.

In most experiments (Cai et al., 2004, Gnauck et al., 2002, Rasmussen et al., 2003, Xu et al., 2004, Zhu et al., 2004a), the DPSK scheme can achieve both high transmission distance and spectral efficiency. Soliton is a good approximation to practical systems using RZ pulses. The optical filter before the receiver is assumed an optical matched filter and an electrical filter after the photodiodes does not further distort the signal.

According to different assumptions, the error performance analysis here includes three different models, which are Gaussain, independent, and dependent model. The Gaussian model in which the linear additive Gaussian noise is projected to the phase of soliton pulse according to first-order perturbation depends solely on the variance of σ_ϕ^2 that also depends solely on SNR. The independent model assumes that additive amplifier noises are independent to nonlinear phase noise.

The dependent model considers the dependence between the amplifier noise induced phase noise and the Gordon-Mollenauer nonlinear phase noise. The two quadrature components of the amplifier noise not only induces the linear phase noise but also contributes to nonlinear phase noise. The dependent model gives the most accurate results by taking into account the dependence between linear and nonlinear phase noise. More important, other models may underestimate the error probability and contradict to the principle of conservative system design.

The error probability of DPSK soliton here ignores the effect of timing jitter. Timing jitter may further degrade the performance of DPSK soliton transmission systems.

Borrowing the idea from timing jitter control using optical filter (Kodama and Hasegawa, 1992, Mecozzi et al., 1991), phase jitter can also be controlled by optical filtering (Boivin et al., 2004, Derevyanko and Turitsyn, 2004, Hanna et al., 1999, 2004).

In summary, the phase statistics of the fiber soliton are derived analytically here based on the first-order perturbation theory. In additional to the well-known Gordon-Mollenauer effect, the nonlinear phase jitter due to the interaction of timing and frequency jitter is also taken into account. The time-frequency effect contributes very little to the error performance of the DPSK soliton signal. The main contribution to the overall phase jitter is always the Gordon-Mollenauer effect and the linear phase noise. For a fixed mean nonlinear phase shift, the contribution of nonlinear phase noise from timing and frequency jitter decreases with pulse propagation distance and SNR.

Based on different assumption, the error performance of DPSK soliton signal is analyzed using Gaussian, independent, and dependent models. The dependent model provides the most accurate results by taking into account the dependency between amplifier noise induced linear phase noise and nonlinear phase jitter. When the linear phase jitter is assumed to be Gaussian distributed, the Gaussian model underestimates the SNR penalty up to 0.32 dB for system without nonlinear phase noise. For mean nonlinear phase shift larger than 0.7 rad, the Gaussian model overestimates the SNR penalty. In all values of mean nonlinear phase shift, the independent model underestimates the error probability compared to the dependent model. For high nonlinear phase shift, the independent and dependent models are almost the same. Both independent and dependent models have the same results for small mean nonlinear phase shift.

Depending mainly on SNR and the mean nonlinear phase shift, the error performance of soliton DPSK signals is insensitive to the pulse propagation distance. The time-frequency effect always contributes little degradation to the performance of the soliton DPSK signals. The distance and the time-frequency effect also contributes little to the p.d.f. of overall phase jitter.

APPENDIX 10.A: Some Deviations

Here, we find the joint characteristic function of

$$\varphi_1 = \int_0^\zeta w_\Omega^2(\zeta_1)\mathrm{d}\zeta_1, \tag{10.A.1}$$

$$\varphi_2 = \int_0^\zeta w_T(\zeta_1)\mathrm{d}w_\Omega(\zeta_1), \tag{10.A.2}$$

$$\varphi_3 = w_\Omega(\zeta)\int_0^\zeta w_\Omega(\zeta_1)\mathrm{d}\zeta_1, \tag{10.A.3}$$

where $w_T(\zeta)$ and $w_\Omega(\zeta)$ are two independent Wiener process for the timing and frequency jitters, respectively, in first-order soliton perturbation.

By changing the integration order, we obtain

$$\varphi_2 = \int_0^\zeta \int_0^{\zeta_1} \mathrm{d}w_T(\zeta_2)\mathrm{d}w_\Omega(\zeta_1) = \int_0^\zeta \left[w_\Omega(\zeta) - w_\Omega(\zeta_2)\right]\mathrm{d}w_T(\zeta_2). \tag{10.A.4}$$

The joint characteristic function of φ_1, φ_2, and φ_3 is

$$\Psi_{\varphi_1,\varphi_2,\varphi_3}(\nu_1,\nu_2,\nu_3) = E\left\{\exp(j\nu_1\varphi_1 + j\nu_2\varphi_2 + j\nu_3\varphi_3)\right\}. \tag{10.A.5}$$

Similar to option pricing with stochastic volatility (Stein and Stein, 1991), the expectation of Eq. (10.A.5) can be evaluated in two steps, first over w_T and than w_Ω. In the average over w_T, it is obvious that φ_2 is a zero-mean Gaussian random variable with a variance of $\sigma_T^2 \int_0^\zeta \left[w_\Omega(\zeta) - w_\Omega(\zeta_1)\right]^2 \mathrm{d}\zeta_1$, we get

$$\begin{aligned}\Psi_{\varphi_1,\varphi_2,\varphi_3}(\nu_1,\nu_2,\nu_3) &= E\Bigg\{-\frac{\sigma_T^2\nu_2^2}{2}\int_0^\zeta \left[w_\Omega(\zeta) - w_\Omega(\zeta_1)\right]^2\mathrm{d}\zeta_1 \\ &\qquad + j\nu_1\int_0^\zeta w_\Omega^2(\zeta_1)\mathrm{d}\zeta_1 + j\nu_3 w_\Omega(\zeta)\int_0^\zeta w_\Omega(\zeta_1)\mathrm{d}\zeta_1\Bigg\} \\ &= E\Bigg\{-\frac{\sigma_T^2\nu_2^2\zeta}{2}w_\Omega^2(\zeta) + (j\nu_3 + \sigma_T^2\nu_2^2)w_\Omega(\zeta)\int_0^\zeta w_\Omega(\zeta_1)\mathrm{d}\zeta_1 \\ &\qquad + \left(j\nu_1 - \frac{\sigma_T^2\nu_2^2}{2}\right)\int_0^\zeta w_\Omega^2(\zeta_1)\mathrm{d}\zeta_1\Bigg\}. \end{aligned} \tag{10.A.6}$$

First of all, we have (Cameron and Martin, 1945, Ho, 2003g, Mecozzi, 1994a,b)

$$\begin{aligned} &E\left\{j\omega_1 w_\Omega(\zeta) + j\omega_2\int_0^\zeta w_\Omega(\zeta_1)\mathrm{d}\zeta_1 + \frac{j\omega_3}{2}\int_0^\zeta w_\Omega^2(\zeta_1)\mathrm{d}\zeta_1\right\} \\ &= \sec^{\frac{1}{2}}\left(\sqrt{j\omega_3}\sigma_\Omega\zeta\right)\exp\Bigg\{-\frac{1}{2}\left(\omega_1^2\sigma_\Omega^2 + \frac{\omega_2^2}{j\omega_3}\right)\frac{\tan\left(\sqrt{j\omega_3}\sigma_\Omega\zeta\right)}{\sqrt{j\omega_3}\sigma_\Omega} \\ &\qquad + j\frac{\omega_1\omega_2}{\omega_3}\left[\sec\left(\sqrt{j\omega_3}\sigma_\Omega\zeta\right) - 1\right] - j\frac{\omega_2^2\zeta}{2\omega_3}\Bigg\} \\ &= \sec^{\frac{1}{2}}\left(\sqrt{j\omega_3}\sigma_\Omega\zeta\right)\exp\left[-\frac{1}{2}\boldsymbol{\omega}_{1,2}^T\mathcal{C}(j\omega_3)\boldsymbol{\omega}_{1,2}\right], \end{aligned} \tag{10.A.7}$$

where $\boldsymbol{\omega}_{1,2} = (\omega_1, \omega_2)^T$ and

$$\mathcal{C}(j\omega_3) = \begin{bmatrix} \dfrac{\sigma_\Omega \tan\left(\sqrt{j\omega_3}\sigma_\Omega\zeta\right)}{\sqrt{j\omega_3}} & \dfrac{1}{j\omega_3}\left[\sec\left(\sqrt{j\omega_3}\sigma_\Omega\zeta\right) - 1\right] \\ \dfrac{1}{j\omega_3}\left[\sec\left(\sqrt{j\omega_3}\sigma_\Omega\zeta\right) - 1\right] & \dfrac{1}{j\omega_3}\left[\dfrac{\tan\left(\sqrt{j\omega_3}\sigma_\Omega\zeta\right)}{\sqrt{j\omega_3}\sigma_\Omega} - \zeta\right] \end{bmatrix}. \quad (10.A.8)$$

As a verification, if ω_3 approaches zero, the covariance matrix is

$$\lim_{\omega_3 \to 0} C(j\omega_3) = \sigma_\Omega^2 \begin{bmatrix} \zeta & \frac{1}{2}\zeta^2 \\ \frac{1}{2}\zeta^2 & \frac{1}{3}\zeta^3 \end{bmatrix}, \quad (10.A.9)$$

for the vector of

$$\mathbf{w}_\zeta = \left(w_\Omega(\zeta), \int_0^\zeta w_\Omega(\zeta_1)\mathrm{d}\zeta_1\right)^T \quad (10.A.10)$$

without any dependence on the random variable φ_1. Note that the equation corresponding to Eq. (10.A.7) in Mecozzi (1994a,b) does not have the limit of Eq. (10.A.9).

The characteristic function of Eq. (10.A.7) is that of a correlated two-dimensional Gaussian random variable of $\mathbf{w}_\zeta$ with dependence to φ_1. The first two terms of Eq. (10.A.6) is a quadratic (or bilinear) function of $\mathbf{w}_\zeta$, i.e., $\frac{1}{2}\mathbf{w}_\zeta^T \mathcal{M}(j\nu_2, j\nu_3)\mathbf{w}_\zeta$, where

$$\mathcal{M}(j\nu_2, j\nu_3) = \begin{bmatrix} -\sigma_T^2\nu_2^2\zeta & j\nu_3 + \sigma_T^2\nu_2^2 \\ j\nu_3 + \sigma_T^2\nu_2^2 & 0 \end{bmatrix}. \quad (10.A.11)$$

The characteristic function of the quadratic function of zero-mean Gaussian random variables is $\det[\mathcal{I} - \mathcal{C}\mathcal{M}]^{-\frac{1}{2}}$ [similar to Eqs. (5.19) and (5.21)], where $\det[\cdot]$ denotes the determinant of a matrix.

The joint characteristic function is

$$\Psi_{\varphi_1 \cdot \varphi_2 \cdot \varphi_3}(\nu_1, \nu_2, \nu_3) = \frac{\sec^{\frac{1}{2}}\left(\sqrt{2j\nu_1 - \sigma_T^2\nu_2^2}\sigma_\Omega\zeta\right)}{\det\left[\mathcal{I} - \mathcal{C}\left(2j\nu_1 - \sigma_T^2\nu_2^2\right)\mathcal{M}(j\nu_2, j\nu_3)\right]^{\frac{1}{2}}}. \quad (10.A.12)$$

where $\mathcal{I}$ is the identity matrix. The substitute of $j\omega_3$ by $2j\nu_1 - \sigma_T^2\nu_2^2$ is obvious by comparing Eqs. (10.A.6) and (10.A.7).

We can get (Cameron and Martin, 1945, Ho, 2003a)

$$\Psi_{\varphi_1}(\nu_1) = \sec^{\frac{1}{2}}\left(\sqrt{2j\nu_1}\sigma_\Omega\zeta\right) \quad (10.A.13)$$

and (Stein and Stein, 1991)

$$\Psi_{\varphi_2}(\nu_2) = \mathrm{sech}^{\frac{1}{2}}\left(\sigma_T\sigma_\Omega\zeta\nu_2\right), \quad (10.A.14)$$

respectively. The statistical properties of

$$\int_0^\zeta w_T(\zeta_1)\mathrm{d}w_\Omega(\zeta_1), \qquad \int_0^\zeta w_\Omega(\zeta_1)\mathrm{d}w_T(\zeta_1), \quad (10.A.15)$$

$$\int_0^\zeta \left[w_\Omega(\zeta) - w_\Omega(\zeta_1)\right]\mathrm{d}w_T(\zeta_1), \qquad \int_0^\zeta \left[w_T(\zeta) - w_T(\zeta_1)\right]\mathrm{d}w_\Omega(\zeta_1) \quad (10.A.16)$$

are the same. We can also obtain

$$\Psi_{\varphi_3}(\nu_3) = \left[1 - j\nu_3\sigma_\Omega^2\zeta^2 + \frac{1}{12}\nu_3^2\sigma_\Omega^4\zeta^4\right]^{-\frac{1}{2}}. \quad (10.A.17)$$

While both random variables φ_1 and φ_2 determine by $\sigma_\Omega\zeta$, the random variable of φ_2 determines by $\sigma_T\sigma_\Omega\zeta$.

Chapter 11

CAPACITY OF OPTICAL CHANNELS

The overall data rate of a wavelength-division-multiplexed (WDM) system can be increased by many methods. When a wider optical bandwidth is used, more channels can be transmitted to increase the overall system throughput. There are many research activities to open up further optical amplifier bands to increase the system throughput (Bigo, 2004, Bromage, 2004, Islam, 2002, Ono et al., 2003), mostly using Raman amplifier to provide optical gain outside the Erbium-doped fiber amplifier (EDFA) gain bandwidth.

The overall data rate also increases linearly with the spectral efficiency. The usage of a wider optical bandwidth typical requires new optical amplifier technologies and further optical components, so raising spectral efficiency is often the more practical and economical alternatives. From Chapter 9, phase-modulated optical communications enable efficient increase of spectral efficiency without a large degradation on receiver sensitivity. When the optical signal is contaminated by noise, the important issue is the ultimate limits of the spectral efficiency that are determined by the information-theoretic capacity per unit bandwidth (Cover and Thomas, 1991, Shannon, 1948, Yeung, 2002). With the allowance of high complexity and long delay, those limits can closely be approached using Turbo or low-density parity check codes (Berrou, 2003, Berrou et al., 1993, Chung et al., 2001). Recently, those advance error-correction codes are implemented for high-speed optical communications (Mizuochi et al., 2004) after the usage was proposed for sometime (Ait Sab and Lemaire, 2001, Bosco et al., 2003, Cai et al., 2003a, Vasic and Djordjevic, 2002).

In this chapter, we calculate the spectral efficiency limits, considering various system design issues, like unconstrained and constant-intensity modulation with coherent or direct detection, and in either linear or nonlinear propagation regime. In most of the cases, optical amplifier noises are assumed to be the dominant noise source. Coherent detection allows information to be encoded in two degrees of freedom per polarization, and its spectral efficiency limits are several b/s/Hz in typical terrestrial systems, even considering nonlinear effects. Using constant-intensity modulation or direct detection, only one degree of freedom per polarization can be exploited, reducing spectral efficiency. Using binary modulation, regardless of detection technique, spectral efficiency cannot exceed 1 b/s/Hz per polarization.

When the number of signal and/or noise photons is small, the channel capacity of optical communication systems is also limited by the particle nature of photons. Coherent communication is equivalent to detecting the real and imaginary parts of the coherent states. Direct detection is equivalent to counting the number of photons in the number states. In the coherent states, if both signal and noise are expressed in terms of photon number, quantum effects add one photon to the noise variance, usually providing a channel capacity slightly smaller than the classical limit. The quantum limit of direct detection is determined by photon statistics and also yields a slightly smaller channel capacity than the classical limit.

While the signal-to-noise ratio (SNR) of a fiber link is proportional to the launched power, fiber nonlinearities induce spurious tones via four-wave-mixing and multiplicative noises via both self- and cross-phase modulation. Fiber nonlinearities certainly also limits the spectral efficiencies. This chapter also reviews the studies on the impact of fiber nonlinearities on the spectral efficiency of lightwave communications.

1. Optical Channel with Coherent Detection

The channel capacity, or the maximum spectral efficiency limit, of a discrete-time channel with X and Y as input and output, respectively, is equal to the maximum mutual information between input and output of

$$C = \max_{p(x)} E\left\{\log \frac{p(y|x)}{p(y)}\right\}, \tag{11.1}$$

where $p(x)$ and $p(y)$ are the probability density function (p.d.f.) of the input of X and output Y, respectively, and $p(y|x)$ is the conditional p.d.f. of the output given the input. The channel capacity can be rewritten as

$$C = \max_{p(x)} \{H(Y) - H(Y|X)\}, \tag{11.2}$$

where the entropy of the output of $H(Y)$ and the conditional entropy of $H(Y|X)$ are

$$H(Y) = -\int p(y)\log p(y)\mathrm{d}y, \tag{11.3}$$

$$H(Y|X) = -\int\int p(x)p(y|x)\log p(y|x)\mathrm{d}x\mathrm{d}y. \tag{11.4}$$

Intuitively, in special case when $H(Y|X)$ is a constant, the channel capacity can be found by using an output density of $p(y)$ to maximize $H(Y)$. However, in general, when the output p.d.f. of $p(y)$ was given to maximize the output entropy of $H(Y)$, an input p.d.f. of $p(x)$ cannot be found with the condition of $p(y) = \int p(y|x)p(x)\mathrm{d}x$. Later in this chapter, the channel capacities of some channels are derived by this special method. Using $\log(\cdot)$ instead of $\log_2(\cdot)$ to calculate entropy, the capacity of Eq. (11.2) has a unit of nat/s/Hz that is 1.44 times less than b/s/Hz.

In additional to Eq. (11.2), the input signal has the constraint of

$$\int g(x)p(x)\mathrm{d}x = A, \tag{11.5}$$

where $g(x) = x$ and $g(x) = x^2$ [or $g(x) = ||x||^2$ for multi-dimensional input] are the most common mean and power constraint. As a p.d.f., we also have the probability constraint of

$$\int p(x)\mathrm{d}x = 1 \text{ and } \int p(y)\mathrm{d}y = 1. \tag{11.6}$$

1.1 Kuhn-Tucker Condition

With constraints of Eqs. (11.5) and (11.6), the optimal problem to find the channel capacity of Eq. (11.2) does not have a simple analytical solution in most cases. Variational principle may be used to derive the Kuhn-Tucker condition for optimality.

Using Lagrange multipliers of λ and μ, the cost function of Eq. (11.2) becomes

$$C_{\mathrm{Lag}} = \int p(x)h_{Y|X}(x)\mathrm{d}x - \int p(y)\log p(y)\mathrm{d}y + \lambda\int g(x)p(x)\mathrm{d}x + \mu\int p(x)\mathrm{d}x, \tag{11.7}$$

where

$$h_{Y|X}(x) = -\int p(y|x)\log p(y|x)\mathrm{d}y \tag{11.8}$$

is a given function determined by the conditional p.d.f. of $p(y|x)$. Using variational method, with $p(x) \to p(x) + \delta_x(x)$ and $p(y) \to p(y) + \delta_y(y)$, where both $\delta_x(x)$ and $\delta_y(y)$ are both very small perturbation with the relationship of $\delta_y(y) = \int p(y|x)\delta_x(x)\mathrm{d}x$, we obtain

$$\begin{aligned}\delta C_{\text{Lag}} = \int h_{Y|X}(x)\delta_x(x)\mathrm{d}x - \int [\log p(y) + 1]\delta_y(y)\mathrm{d}y \\ + \lambda \int g(x)\delta_x(x)\mathrm{d}x + \mu \int \delta_x(x)\mathrm{d}x. \\ = \int \left\{ h_{Y|X}(x) + \lambda g(x) + \mu - \int [\log p(y) + 1] p(y|x)\mathrm{d}y \right\} \delta_x(x)\mathrm{d}x. \quad (11.9)\end{aligned}$$

If $\delta_x(x)$ is a continuous function, we need to solve the integral equation of $\int p(y|x) \log p(y)\mathrm{d}y = h_{Y|X}(x) + \lambda g(x) + \mu - 1$. However, $\delta_x(x)$ is not necessary a continuous function but may be a discrete function, we obtain the Kuhn-Tucker condition of

$$\int p(y|x) \log p(y)\mathrm{d}y + h_{Y|X}(x) + \lambda [g(x) - A] + C \geq 0, \quad (11.10)$$

with $\lambda > 0$, where the equal sign is satisfied at the locations when $p(x) \neq 0$ [or $\delta_x(x) \neq 0$]. When the optimal $p(x)$ is multiplied to Eq. (11.10) and integrates over the whole region, we obtain $C = H(Y) - H(Y|X)$ for the optimal $p(x)$, conform to the definition of Eq. (11.2).

1.2 Unconstrained Channel

For an optical channel with only amplifier noise and power constrained on the input signal of X, the output of the channel is given by

$$Y = X + N, \quad (11.11)$$

where N is two-dimensional Gaussian distributed noise. For zero mean X and Y, the variance of Y is the summation of $\sigma_y^2 = \sigma_x^2 + 2\sigma_n^2$, where σ_n^2 is the noise variance per dimension and σ_x^2 and σ_y^2 are the variance or power of the input and output, respectively, where the input constraint of Eq. (11.5) is $\int ||x||^2 p(x)\mathrm{d}x = \sigma_x^2$. The unconstrained channel of Eq. (11.11) requires the usage of coherent detection to recover the two-dimensional component of Y. The conditional p.d.f. of the channel is

$$p(y|x) = \frac{1}{2\pi\sigma_n^2} \exp\left[-\frac{||x - y||^2}{2\sigma_n^2} \right], \quad (11.12)$$

with x and y as two-dimensional vectors. The conditional entropy of $H(Y|X) = h_{Y|X}(x) = \log\left(2\pi\sigma_n^2\right) + 1$ is independent of the channel input.

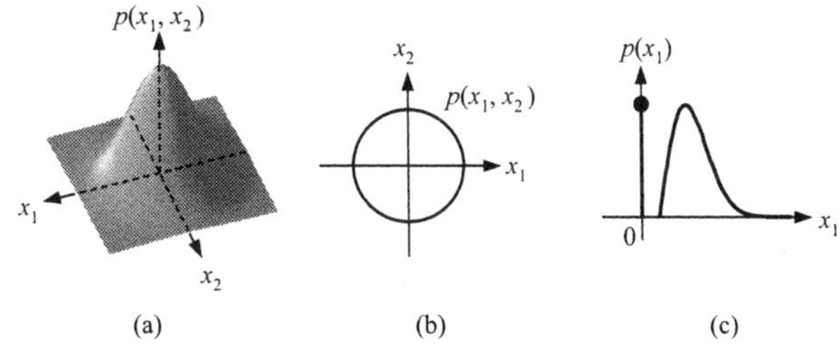

Figure 11.1. The p.d.f. of the input signal to maximize the spectral efficiency for (a) unconstrained, (b) constant-intensity, (c) intensity-modulation/direct-detection (IMDD) signal. [Adapted from Kahn and Ho (2004)]

The output density that maximizes $H(Y)$ of Eq. (11.3) is found to be zero-mean two-dimensional Gaussian distribution with overall variance of σ_y^2 and $H(Y) = \log(\pi\sigma_y^2) + 1$. The Gaussian distribution of the input signal is shown in Fig. 11.1(a).

With only power constraint, the channel capacity is

$$C = \log\left(1 + \rho_s\right), \tag{11.13}$$

where $\rho_s = \sigma_x^2/2\sigma_n^2$ is the SNR of the channel.

The unconstrained spectral efficiency limit of Eq. (11.13) was derived by Shannon (1948) and can be found in most textbooks on information theory (Cover and Thomas, 1991, Yeung, 2002).

For continuous $p(x)$, because $\int ||y||^2 p(y|x)\mathrm{d}y = ||x||^2 + 2\sigma_n^2$, for $p(y) = e^{-||y||^2/2\sigma_y^2}/2\pi\sigma_y^2$, the Kuhn-Tucker condition becomes

$$\int p(y|x)\log p(y)\mathrm{d}y = -\frac{||x||^2}{2\sigma_y^2} - \frac{\sigma_n^2}{\sigma_y^2} - \log(2\pi\sigma_y^2). \tag{11.14}$$

The Kuhn-Tucker condition is conformed by the capacity of Eq. (11.13) with a Lagrange multiplier of $\lambda = 1/2\sigma_y^2$.

1.3 Constant-Intensity Modulation

The input of X in the optical channel of Eq. (11.11) may be a constant-intensity signal similar to a phase- or frequency-modulated signal. In wireless communications, constant-intensity signal is used such that non-linear amplifiers can be used in the transmitter. When constant-intensity signal is used in optical fiber, both self- and cross-phase modulation gives a constant phase shift to the channel itself or all other WDM channels. Constant-intensity signal may increase the spectral efficiency of an optical signal if self- or cross-phase modulation is the dominant impairment. However, constant-intensity modulation cannot solve the problem of nonlinear phase noise of Chapter 5. If self-phase modulation induced nonlinear phase noise is the dominant impairment, constant-intensity modulation should not be used.

For constant intensity signal, the input of X should uniformly distributed as a circle with a radius of A as shown in Fig. 11.1(b). The SNR of the channel is $\rho_s = A^2/2\sigma_n^2$. While $H(Y|X)$ is the same as that of the unconstrained channel of Eq. (11.11), the output entropy of $H(Y)$ must calculate differently. The two-dimensional output density is equal to

$$p(y) = \frac{1}{4\pi^2\sigma_n^2}\int_{-\pi}^{+\pi}\exp\left[-\frac{(y_1 - A\cos\theta)^2 + (y_2 - A\sin\theta)^2}{2\sigma_n^2}\right]\mathrm{d}\theta, \tag{11.15}$$

or

$$p(y) = \frac{1}{2\pi\sigma_n^2}\exp\left[-\frac{y_1^2 + y_2^2 + A^2}{2\sigma_n^2}\right] I_0\left(\frac{\sqrt{y_1^2 + y_2^2}A}{\sigma_n^2}\right). \tag{11.16}$$

The channel capacity is equal to

$$\begin{aligned} C &= -\int_{-\infty}^{+\infty}\int_{-\infty}^{+\infty} p(y)\log p(y)\mathrm{d}y_1\mathrm{d}y_2 - \log\left(2e\pi\sigma_n^2\right) \\ &= -2\pi\int_0^{+\infty} rf(r)\log f(r)\mathrm{d}r - \log\left(2e\pi\sigma_n^2\right) \end{aligned} \tag{11.17}$$

where

$$f(r) = \frac{1}{2\pi\sigma_n^2}\exp\left[-\frac{r^2 + A^2}{2\sigma_n^2}\right] I_0\left(\frac{rA}{\sigma_n^2}\right). \tag{11.18}$$

is the substitute of $r = \sqrt{y_1^2 + y_2^2}$ to $p(y)$ of Eq. (11.16). The integration of Eq. (11.17) has changed from rectangular to polar coordinate and finds that the p.d.f. of $p(y)$ is independent of angle. With large SNR of ρ_s, the spectral limit of Eq. (11.17) can be simplified using the asymptotic expression of

$$I_0(x) \sim \frac{e^x}{\sqrt{2\pi x}} \tag{11.19}$$

and

$$f(r) \sim \frac{1}{(2\pi)^{3/2}A\sigma}\exp\left[-\frac{(r - A)^2}{2\sigma_n^2}\right]. \tag{11.20}$$

In Eq. (11.20), outside the exponential, we also approximate r with A. The asymptotic result for $-2\pi\int_0^{+\infty} rf(r)\log f(r)\mathrm{d}r$ is

$$\log\left[(2\pi)^{3/2}A\sigma\right] + \frac{1}{2}. \tag{11.21}$$

The asymptotic spectral efficiency under a constant-intensity constraint is

$$C \sim \log \frac{A}{\sigma_n} + \frac{1}{2} \log \frac{2\pi}{e} = \frac{1}{2} \log \rho_s + 1.10 \log 2. \tag{11.22}$$

The asymptotic limit of Eq. (11.22) is half of the unconstrained limit of Eq. (11.13) plus 1.10 b/s/Hz. While the unconstrained signal can use two dimensions, the constant-intensity modulation can only use a single dimension that gives the factor of 1/2 in Eq. (11.22). The factor of 1.10 may come from the usage of a circle instead of just only the x-axis of the signal.

In general, constant-intensity signal requires coherent detection or interferometric detection. The constant-intensity spectral efficiency of Eq. (11.22) was first derived by Geist (1990) and re-derived independently by Aldis and Burr (1993) and Ho and Kahn (2002). The derivation here is mostly based on Ho and Kahn (2002) and Kahn and Ho (2004).

2. Intensity-Modulation/Direct-Detection Channel

While this book focuses on phase-modulated optical communications, the IMDD channel is very popular as a low-cost solution. The previous section derives the capacity for system with coherent detection, for comparison propose, we derive the capacity of IMDD system in this section.

In IMDD channel limited by amplifier noise, the discrete-time model of the channel is

$$Y = |X + N|^2, \tag{11.23}$$

where the additive noise is the same as that for unconstrained channel of Eq. (11.11). We assume that the random variables of X and N are all complex numbers but Y is positive real random variable. The mean of the output is $m_y = \sigma_x^2 + 2\sigma_n^2$. The conditional p.d.f. of $p(y|x)$ is a noncentral chi-square (χ^2) distribution with two degrees of freedom of

$$p(y|x) = \frac{1}{2\sigma_n^2} \exp\left[-\frac{y + x_1^2 + x_2^2}{2\sigma_n^2}\right] I_0\left(\frac{\sqrt{y(x_1^2 + x_2^2)}}{\sigma_n^2}\right) \tag{11.24}$$

with noncentrality parameter of $x_1^2 + x_2^2$ from the two components of the two-dimensional input of X.

2.1 Some Approximated Results

The channel capacity of the IMDD channel of Eq. (11.23) is difficult to derive analytically. Some approximated capacities for IMDD channel are given here.

One-Dimensional Gaussian Channel Approximation

If the channel of Eq. (11.23) is rewritten as $Y = |X|^2 + X \cdot N^* + X^* \cdot N + |N|^2$, in high SNR, the signal is of $|X|^2$ is a χ^2 random variable with two degrees of freedom with a variance of $2\sigma_x^4$. Ignored the small noise of $|N|^2$, the noise of $X^* \cdot N + X \cdot N^*$ has a variance of $4\sigma_x^2\sigma_n^2$. With a SNR of $\sigma_x^2/2\sigma_n^2$ as a one-dimensional Gaussian channel, the channel capacity is approximately equal to

$$C \sim \frac{1}{2} \log \rho_s, \tag{11.25}$$

where ρ_s is the same as that for Eq. (11.13). Compared with the Shannon limit of Eq. (11.13), in this approximation, the capacity for IMDD channel is approximately half of the Shannon limit.

The channel capacity of Eq. (11.25) is first derived by Desurvire (2000) and also used in Wegener et al. (2004). The approximated capacity of Eq. (11.25) is very simple. Here, we assume that X is Gaussian distribution but Desurvire (2000) does not assume an input distribution of X. Note that $|X|^2 > 0$ as a positive random variable. Gaussian distribution with a variance of $2\sigma_x^2$ may have negative value, giving larger entropy than $|X|^2$. This approximation gives larger channel capacity than the exact results.

Maximum Output Entropy

For an output of $Y > 0$ with a mean constraint of m_y, the output entropy of $H(Y)$ is maximized by the exponential distribution of

$$p(y) = \frac{1}{m_y} \exp\left(-\frac{y}{m_y}\right), \quad y > 0, \tag{11.26}$$

with $H(Y) = \log m_y + 1$. In order to obtain the exponential distribution of $p(y)$ in Eq. (11.26), $X + N$ may have a two-dimensional Gaussian distribution with overall variance of m_y. Because the noise of N is a two-dimensional Gaussian distribution with per dimension variance of σ_n^2, the input X may also have a two-dimensional Gaussian distribution with an overall variance of $\sigma_x^2 = m_y - 2\sigma_n^2$.

If the input of X is two-dimensional Gaussian distributed with overall variance of σ_x^2, its intensity of $x_1^2 + x_2^2$ has the exponential distribution of $p(y_i) = \exp(-y_i/\sigma_x^2)/\sigma_x^2$. The channel capacity is

$$C = \log m_y + 1 + \int_0^{+\infty} p(y_i)\mathrm{d}y_i \int_0^{+\infty} g(y, y_i)\log g(y, y_i)\mathrm{d}y, \quad (11.27)$$

where

$$g(y, y_i) = \frac{1}{2\sigma_n^2}\exp\left[-\frac{y + y_i}{2\sigma_n^2}\right] I_0\left(\frac{\sqrt{y y_i}}{\sigma_n^2}\right). \quad (11.28)$$

Using the asymptotic expression of Eq. (11.19) for large SNR, we obtain

$$g(y, y_i) \sim \frac{1}{2\sqrt{2\pi\sigma_n^2 y_i}}\exp\left[-\frac{(\sqrt{y} - \sqrt{y_i})^2}{2\sigma_n^2}\right], \quad (11.29)$$

and

$$-\int_0^{+\infty} g(y, y_i)\log g(y, y_i)\mathrm{d}y \sim \frac{1}{2}\log\left(8\pi e\sigma_n^2 y_i\right) \quad (11.30)$$

for a one-dimensional Gaussian p.d.f. with variance of $4\sigma_n^2 y_i$ that is approximately equal to the variance of the noncentral χ^2 distribution of Eq. (11.24) or (11.28) with $y_i = x_1^2 + x_2^2$. In the integration of Eq. (11.30), we use the approximation of $y \sim y_i$ for large SNR. We obtain

$$H(Y|X) \sim \frac{1}{2}\log\left(8\pi e\sigma_n^2\right) + \frac{1}{2}\log\sigma_x^2 - \frac{1}{2}\gamma_e, \quad (11.31)$$

where $\gamma_e = 0.577$ is the Euler gamma constant. We obtain

$$\begin{aligned} C &\sim \frac{1}{2}\log\frac{\sigma_x^2}{2\sigma_n^2} + \frac{1}{2}(1+\gamma_e) - \frac{1}{2}\log 4\pi \\ &= \frac{1}{2}\log\rho_s - 0.688\log 2. \end{aligned} \quad (11.32)$$

The channel capacity of Eq. (11.32) was derived in Hall (1994), Hall and O'Routke (1993), and Kahn and Ho (2004) as an approximation. Because $h_{Y|X}(x)$ is not a constant for IMDD channel, the p.d.f. that maximizes the output entropy does not necessary also give the channel capacity.

The channel model of Eq. (11.23) assumes that a polarizer precede the receiver to filter out the noise from the polarization orthogonal to the signal. Without the polarizer, the output signal is equal to

$$Y = |X + N_1|^2 + |N_2|^2, \quad (11.33)$$

where N_1 and N_2 are two independent two-dimensional Gaussian distributed random variables from both polarizations. Unlike the model of Eq. (11.23), we are not able to find an input density of X for the model of Eq. (11.33) to give the output density of Eq. (11.26) for Y. We assume a Gaussian input density for X to give a lower bound for the ultimate spectral limit. The characteristic function of the output Y for the model of Eq. (11.33) is

$$\Psi_Y(\nu) = \frac{1}{[1 - j\nu(\sigma_x^2 + 2\sigma_n^2)](1 - j2\nu\sigma_n^2)}, \tag{11.34}$$

where the first factor is the characteristic function of $|X + N_1|^2$ and the second factor is that for $|N_2|^2$. Taking an inverse Fourier transform, the p.d.f. of the output Y is

$$p(y) = \frac{1}{\sigma_x^2}\left[\exp\left(-\frac{y}{\sigma_x^2 + 2\sigma_n^2}\right) - \exp\left(-\frac{y}{2\sigma_n^2}\right)\right]. \tag{11.35}$$

The conditional entropy of $H(Y|X)$ can be calculated using the conditional p.d.f. of

$$g_2(y, y_i) = \frac{1}{2\sigma_n^2}\sqrt{\frac{y}{y_i}}\exp\left(-\frac{y + y_i}{2\sigma_n^2}\right) I_1\left(\frac{\sqrt{yy_i}}{\sigma_n^2}\right) \tag{11.36}$$

of noncentral χ^2 distribution with four degrees of freedom. In the channel model of Eq. (11.33), both $H(Y)$ and $H(Y|X)$ must be evaluated numerically. Without going into detail, with Gaussian input, the asymptotic limit of Eq. (11.32) is valid for both the one or two polarization noise model of Eqs. (11.23) and (11.33), respectively.

Half-Gaussian Input Distribution

The Gaussian input distribution that maximizes the output entropy cannot give the maximum spectral efficiency. As a counter example, if the input electric field is one-dimensionally Gaussian distributed with variance of σ_x^2, the input intensity of $y_i = x_1^2$ has a p.d.f. of $p(y_i) = e^{-y_i/2\sigma_x^2}/\sqrt{2\pi\sigma_x^2 y_i}$. Using the asymptotic results of Eq. (11.30) and similar to Eq. (11.31) but using different $p(y_i)$, we obtain

$$H(Y|X) \sim \frac{1}{2}\log\left(4\pi\sigma_n^2\sigma_x^2\right) + \frac{1}{2} - \frac{1}{2}\gamma_e. \tag{11.37}$$

The output distribution can be found analytically as

$$p(y) = \frac{1}{2\sigma_n\sqrt{\sigma_n^2 + \sigma_x^2}}\exp\left[-\frac{y(2\sigma_n^2 + \sigma_x^2)}{4\sigma_n^2(\sigma_n^2 + \sigma_x^2)}\right] I_0\left[\frac{y\sigma_x^2}{4\sigma_n^2(\sigma_n^2 + \sigma_x^2)}\right] \tag{11.38}$$

with asymptotic entropy of $H(Y) \sim \log\sigma_x^2 + \frac{1}{2}(\log\pi - 1 - \gamma_e)$. The asymptotic channel capacity is

$$C \sim \frac{1}{2}\log\rho_s - \frac{1}{2}\log 2, \tag{11.39}$$

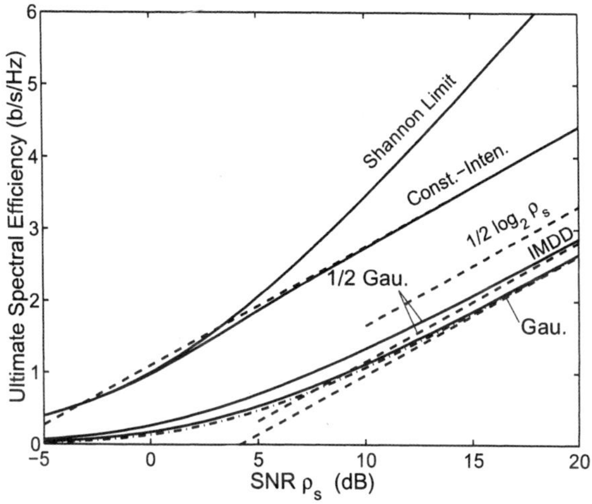

Figure 11.2. The maximum spectral efficiency of optical channel in linear regime. Those for IMDD channels are approximation using various input distributions. The dashed lines are asymptotic limits for constant-intensity and IMDD signal. The two curves with Gaussian input for IMDD signal include noise from one or two polarizations.

that is 0.5 b/s/Hz worse than half of the Shannon limit.

Mecozzi and Shtaif (2001) uses the above approximation to find the maximum spectral efficiency of IMDD channel. This is an example to shown that to maximize the output entropy does not necessary give the channel capacity.

Figure 11.2 shows the ultimate spectral efficiency of optical channel in linear regime as a function of SNR ρ_s. The unconstrained capacity is directly calculated from Eq. (11.13). The capacity of constant-intensity modulation is calculated by Eq. (11.17) by numerical integration. For Gaussian input, the capacity of IMDD channel is calculated using Eq. (11.27) using numerical integration including and excluding the noise from the polarization orthogonal to the signal. For half-Gaussian input, the channel capacity is calculated directly using Eqs. (11.3) and (11.4).

The asymptotic limits of constant-intensity modulation of Eq. (11.22) and IMDD channel of Eqs. (11.32) and (11.39) are both plotted as dashed lines. From Fig. 11.2, the asymptotic limit of Eq. (11.22) is very accurate for constant-intensity modulation in a wide range of SNR. The asymptotic limit for IMDD is valid for a SNR larger than 10 dB. Figure 11.2 also shows the limit of $\frac{1}{2}\log_2 \rho_s$ that is larger than other limits.

2.2 Exact Capacity by Numerical Calculation

For the IMDD channel of Eq. (11.23) with the conditional p.d.f. of Eq. (11.24), the channel output is the intensity $Y \geq 0$ but monotonic one-to-one transfer to the amplitude of $R = \sqrt{Y} \geq 0$ does not change the channel capacity. Using the input and output amplitude random variables of S and R, respectively, the conditional p.d.f. becomes a Rice distribution of

$$p(r|s) = \frac{r}{\sigma_n^2} \exp\left[-\frac{r^2+s^2}{2\sigma_n^2}\right] I_0\left(\frac{rs}{\sigma_n^2}\right), \quad r, s \geq 0. \tag{11.40}$$

The channel capacity, or the maximum spectral efficiency limit, is also equal to the maximum mutual information of

$$C = \max_{p(s)} E\left\{\log_2 \frac{p(r|s)}{p(r)}\right\}, \tag{11.41}$$

where $E\{\cdot\}$ denotes expectation, $p(s)$ and $p(r) = \int_0^\infty p(r|s)p(s)\mathrm{d}s$ are the p.d.f. of the input and output amplitudes, respectively. The channel capacity is

$$C = \max_{p(s)} \left\{H(R) - H(R|S)\right\}, \tag{11.42}$$

where the entropy of the output of $H(R)$ and the conditional entropy of $H(R|S)$ are

$$H(R) = -\int p(r)\log_2 p(r)\mathrm{d}r, \tag{11.43}$$

$$H(R|S) = -\int\int p(s)p(r|s)\log p(r|s)\mathrm{d}r\mathrm{d}s, \tag{11.44}$$

where all integrations are from 0 to $+\infty$. The capacity of Eq. (11.42) should be evaluated together with the average power and probability constraints of

$$\int s^2 p(s)\mathrm{d}s = \sigma_x^2, \quad \int p(s)\mathrm{d}s = 1. \tag{11.45}$$

Based on different assumptions, three algorithms are used to find the optimal input distribution to maximize the channel capacity given by Eq. (11.42).

At large amplitude of $r, s \gg \sigma_n$, the conditional p.d.f. of $p(r|s)$ is approximately a one-dimensional Gaussian distribution with a variance of σ_n^2. In the Kuhn-Tucker condition of Eq. (11.10), the corresponding function of $h_{R|S}(s) = \frac{1}{2}\log(2\pi e\sigma_n^2)$ is a constant at large amplitude. At large amplitude of r and s, similar to Gaussian channel, the output amplitude may have a tail distribution of $p(r) \sim e^{-\lambda r^2}$ where $\lambda > 0$ is the Lagrange multiplier. The input amplitude also has a tail distribution of $p(s) \sim e^{-\kappa_s s^2}$ where λ and κ_s has the relationship of $\lambda^{-1} = 2\sigma_n^2 + \kappa_s^{-1}$. With a constant $h_{R|S}(s)$, the tail distribution of both $p(s)$ and $p(r)$ approaches a continuous distribution. However, the above argument is not sufficient to prove that both the input and output amplitude is continuously distributed with a Gaussian tail. Alternatively, the input and output may have many points very close to each other at large amplitude. In practice, the small probability at large amplitude does not affect the capacity of a practical channel.

At small amplitude of s approaches zero, the function of $H_{R|S}(s)$ approaches $H_{R|S}(0) = \frac{1}{4}\log(2\pi e\sigma_n^2)$. At low intensity, the input p.d.f. has discrete points as shown in Fig. 11.1(c).

Arimoto Algorithm

The Arimoto algorithm can calculate the channel capacity iteratively (Arimoto, 1972, Blahut, 1972, Cover and Thomas, 1991). The single optimization of Eq. (11.41) can change to double iterative optimization of

$$C = \max_{p(s)} \max_{q(s|r)} \int\int p(s)p(r|s)\log\frac{q(s|r)}{p(s)}\mathrm{d}r\mathrm{d}s. \tag{11.46}$$

Given an input distribution of $p(s)$, the optimal conditional p.d.f. for $q(s|r)$ is

$$q(s|r) = \frac{p(s)p(r|s)}{\int p(s)p(r|s)\mathrm{d}s}. \tag{11.47}$$

Given $q(s|r)$, with the condition of Eq. (11.45), the optimal input distribution of $p(s)$ is

$$p(s) = \frac{\exp\left[\int p(r|s)\log q(s|r)\mathrm{d}r - \lambda s^2\right]}{\int \exp\left[\int p(r|s)\log q(s|r)\mathrm{d}r - \lambda s^2\right]\mathrm{d}s}, \tag{11.48}$$

in which the multiplier of λ can be found by the power constraint of $\int s^2 p(s)\mathrm{d}s = \sigma_x^2$. In the Arimoto algorithm, the two procedures of Eqs. (11.47) and (11.48) should be operated iteratively to find the optimal distribution. In practice, with an original input amplitude distribution of $p_{\mathrm{old}}(s)$, the two steps of Eqs. (11.47) and (11.48) can be combined into a single step to give a new input amplitude distribution of $p_{\mathrm{new}}(s)$

by

$$r(s) = \exp\left[- \int p(r|s) \log\left(\int p_{\mathrm{old}}(s')p(r|s')\mathrm{d}s' \right) \mathrm{d}r - h_{R|S}(s) - \lambda s^2 \right], \tag{11.49}$$

and

$$p_{\mathrm{new}}(s) = \frac{p_{\mathrm{old}}(s)r(s)}{\int p_{\mathrm{old}}(s)r(s)\mathrm{d}s}. \tag{11.50}$$

In the expression of Eq. (11.49), inside the exponential is a factor the same as Kuhn-Tucker condition of Eq. (11.10) with $\int p(r|s)\log p(r)\mathrm{d}r + h_{R|S}(s)$. Comparing Eq. (11.49) with the condition Eq. (11.10), the ratio of $r(s)$ decreases the probability where the condition of Eq. (11.10) is greater than zero and increases the probability where the condition of Eq. (11.10) is small than zero. At locations in which the Kuhn-Tucker condition is conformed, the input probability is converged to a fixed value. The Arimoto algorithm can operate for an initial p.d.f. of $p_{\mathrm{ini}}(s)$ either discrete or continuous. For example, for the case of Fig. 11.1(c) with a discrete point at $s = 0$, the initial p.d.f. can be $p_{\mathrm{ini}}(s) = p_0\delta(s) + (1 - p_0)p_{\mathrm{ini},1}(s)$. However, in the region that $p_{\mathrm{ini}}(s) = 0$, the algorithm cannot obtain a nonzero probability of $p(s) > 0$ afterward. With initial nonzero probability of $p_{\mathrm{ini}}(s) > 0$, the algorithm can converge to a very small probability of $p(s)$ approaching zero.

To implement the Arimoto algorithm of Eqs. (11.49) and (11.50) for continuous amplitude p.d.f., the channel needs to be first discretized to $p(s_i)\Delta s$ for the input and $p(r_j)\Delta r$ for the output. The Gaussian tail is also given by the multiplier of λ and κ_s.

Numerical Optimization

If an artificial peak-power (or equivalently peak-amplitude) constraint is imposed to the IMDD channel, the optimal input distribution is discrete (Abou-Faycal et al., 2001, Smith, 1971). Practical system should have a peak-amplitude constraint of $s \leq s_{\max}$ limited by the maximum rating of the transmitter or fiber nonlinearities. Optical amplifiers also cannot provide an infinitely large amplitude, even with very small probability. Given certain number of discrete points, numerical nonlinear programming algorithm can be used to find the optimal distribution. The input distribution of $p(s) = \sum_{k=1}^{K} p_k\delta(s - s_k)$ is fully determined by $2K$ parameters of p_k and s_k, $k = 1, \ldots, K$. The channel capacity of Eq. (11.42) can be maximized for those $2K$ parameters with the constraints of

$$0 \leq s_1 < s_2 < \cdots < s_{K-1} < s_K \leq s_{\max}, \tag{11.51}$$

and

$$\sum_k p_k = 1, \quad \text{and} \quad \sum_k p_k s_k^2 = \sigma_x^2. \tag{11.52}$$

To determine whether those $2K$ parameters are the global optimum of the channel capacity of Eq. (11.42), the Kuhn-Tucker condition of Eq. (11.10) can be used to verify the optimality of the parameters. If the Kuhn-Tucker condition cannot be conformed, an additional discrete points can be added to the optimization procedure until the conformance of Eq. (11.10). In the condition of Eq. (11.10), equality must be satisfied at s_k and the inequality in $s \neq s_k$. In all cases, $s_1 = 0$ is one of the solution. Depending on the ratio of $s_{\text{max}}^2/\sigma_x^2$ and SNR of ρ_s, the maximum point of s_K is usually but not always equal to s_{max}.

The nonlinear programming algorithm can be initiated with two discrete points of $K = 2$ with the number of discrete points increasing until the conformance of the Kuhn-Tucker condition. The channel capacity is also increased with the number of discrete points and the nonlinear programming algorithm can stop with a stable capacity or one of the discrete point has zero probability. At high SNR, the Kuhn-Tucker condition of Eq. (11.10) is difficult to ideally verify. A convergent capacity can be used instead at high SNR.

With fixed positions of $s_k, k = 1, \ldots, K$, the Arimoto algorithm can iteratively find the optimal probability of $p_k, k = 1, \ldots, K$. While possible, the Arimoto algorithm requires lengthy computation to find the optimal positions of s_k.

Combined Numerical Optimization and Arimoto Algorithm

As discussed earlier, the Arimoto algorithm can be modified to include some discrete points, especially for a single discrete point at zero intensity. Instead of using continuous distribution as initial assumption, the algorithm is modified with discrete probability at zero intensity. With more than one discrete point, other than the single point at zero intensity, the positions of other discrete points must be optimized accordingly. With the prior assumption that there are several discrete points at small input amplitude, numerical optimization can be used to find the locations of those optimal discrete points and Arimoto algorithm is used to find the optimal probability (both continuous and discrete parts). The two procedures are used alternatively with increase channel capacity in each iteration.

Based on the above three algorithms, the optimal input distribution is evaluated to maximize the channel capacity of Eq. (11.42). Figure 11.3 shows the channel capacity as a function of SNR for IMDD channel of Eq. (11.23). Different algorithms give different input distributions as

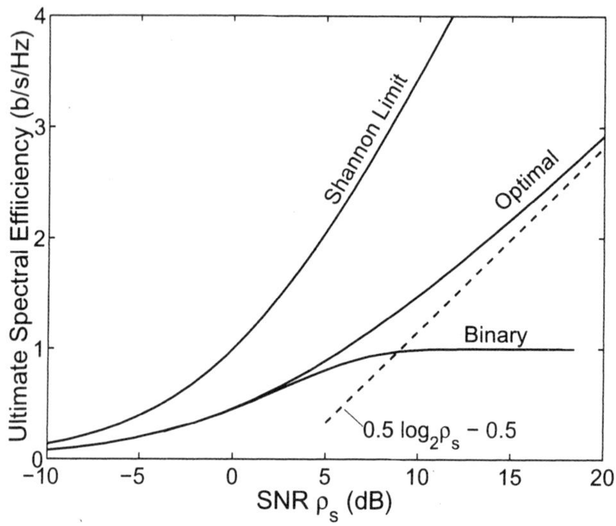

Figure 11.3. The channel capacity as a function of SNR ρ_s for IMDD channel using the optimal input distribution. Also shown the spectral efficiency of binary signal and the asymptotic limit of $\frac{1}{2}\log_2 \rho_s - \frac{1}{2}$.

shown in Fig. 11.4 for $\rho_s = 10$ dB but the same channel capacity in Fig. 11.3. The Shannon limit of Eq. (11.13) is also shown in Fig. 11.3 for comparison.

Calculated using numerical optimization, Figure 11.4 shows the optimal 9 discrete points with the corresponding probability for a peak-power constraint 10 times the average power. Theoretically, the larger is the peak-power constraint, the larger is the channel capacity. Numerically, the usage of 10 times the average power as peak-power constraint gives a channel capacity virtually the same as other algorithms.

Without the artificial peak-power constraint, the optimal input distribution has continuous component or infinite number of points very close to each other in its tail. The Arimoto algorithm gives only a single discrete point at zero intensity. Instead of using continuous distribution as initial assumption, the algorithm is modified with discrete probability at zero intensity. Figure 11.4 shows the optimal input distribution with discrete probability at zero intensity (overlapped with the square there) and continuous-distribution as dash-dotted line. Note that unlike Rayleigh channel in Abou-Faycal et al. (2001), the Arimoto algorithm converges very fast for IMDD channel. Figure 11.4 also shows the optimal input distribution with two discrete points at low intensity (empty

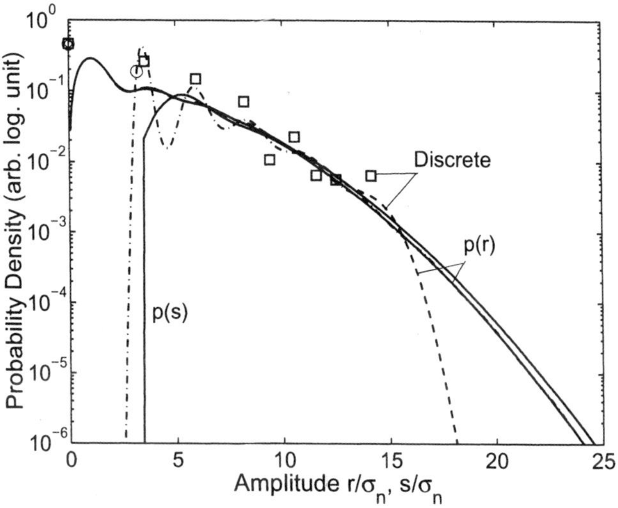

Figure 11.4. The input and output probability density as a function of normalized input and output amplitude of $r/\sigma_n, s/\sigma_n$ for SNR of $\rho_s = 10$ dB. [Adapted from Ho (2005b)]

circles, the circle at zero intensity overlapped with the square) and the continuous distribution at large amplitude (solid line). Not shown in Fig. 11.4, more than two discrete points are used for low intensity in further calculations that do not give a channel capacity with observable difference with that in Fig. 11.3.

All three algorithms converge to the same channel capacity without observable difference. However, Figure 11.4 shows that the input distribution of $p(s)$ has significant difference from one algorithm to others. At $\rho_s = 10$ dB, the three algorithms give a channel capacity within $\pm 0.05\%$ of each other. The output distributions of $p(r)$ in Fig. 11.4 from the three input distributions also have no significant difference at small amplitude. Only the tail distribution of $p(r)$ has major difference when the input distribution is totally or partially discrete.

Figure 11.3 also shows the channel capacity for binary signal (two discrete points in the input distribution) calculated by numerical optimization. Binary signal achieves the channel capacity for SNR less than about 5 dB. The channel capacity of binary signal was also calculated in Mecozzi and Shtaif (2001) with the assumption that the two levels are equally probable. Except for large SNR ($\rho_s > 12$ dB), the optimal binary signal has larger probability at zero-intensity and smaller probability at nonzero-intensity. For example, if only 10% probability at nonzero-intensity, the nonzero-intensity is 10 times the averaged intensity as compared with twice the averaged intensity for equal-probable case. Compared to similar curve in Mecozzi and Shtaif (2001), the binary signal can achieve better channel capacity at low SNR. By sending occasional pulses with large intensity above the noise, the system is similar to the essence of return-to-zero (RZ) signaling. Systems with powerful forward error correction (Mizuochi et al., 2004) can operate around $\rho_s = 5$ to 7 dB. Binary instead of multilevel signals may be sufficient for those systems.

Based on the same channel model of Eq. (11.23), the half-Gaussian distribution of Mecozzi and Shtaif (2001) and Fig. 11.2 provides a lower bound and is 0.07 to 0.21 b/s/Hz worse than the optimal distribution calculated by numerical algorithms. The optimal distribution has a tail profile of $e^{-\kappa_1 r^2}$, similar to that of half-Gaussian distribution. For very large SNR ($\rho_s > 30$ dB), the discrete region of Fig. 11.4 at low intensity becomes insignificant and the half-Gaussian distribution should be very close to the optimal distribution.

2.3 Thermal Noise Dominated IMDD Channel

For an IMDD system without optical amplifiers, the system is limited by thermal noise of N_{th} from the receiver circuitry. The photodetector gives the intensity of the signal as $|X|^2$, with additive thermal noise, the receiver output is

$$Y = |X|^2 + N_{\text{th}}, \tag{11.53}$$

where $|X|^2 \geq 0$ because optical intensity is always positive. Naturally, there is a peak constraint of the instantaneous optical power of $|X|^2 \leq P_{\text{max}}$. From Smith (1971), the optimal input distribution to give the ultimate spectral efficiency is a discrete distribution.

Using numerical optimization, Figure 11.5 shows the channel capacity of thermal noise limited IMDD channel with peak intensity constraint. The SNR of Fig. 11.5 is defined as $m_{|X|^2}/\sigma_{\text{th}}$, where $m_{|X|^2} = E\{|X|^2\}$ is the average of the optical intensity and $\sigma_{\text{th}}^2 = E\{N_{\text{th}}^2\}$ is the variance of the Gaussian distributed zero-mean thermal noise. This definition of SNR is consistent with the definition of Q factor for binary equal probable signal in Eq. (3.139). Figure 11.5 shows the channel capacity when the peak intensity of P_{max} is either 3 or 10 times larger than the average optical intensity of $m_{|X|^2}$.

At high SNR, the output of Y may be considered as confined to $Y \geq -\sigma_{\text{th}}$, the optimal distribution to maximize the output entropy is exponential distribution with p.d.f. of $p(y) = \exp[-(y + \sigma_{\text{th}})/m_y]/m_y, y \geq -\sigma_{\text{th}}$, where $m_y = m_{|X|^2} + \sigma_{\text{th}}$, where σ_{th} is added to take into ac-

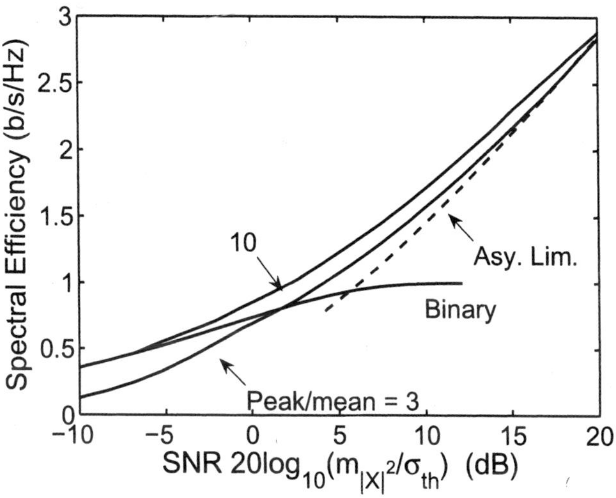

Figure 11.5. The maximum spectral efficiency of an IMDD channel limited by thermal noise

count the small negative value of the output Y. The output entropy is $H(Y) \sim \log(m_y) + 1$ and the channel capacity has an asymptotic limit of

$$\begin{aligned} C &\sim \log(m_y) + 1 - \frac{1}{2}\left[\log\left(2\pi\sigma_{\text{th}}^2\right) + 1\right] \\ &= \log\left(1 + \frac{m_{|X|^2}}{\sigma_{\text{th}}}\right) - 0.604\log 2. \end{aligned} \tag{11.54}$$

This asymptotic limit is also shown in Fig. 11.5. Comparing the channel capacity of Fig. 11.5 limited by thermal noise with similar channel capacity of Fig. 11.3 limited by amplifier noise, the channel capacity limited by thermal noise is significantly larger at low SNR and slightly smaller at high SNR. At low SNR, IMDD channel limited by thermal noise can have negative output but that limited by amplifier noise always has positive output. The channel capacity improves with the possibility of negative output.

The asymptotic limit for the channel capacity limited by amplifier noise is 0.1 b/s/Hz larger than that limited by thermal noise at high SNR. The conditional entropy of thermal-noise-limited channel is a constant independent of input signal. The conditional entropy $H(R|S)$ of IMDD channel limited by amplifier noise at small input is half of that at large input. The channel capacity increases slightly for channel limited by amplifier noise due to the reduction of conditional entropy.

The discrete nature of the optimal input distribution was first shown in Smith (1971) for Gaussian channel with peak-amplitude constraint and later used by Abou-Faycal et al. (2001) for Rayleigh channel. Many channels have discrete optimal input distribution (Huang and Meyn, 2003, Shamai and Bar-David, 1995).

Most short-distance optical communication systems without optical amplifiers are primary limited by thermal noise. Those systems usually use single-channel on-off keying and the channel capacity is usually not a major issue.

3. Quantum-Limited Capacity

When the number of signal and/or noise photons is small, quantum effects must be considered to compute the capacity of an optical channel. While a classical continuous-time channel can be converted to a discrete-time channel using the sampling theorem (Shannon, 1948), the particle-based quantum channel does not have the corresponding sampling theorem. However, one may assume that the measurement is made within a time interval limited by the channel bandwidth. Hence, most studies of quantum-limited capacity assume a discrete-time channel. Corresponding to a narrow-band WDM channel, this type of channel is generally referred to as narrow-band channel.

In coherent detection of an optical signal, the input signal can be assumed as a coherent state (Caves and Drummond, 1994, Gardiner, 1985, Yamamoto and Haus, 1986). If there is an average of $\bar{n}_S$ signal photon and $\bar{n}_N$ noise photon, the channel capacity is the same as that of Eq. (11.13) with one additional noise photon of

$$C = \log\left(1 + \frac{\bar{n}_S}{1 + \bar{n}_N}\right). \tag{11.55}$$

The classic SNR of ρ_s is the ratio of $\bar{n}_S$ to $\bar{n}_N$. If a quantum SNR is defined by the ratio of $\bar{n}_S/(1+\bar{n}_N)$, the quantum capacity of Eq. (11.55) is the same as that of the classic limit of Eq. (11.13). Intuitively, there is a minimum of one noise photon in coherent state. Here, the SNR is expressed as the ratio of photons instead of power like Eq. (3.36). The relationship of SNR to number of photons is very obvious from Table 3.1. For typical optical communication systems with amplifier noise, the quantum-limited capacity is slightly less than the classic limit of Eq. (11.13). However, for large signal and noise photons of both $\bar{n}_S$ and $\bar{n}_N$, the difference between quantum and classic limit is small. However, for "quantum" thermal noise limited system, the number of noise photons of $\bar{n}_N$ is very small (Hall, 1994), usually in the other of

10^{-3} or less. Of course, the quantum thermal noise is not the same as receiver thermal noise.

The IMDD channel considered in previous section is equivalent to the classical limit of a photon-counting channel. The continuous-time photon-counting channel is modeled by information theorists as a Poisson channel with unlimited bandwidth but peak and average power constraints (Davis, 1980, Kabanov, 1978, Massey, 1981, Wyner, 1988). A discrete-time Poisson channels can be used to model a quantum-limited, band-limited photon-counting channel (Caves and Drummond, 1994, Gordon, 1962, Hall, 1994, Stern, 1960, Yamamoto and Haus, 1986).

Optical amplification alters the photon statistics of amplified light. While an amplified signal has Poisson statistics, amplified spontaneous emission (ASE) noise obeys Bose-Einstein statistics. For a signal having n_S signal photons and an average of $\bar{n}_{\text{ASE}}$ ASE photons, the ASE noise is equivalent to Poisson-distributed light where the mean number of photons has an exponential distribution of

$$p_{n_N}(\bar{n}) = \frac{1}{\bar{n}_N} \exp\left(-\frac{\bar{n}}{\bar{n}_N}\right), \quad \bar{n} \geq 0. \tag{11.56}$$

With n_S of signal photon, the average number of photons has an distribution of $p_{n_N}(\bar{n} - n_S)$ with $\bar{n} \geq n_S$. The output photon-number distribution is equal to

$$\begin{aligned} p_{n_S}(n) &= \frac{1}{n!} \int_{n_S}^{+\infty} p_{n_N}(\bar{n} - n_S) e^{-\bar{n}} \bar{n}^n \mathrm{d}\bar{n} \\ &= \frac{e^{-n_S}}{n!} \int_0^{+\infty} p_{n_N}(\bar{n}) e^{-\bar{n}} (\bar{n} + n_S)^n \mathrm{d}\bar{n}, \end{aligned} \tag{11.57}$$

or

$$p_{n_S}(n) = \frac{\bar{n}^n}{(1+\bar{n}_N)^{n+1}} \exp\left(-\frac{n_S}{1+\bar{n}_N}\right) L_n\left(-\frac{n_S/\bar{n}_N}{1+\bar{n}_N}\right), \tag{11.58}$$

where $L_n(\cdot)$ is the Laguerre polynomial of

$$L_n(x) = \sum_{k=0}^{n} \binom{n}{k} \frac{(-x)^k}{k!}. \tag{11.59}$$

The distribution of Eq. (11.58) includes only the ASE photons from the same polarization of the signal.

If the input distribution is $p_S(n_S)$, the output photon-number distribution is $p_N(n) = \int_0^{+\infty} p_S(n_S) p_{n_S}(n) \mathrm{d}n_S$. The maximum spectral efficiency is

$$C = \max_{p_S(n_S)} \left\{ H(N) - H(N|S) \right\}, \tag{11.60}$$

where

$$H(N) = -\sum_{n=0}^{\infty} p_N(n) \log p_N(n), \tag{11.61}$$

and

$$H(N|S) = -\int_0^{\infty} p_S(n_S) \sum_{n=0}^{\infty} p_{n_S}(n) \log p_{n_S}(n) \mathrm{d}n_S. \tag{11.62}$$

As an approximation, the maximum spectral efficiency can be derived by maximizing the output entropy of $H(N)$. Previous section already shows that the maximizing of the output entropy of $H(N)$ does not necessary give the channel capacity. However, the calculation here gives a lower bound of the channel capacity. If the input photon is exponential distributed with $p_S(n_s) = \exp\left(-n_S/\bar{n}_S\right)/\bar{n}_S$, the output photon distribution is $p_N(n) = \bar{n}^n/(1+\bar{n})^{n+1}$ and $H(N)$ is

$$H(N) = \log(1+\bar{n}) + \bar{n}\log(1+1/\bar{n}), \tag{11.63}$$

where $\bar{n} = \bar{n}_S + \bar{n}_N$. With an input exponential distribution of $p_S(n_s)$ and the conditional p.d.f. of Eq. (11.58), the conditional entropy of Eq. (11.62) can be calculated numerically.

For a large number of photons and high SNR, the summation of $\sum_{n=0}^{\infty} p_{n_S}(n) \log p_{n_S}(n)$ can be approximated by integration. If the conditional probability of $p_{n_S}(n)$ is assumed to be Gaussian distributed with variance of $\sigma_s^2(n_S) = n_S + \bar{n}_N + 2n_S\bar{n}_N + n\bar{n}_N^2$ for the signal shot noise, ASE shot noise, signal-spontaneous beat noise, and spontaneous-spontaneous beat noise, respectively (Desurvire, 1994). Based on the Gaussian approximation, we obtain

$$\sum_{n=0}^{\infty} p_{n_S}(n) \log p_{n_S}(n) \approx -\frac{1}{2}\log\left[2\pi e \sigma_s^2(n_S)\right]. \tag{11.64}$$

Using Eq. (11.64) to calculate $H(N|S)$, the asymptotic limit is

$$\begin{aligned} C &\sim \frac{1}{2}\log\left(\frac{\bar{n}_S e^{1+\gamma_e}}{\bar{n}_N \pi}\right) - 1 \\ &= \frac{1}{2}\log\frac{\bar{n}_S}{\bar{n}_N} - 0.688\log 2, \end{aligned} \tag{11.65}$$

the same as that of Eq. (11.32).

Figures 11.6 present spectral efficiency limits given by Eq. (11.60) for a photon-counting channel with ASE noise. Figure 11.6(a) shows the spectral efficiency as a function of the classical SNR ρ_s for various values

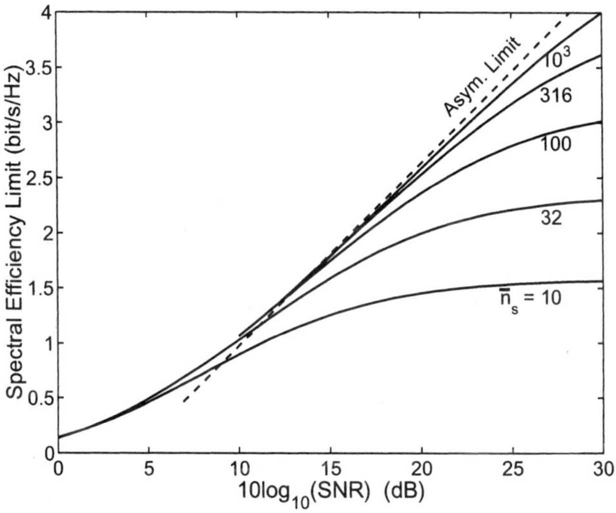

(a)

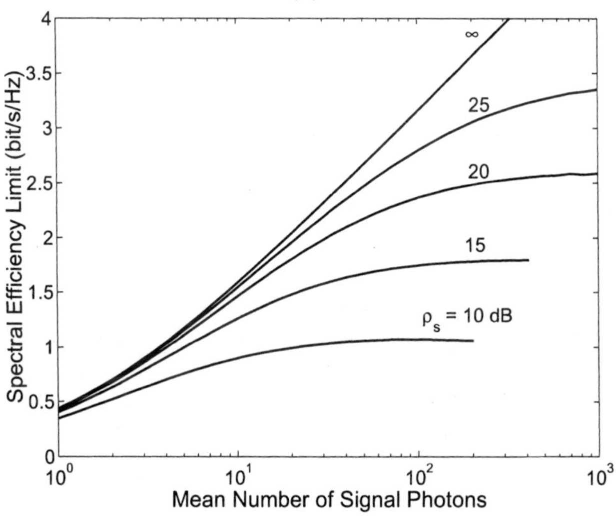

(b)

Figure 11.6. The quantum-limited maximum spectral efficiency of photon-counting optical channel as a function of (a) the SNR of $\bar{n}_S/\bar{n}_N$ (b) the average number of signal photons $\bar{n}_S$.

of the mean number of signal photons $\bar{n}_S$. Figure 11.6(a) also shows the asymptotic limit of Eq. (11.65) for large number of signal photons and high SNR. Figures 11.6(a) and (b) show the spectral efficiency of Eq. (11.60) as a function of the mean signal photon number $\bar{n}_S$ for various values of the SNR ρ_s. The case with infinite SNR corresponds to the spectral efficiency for Poisson-distributed photons (Gordon, 1962, Stern, 1960).

Unlike the classical case in Fig. 11.2 in which the spectral efficiency depends only on the SNR, the quantum-limited spectral efficiency depends on both the SNR and the number of signal photons. Even at high SNR, high spectral efficiency cannot be achieved with a small number of signal photons. As shown in Desurvire (2002, 2003), the effect of fiber nonlinearity upon quantum-limited spectral efficiency is equivalent to an increase in the number of ASE photons.

4. Channel Capacity in Nonlinear Regime

Fiber nonlinearities limit the transmission distance and the overall capacity of a WDM system. The major fiber nonlinearities are the Kerr effect, stimulated Raman scattering, and stimulated Brillouin scattering (Agrawal, 2001). The Kerr effect leads to self- and cross-phase modulation, and four-wave mixing. In the Kerr effect, from Eq. (5.1), the intensity of the aggregated optical signal perturbs the fiber refractive index, thereby modulating the signal phase. In WDM systems, self- and cross-phase modulations arise when the phase of a channel is modulated by its own intensity and by the intensity of other channels, respectively. Four-wave-mixing arises when two channels beat with each other, causing intensity modulation at the different frequency, thereby phase-modulating all the channels and generating new frequency components. Previous chapter considers both self- and cross-phase modulation but not four-wave-mixing. The new frequency components from four-wave mixing can be considered as additive noise in additional to amplifier noise.

Fiber propagation with the Kerr effect is modeled using the nonlinear Schrödinger equation of Eq. (7.9) for single channel systems and coupled nonlinear Schrödinger equations for WDM systems. Among the various nonlinearities, the Kerr effect has the greatest impact on a WDM system and thus its channel capacity. Early studies focused on the effect of fiber nonlinearity on specific modulation and detection techniques, including on-off keying with direct detection (Chraplyvy, 1990, Chraplyvy and Tkach, 1993, Forghieri et al., 1997, Wu and Way, 2004) or simple modulations with coherent detection (Shibata et al., 1990, Waarts et al., 1990). Recently, the combined effect of amplifier noise and Kerr nonlinearity on the Shannon capacity has been studied.

Mitra and Stark (2001) argued that the capacity of WDM systems is fundamentally limited mostly by cross-phase modulation. As a signal propagates, chromatic dispersion converts cross-phase-modulation-induced phase modulation to intensity noise. Capacity limitations caused by cross-phase modulation are further studied in Green et al. (2002), Stark et al. (2001), and Wegener et al. (2004). In fibers with nonzero dispersion, cross-phase modulation has a much greater impact than four-wave-mixing. WDM systems with many channels are likely to be limited by cross-phase modulation, perhaps allowing self-phase modulation to be ignored to first order for systems with many channels. However, the methods of Mitra and Stark (2001), Stark et al. (2001), and Green et al. (2002) do not quantify the importance of self- relative to cross-phase modulation, and cannot be applied to single-channel systems limited primarily by self-phase modulation.

With constant-intensity modulation, such as phase or frequency modulation (Ho and Kahn, 2002), ideally, both self- and cross-phase modulation cause only time-invariant phase shifts, eliminating both phase and intensity distortion. If one could further neglect four-wave-mixing, propagation would be linear; the capacity would be given by the expressions of previous section, and increasing the launched power would lead to a monotonic increase in spectral efficiency. In reality, chromatic dispersion converts phase or frequency modulation to intensity modulation (Norimatsu and Iwashita, 1993, Wang and Petermann, 1992), and laser intensity noise and imperfect modulation cause additional intensity fluctuations. Hence, it is difficult to maintain constant intensity along an optical fiber. Furthermore, in reality, constant-intensity modulation is subject to four-wave-mixing. As shown earlier, constant-intensity modulation is also fundamentally limited by nonlinear phase noise.

Tang (2001a,b, 2002) solved the nonlinear Schrödinger equation using a series expansion, similar to the Volterra series in Peddanarappagari and Brandt-Pearce (1997, 1998). The linear term is considered to be signal and all higher-order terms are considered to be noise. If sufficient number of terms is included, methods based on series expansion are very accurate. Tang (2001a,b, 2002) has included many terms, yielding a quasi-exact closed-form treatment.

In a single-channel system, the channel capacity is limited by self-phase modulation. In a WDM system, when all channels are detected together, the impact of cross-phase modulation can be reduced using a multi-user detection or interference-cancellation scheme. Using perturbation methods, Narimanov and Mitra (2002) found the channel capacity of single-channel systems. For a single-channel system with zero average dispersion, Turitsyn et al. (2003) solves the nonlinear Schrödinger equation analytically to find the channel capacity. Basically, Turitsyn et al. (2003) finds the p.d.f. of signal with nonlinear phase noise of Eq. (6.66) and uses it to find the asymptotic channel capacity of the channel. With large nonlinear phase noise, IMDD signal without the usage of phase information can be used instead.

To quantify the SNR in the presence of cross-phase modulation, Mitra and Stark (2001) introduced a nonlinear intensity scale I_0. For transmitted power per channel well below I_0, increasing the power raises the SNR, increasing capacity. As the transmitted power approaches I_0, cross-phase modulation noise increases rapidly, causing capacity to decrease precipitously. This nonlinear intensity scale I_0 is also applicable to the models of Tang (2001a,b, 2002), Narimanov and Mitra (2002), constant-intensity modulation of Ho and Kahn (2002), and even the nonlinear quantum limit of Desurvire (2002, 2003). In each of those models, the launched power that maximizes the channel capacity increase with I_0. In WDM systems, the nonlinear intensity scale I_0, and thus the capacity, increases with fiber dispersion, channel spacing and signal bandwidth, and decreases with the total number of spans and the total number of channels. In Mitra and Stark (2001), the nonlinear intensity scale for cross-phase modulation was found to be

$$I_0 = \left[\frac{BD\Delta\lambda}{2\gamma^2 \log(M) L_{\text{eff}}}\right]^{1/2} \tag{11.66}$$

and the maximum spectral efficiency is lowered bound by

$$C = \log\left[1 + \frac{e^{-(I_t/I_0)^2} I_t}{I_n + (1 - e^{-(I_t/I_0)^2}) I_t}\right] \tag{11.67}$$

where B is the number of symbol per second, D is the dispersion coefficient and $\Delta\lambda$ is the channel spacing of the WDM system, $2M+1$ is the overall number of channels, γ is the fiber nonlinear coefficient, I_t and I_n are the signal and noise power per channel, respectively. For a system with N_A spans, the overall effective nonlinear length is approximately equal to $L_{\text{eff}} = N_A/\alpha$ where α is the fiber attenuation coefficient.

Using the spectral efficiency lower bound of Eq. (11.67), the power per channel that maximizes spectral efficiency is approximately equal to $(I_0^2 I_n/2)^{1/3}$, and the maximum spectral efficiency is approximately equal to

$$\frac{2}{3}\log\left(\frac{2I_0}{I_n}\right) \tag{11.68}$$

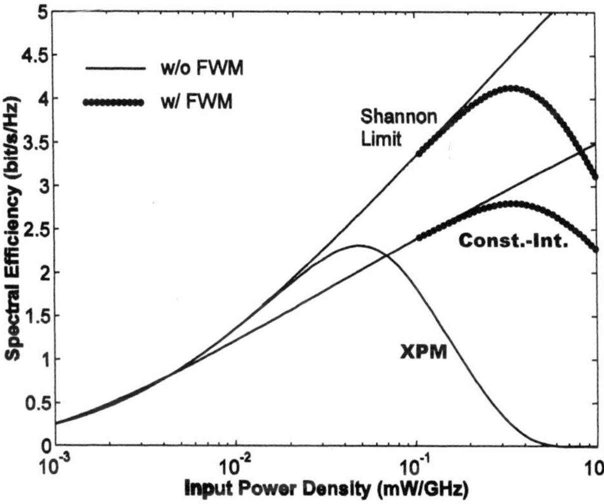

Figure 11.7. The maximum spectral efficiency of optical channel in nonlinear regime for both unconstrained or constant-intensity signal. The unconstrained signal is limited by cross-phase modulation (XPM) but constant-intensity signal is limited by four-wave-mixing (FWM).

In a WDM system limited by four-wave-mixing instead of cross-phase modulation, the spectral efficiency bound is also given by Eq. (11.67), and the nonlinear intensity scale is given by:

$$I_0^{-1} = \frac{N_A \gamma^2}{9} \sum_{\substack{p,q,p\neq 0,q\neq 0 \\ |p+q|\leq M}} \frac{D_{pq}^2}{\alpha^2 + \Delta k_{pq}^2}, \tag{11.69}$$

where $-M \leq p, q \leq M$, $D_{pq} = 3$ if $p = q$ and $D_{pq} = 6$ if $p \neq q$, and $\Delta k_{pq} = 2\pi\lambda^2 D \Delta f^2 qp/c$, λ is the optical wavelength, and c is the speed of light. The expression Eq. (11.69) has been derived for the center (worst-case) channel. The additive noise from four-wave-mixing was studied by Eiselt (1999), Tkach et al. (1995), and Forghieri et al. (1997).

Figure 11.7 shows the spectral efficiency as a function of input power density, including the Shannon limit of Eq. (11.13), the numerical expression of Eq. (11.17) for constant-intensity signal, and the results of Mitra and Stark (2001) limited by cross-phase modulation. In the absence of four-wave-mixing, as input power is increased, spectral efficiency increases monotonically for the Shannon limit and for constant-intensity modulation, but the spectral efficiency computed following Mitra and Stark (2001) reaches a maximum value limited by cross-phase modulation. When four-wave-mixing is modeled as extra additive Gaussian noise, spectral efficiencies for the Shannon limit and for constant-intensity modulation reach maximum values limited by four-wave-mixing, while the spectral efficiency computed following Mitra and Stark (2001) remains unchanged. The maximum spectral efficiency of constant-intensity signal is about 2.8 bit/s/Hz, compared with 2.3 bit/s/Hz computed following Mitra and Stark (2001).

The system of Fig. 11.7 has $2M + 1 = 101$ WDM channels and $N_A = 10$ fiber spans; uses optical fiber having attenuation coefficient $\alpha = 0.2$ dB/km, nonlinear coefficient of $\gamma = 1.24$ rad/W/km, and dispersion coefficient $D = 17$ ps/km/nm; operates around the wavelength of $\lambda = 1.55$ μm with channel bandwidth $B = 40$ GHz, and channel separation $\Delta f = 1.5B$; uses optical amplifiers with noise figure of 4 dB and gain of 30 dB. Using an overall effective length of $L_{\text{eff}} = N_A/\alpha$, the nonlinear intensity scale of Eq. (11.66) is $I_0 = 11.2$ mW. In Fig. 11.7, we assume that all four-wave-mixing components from the same span and from each fiber span combine incoherently by ignoring the phase dependence between four-wave-mixing components.

Limited by four-wave-mixing, constant-intensity modulation may provide better spectral efficiencies than those of Mitra and Stark (2001) in the regime in which cross-phase modulation dominates over four-wave-mixing. Because four-wave-mixing decreases more rapidly than cross-phase modulation as channel spacing is increased, the improvement obtained using constant-intensity modulation is more significant for systems having large channel spacing. Both four-wave-mixing and cross-phase modulation decrease with an increase of fiber dispersion. Four-wave-mixing dominates over cross-phase modulation for zero-dispersion optical fiber. Of course, although the system for Fig. 11.7 shows that constant-intensity modulation has better maximum spectral efficiency than unconstrained modulation. Depending on system parameters, unconstrained modulation may have maximum spectral efficiency better than constant-intensity modulation (Kahn and Ho, 2004).

While the nonlinear Schrödinger equation with noise provides a very accurate model for nonlinear propagation in fiber, the equation does not have an analytical solution except in some special cases (Turitsyn et al., 2003). While all of the works are based on this accurate formulation, they make different assumptions and approximations, leading to different estimation of the channel capacity.

In order to illustrate the major qualitative differences between the various models, we consider a simplified memoryless monotonic transfer characteristic of $y = f(x) = x + \epsilon x^3$, where x and y are the input and output, respectively, and ϵ is a small number. While there is no nonlinear fiber channel, or other channel type, having transfer characteristic of $f(x)$, this simple function with linear term x and nonlinear term of ϵx^3 yields insight into fiber systems. For a monotonic, one-to-one function such as $f(x)$, if we interpret both terms x and ϵx^3 as signal, then in the absence of any noise, the entropy of the output given the input, $H(Y|X)$, is equal to zero. The mutual information between the input and the output, and thus the channel capacity, equals the entropy of the input x.

We draw an equivalence to the most models by considering the linear term x to be signal and the nonlinear term ϵx^3 to be noise. In WDM systems with many channels, the nonlinear term ϵx^3 includes "intermodulation products" corresponding to the cross-phase modulation and four-wave-mixing caused by other channels. As all channels are typically independent from one another, the models concerning cross-phase-modulation-induced distortion can indeed model cross-phase modulation as noise independent from the signal (Green et al., 2002, Mitra and Stark, 2001, Stark et al., 2001, Wegener et al., 2004). Likewise, the model concerning four-wave-mixing components from other channels for constant-intensity modulation can model four-wave-mixing to be noise independent from the signal (Ho and Kahn, 2002). In a single-channel system (Turitsyn et al., 2003), the nonlinear distortion caused by self-phase modulation depends on the signal and cannot be modeled as signal-independent noise. While only contributions from cross-phase modulation are modeled as signal-independent noise in Green et al. (2002), Mitra and Stark (2001), Stark et al. (2001), and Wegener et al. (2004), the series expansion model of Tang (2001a,b, 2002) regards all higher-order terms as noise independent of the signal. In fact, the two main groups of models are not complete because Tang (2001a,b, 2002) cannot account for the dependence of higher-order terms on the signal and Green et al. (2002), Mitra and Stark (2001), Stark et al. (2001), and Wegener et al. (2004) cannot include the higher-order terms caused by self-phase modulation. In WDM systems with many channels, the methods of Green et al. (2002), Mitra and Stark (2001), Stark et al. (2001), Tang (2001a,b), and Wegener et al. (2004) can be considered to be equivalent if the effect of self-phase modulation is negligible compared to cross-phase modulation. In single-channel systems, where self-phase modulation must be considered, only the method of Turitsyn et al. (2003) yields the probability density of the channel output including nonlinearity and uses it to calculate the channel capacity. In all cases, if the nonlinear term of ϵx^3 is considered as noise, the channel capacity decreases at high launched power and a nonlinear intensity scale similar to Eqs. (11.66) and (11.69) can be evaluated. In the single-channel system of Narimanov and Mitra (2002), the channel capacity curve behaves like the curves in Fig. 4 with four-wave mixing.

The single-channel capacity of Turitsyn et al. (2003) has been evaluated for fiber links with zero average dispersion. Only nonlinear phase noise of Chapter 5 modulates the signal phase. The channel capacity is calculated using the probability density of the signal with nonlinear phase noise of Eq. (6.66). In the limit of very high nonlinear phase noise, the capacity degenerates to that of direct detection in the last section, which increases logarithmically with launched power. The nonlinear phase noise causes no amplitude noise.

The impact of Kerr nonlinearity can be reduced or canceled using phase conjugation (Brener et al., 2000, Pepper and Yariv, 1980). In WDM systems with many channels, Kerr effect compensation reduces or cancels the nonlinear terms originating from other channels. In such a case, Kerr effect compensation yields the obvious benefit of reducing the "noise". In single-channel systems, Kerr effect compensation changes the statistics of the signal with noise. While mid-span or distributed Kerr effect compensation can improve the capacity, Kerr effect compensation just before the receiver does not improve capacity, and may actually reduce capacity by adding more noise.

5. Summary

Increasing spectral efficiency is often the most economical means to increase WDM system capacity. In this chapter, we find the information-theoretical spectral efficiency limits for various modulation and detection techniques in both classical and quantum regimes, considering both linear and nonlinear fiber propagation regimes. Spectral efficiency limits for unconstrained modulation with coherent detection are several b/s/Hz in terrestrial WDM systems, even considering nonlinear effects. Spectral efficiency limits are reduced significantly using either constant-intensity modulation or direct detection. Using binary modulation, regardless of detection technique, spectral efficiency cannot exceed 1 b/s/Hz per polarization.

Optical signals propagating in fibers offer several degrees of freedom, including time, frequency and polarization. The combined coding over these degrees of freedom has been seldom explored as a means to increase transmission capacity in fibers, especially as a way to combat or benefit from fiber nonlinearity and polarization-mode dispersion.

Both fiber chromatic dispersion and polarization-mode dispersion does not limit the maximum spectral efficiency of the optical channel. In the ultimate limit, the bandwidth per channel can be very small to reduce the impacts of both chromatic and polarization-mode dispersion. The spectral efficiency can also be doubled using polarization-division multiplexing (PDM). For fiber with polarization-mode dispersion, the two orthogonal polarized signals can propagate along the two principle states of polarization. Of course, the transmitter requires active tracking such that the signal can follow the two principle states of polarization for a time varying fiber channel.

Bibliography

Aarts, W. H. J. and Khoe, G. D. (1989). New endless polarization control method using three fiber squeezers. *J. Lightwave Technol.*, 7(7):1033–1043.

Abbas, G. L., Chan, V. W. S., and Yee, T. K. (1985). A dual-detector optical heterodyne receiver for local oscillator noise suppression. *J. Lightwave Technol.*, LT-3(5):1110–1122.

Abidi, A. A. (1995). Direct-conversion radio transceivers for digital communications. *IEEE J. Solid-State Circuits*, 30(12):1399–1410.

Ablowitz, M. J. and Hirooka, T. (2001). Intrachannel pulse interaction in dispersion-managed transmission systems: Timing shifts. *Opt. Lett.*, 26(23):1846–1848.

Ablowitz, M. J. and Hirooka, T. (2002). Resonant intrachannel pulse interaction in dispersion-managed transmission systems. *IEEE J. Sel. Top. Quantum Electron.*, 8(3):603–614.

Abou-Faycal, I. C., Trott, M. D., and Shamai (Shitz), S. (2001). The capacity of discrete-time memoryless Rayleigh-fading channels. *IEEE Trans. Info. Theory*, 47(4):1290–1301.

Abrams, M., Becker, P. C., Fujimoto, Y., O'Byrne, V., and Piehler, D. (2005). FTTP deployments in the United States and Japan-equipment choices and service provider imperatives. *J. Lightwave Technol.*, 23(1):236–246.

Agrawal, G. P. (2001). *Nonlinear Fiber Optics.* Academic Press, San Diego, 3 edition.

Agrawal, G. P. (2002). *Fiber-Optic Communication Systems.* Wiley-Interscience, New York, 3 edition.

Agrawal, G. P. and Dutta, N. K. (1986). *Long-Wavelength Semiconductor Lasers.* Van Norstrand-Reinhold, Princeton, NJ.

Ait Sab, O. and Lemaire, V. (2001). Block Turbo code performance for long-haul DWDM optical transmission systems. In *Optical Fiber Commun. Conf.* paper TuF1.

Akage, Y., Kawano, K., Oku, S., Iga, R., Okamoto, H., Miyamoto, Y., and Takeuchi, H. (2001). Wide bandwidth of over 50 GHz traveling-wave electrode electroabsorption modulator integrated DFB lasers. *Electron. Lett.*, 37(5):299–300.

Akiba, S., Usami, M., and Utaka, K. (1987). 1.5 μm $\lambda/4$-shifted InGaAsP/InP DFB lasers. *J. Lightwave Technol.*, LT-5(11):1564–1573.

Akulova, Y. A., Fish, G. A., Koh, P.-C., Schow, C. L., Kozodoy, P., Dahl, A. P., Nakagawa, S., Larson, M. C., Mack, M. P., Strand, T. A., Coldren, C. W., Hegblom, E., Penniman, S. K., Wipiejewski, T., and Coldren, L. A. (2002). Widely tunable electroabsorption-modulated sampled-grating DBR laser transmitter. *IEEE J. Sel. Top. Quantum Electron.*, 8(6):1349–1357.

Aldis, J. P. and Burr, A. G. (1993). The channel capacity of discrete time phase modulation in AWGN. *IEEE Trans. Info. Theory*, 39(1):184–185.

Alexander, S. B. (1987). Design of wide-band optical heterodyne balanced mixer receivers. *J. Lightwave Technol.*, LT-5(4):523–537.

Alexander, S. B., Welford, D., and Marquis, D. V. L. (1989). Passive equalization of semiconductor diode laser frequency modulation. *J. Lightwave Technol.*, 7(1):11–23. Comments by S. Dilwali and G. S. Pandian, 10(7):1000-1003.

Alexander, S. B., Barry, R., Castagnozzi, D. M., Chan, V. W. S., Hodsdon, D. M., Jeromin, L. L., Kaufmann, J. E., Materna, D. M., Parr, R. J., Stevens, M. L., and White, D. W. (1990). 4-ary FSK coherent coherent optical communication system. *Electron. Lett.*, 26(17):1346–1347.

Alferness, R. C. (1981). Guided-wave devices for optical communication. *IEEE J. Quantum Electron.*, QE-17(6):946–959.

Alferness, R. C. (1982). Waveguide electrooptic modulators. *IEEE Trans. Microwave Theory & Tech.*, MTT-30(8):1121–1137. Erratum: MTT-31(3):315.

Alferness, R. C. (1990). Titanium-diffused Lithium Niobate waveguide devices. In Tamir, T., editor, *Guided-Wave Optoelectronics*, pages.145–210. Springer-Verlag, Berlin, 2 edition.

Alic, N. and Fainman, Y. (2004). Data-dependent phase coding for suppression of ghost pulses in optical fibers. *IEEE Photon. Technol. Lett.*, 16(4):1041–1135.

Altas, D. A. and Kazovsky, L. G. (1990). An optical PSK homodyne transmission experiment using 1320 nm diode-pumped Nd:YAG lasers. *IEEE Photon. Technol. Lett.*, 2(5):367–370.

Amos, D. E. (1986). A portable package for Bessel functions of a complex argument and nonnegative order. *ACM Trans. on Math. Software*, 12(3):265–273.

Anthon, D., Berger, J. D., Drake, J., Dutta, S., Fennema, A., Grade, J. D., Hrinya, S., Ilkov, F., Jerman, H., King, D., Lee, H., and Tselikov, A.and Yasumura, K. (2002). External cavity diode lasers tuned with silicon MEMS. In *Optical Fiber Commun. Conf.*, paper TuO7.

Aoki, K., Kondo, J., Kondo, A., Mori, T., Mizuno, Y., Shimodaira, S., Imaeda, M., Kozuka, Y., Mitomi, O., and Minakata, M. (2004). High-performance optical modulator with a wide center electrode and thin x-cut $LiNbO_3$ substrate. *IEEE Photon. Technol. Lett.*, 16(12):2610–2612.

Appathurai, S., Mikhailov, V., Killey, R. I., and Bayvel, P. (2004). Investigation of the optimum alternate-phase RZ modulation format and its effectiveness in the suppression of intrachannel nonlinear distortion in 40-Gbit/s transmission over standard single-mode fiber. *IEEE J. Sel. Top. Quantum Electron.*, 10(2):239–249.

Arbore, M. A., Zhou, Y., Thiele, H., Bromage, J., and Nelson, L. (2003). S-band Erbium-doped fiber amplifiers for WDM transmission between 1488 and 1508 nm. In *Optical Fiber Commun. Conf.*, paper WK2.

Arimoto, S. (1972). An algorithm for computing the capacity of arbitrary discrete memoryless channels. *IEEE Trans. Info. Theory*, IT-18(1):14–20.

Atia, W. A. and Bondurant, R. S. (1999). Demonstration of return-to-zero signaling in both OOK and DPSK formats to improve receiver sensitivity in an optically preamplifier receiver. In *Proc. LEOS '99*. paper TuM3.

Azizoğlu, M. and Humblet, P. A. (1991). Envelope detection of orthogonal signals with phase noise. *J. Lightwave Technol.*, 9(10):1398–1410.

Bakhshi, B., Manna, M., Mohs, G., Kovsh, D. I., Lynch, R. L., M.Vaa, Golovchenko, E. A., Patterson, W. W., Anderson, W. T., Corbett, P., Jiang, S., Sanders, M. M., Li, H., Harvey, G. T., Lucero, A., and Abbott, S. M. (2004). First dispersion-flattened transpacific undersea system: From design to Terabit/s field trail. *J. Lightwave Technol.*, 22(1):233–241.

Barry, J. R. and Kahn, J. M. (1992). Carrier synchronization for homodyne and heterodyne detection of optical quadriphase-shift-keying. *J. Lightwave Technol.*, 10(12):1934–1951.

Becker, P. C., Olsson, N. A., and Simpson, J. R. (1999). *Erbium-Doped Fiber Amplifiers: Fundamental and Technology.* Academic Press, San Diego.

Becouarn, L., Vareille, G., Pecci, P., and Marcerou, J.F. (2003). 3 Tbit/s transmission (301 DPSK channels at 10.709 Gb/s) over 10270 km with a record efficiency of 0.65(bit/s)/Hz. In *European Conf. on Optical Commun.* postdeadline paper Th4.3.2.

Becouarn, L., Vareille, G., Dupont, S., Plantady, P., Marcerou, J.F., Klekamp, A., Dischler, R., Idler, W., and Charlet, G. (2004). 42 × 42.7 Gb/s RZ-DPSK transmission over a 4820 km long NZDSF deployed line using C-band-only EDFAs. In *Optical Fiber Commun. Conf.*, postdeadline paper PDP37.

Beling, A., Bach, H.-G., Schmidt, D., Mekonnen, G. G., Ludwig, R., Ferber, S., Schubert, C., Boerner, C., Schmauss, B., Berger, J., Schmidt, C., Troppenz, U., and Weber, H. G. (2003). Monolithically integrated balanced photodetector and its application in OTDM 160 Gbit/s DPSK transmission. *Electron. Lett.*, 29(16):1204–1205.

Bellotti, G. and Bigo, S. (2000). Cross-phase modulation suppressor for multispan dispersion-managed WDM transmissions. *IEEE Photon. Technol. Lett.*, 12(6):726–728.

Bellotti, G., Bigo, S., Cortes, P.-Y., Gauchard, S., and LaRochelle, S. (2000). 10 × 10 Gb/s cross-phase modulation suppressor for multispan transmissions using WDM narrow-band fiber Bragg gratings. *IEEE Photon. Technol. Lett.*, 12(10):1403–1405.

Benedetto, S., Gaudino, R., and Poggiolini, P. (1995a). Direct detection of optical digital transmission based on polarization shift keying modulation. *IEEE J. Sel. Areas Commun.*, 13(3):531–542.

Benedetto, S., Gaudino, R., and Poggiolini, P. (1995b). Performance of coherent optical polarization shift keying modulation in the presence of phase noise. *IEEE Trans. Commun.*, 43:1603–1612.

Benedetto, S. and Poggiolini, P. (1992). Theory of polarization shift keying modulation. *IEEE Trans. Commun.*, 40(4):708–721.

Benedetto, S. and Poggiolino, P. (1990). Performance evaluation of polarisation shift keying modulation schemes. *Electron. Lett.*, 26(4):256–258.

Bennett, B. R. and Soref, R. A. (1987). Electrorefraction and electroabsorption in InP, GaAs, GaSb, InAs, and InSb. *IEEE J. Quantum Electron.*, QE-23(12):2159–2166.

Bergano, N. S. and Davidson, C. R. (1996). Wavelength division multiplexing in long-haul transmission systems. *J. Lightwave Technol.*, 14(6):1299–1308.

Berrou, C. (2003). The ten-year-old Turbo codes are entering into service. *IEEE Commun. Mag.*, 41(8):110–116.

Berrou, C., Glavieux, A., and Thitimajshima, P. (1993). Near Shannon limit error-correcting coding and decoding: Turbo-codes (1). In *Proc. of IEEE Int'l Conf. on Commun.*, volume 2, pages 1064–1070, Geneva.

Betti, S., Curti, T., Daino, B., de Marchis, G., and Iannone, E. (1988). State of polarisation and phase noise independent coherent optical transmission system based on Stokes parameter detection. *Electron. Lett.*, 24(23):1460–1461.

Betti, S., de Marchis, G., Iannone, E., and Lazzaro, P. (1991). Homodyne optical coherent systems based on polarization modulation. *J. Lightwave Technol.*, 9(10):1314–1320.

Betti, S., de Marchis, G., and Iannone, E. (1995). *Coherent Optical Communication Systems.* John Wiley & Sons, New York.

Bigo, S. (2004). Multiterabit/s DWDM terrestrial transmission with bandwidth-limited optical filtering. *IEEE J. Sel. Top. Quantum Electron.*, 10(2):329–340.

Bigo, S., Frignac, Y., Charlet, G., Idler, W., Borne, S., Gross, H., Dischler, R., Poehlmann, W., Tran, P., Simonneau, C., Bayart, D., Veith, G., Jourdan, A., , and Hamaide, J.-P. (2001). 10.2 Tbit/s (256 × 42.7 Gbit/s PMD/WDM) transmission over 100 km TeraLight fiber with 1.28 bit/s/Hz spectral efficiency. In *Optical Fiber Commun. Conf.*, postdeadline paper PD25.

Bissessur, H., Charlet, G., Gohin, E., Simonneau, C., Pierre, L., and Idler, W. (2003). 1.6 Tbit/s (40 × 40 Gbit/s) DPSK transmission over 3 × 100 km of TeraLight fibre with direct detection. *Electron. Lett.*, 39(2):192–193.

Blachman, N. M. (1981). The effect of phase error on DPSK error probability. *IEEE Trans. Commun.*, COM-29(3):364–465.

Blachman, N. M. (1988). Gaussian noise - part II: Distribution of phase change of narrow-band noise plus sinusoid. *IEEE Trans. Info. Theory*, 34(6):1401–1405.

Blahut, R. (1972). Computation of channel capacity and rate-distortion functions. *IEEE Trans. Info. Theory*, IT-18(4):460–473.

Blow, K., Doran, N., and Phoenix, S. (1992). The soliton phase. *Opt. Commun.*, 88(2-3):137–140.

Bødker, E., Christensen, B., Jacobsen, G., Olsen, H. J., Madsen, P. W., Nielsen, S. K., Skipper, B. F., Thorsen, P., Hoffmann, B., and Jensen, T. (1991). Field test of fully implemented, highly sensitive 636 Mbit/s polarization diversity SF-FSK system. In *European Conf. Opt. Commun.*, pages 401–403.

Boivin, D., Hanna, M., Lacourt, P.-A., and Goedgebuer, J.-P. (2004). Reduction of phase jitter in dispersion-managed systems by in-line filtering. *Opt. Lett.*, 29(7):688–690.

Bosco, G. and Poggiolini, P. (2003). The effect of receiver imperfections on the performance of direct-detection optical systems using DPSK modulation. In *Optical Fiber Commun. Conf.*, paper ThE6.

Bosco, G. and Poggiolini, P. (2004a). On the Q factor inaccuracy in the performance analysis of optical direct-detection DPSK systems. *IEEE Photon. Technol. Lett.*, 16(2):665–667.

Bosco, G. and Poggiolini, P. (2004b). Analysis of impact of receiver imperfections on performance of optical DQPSK systems *Electron. Lett.*, 40(18):1147–1149.

Bosco, G., Carena, A., Curri, V., Gaudino, R., Poggiolini, P., and Benedetto, S. (2001). A novel analytical approach to the evaluation of the impact of fiber parametric gain on the bit error rate. *IEEE Trans. Commun.*, 49(12):2154–2163.

Bosco, G., Montorsi, G., and Benedetto, S. (2003). Soft decoding in optical systems. *IEEE Trans. Commun.*, 51(8):1258–1265.

Boyd, R. W. (2003). *Nonlinear Optics*. Academic Press, Amsterdam, 2 edition.

Brain, M. C., Creaner, M. J., Steele, R. C., Walker, N. G., Walker, G. R., Mellis, J., Al-Chalabi, S., Davidson, J., Rutherford, M., and Sturgess, I.C. (1990). Progress towards the field deployment of coherent optical fiber systems. *J. Lightwave Technol.*, 8(3):423–437.

Brener, I., Mikkelsen, B., Rottwitt, K., Burkett, W., Raybon, G., Stark, J. B., Parameswaran, K., Chou, M. H., Fejer, M. M., Chaban, E. E., Harel, R., Philen, D. L., and Kosinski, A. (2000). Cancellation of all Kerr nonlinearities in long fiber spans using a $LiNbO_3$ phase conjugator and Raman amplification. In *Optical Fiber Commun. Conf.*, postdeadline paper PD33.

Brittain, J. E. (2004). Electrical engineering hall of fame-Edwin H. Armstrong. *Proc. IEEE*, 92(3):575–578.

Bromage, J. (2004). Raman amplification for fiber communications systems. *J. Lightwave Technol.*, 22(1):79–93.

Bruyère, F. (1996). Impact of first- and second-order PMD in optical digital transmission systems. *Opt. Fiber Technol.*, 2(4):269–280.

Bruyère, F. and Audouin, O. (1994). Penalties in long-haul optical amplifier systems due to polarization dependent loss and gain. *IEEE Photon. Technol. Lett.*, 6(5):654–656.

Buchali, F. and Bülow, H. (2004). Adaptive PMD compensation by electrical and optical techniques. *J. Lightwave Technol.*, 22(4):1116–1126.

Bülow, H. (1998). System outage probability due to first- and second-order PMD. *IEEE Photon. Technol. Lett.*, 10(5):696–698.

Burns, W. K., Howerton, M. M., Moeller, R. P., R. Krähenbühl, R., McElhanon, R. W., and Greenblatt, A. S. (1999). Low drive voltage, broad-band $LiNbO_3$ modulators with and without etched ridges. *J. Lightwave Technol.*, 17(12):2551–2555.

Cahn, C. R. (1959). Performance of digital phase-modulation communication systems. *IRE Trans. Commun. Sys.*, CS-7(1):3–6.

Cai, J.-X., Nissov, M., Davidson, C. R., Pilipetskii, A. N., Mohs, G., Li, H., Cai, Y., Golovchenko, E. A., Lucero, A. J., Foursa, D. G., and Bergano, N. S. (2002). Long-haul 40 Gb/s DWDM transmission with aggregate capacities exceeding 1 Tb/s. *J. Lightwave Technol.*, 20(12):2247–2258.

Cai, Y., Morris, J. M., Adali, T., and Menyuk, C. R. (2003a). On Turbo code decoder performance in optical-fiber communication systems with dominating ASE noise. *J. Lightwave*

Technol., 20(9):1653–1663.

Cai, J.-X., Foursa, D. G., Davidson, C. R., Cai, Y., Domagala, G., Li, H., Liu, L., Patterson, W. W., Pilipetskii, A. N., Nissov, M., and Bergano, N. S. (2003b). A DWDM demonstration of 3.73 Tb/s over 11,000 km using 373 RZ-DPSK channels at 10 Gb/s. In *Optical Fiber Commun. Conf.*, postdeadline paper PD22.

Cai, J.-X., Foursa, D. G., Liu, L., Davidson, C. R., Cai, Y., Patterson, W. W., Lucero, A. J., Bakhshi, B., Mohs, G., Corbett, P. C., Gupta, V., Anderson, W., Vaa, M., Domagala, G., Mazurczyk, M., Li, H., Jiang, S., Nissov, M., Pilipetskii, A. N., and Bergano, N. S. (2004). RZ-DPSK field trial over 13,100 km of installed non slope-matched submarine fibers. In *Optical Fiber Commun. Conf.*, postdeadline paper PDP34.

Cai, J.-X., Foursa, D. G., Liu, L., Davidson, C. R., Cai, Y., Patterson, W. W., Lucero, A. J., Bakhshi, B., Mohs, G., Corbett, P. C., Gupta, V., Anderson, W., Vaa, M., Domagala, G., Mazurczyk, M., Li, H., Jiang, S., Nissov, M., Pilipetskii, A. N., and Bergano, N. S. (2005). RZ-DPSK field trial over 13,100 km of installed non slope-matched submarine fibers. *J. Lightwave Technol.*, 23(1):95–103.

Calderbank, A. R. and Ozarow, L. H. (1990). Nonequiprobable signaling on the Gaussian channel. *IEEE Trans. Info. Theory*, 36(4):726–740.

Calvani, R., Caponi, R., and Cisternino, F. (1988). Polarization phase-shift keying: A coherent transmission technique with differential heterodyne detection. *Electron. Lett.*, 24(10):642–643.

Cameron, R. H. and Martin, W. T. (1945). Evaluation of various Wiener integrals by use of certain Sturm-Liouville differential equations. *Bull. Am. Math. Soc.*, 51(2):73–90.

Cantrell, P. E. and Ojha, A. K. (1987). Comparison of generalized Q-function algorithms. *IEEE Trans. Info. Theory*, IT-33(7):591–596.

Caplan, D. O. and Atia, W. A. (2001). A quantum-limited optically-matched communication link. In *Optical Fiber Commun. Conf.*, paper MM2.

Caponio, N., Delpiano, F., Gambini, P., Puleo, M., Seano, V., and Vezzoni, E. (1991). Polarisation insensitive coherent transmission by synchronous intrabit polarisation spreading. *Electron. Lett.*, 27(4):337 – 338.

Cartaxo, A. V. T. (1999). Cross-phase modulation in intensity modulation-direct detection WDM systems with multiple optical amplifiers and dispersion compensators. *J. Lightwave Technol.*, 17(2):178–190.

Cartaxo, A. V. T., Wedding, B., and Idler, W. (1999). Influence of fiber nonlinearity on the fiber transfer function: Theoretical and experimental analysis. *J. Lightwave Technol.*, 17(10):1806–1813.

Cartledge, J. C. and Srinivasan, R. C. (1997). Extraction of DFB laser rate equation parameters for system simulation purposes. *J. Lightwave Technol.*, 15(5):852–860.

Casadevall, F. J. and Valdovinos, A. (1995). Performance analysis of QAM modulations applied to the LINC transmitter. *IEEE Trans. Vehicular Technol.*, 42(4):399–406.

Casey, H. C. and Panish, M. B. (1978). *Heterostructure Lasers.* Academic Press, San Diego.

Caves, C. M. and Drummond, P. D. (1994). Quantum limits on Bosonic communication rates. *Rev. Mod. Phys.*, 66(2):481–537.

Chan, B. and Conradi, J. (1997). On the non-Gaussian noise in Erbium-doped fiber amplifiers. *J. Lightwave Technol.*, 15(4):680–687.

Chan, V. W. S. (1987). Space coherent optical communication systems - an introduction. *J. Lightwave Technol.*, 5(4):633–637.

Chan, V. W. S. (2000). Optical space communications. *IEEE J. Sel. Top. Quantum Electron.*, 6(6):959–975.

Chan, V. W. S. (2003). Optical satellite networks. *J. Lightwave Technol.*, 21(11):2811–2827.

Charlet, G., Lazaro, J., Corbel, E., Tran, P., A.Klekamp, Lopez, T., Mardoyan, H., Idler, W., Konczykowska, A., Thiery, J.-P., Dischlet, R., and Bigo, S. (2003). One-hundred WDM-channel transatlantic transmission experiment at 43 Gbit/s using Raman repeaters with large 65 km spacing. In *European Conf. on Optical Commun.* postdeadline paper Th4.3.4.

Charlet, G., Corbel, E., Lazaro, J., Klekamp, A., Dischler, R., Tran, P., Idler, W., Mardoyan, H., Konczykowska, A., Jorge, F., and Bigo, S. (2004a). WDM transmission at 6 Tbit/s capacity over transatlantic distance, using 42.7 Gb/s differential phase-shift keying without pulse carver. In *Optical Fiber Commun. Conf.*, postdeadline paper PDP36.

Charlet, G., Dischler, R., Klekamp, A., Tran, P., Mardoyan, H., Pierre, L., Idler, W., and Bigo, S. (2004b). WDM bit-to-bit alternate-polarisation RZ-DPSK transmission at 40 × 42.7 Gbit/s over transpacific distance with large Q-factor margin. In *European Conf. on Optical Commun.* postdeadline paper Th4.4.5.

Charlet, G., Corbel, E., Lazaro, J., Klekamp, A., Dischler, R., Tran, P., Idler, W., Mardoyan, H., Konczykowska, A., Jorge, F., and Bigo, S. (2005). WDM transmission at 6 Tbit/s capacity over transatlantic distance, using 42.7 Gb/s differential phase-shift keying without pulse carver. *J. Lightwave Technol.*, 23(1):104–107.

Chen, W. H. and Way, W. I. (2004). Multichannel single-sideband SCM/DWDM transmission systems. *J. Lightwave Technol.*, 22(7):1679–1693.

Chen, Y.-K. and Lee, C.-C. (1998). Fiber Bragg grating-based large nonblocking multiwavelength cross-connects. *J. Lightwave Technol.*, 16(10):1746–1756.

Chen, Y., Zucker, J. E., Sauer, N. J., and Chang, T. Y. (1992). Polarization-independent strained InGaAs/InGaAlAs quantum-well phase modulators. *IEEE Photon. Technol. Lett.*, 4(10):1992.

Chen, C.-C., Forte, H., Carenco, A., Goedgebuer, J.-P., and Armbruster, V. (1997a). Phase correction by laser ablation of a polarization independent $LiNbO_3$ Mach-Zehnder modulator. *IEEE Photon. Technol. Lett.*, 9(10):1361–1363.

Chen, D., Fetterman, H. R., Chen, A., Steier, W. H., Dalton, L. R., Wang, W., and Shi, Y. (1997b). Demonstration of 110 GHz electro-optic polymer modulators. *Appl. Phys. Lett.*, 70(25):3335–3337.

Cheng, K. S. and Conradi, J. (2002). Reduction of pulse-to-pulse interaction using alternative RZ formats in 40-Gb/s systems. *IEEE Photon. Technol. Lett.*, 14(1):98–100.

Cheng, Y. H. and Okoshi, T. (1989). Phase-noise-cancelling dual-frequency heterodyne optical fibre communication systems. *Electron. Lett.*, 25(13):395–396.

Cheng, Y. H., Okoshi, T., and Ishida, O. (1989). Performance analysis and experiment of a homodyne receiver insensitive to both polarization and phase fluctuations. *J. Lightwave Technol.*, 7(2):368–374.

Chi, N., Mikkelsen, C., Xu, L., Zhang, J., Holm-Nielsen, P. V., Ou, H., Seoane, J., Peucheret, C., and Jeppesen, P. (2003). Transmission and label encoding/erasure of orthogonally labelled signal using 40 Gbit/s RZ-DPSK payload and 2.5 Gbit/s IM label. *Electron. Lett.*, 39(18):1335–1337.

Chiang, T.-K., Kagi, N., Fong, T. K., Marhic, M. E., and Kazovsky, L. G. (1994). Cross-phase modulation in dispersive fibers: Theoretical and experimental investigation of the impact of modulation frequency. *IEEE Photon. Technol. Lett.*, 6(6):733–736.

Chiang, T.-K., Kagi, N., Marhic, M. E., and Kazovsky, L. G. (1996). Cross-phase modulation in fiber links with multiple optical amplifiers and dispersion compensators. *J. Lightwave Technol.*, 14(3):249–260.

Chikama, T., Naito, T., Onaga, H., Kiyonaga, T., Watanabe, S., Suyama, M., and Wenke, G. (1988). 1.2 Gbit/s, 201 km optical DPSK heterodyne transmission experiment using a compact, stable external fibre cavity DFB laser module. *Electron. Lett.*, 24(10):636–637.

Chikama, T., Naito, T., Watanabe, S., Kiyonaga, T., Suyama, M., and Kuwahara, H. (1990a). Optical heterodyne image-rejection receiver for high-density optical frequency division multiplexing system. *J. Lightwave Technol.*, 8(6):1087–1093.

Chikama, T., Watanabe, S., Naito, T., Onaka, H., Kiyonaga, T., Onoda, Y., Miyata, H., Suyama, M., Seino, M., and Kuwahara, H. (1990b). Modulation and demodulation techniques in optical heterodyne PSK transmission systems. *J. Lightwave Technol.*, 8(3):309–322.

Chinn, S. R., Boroson, D. M., and Livas, J. C. (1996). Sensitivity of optically preamplified DPSK receivers with Fabry-Perot filters. *J. Lightwave Technol.*, 14(3):370–376.

Cho, P. S., Grigoryan, V. S., Godin, Y. A., Salamon, A., and Achiam, Y. (2003). Transmission of 25-Gb/s RZ-DQPSK signals with 25-GHz channel spacing over 1000 km of SMF-28 fiber. *IEEE Photon. Technol. Lett.*, 15(3):473–475.

Cho, P. S., Achiam, Y., Levy-Yurista, G., Margalit, M., Gross, Y., and Khurgin, J. B. (2004a). Investigation of SOA nonlinearities on the amplification of DWDM channels with spectral efficiency up to 2.5 b/s/Hz. *IEEE Photon. Technol. Lett.*, 16(3):918–920.

Cho, P. S., Harston, G., Kerr, C. J., Greenblatt, A. S., Kaplan, A., Achiam, Y., Levy-Yurista, G., Margalit, M., Gross, Y., and Khurgin, J. B. (2004b). Investigation of 2-b/s/Hz 40-Gb/s DWDM transmission over 4 × 100 km SMF-28 fiber using RZ-DQPSK and polarization multiplexing. *IEEE Photon. Technol. Lett.*, 16(2):656–658.

Cho, P. S., Harston, G., Kerr, C. J., Greenblatt, A. S., Kaplan, A., Achiam, Y., and Shpantzer, I. (2004c). Coherent homodyne detection of BPSK signals using time-gated amplification and $LiNbO_3$ optical 90° hybrid. *IEEE Photon. Technol. Lett.*, 16(7):1727–1729.

Choi, W. J., Bond, A. E., Kim, J., Zhang, J., Jambunathan, R., Foulk, H., O'Brien, S., van Norman, J., Vandegrift, D., Wanamaker, C., Shakespeare, J., and Cao, H. (2002). Low insertion loss and low dispersion penalty InGaAsP quantum-well high-speed electroabsorption modulator for 40-Gb/s very-short-reach, long-reach, and long-haul applications. *J. Lightwave Technol.*, 20(12):2052–2056.

Chowdhury, A. and Essiambre, R.-J. (2004). Optical phase conjugation and pseudolinear transmission. *Opt. Lett.*, 29(10):1105–1107.

Chowdhury, A., Raybon, G., Essiambre, R.-J., Sinsky, J., Adamiecki, A., Leuthold, J., Doerr, C. R., and Chandrasekhar, S. (2005). Compensation of intra-channel nonlinearities in 40 Gb/s pseudo-linear systems using optical phase conjugation. *J. Lightwave Technol.*, 23(1):172–177.

Chraplyvy, A. R. (1990). Limitation on lightwave communications imposed by optical-fiber nonlinearities. *J. Lightwave Technol.*, 8(12):1548–1557.

Chraplyvy, A. R. and Tkach, R. W. (1993). What is the actual capacity of single-mode fibers in amplified lightwave systems? *IEEE Photon. Technol. Lett.*, 5(6):666–668.

Chraplyvy, A. R., Tkach, R. W., Gnauck, A. H., Kasper, B. L., and Derosier, R.M. (1989). 8 Gbit/s FSK modulation of DFB lasers with optical demodulation. *Electron. Lett.*, 25(5):319–321.

Chung, S.-Y., Forney, G. D., Richardson, T. J., and Urbanke, R. (2001). On the design of low-density parity-check codes within 0.0045 dB of the Shannon limit. *IEEE Commun. Lett.*, 5(1):58–60.

Cimini Jr., L. J. and Foschini, G. J. (1993). Can multilevel signaling improve the spectral efficiency of ASK optical FDM systems? *IEEE Trans. Commun.*, 41(7):1084–1090.

Cimini Jr., L. J., Habbab, I. M. I., Yang, S., Rustako, A. J., Liou, K. Y., and Burrus, C. A. (1988). Polarisation-insensitive coherent lightwave system using wide-deviation FSK and data-induced polarisation switching. *Electron. Lett.*, 24(6):358–360.

Cites, J. S. and Ashley, P. R. (1994). High-performance Mach-Zehnder modulators in multiple quantum well GaAs/AlGaAs. *J. Lightwave Technol.*, 12(7):1167–1173.

Cline, T. W., Delavaux, J.-P. M, Dutta, N. K., Eijk, P. V., Kuo, C. Y., Owen, B., Park, Y. K., Pleiss, T. C., Riggs, R. S., Tench, R. E., Twu, Y., Tzeng, L. D., and Wagner, E. J. (1990). A field demonstration of 1.7 Gb/s coherent lightwave regenerators. *IEEE Photon. Technol. Lett.*, 2(6):425–427.

Coldren, L. A. (2000). Monolithic tunable diode lasers. *IEEE J. Sel. Top. Quantum Electron.*, 6(6):988–999.

Coldren, L. A. and Corzine, C. W. (1995). *Diode Lasers and Photonic Integrated Circuits.* John Wiley & Sons, New York.

Coldren, L. A., Fish, G. A., Akulova, Y., Barton, J. S., Johansson, L., and Coldren, C. W. (2004). Tunable semiconductor lasers: A tutorial. *J. Lightwave Technol.*, 22(1):193–202.

Cornwell, W. D. and Andonovic, I. (1996). interferometeric noise for a single interferer: Comparison between theory and experiment. *Electron. Lett.*, 32(15):1501–1502.

Corvaja, R., Pierobon, G. L., and Tomba, L. (1992). Accurate performance evaluation of weakly coherent optical systems. *J. Lightwave Technol.*, 10(11):1665–1673.

Corvini, P. J. and Koch, T. L. (1987). Computer simulation of high-bit-rate optical fiber transmission using single-frequency lasers. *J. Lightwave Technol.*, LT-5(11):1591–1595.

Cover, T. M. and Thomas, J. A. (1991). *Elements of Information Theory.* John Wiley & Sons, New York.

Cox, D. C. (1974). Linear amplification with nonlinear components. *IEEE Trans. Commun.*, COM-22(12):1942–1945.

Cox, D. C. and Leck, R. P. (1975). Component signal separation and recombination for linear amplification with nonlinear components. *IEEE Trans. Commun.*, COM-23(11):1281–1287.

Creaner, M. J., Steele, R. C., Marshall, I., Walker, G. R., Walker, N. G., Mellis, J., Al Chalabi, S., Sturgess, I., Rutherford, M., Davidson, J., and Brain, M. (1988). Field demonstration of 565 Mbit/s DPSK coherent transmission system over 176 km of installed fibre. *Electron. Lett.*, 24(22):1354–1356.

Cvijetic, M. (1996). *Coherent and Nonlinear Lightwave Communications.* Artech House, Boston.

Dadap, J. I., Chu, P. B., Brener, I., Pu, C., Lee, C. D., Bergman, K., Bonadeo, N., Chau, T., Chou, M., Doran, R., Gibson, R., Harel, R., Johnson, J. J., Lee, S. S., Park, S., Peale, D. R., Rodriguez, R., Tong, D., Tsai, M., Wu, C., Zhong, W., Goldstein, E. L., Lin, L. Y., and Walker, J. A. (2003). Modular MEMS-based optical cross-connect with large port-count. *IEEE Photon. Technol. Lett.*, 15(12):1773–1775.

Dainesi, P., Küng, A., Chabloz, M., Lagos, A., Flückiger, Ph., Ionescu, A., Fazan, P., Declerq, M., Renaud, Ph., and Robert, Ph. (2000). CMOS compatible fully integrated Mach-Zehnder interferometer in SOI technology. *IEEE Photon. Technol. Lett.*, 12(6):660–662.

Daino, B., Galeotti, M., and Sette, D. (1974). Error-rate measurements in an atmospheric twin-channel optical link. *IEEE J. Quantum Electron.*, QE-10(1):86–87.

Dalton, L., Harper, A., Ren, A., Wang, F., Todorova, G., Zhang, J. Chenand C., and Lee, M. (1999). Polymeric electro-optic modulators: From chromophore design to integration with semiconductor very large scale integration electronics and silica fiber optics. *Ind. Eng. Chem. Res.*, 38(1):8–33.

Danielsen, S. L., Joergensen, C., Mikkelsen, B., and Stubkjaer, K. E. (1998). Analysis of interferometric crosstalk in optical switch blocks using moment generating functions. *IEEE Photon. Technol. Lett.*, 10(11):1635–1637.

Darcie, T. E. and Glance, B. (1986). Optical heterodyne image-rejection mixer. *Electron. Lett.*, 22(15):825–826.

Davis, A. W., Pettitt, M. J., King, J. P., and Wright, S. (1987). Phase diversity techniques for coherent optical receivers. *J. Lightwave Technol.*, LT-5(4):561–572.

Davis, A. W. and Wright, S. (1986). Coherent optical receiver for 680 Mbit/s using phase diversity. *Electron. Lett.*, 22(1):9–11.

Davis, M. H. A. (1980). Capacity and cutoff rate for Poisson-type channels. *IEEE Trans. Info. Theory*, IT-26(6):710–715.

DeLange, O. E. (1972). Wideband optical communications: Part II-frequency division multiplexing. *Proc. IEEE*, 58(10):1683–1690.

Delavaux, J.-M. P. and Riggs, R. (1990). High performance packaged optical coherent hybrid. *Electron. Lett.*, 16(14):1022–1024.

Delavaux, J.-M. P., Park, Y. K., Barski, S., Lefevre, H., Failes, M., and Warner, C. (1990). All-fibre optical hybrid for coherent polarisation diversity receivers. *Electron. Lett.*, 16(16):1303–1305.

Derevyanko, S. A. and Turitsyn, S. K. (2004). Reduction of the phase jitter in differential phase-shift-keying soliton transmission systems by in-line Butterworth filters. *Opt. Lett.*, 29(1):35–37.

Derr, F. (1990). Optical QPSK homodyne transmission of 280 Mbit/s. *Electron. Lett.*, 26(6):401–403.

Derr, F. (1992). Coherent optical QPSK intradyne system: Concept and digital receiver realization. *J. Lightwave Technol.*, 10(9):1290–1296.

Desurvire, E. (1994). *Erbium-Doped Fiber Amplifiers.* John Wiley & Sons, New York.

Desurvire, E. (2000). Fundamental information-density limits in optically amplified transmission: An entropy analysis. *Opt. Lett.*, 25(10):701–703.

Desurvire, E. (2002). A quantum model for optically amplified nonlinear transmission systems. *Opt. Fiber Technol.*, 8(3):210–230.

Desurvire, E. (2003). Quantum and nonlinearity limitations of the optical communication channel. *C. R. Physique*, 4(1):11–28.

Desurvire, E., Simpson, J. R., and Becker, P. C. (1987). High gain Erbium-doped traveling wave fiber amplifier. *Opt. Lett.*, 12(11):888–890.

Dietrich, E., Enning, B., Gross, B., and Knupke, E. (1987). Heterodyne transmission of a 560 Mbit/s optical signal by means of polarisation shift keying. *Electron. Lett.*, 23:421–422.

Dods, S. D. and Tucker, R. S. (2001). A comparison of the homodyne crosstalk characteristics of optical add-drop multiplexers. *J. Lightwave Technol.*, 19(12):1829–1838.

Dods, S. D. and Tucker, R. S. (2002). Analysis of incoherent homodyne crosstalk in optical networks. *J. Opt. Commun.*, 23(4):178–181.

Dolfi, D. W., Nazarathy, M., and Jungerman, R. L. (1988). 40 GHz electro-optic modulator with 7.5 V drive voltage. *Electron. Lett.*, 24(9):528–529.

Domash, L., Wu, M., Nemchuk, N., and Ma, E. (2004). Tunable and switchable multiple-cavity thin film filters. *J. Lightwave Technol.*, 22(1):126–135.

Dorrer, C., Kilper, D. C., Stuart, H. R., Raybon, G., and Raymer, M. G. (2003). Linear optical sampling. *IEEE Photon. Technol. Lett.*, 15(12):1746–1748.

Dorrer, C., Doerr, C. R., Kang, I., Ryf, R., Leuthold, J., and Winzer, P. J. (2005). Measurement of eye diagrams and constellation diagrams of optical sources using linear optics and waveguide technology. *J. Lightwave Technol.*, 23(1):178–186.

Douglas, A. (1990). Who invented the superheterodyne? *Proc. of the Radio Club of America*, 64(3):123–142.

Dragone, C. (1991). An $N \times N$ optical multiplexer using a planar arrangement of two star couplers. *IEEE Photon. Technol. Lett.*, 3(9):812–815.

Dragone, C. (2005). Low-loss wavelength routers for WDM optical networks and high-capacity IP routers. *J. Lightwave Technol.*, 23(1):66–79.

Einarsson, G. H. (1996). *Principles of Lightwave Communications.* John Wiley & Sons, New York.

Eiselt, M. (1999). Limits on WDM systems due to four-wave mixing: A statistical approach. *J. Lightwave Technol.*, 17(11):2261–2267.

Elrefaie, A. E. and Wagner, R. E. (1991). Chromatic dispersion limitations for FSK and DPSK systems with direct detection receivers. *IEEE Photon. Technol. Lett.*, 3(1):71–73.

Elrefaie, A. E., Wagner, R. E., Atlas, D. A., and Daut, D. G. (1988). Chromatic dispersion limitations in coherent lightwave transmission systems. *J. Lightwave Technol.*, 6(5):704–709.

Emura, K., Shikada, M., Fujita, S., Mito, L., Honmou, H., and Minemura, K. (1984). Novel optical FSK heterodyne single filter detection system using a directly modulated DFB-laser diode. *Electron. Lett.*, 20(24):1022–1023.

Emura, K., Sato, K., Yamazaki, S., Murata, S., Shikada, M., and Minemura, K. (1990a). Optimum system design for CPFSK heterodyne delay demodulation system with DFB LDs. *J. Lightwave Technol.*, 8(2):251–258.

Emura, K., Yamazaki, S., Yamaguchi, M., Shikada, M., and Minemura, K. (1990b). Optical FSK heterodyne dual filter detection system for taking advantage of DFB LD applications. *J. Lightwave Technol.*, 8(2):243–250.

Essiambre, R.-J., Mikkelsen, B., and Raybon, G. (1999). Intra-channel cross-phase modulation and four-wave mixing in high-speed TDM systems. *Electron. Lett.*, 35(19):1576–1578.

Essiambre, R.-J., Raybon, G., and Mikkelsen, B. (2002). Pseudo-linear transmission of high-speed TDM signals: 40 and 160 Gb/s. In Kaminow, I. P. and Li, T., editors, *Optical Fiber*

Telecommunications, volume IV. Academic Press, San Diego.

Favre, F., Jeunhomme, L., Joindot, I., Monerie, M., and Simon, J. C. (1981). Progress towards heterodyne-type single-mode fiber communication systems. *IEEE J. Quantum Electron.*, QE-17(6):897–906.

Favre, F. and LeGuen, D. (1980). High frequency stability of laser diode for heterodyne communication systems. *Electron. Lett.*, 16(18):709–710.

Favre, F. and LeGuen, D. (1982). Effect of semiconductor phase noise on BER performance in an optical DPSK heterodyne-type experiment. *Electron. Lett.*, 18(22):964–965.

Fleming, M. W. and Mooradian, A. (1981a). Fundamental line broadening of single-mode (GaAl)As diode lasers. *Appl. Phys. Lett.*, 23(7):511–513.

Fleming, M. W. and Mooradian, A. (1981b). Spectral characteristics of external-cavity controlled semiconductor lasers. *IEEE J. Quantum Electron.*, QE-17(1):44–59.

Forestieri, E. (2000). Evaluating the error probability in lightwave systems with chromatic dispersion, arbitrary pulse shape and pre- and postdetection filtering. *J. Lightwave Technol.*, 18(11):1493–1503.

Forghieri, F., Tkach, R. W., and Chraplyvy, A. R. (1995). Effect of modulation statistics on Raman crosstalk in WDM systems. *IEEE Photon. Technol. Lett.*, 7(1):101–103.

Forghieri, F., Tkach, R. W., and Chraplyvy, A. R. (1997). Fiber nonlinearities and their impact on transmission systems. In Kaminow, I. P. and Koch, T. L., editors, *Optical Fiber Telecommunications*, volume IIIA, pages 196–264. Academic Press, San Diego, CA.

Forzati, M., Mårtensson, J., Berntson, A., Djupsjöbacka, A., and Johannisson, P. (2002). Reduction of intrachannel four-wave mixing using the alternate-phase RZ modulation format. *IEEE Photon. Technol. Lett.*, 14(9):1285–1287.

Foschini, G. J. and Poole, C. D. (1991). Statistical theory of polarization dispersion in single mode fibers. *J. Lightwave Technol.*, 9(11):1439–1456.

Foschini, G. J. and Vannucci, G. (1988). Characteristing filtered light waves corrupted by phase noise. *IEEE Trans. Info. Theory*, 34(6):1438–1448.

Foschini, G. J., Gitlin, R. D., and Weinstein, S. B. (1974). Optimization of two-dimensional signal constellations in the presence of Gaussian noise. *IEEE Trans. Commun.*, COM-22(1):28–38.

Foschini, G. J., Greenstein, L. J., and Vannucci, G. (1988). Noncoherent detection of coherent lightwave signals corrupted by phase noise. *IEEE Trans. Commun.*, 36(3):306–314.

Foschini, G. J., Vannucci, G., and Greenstein, L. J. (1989). Envelope statistics for filtered optical signals corrupted by phase noise. *IEEE Trans. Commun.*, 37(12):1293–1302.

Foschini, G. J., Jopson, R. M., Nelson, L. E., and Kogelnik, H. (1999). The statistics of PMD-induced chromatic fiber dispersion. *J. Lightwave Technol.*, 17(9):1560–1565.

Foschini, G. J., Jopson, R. M., Nelson, L. E., and Kogelnik, H. (2001). Statistics of second-order PMD depolarization. *J. Lightwave Technol.*, 19(12):1882–1886.

Francia, C., Bruyère, F., Penninckx, D., and Chbat, M. (1998). PMD second-order effects on pulse propagation in single-mode optical fibers. *IEEE Photon. Technol. Lett.*, 10(12):1739–1741.

Frateschi, N. C., Zhang, J., Choi, W. J., Gebretsadik, H., Jambunathan, R., and Bond, A. E. (2004). High performance uncooled C-band, 10 Gbit/s InGaAlAs MQW electroabsorption modulator integrated to semiconductor amplifier in laser-integrated modules. *Electron. Lett.*, 40(2):140–141.

Frignac, Y., Charlet, G., Idler, W., Dischler, R., Tran, P., Lanne, S., Borne, S., Martinelli, C., Veith, G., Jourdan, A., Hamaide, J.-P., and Bigo, S. (2002). Transmission of 256 wavelength-division and polarization-division-multiplexed channels at 42.7 Gb/s (10.2 Tb/s capacity) over 3 × 100 km of TeraLight^{TM} fiber. In *Optical Fiber Commun. Conf.*, postdeadline paper FC5.

Fujiwara, T., Watanabe, A., and Mori, H. (1990). Measurement of uniformity of driving voltage in Ti:LiNbO3 waveguides using Mach-Zehnder interferometers. *IEEE Photon. Technol. Lett.*, 2(4):260–261.

Fukada, Y. (2002). Probability density function of polarization dependent loss (PDL) in optical transmission system composed of passive devices and connecting fibers. *J. Lightwave Technol.*, 20(6):953–964.

Fukuchi, K. (2002). Wideband and ultra-dense WDM transmission technologies toward over 10-Tb/s capacity. In *Optical Fiber Commun. Conf.*, paper ThX5.

Fukuchi, K., Kasamatsu, T., Morie, M., Ohhira, R., Ito, T., Sekiya, K., Ogasahara, D., and Ono, T. (2001). 10.92-Tb/s (273 × 40-Gb/s) triple-band/ultra-dense WDM optical-repeatered transmission experiment. In *Optical Fiber Commun. Conf.*, postdeadline paper PD24.

Funabashi, M., Nasu, H., Mukaihara, T., Kimoto, T., Shinagawa, T., Kise, T., Takaki, K., Takagi, T., Oike, M., Nomura, T., and Kasukawa, A. (2004). Recent advances in DFB lasers for ultradense WDM applications. *IEEE J. Sel. Top. Quantum Electron.*, 10(2):312–320.

Galtarossa, A. and Palmieri, L. (2004). Reflectometric measurements of polarization properties in optical-fiber links. *IEEE Trans. Instr. & Measur.*, 53(1):86–94.

Galtarossa, A., Palmieri, L., Schiano, M., and Tambosso, T. (2001). Measurement of birefringence correlation length in long, single-mode fibers. *Opt. Lett.*, 26(13):962–964.

Garcia, J. S., Gonzalez, A. G., and Iribas, M. L. (1996). Polarization mode dispersion power penalty: Influence of rise/fall times, receiver Q and amplifier noise. *IEEE Photon. Technol. Lett.*, 8(12):1719–1721.

Gardiner, C. W. (1985). *Handbook of Stochastic Methods.* Springer, Berlin, second edition.

Garrett, I., Bond, D. J., Waite, J. B., Lettis, D. S. L., and Jacobsen, G. (1990). Impact of phase noise in weakly coherent systems: A new and accurate approach. *J. Lightwave Technol.*, 8(3):329–337.

Geist, J. M. (1990). Capacity and cutoff rate for dense M-ary PSK constellations. In *Proc. of IEEE Mil. Commun. Conf.*, pages 168–170, Monterey, CA.

Gene, J. M., Soler, M., Killey, R. I., and Prat, J. (2004). Investigation of 10-Gb/s optical DQPSK systems in presence of chromatic dispersion, fiber nonlinearities, and phase noise. *IEEE Photon. Technol. Lett.*, 16(3):924–926.

Georges, T. (1995). Perturbation theory for the assessment of soliton transmission control. *Opt. Fiber Technol.*, 1(2):97–116.

Giles, C. R., Kahn, J. M., Korotky, S. K., Veselka, J. J., Burrus, C. A., Perino, J., and Presby, H. M. (1991). Polarization effects on ultralong distance signal transmission in amplified optical-fiber loops. *IEEE Photon. Technol. Lett.*, 3(10):948–951.

Gill, D. M., Liu, X., Wei, X., Banerjee, S., and Su, Y. (2003). $\pi/2$ alternate-phase ON-OFF keyed 40-Gb/s transmission on standard single-mode fiber. *IEEE Photon. Technol. Lett.*, 15(12):1776–1778.

Gimlett, J. L. and Cheung, N. K. (1989). Effects of phase-to-intensity noise conversion by multiple reflections on Gigabit-per-second DFB laser transmission systems. *J. Lightwave Technol.*, 7(6):888–895.

Gimlett, J. L., Vodhanel, R. S., Choy, M. M., Elrefaie, A. F., Cheung, N. K., and Wagner, R. E. (1987). A 2-Gbit/s optical FSK heterodyne transmission experiment using a 1520-nm DFB laser transmitter. *J. Lightwave Technol.*, LT-5(9):1315–1324.

Glance, B. (1986a). Performance of homodyne detection of binary PSK optical signals. *J. Lightwave Technol.*, LT-4(2):228–235.

Glance, B. S. (1986b). An optical heterodyne mixer providing image-frequency rejection. *J. Lightwave Technol.*, LT-4(11):1722–1725.

Glance, B. (1987). Polarization independent coherent optical receiver. *J. Lightwave Technol.*, LT-5(2):274–276.

Glance, B., Eisenstein, G., Fitzgerald, P. J., Pollock, K. J., and Raybon, G. (1988). Sensitivity of an optical heterodyne receiver in presence of an optical preamplifier. *Electron. Lett.*, 24(19):1229–1230.

Gnauck, A. H. and Winzer, P. J. (2005). Optical phase-shift-keyed transmission. *J. Lightwave Technol.*, 23(1):115–130.

Gnauck, A. H., Reichmann, K. C., Kahn, J. M., Korotky, S. K., Veselka, J. J., and Koch, T.L. (1990). 4-Gb/s heterodyne transmission experiments using ASK, FSK and DPSK modulation. *IEEE Photon. Technol. Lett.*, 2(12):908–910.

Gnauck, A. H., Korotky, S. K., Veselka, J. J., Nagel, J., Kemmerer, C. T., Minford, W. J., and Moser, D. T. (1991). Dispersion penalty reduction using an optical modulator with adjustable chirp. *IEEE Photon. Technol. Lett.*, 3(10):916–918.

Gnauck, A. H., Raybon, G., Chandrasekhar, S., Leuthold, J., Doerr, C., Stulz, L., Agrawal, A., Banerjee, S., Grosz, D., Hunsche, S., Kung, A., Marhelyuk, A., Maymar, D., Movassaghi, M., Liu, X., Xu, C., Wei, X., and Gill, D. M. (2002). 2.5 Tb/s (64 × 42.7 Gb/s) transmission over 40 × 100 km NZDSF using RZ-DPSK format and all-Raman-amplified spans. In *Optical Fiber Commun. Conf.*, postdeadline paper FC2.

Gnauck, A. H., Chandrasekhar, S., Leuthold, J., and Stulz, L. (2003a). Demonstration of 42.7-Gb/s DPSK receiver with 45 photon/bit sensitivity. *IEEE Photon. Technol. Lett.*, 15(1):99–101.

Gnauck, A. H., Raybon, G., Bernasconi, P. G., Leuthold, J., Doerr, C. R., and Stulz, L. W. (2003b). 1-Tb/s (6 × 170.6 Gb/s) transmission over 2000-km NZDF using OTDM and RZ-DPSK format. *IEEE Photon. Technol. Lett.*, 15(11):1618–1620.

Gnauck, A. H., Raybon, G., Chandrasekhar, S., Leuthold, J., Doerr, C., Stulz, L., and Burrows, E. (2003c). 25 40-Gb/s copolarized DPSK transmission over 12 100-km NZDF with 50-GHz channel spacing. *IEEE Photon. Technol. Lett.*, 15(3):467–469.

Gnauck, A. H., Leuthold, J., Xie, C., Kang, I., Chandrasekhar, S., Bernasconi, P., Doerr, C., Buhl, L., Bull, J. D., Jaeger, N. A. F., Kato, H., and Guest, A. (2004a). 6 × 42.7-Gb/s transmission over ten 200-km EDFA-amplified SSMF spans using polarization-alternating RZ-DPSK. In *Optical Fiber Commun. Conf.*, postdeadline paper PDP35.

Gnauck, A. H., Winzer, P. J., Chandrasekhar, S., and Dorrer, C. (2004b). Spectrally efficient (0.8 b/s/Hz) 1-Tb/s (25 × 42.7 Gb/s) RZ-DQPSK transmission over 28 100-km SSMF spans with 7 optical add/drops. In *European Conf. on Optical Commun.* postdeadline paper Th.4.4.1.

Gockenbach, M. S. and Kearsley, A. J. (1999). Optimal signal sets for non-Gaussian detectors. *SIAM J. on Optimization*, 9(2):316–326.

Goldstein, E. L., Eskildsen, L., and Elrefaie, A. F. (1994). Performance implications of component crosstalk in transparent lightwave networks. *IEEE Photon. Technol. Lett.*, 6(5):657–660.

Golovchenko, E. A., Pilipetskii, A. N., Bergano, N. S., Davidson, C. R., Khatri, F. I., Kimball, R. M., and Mazurczyk, V. J. (2000). Modeling of transoceanic fiber-optic WDM communication systems. *IEEE J. Sel. Top. Quantum Electron.*, 6(2):337–347.

Goodwin, F. E. (1967). A 3.39-micron infrared optical heterodyne communication system. *IEEE J. Quantum Electron.*, QE-3(11):524–531.

Gopalakrishnan, G. K., Burns, W. K., McElhanon, R. W., Bulmer, C. H., and Greenblatt, A. S. (1994). Performance and modeling of broadband $LiNbO_3$ traveling wave optical intensity modulators. *J. Lightwave Technol.*, 12(10):1807–1819.

Gordon, J. P. (1962). Quantum effects in communication systems. *Proc. IRE*, 50(9):1898–1908.

Gordon, J. P. (1983). Interaction forces among solitons in optical fibers. *Opt. Lett.*, 8(11):596–598.

Gordon, J. P. and Haus, H. A. (1986). Random walk of coherently amplified solitons in optical fiber transmission. *Opt. Lett.*, 11(10):865–867.

Gordon, J. P. and Kogelnik, H. (2000). PMD fundamentals: Polarization mode dispersion in optical fibers. *Proc. Nat. Acad. Sci.*, 97(9):4541–4550.

Gordon, J. P. and Mollenauer, L. F. (1990). Phase noise in photonic communications systems using linear amplifiers. *Opt. Lett.*, 15(23):1351–1353.

Gradshteyn, I. S. and Ryzhik, I. M. (1980). *Table of Integrals, Series, and Products.* Academic Press, San Diego.

Grant, M. A., Michie, W. C., and Fletcher, M. J. (1987). The performance of optical phase-locked loops in the presence of nonnegligible loop propagation delay. *J. Lightwave Technol.*, LT-5(4):592–597.

Gray, R. M. (1973). On the asymptotic eigenvalue distribution of Toeplitz matrices. *IEEE Trans. Info. Theory*, 18(6):725–730.

Green, A. G., Littlewood, P. B., and P. P. Mitra, and L. G. L. Wegener (2002). Schrödinger equation with a spatially and temporally random potential: Effects of cross-phase modulation in optical communication. *Phys. Rev. E*, 66(046627).

Green, A. G., Mitra, P. P., and Wegener, L. G. L. (2003). Effect of chromatic dispersion on nonlinear phase noise. *Opt. Lett.*, 28(24):2455–2457.

Greenstein, L. J. (1977). Spectra of PSK signals with overlapping baseband pulses. *IEEE Trans. Commun.*, COM-25(5):523–530.

Greenstein, L. J., Vannucci, G., and Foschini, G. J. (1989). Optical power requirements for detecting OOK and FSK signals corrupted by phase noise. *IEEE Trans. Commun.*, 37(12):1275–1281.

Griffin, R. A. and Carter, A. C. (2002). Optical differential quadrature phase-shift key (oDQPSK) for high capacity optical transmission. In *Optical Fiber Commun. Conf.*, paper WX6.

Griffin, R. A., Johnstone, R. I., Walker, R. G., Hall, J., Wadsworth, S. D., Berry, K., Carter, A. C., Wale, M. J., Jerram, P. A., and Parsons, N. J. (2002). 10 Gb/s optical differential quadrature phase shift key (DQPSK) transmission using GaAs/AlGaAs integration. In *Optical Fiber Commun. Conf.*, postdeadline paper FD6.

Griffin, R., Johnstone, R., Walker, R., Wadsworth, S., Carter, A., and Wale, M. (2003). Integrated DQPSK transmitter for dispersion-tolerant and dispersion-managed DWDM transmission. In *Optical Fiber Commun. Conf.*, paper FP6.

Griffin, R. A., Walker, R. G., and Johnstone, R. I. (2004). Integrated devices for advanced modulation formats. In *IEEE/LEOS Workshop on Advanced Modulation Formats.* paper FC1.

Grüner-Nielsen, L., Edvold, B., Magnussen, D., Peckham, D., Vengsarkar, A., Jacobsen, D., Veng, T., Larsen, C. C., and Damsgaard, H. (1998). Large volume manufacturing of dispersion compensating fibers. In *Optical Fiber Commun. Conf.*, paper TuD5.

Habbab, I. M. I. and Cimini Jr., L. J. (1988). Polarization-switching techniques for coherent optical communications. *J. Lightwave Technol.*, 6(10):1537–1548.

Hakki, B. W. (1997). Polarization mode dispersion compensation by phase diversity detection. *IEEE Photon. Technol. Lett.*, 9(1):121–123.

Hall, M. J. W. (1994). Gaussian-noise and quantum-optical communication. *Phys. Rev. A*, 50(10):3295–3303.

Hall, M. J. W. and O'Routke, M. J. (1993). Realistic performance of the maximum information channel. *Quantum Opt.*, 5(1):161–180.

Hall, A. N., Fenner, G. E., Kingsley, T. D., Soltyo, T. I., and Carlson, R. O. (1962). Stimulated emission of radiation from GaAs *p-n* junctions. *Phys. Rev. Lett.*, 9(11):366–378.

Hamdy, W. M. and Humblet, P. A. (1993). Sensitivity analysis of direct detection optical FDMA networks with OOK modulation. *J. Lightwave Technol.*, 11(6):783–794.

Han, Y., Kim, C., and Li, G. (2004). Simplified receiver implementation for optical differential 8-level phase-shift keying. *Electron. Lett.*, 40(21):1372–1373.

Hanna, M., Porte, H., Goedgebuer, J.-P., and Rhodes, W. T. (1999). Soliton optical phase control by use of in-line filters. *Opt. Lett.*, 24(11):732–734.

Hanna, M., Porte, H., Goedgebuer, J.-P., and Rhodes, W. T. (2000). Experimental investigation of soliton optical phase jitter. *IEEE J. Quantum Electron.*, 36(11):1333–1338.

Hanna, M., Porte, H., Goedgebuer, J.-P., and Rhodes, W. T. (2001). Performance assessment of DPSK soliton transmission system. *Electron. Lett.*, 37(10):644–646.

Hanna, M., Boivin, D., Lacourt, P.-A., and Goedgebuer, J.-P. (2004). Effect of sliding filters on the soliton optical phase jitter in constant-dispersion systems. *Opt. Commun.*, 231:181–185.

Hansryd, J., van Howe, J., and Xu, C. (2004). Nonlinear crosstalk and compensation in ODPASK optical communication systems. *IEEE Photon. Technol. Lett.*, 16(8):1975–1977.

Hansryd, J., van Howe, J., and Xu, C. (2005). Experimental demonstration of nonlinear phase jitter compensation in DPSK modulated fiber links. *IEEE Photon. Technol. Lett.*, 17(1):232–234.

Hao, M.-J. and Wicker, S. B. (1995). Performance evaluation of FSK and CPFSK optical communication systems: A stable and accurate method. *J. Lightwave Technol.*, 13(8):1613–1623.

Hasegawa, A. and Tappert, F. D. (1973). Transmission of stationary nonlinear optical pulses in dispersive dielectric fibers: I. Anomalous dispersion. *Appl. Phys. Lett.*, 23(3):142–144.

Hatami-Hanza, H., Mostofi, A., and Chu, P. L. (1997). A multilevel soliton communication system. *J. Lightwave Technol.*, 15(1):6–19.

Haunstein, H. F., Sauer-Greff, W., Dittrich, A., Sticht, K., and Urbansky, R. (2004). Principles for electronic equalization of polarization-mode dispersion. *J. Lightwave Technol.*, 22(4):1169–1182.

Haus, H. A., Wong, W. S., and Khatri, F. I. (1997). Continuum generation by perturbation of soliton. *J. Opt. Soc. Amer. B*, 14(2):304–313.

Hayase, S., Kikuchi, N., Sekine, K., and Sasaki, S. (2004). Chromatic dispersion and SPM tolerance of 8-state/symbol (binary ASK and QPSK) modulated signal. In *Optical Fiber Commun. Conf.*, paper ThM3.

Hayashi, I., Panish, M. B., Foy, P. W., and Sumelay, S. (1970). Junction lasers which operate continuously at room temperature. *Appl. Phys. Lett.*, 17(3):109–111.

Haykin, S. (1988). *Digital Communications*. John Wiley & Sons, New York.

Helstrom, C. W. (1955). The resolution of signals in white Gaussian noise. *Proc. IRE*, 43(9):1111–1118.

Henry, C. H. (1982). Theory of the linewidth of semiconductor lasers. *IEEE J. Quantum Electron.*, 18(2):259–264.

Henry, C. H. (1983). Theory of the phase noise and power spectrum of a single mode injection laser. *IEEE J. Quantum Electron.*, QE-19(9):1391–1397.

Henry, C. H. (1986). Phase noise in semiconductor lasers. *J. Lightwave Technol.*, 4(3):298–311.

Henry, P. S. and Personick, S. D., editors (1990). *Coherent Lightwave Communications*. IEEE Press, New York.

Henry, C. H., Logan, R. A., and Bertness, K. A. (1981). Spectral dependence of the change in refractive index due to carrier injection in GaAs lasers. *J. Appl. Phys.*, 52(7):4457–4461.

Hibino, Y. (2002). Recent advances in high-density and large-scale AWG multi/demultiplexers with higher index-contrast silica-based PLCs. *IEEE J. Sel. Top. Quantum Electron.*, 8(6):1090–1101.

Hiew, C. C., Abbou, F. M., Chuah, H. T., Majumder, S.P., and Hairul, A. A. R. (2004). BER estimation of optical WDM RZ-DPSK systems through the differential phase Q. *IEEE Photon. Technol. Lett.*, 16(12):2619–2621.

Hill, A. M. and Payne, D. B. (1985). Linear crosstalk in wavelength-division-multiplexed optical-fiber transmission systems. *J. Lightwave Technol.*, LT-3(3):643–651.

Hill, P. M., Olshansky, R., and Burns, W. K. (1992). Optical polarization division multiplexing at 4 Gb/s. *IEEE Photon. Technol. Lett.*, 4(5):500–502.

Ho, C.-L. and Wang, J.-N. (1995). Calculating the performance of optical CPFSK phase diversity receivers. *J. Lightwave Technol.*, 13(5):971–976.

Ho, K.-P. (1998). Analysis of co-channel crosstalk interference in optical networks. *Electron. Lett.*, 34(4):383–385.

Ho, K.-P. (1999a). Analysis of homodyne crosstalk in optical networks using Gram-Charlier series. *J. Lightwave Technol.*, 17(2):149–154.

Ho, K.-P. (1999b). Spectral density of cross-phase modulation induced phase noise. *Opt. Commun.*, 169:63–68.

Ho, K.-P. (2000). Statistical properties of simulated Raman crosstalk in WDM systems. *J. Lightwave Technol.*, 18(7):915–921.

Ho, K.-P. (2003a). Asymptotic probability density of nonlinear phase noise. *Opt. Lett.*, 28(15):1350–1352.

Ho, K.-P. (2003b). Compensation improvement of DPSK signal with nonlinear phase noise. *IEEE Photon. Technol. Lett.*, 15(9):1216–1218.

Ho, K.-P. (2003c). Non-Gaussian statistics of the soliton timing jitter due to amplifier noise. *Opt. Lett.*, 28(22):2165–2167.

Ho, K.-P. (2003d). The optimal compensator for nonlinear phase noise. *Opt. Commun.*, 211(4-6):419–425.

Ho, K.-P. (2003e). Performance degradation of phase-modulated systems with nonlinear phase noise. *IEEE Photon. Technol. Lett.*, 15(9):1213–1215.

Ho, K.-P. (2003f). Probability density of nonlinear phase noise. *J. Opt. Soc. Amer. B*, 20(9):1875–1879.

Ho, K.-P. (2003g). Statistical properties of nonlinear phase noise. In Arkin, W. T., editor, *Advances in Optics and Laser Research*, volume 3. Nova Science Publishers, Hauppauge, NY. E-print: `http://arXiv.org/physics/0303090`.

Ho, K.-P. (2004a). The effect of interferometer phase error on direct-detection DPSK and DQPSK signals. *IEEE Photon. Technol. Lett.*, 16(1):308–310.

Ho, K.-P. (2004b). Error probability of DPSK signals with cross-phase modulation induced nonlinear phase noise. *IEEE J. Sel. Top. Quantum Electron.*, 10(2):421–427.

Ho, K.-P. (2004c). Impact of nonlinear phase noise to DPSK signals: A comparison of different models. *IEEE Photon. Technol. Lett.*, 16(5):1403–1405.

Ho, K.-P. (2004d). Performance of DPSK signals with nonlinear phase noise for systems with small number of fiber spans. In *IEEE/LEOS Workshop on Advanced Modulation Formats*, pages 19–20. paper ThC4.

Ho, K.-P. (2004e). Phase statistics of the soliton. *J. Opt. Soc. Amer. B*, 21(2):266–272.

Ho, K.-P. (2005a). Mid-span compensation of nonlinear phase noise. *Opt. Commun.*, 245(1-6):391–398.

Ho, K.-P. (2005b). Exact evaluation of the channel capacity for intensity-modulated direct-detection channels with optical amplifier noises. *IEEE Photon. Technol. Lett.*, 17(4).

Ho, K.-P. (2005c). Error probability of DPSK signals with intrachannel four-wave-mixing in highly dispersive transmission systems. *IEEE Photon. Technol. Lett.*, 17(4).

Ho, K.-P. and Cuei, H.-W. (2005). Generation of arbitrary quadrature signals using one dual-drive modulator. *J. Lightwave Technol.*, 23(2).

Ho, K.-P. and Kahn, J. M. (1996). Methods for crosstalk measurement and reduction in dense WDM systems. *J. Lightwave Technol.*, 14(6):1127–1135.

Ho, K.-P. and Kahn, J. M. (2002). Channel capacity of WDM systems using constant-intensity modulation formats. In *Optical Fiber Commun. Conf.*, paper ThGG85.

Ho, K.-P. and Kahn, J. M. (2004a). Electronic compensation technique to mitigate nonlinear phase noise. *J. Lightwave Technol.*, 22(3):779–783.

Ho, K.-P. and Kahn, J. M. (2004b). Multilevel optical signals optimized for systems having signal-dependent and signal-independent noises, finite transmitter extinction and inter-symbol interference. US Patent 6,690,894.

Ho, K.-P. and Kahn, J. M. (2004c). Spectrum of externally modulated optical signals. *J. Lightwave Technol.*, 22(2):658–663.

Ho, K.-P. and Liaw, S.-K. (1998a). Demultiplexer crosstalk rejection requirements for hybrid WDM system with analog and digital channels. *IEEE Photon. Technol. Lett.*, 10(5):737–739.

Ho, K.-P. and Liaw, S.-K. (1998b). Eight-channel bidirectional WDM add/drop multiplexer. *Electron. Lett.*, 34(10):947–948.

Ho, K.-P. and Lin, C. (1996). Electronic crosstalk cancellation in high density WDM systems using linear cancellation. *Electron. Lett.*, 32(12):1119–1120.

Ho, K.-P. and Lin, C. (1997). Performance analysis of optical transmission system with polarization-mode dispersion and forward error correction. *IEEE Photon. Technol. Lett.*, 9(9):1288–1290.

Ho, K.-P., Walker, J. D., and Kahn, J. M. (1993). External optical feedback effects on intensity noise of vertical-cavity surface-emitting lasers. *IEEE Photon. Technol. Lett.*, 5(8):892–895.

Ho, K.-P., Liaw, S.-K., and Lin, C. (1997). Efficient photonic mixer with frequency doubling. *IEEE Photon. Technol. Lett.*, 9(4):511–513.

Ho, K.-P., Chan, C.-K., Tong, F., and Chen, L. K. (1998a). Exact analysis of homodyne crosstalk induced penalty in WDM networks. *IEEE Photon. Technol. Lett.*, 10(3):457–458.

Ho, K.-P., Dai, H., Lin, C., Liaw, S.-K., Gysel, H., and Ramachandra, M. (1998b). Hybrid wavelength-division-multiplexing systems for high-capacity digital and analog video trunking applications. *IEEE Photon. Technol. Lett.*, 10(2):297–299.

Ho, K.-P., Kong, E. T.-P., Chan, L. Y., Chan, L.-K., and Tong, F. (1999). Analysis and measurement of root-mean-squared bandwidth of cross-phase-modulation-induced spectral broadening. *IEEE Photon. Technol. Lett.*, 11(9):1126–1128.

Ho, K.-P., Chen, L.-K., and Tong, F. (2000a). Modeling of waveform distortion due to optical filtering. *IEEE J. Sel. Top. Quantum Electron.*, 6(2):223–226.

Ho, K.-P., Liaw, S.-K., and Tong, F. (2000b). Bidirectional single-fiber multiwavelength ring networks. *IEICE Trans. Commun.*, E83B(10):2245–2252.

Ho, K.-P., Yu, H., Chen, L.-K., and Tong, F. (2000c). High-resolution measurement and spectral overlap of cross-phase modulation induced spectral broadening. *IEEE Photon. Technol. Lett.*, 12(11):1534–1536.

Hodgkinson, T. (1987). Receiver analysis for synchronous coherent optical fiber transmission systems. *J. Lightwave Technol.*, LT-5(4):573–586.

Hodgkinson, T. G., Harmon, R. A., and Smith, D. W. (1985). Demodulation of optical DPSK using in-phase and quadrature detection. *Electron. Lett.*, 21(21):867–868.

Hodgkinson, T. G., Harmon, R. A., and Smith, D. A. (1987). Polarisation-insensitive heterodyne detection using polarisation scrambling. *Electron. Lett.*, 23(10):513–514.

Hodgkinson, T. G., Harmon, R. A., Smith, D. W., and Chidgey, P. J (1988). In-phase and quadrature detection using 90° optical hybrid receiver: Experiments and design considerations. *IEE Proc. J*, 135(3):260–267.

Hoffman, D., Heidrich, H., Wenke, G., Langenhorst, R., and Dietrich, E (1989). Integrated optics eight-port 90° hybrid on $LiNbO_3$. *J. Lightwave Technol.*, 7(5):794–798.

Holzlohner, R., Grigoryan, V. S., Menyuk, C. R., and Kath, W. L. (2002). Accurate calculation of eye diagrams and bit error rates in optical transmission systems using linearization. *J. Lightwave Technol.*, 20(3):389–400.

Hooijmans, P. W. (1994). *Coherent Optical System Design.* John Wiley & Sons, New York.

Hooijmans, P. W., Tomesen, M. T., and van de Grijp, A. (1990). Penalty free biphase linecoding for pattern independent FSK coherent transmission systems. *J. Lightwave Technol.*, 8(3):323–328.

Howerton, M. M., Moeller, R. P., Greenblatt, A. S., and Krähenbühl, R. (2000). Fully packaged, broad-band $LiNbO_3$ modulator with low drive voltage. *IEEE Photon. Technol. Lett.*, 12(7):792–794.

Huang, J. and Meyn, S. (2003). Characterization and computation of optimal distributions for channel code. In *Conf. on Info. Sci. and Sys.*

Huang, J.-A. and Ho, K.-P. (2003). Probability density of signal with nonlinear phase noise. In *Proc. of the 5th Pacific Rim Conf. on Lasers and Electro-Optics, CLEO/PR '03*, Taipei, Taiwan. paper THP-(5)-16.

Huang, S. and Wang, L. (1995). Optimum phase deviations between mark-and space-state bits in long-haul coherent communication systems with balanced PLL receivers. *J. Lightwave Technol.*, 13(10):1963–1970.

Huang, S. and Wang, L. (1996). Exact evaluation of laser linewidth requirements for optical PSK homodyne communication systems with balanced PLL receivers. *J. Lightwave*

Technol., 14(5):661–664.

Hui, R., O'Sullivan, M., Robinson, A., and Taylor, M. (1997). Modulation instability and its impact in multispan optical amplified IMDD systems: Theory and experiments. *J. Lightwave Technol.*, 15(7):1071–1082.

Hui, R., Demarest, K. R., and Allen, C. T. (1999). Cross-phase modulation in multispan WDM optical fiber systems. *J. Lightwave Technol.*, 17(6):1018–1026.

Hui, R., Zhu, B., Huang, R., Allen, C. T., Demarest, K. R., and Richards, D. (2002). Subcarrier multiplexing for high-speed optical transmission. *J. Lightwave Technol.*, 13(8):896–898.

Humblet, P. A. and Azizoğlu, M. (1991). On the bit error rate of lightwave systems with optical amplifiers. *J. Lightwave Technol.*, 9(11):1576–1582.

Humblet, P. A. and Hamdy, W. M. (1990). Crosstalk analysis and filter optimization of single- and double-cavity Fabry-Perot filters. *IEEE J. Sel. Areas Commun.*, 8(6):1095–1107.

Hung, W., Chan, C.-K., and Chen, L.-K. (2004). A novel optical packet labeling scheme using interleaved low-speed DPSK labels. *IEEE Photon. Technol. Lett.*, 16(2):698–700.

Huttner, B., Reecht, J., Gisin, N., Passy, R., and von der Weid, J. P. (1998). Local birefringence measurements in single-mode fibers with coherent optical frequency-domain reflectometry. *IEEE Photon. Technol. Lett.*, 10(10):1458–1460.

Huttner, B., Geiser, C., and Gisin, N. (2000). Polarization-induced distortions in optical fiber networks with polarization-mode dispersion and polarization-dependent losses. *IEEE J. Sel. Top. Quantum Electron.*, 6(2):317–329.

Iannone, E., Matera, F., Mecozzi, A., and Settembre, M. (1998). *Nonlinear Optical Communication Networks.* John Wiley & Sons, New York.

Idler, W., Klekamp, A., Dischler, R., and Wedding, B. (2004). Advantages of frequency shift keying in 10 Gb/s systems. In *IEEE/LEOS Workshop on Advanced Modulation Formats.* paper FD3.

Iga, K., Koyama, F., and Kinoshita, S. (1988). Surface emitting semiconductor lasers. *IEEE J. Quantum Electron.*, 24(9):1845–1855.

Imai, T., Hayashi, Y., Ohkawa, N., Sugie, T., Ichihashi, Y., and Ito, T. (1990a). Field demonstration of 2.5 Gbit/s coherent optical transmission through installed submarine fibre cable. *Electron. Lett.*, 26(17):1407–1409.

Imai, Y., Iizuka, K., and James, R. T. B. (1990b). Phase-noise-free coherent optical communication system utilizing differential polarization shift keying (DPolSK). *J. Lightwave Technol.*, 8(5):691–698.

Imai, T., Ohkawa, N., Hayashi, Y., and Ichihashi, Y. (1991). Polarization diversity detection performance of 2.5-Gb/s CPFSK regenerators intended for field use. *J. Lightwave Technol.*, 9(6):761–769.

Ishida, K., Kobayashi, T., Abe, J., Kinjo, K., Kuroda, S., and Mizuochi, T. (2003). A comparative study of 10 Gb/s RZ-DPSK and RZ-ASK WDM transmission over transoceanic distances. In *Optical Fiber Commun. Conf.*, paper ThE2.

Islam, M. N. (2002). Raman amplifiers for telecommunications. *IEEE J. Sel. Top. Quantum Electron.*, 8(3):548–559.

Iwashita, K. and Matsumoto, T. (1987). Modulation and detection characteristics of optical continuous phase FSK transmission system. *J. Lightwave Technol.*, LT-5(4):452–460.

Iwashita, K. and Takachio, N. (1988). Compensation of 202 km single-mode fibre chromatic dispersion in 4 Gbit/s optical CPFSK transmission experiment. *Electron. Lett.*, 24(12):36.

Iwashita, K. and Takachio, N. (1990). Chromatic dispersion compensation in coherent optical communications. *J. Lightwave Technol.*, 8(3):367–375.

Iwashita, K., Imai, T., and Matsumoto, T. (1986). 400 Mbit/s optical FSK transmission experiment over 270 km of single-mode fibre. *Electron. Lett.*, 22(3):164–165.

Jackson, S. M., Hewitt, P. D., Reed, G. T., Tang, C. K., Evans, A. G. R., Clark, J., Aveyard, C., and Namavar, F. (1998). A novel optical phase modulator design suitable for phased arrays. *J. Lightwave Technol.*, 16(11):2016–2019.

Jacobsen, G (1993). Performance of DPSK and CPFSK systems with significant post-detection filtering. *J. Lightwave Technol.*, 11(10):1622–163.

Jacobsen, G. and Garrett, I. (1987). Theory for optical heterodyne DPSK receivers with post-detection filtering. *J. Lightwave Technol.*, 5(4):478–484.

Jain, P. C. (1974). Error probabilities in binary angle modulation. *IEEE Trans. Info. Theory*, IT-20(1):36–42.

Jain, P. C. and Blachman, N. M. (1973). Detection of a PSK signal transmitted through a hard-limited channel. *IEEE Trans. Info. Theory*, IT-19(5):623–630.

Janssen, A. (2001). The evolution of passive devices in WDM networks. *IEEE Trans. Adv. Packaging*, 24(4):590–594.

Jeromin, L. L. and Chan, V. W. S. (1986). M-ary FSK performance for coherent optical communication system using semiconductor lasers. *IEEE Trans. Commun.*, COM-34(4):375–381.

Jeruchim, M. C. (1984). Techniques for estimating the bit error rate in the simulation of digital communication systems. *IEEE J. Sel. Areas Commun.*, SAC-2(1):153–170.

Jiang, X. and Roudas, I. (2001). Asymmetric probability density function of a signal with interferometric crosstalk. *IEEE Photon. Technol. Lett.*, 13(2):160–162.

Johansson, L. A., Akulova, Y. A., Fish, G. A., and Coldren, L. A. (2004). Sampled-grating DBR laser integrated with SOA and tandem electroabsorption modulator for chirp-control. *Electron. Lett.*, 40(1):70–71.

Jørgensen, B. F., Mikkelsen, B., and Mahan, C. J. (1992). Analysis of optical amplifier noise in coherent communication systems with optical image rejection receivers. *J. Lightwave Technol.*, 10(5):660–670.

Kabanov, Y. M. (1978). The capacity of a channel of Poisson type. *Theory Prob. & Its Appl.*, 23(1):143–147.

Kac, M. and Siegert, A. J. F. (1947). On the theory of noise in radio receivers with square law detectors. *J. Appl. Phys.*, 18(4):383–397.

Kahn, J. M. (1989). 1 Gbit/s PSK homodyne transmission system using phase-locked semiconductor lasers. *IEEE Photon. Technol. Lett.*, 1(10):340–342.

Kahn, J. M. (1990). BPSK homodyne detection experiment using balanced optical phase-locked loop with quantized feedback. *IEEE Photon. Technol. Lett.*, 2(11):840–843.

Kahn, J. M and Ho, K.-P. (2001). A bottleneck for optical fibres. *Nature*, 411:1007–1010.

Kahn, J. M. and Ho, K.-P. (2004). Spectral efficiency limits and modulation/detection techniques for DWDM systems. *IEEE J. Sel. Top. Quantum Electron.*, 10(2):259–272.

Kahn, J. M., Gnuack, A. H., Veselka, J. J., Korotky, S. K., and Kasper, B. L. (1990). 4-Gb/s PSK homodyne transmission system using phase-locked semiconductor lasers. *IEEE Photon. Technol. Lett.*, 2(4):285–287.

Kahn, J. M., Porter, A. M., and Padan, U. (1992). Heterodyne detection of 310-Mb/s quadriphase-shift keying using fourth-power optical phase-locked loop. *IEEE Photon. Technol. Lett.*, 4(12):1397–1400.

Kaiser, C. P., Shafi, M., and Smith, P. J. (1993). Analysis methods for optical heterodyne DPSK receivers corrupted by laser phase noise. *J. Lightwave Technol.*, 11(11):1820–1830.

Kaiser, P. and Keck, D. B. (1988). Fiber types and their status. In Miller, S. E. and Kaminow, I. P., editors, *Optical Fiber Telecommunications*, volume II, pages 29–54. Academic Press, San Diego.

Kajikawa, M. and Kataoka, I. (1997). A design method of optical bandpass filters. *J. Lightwave Technol.*, 15(9):1720–1727.

Kamalakis, T. and Sphicopoulos, T. (2003). Asymptotic behavior of in-band crosstalk noise in WDM networks. *IEEE Photon. Technol. Lett.*, 15(3):476–478.

Kamalakis, T., Sphicopoulos, T., and Sagriotis, M. (2003). Accurate estimation of the error probability in the presence of in-band crosstalk noise in WDM networks. *J. Lightwave Technol.*, 21(10):2172–2181.

Kaminow, I. P. (1981). Polarization in optical fiber. *IEEE J. Quantum Electron.*, 17(1):15–22.

Kaminow, I. P. (1990). FSK with direct detection in optical multiple-access FDM networks. *IEEE J. Sel. Areas Commun.*, 8(6):1005–1014.

Kaminow, I. P., Iannone, P. P., Stone, J., and Stulz, L. W. (1988). FDMA-FSK star network with a tunable optical filter demultiplexer. *J. Lightwave Technol.*, 6(9):1406–1461.

Kanamori, H., Yokota, H., Tanaka, G., Watanabe, M., Ishiguro, Y., Yoshida, I., Kakii, T., Itou, S., Asano, Y., and Tanaka, S. (1986). Transmission characteristics and reliability of pure silica core single mode fibers. *J. Lightwave Technol.*, LT-4(8):1144–1150.

Kao, K. C. and Hockham, G. A. (1966). Dielectric fiber surface waveguides for optical frequencies. *Proc. IEE*, 133(7):1151–1158.

Kapron, F. P., Keck, D. B., and Maurer, R. D. (1970). Radiation losses in glass optical waveguides. *Appl. Phys. Lett.*, 17(11):423–425.

Karpman, V. I. and Maslov, E. M. (1977). Perturbation theory for solitons. *JETP*, 73(2):537–559.

Kaup, D. J. (1976). A perturbation expansion for the Zakharov-Shabat inverse scattering transform. *SIAM J. Appl. Math.*, 31(1):121–133.

Kaup, D. J. (1990). Perturbation theory for solitons in optical fibers. *Phys. Rev. A*, 42(9):5689–5694.

Kaup, D. J. (1991). Second-order perturbation for solitons in optical fibers. *Phys. Rev. A*, 44(7):4582–4590.

Kaup, D. J. and Newell, A. C. (1978). Solitons as particles, oscillators, and in slowly changing media: A singular perturbation theory. *Proc. R. Soc. London A*, 361(1707):413–446.

Kavehrad, M. and Glance, B. S. (1988). Polarization-insensitive frequency-shift-keying optical heterodyne receiver using discriminator demodulation. *J. Lightwave Technol.*, 6(9):1386–1394.

Kawanishi, H., Yamauchi, Y., Mineo, N., Shibuya, Y., Mural, H., Yamada, K., and Wada, H. (2001). EAM-integrated DFB laser modules with more than 40-GHz bandwidth. *IEEE Photon. Technol. Lett.*, 13(9):954–956.

Kawano, K., Kitoh, T., Mitomi, O., Nozawa, T., and Jumonji, H. (1989). A wide-band and low-driving-power phase modulator employing a Ti:LiNbO$_3$ optical waveguide at 1.5 μm. *IEEE Photon. Technol. Lett.*, 1(2):33–34.

Kawano, K., Noguchi, K., Kitoh, T., and Miyazawa, H. (1991). A finite element method (FEM) analysis of a shielded velocity-matched Ti:LiNbO$_3$ optical modulator. *IEEE Photon. Technol. Lett.*, 3(10):919–921.

Kazovsky, L. G. (1985). Decision-driven phase-locked loop for optical homodyne receivers: Performance analysis and laser linewidth requirements. *J. Lightwave Technol.*, LT-3(12):1238–1247.

Kazovsky, L. G. (1986a). Balanced phase-locked loops for optical homodyne receivers: Performance analysis, design considerations, and laser linewidth requirements. *J. Lightwave Technol.*, LT-4(2):182–195.

Kazovsky, L. G. (1986b). Performance analysis and laser linewidth requirements for optical PSK heterodyne communications systems. *J. Lightwave Technol.*, LT-4(4):415–425.

Kazovsky, L. G. (1989). Phase- and polarization-diversity coherent optical techniques. *J. Lightwave Technol.*, 7(2):279–292.

Kazovsky, L. G. and Atlas, D. A. (1990). A 1320-nm experimental optical phase-locked loop: Performance investigation and PSK homodyne experiments at 140 Mb/s and 2 Gb/s. *J. Lightwave Technol.*, 8(9):1414–1425.

Kazovsky, L. G. and Tonguz, O. K. (1990). ASK and FSK coherent lightwave systems: A simplified approximate analysis. *J. Lightwave Technol.*, 8(3):338–352.

Kazvosky, L. G., Meissner, P., and Patzak, E. (1987). ASK multiport optical homodyne receivers. *J. Lightwave Technol.*, LT-5(6):770–791.

Kazovsky, L. G., Curtis, I., Young, W. C., and Cheung, N. K. (1987). All-fiber 90° optical hybrid for coherent communications. *Appl. Opt.*, 26(3):437–439.

Kazovsky, L. G., Atlas, D. A., and Smith, R. W. (1990). Optical phase-locked PSK heterodyne experiment at 4 Gb/s. *IEEE Photon. Technol. Lett.*, 2(8):588–590.

Kazovsky, L. G., Benedetto, S., and Willner, A. E. (1996). *Optical Fiber Communication Systems*. Artech House, Boston.

Keiser, G. (1999). *Optical Fiber Communications*. McGraw Hill, New York, 3 edition.

Kieckbusch, S., Ferber, S., Rosenfeldt, H., Ludwig, R., Boerner, C., Ehrhardt, A., Brinkmeyer, E., and Weber, H.-G. (2005). Automatic PMD compensator in a 160-Gb/s OTDM transmission over deployed fiber using RZ-DPSK modulation format. *J. Lightwave Technol.*, 23(1):165–171.

Kikuchi, K. and Katoh, K. (2002a). Differential detection of single modulation sideband for ultra-dense optical frequency-division multiplexed systems. *Electron. Lett.*, 38(17):980–981.

Kikuchi, K. and Katoh, K. (2002b). Optical heterodyne receiver for selecting densely frequency division multiplexed signals. *Electron. Lett.*, 38(6):283–285.

Kikuchi, K. and Okoshi, T. (1985). Estimation of linewidth enhancement factor of AlGaAs lasers by correlation measurement between FM and AM noises. *IEEE J. Quantum Electron.*, QE-21(6):669–673.

Kikuchi, K., Okoshi, T., and Kitano, J. (1981). Measurement of bit-error rate of heterodyne-type optical communication system–a simulation experiment. *IEEE J. Quantum Electron.*, QE-17(12):2266–2267.

Kikuchi, K., Okoshi, T., Nagamatsu, M., and Henmi, N. (1983). Bit-error rate of PSK heterodyne optical communication system and its degradation due to spectral spread of transmitter and local oscillator. *Electron. Lett.*, 19(11):417–418.

Kikuchi, K., Okoshi, T., Nagamatsu, M., and Henmi, N. (1984). Degradation of bit-error rate in coherent optical communications due to spectral spread of the transmitter and the local oscillator. *J. Lightwave Technol.*, LT-2(6):1024–1033.

Kim, C. H., Lee, C.-H., and Chung, Y. C. (1998). A novel bidirectional add/drop amplifier (BADA). *IEEE Photon. Technol. Lett.*, 10(8):1118–1120.

Kim, H. (2003). Cross-phase-modulation-induced nonlinear phase noise in WDM direct-detection DPSK systems. *J. Lightwave Technol.*, 21(8):1770–1774.

Kim, H. and Essiambre, R.-J. (2003). Transmission of 8 × 20 Gb/s DQPSK signals over 310-km SMF with 0.8 b/s/Hz spectral efficiency. *IEEE Photon. Technol. Lett.*, 15(5):769–771.

Kim, H. and Gnauck, A. H. (2002). Chirp characteristics of dual-drive, Mach-Zehnder modulator with a finite DC extinction ratio. *IEEE Photon. Technol. Lett.*, 12(3):298–300.

Kim, H. and Gnauck, A. H. (2003). Experimental investigation of the performance limitation of DPSK systems due to nonlinear phase noise. *IEEE Photon. Technol. Lett.*, 15(2):320–322.

Kim, H. and Winzer, P. J. (2003). Robustness to laser frequency offset in direct-detection DPSK and DQPSK systems. *J. Lightwave Technol.*, 21(9):1887–1891.

Kim, H., Doerr, C. R., Pafchek, R., Stulz, L. W., and Bernasconi, P. (2002). Polarisation-mode dispersion impairments in direct-detection differential phase-shift-keying systems. *Electron. Lett.*, 38(18):1047–1048.

Kivshar, Yu. S. and Malomed, B. A. (1989). Dynamics of solitons in nearly integrable systems. *Rev. Mod. Phys.*, 61(4):763–915. Addendum: 63(1):211, 1993.

Kjebon, O., Schatz, R., Lourdudoss, S., Nilsson, N., Stålnacke, B., and Bäckborn, L. (1997). 30 GHz direct modulation bandwidth in detuned loaded InGaAsP DBR lasers at 1.55 μm wavelength. *Electron. Lett.*, 33(6):488–489.

Kobayashi, S. and Kimura, T. (1982). Optical phase modulation in an injection locked AlGaAs semiconductor laser. *IEEE J. Quantum Electron.*, QE-18(10):1662–1669.

Kobayashi, S., Yamamoto, Y., Ito, M., and Kimura, T. (1982). Direct frequency modulation in AlGaAs semiconductor lasers. *IEEE J. Quantum Electron.*, QE-18(4):582–595.

Kobayashi, S., Yamamoto, Y., and Kimura, T. (1991). Frequency and phase modulation of semiconductor lasers. In Li, T., editor, *Topics in Lightwave Transmission Systems*, pages 151–202. Academic Press, San Diego.

Kodama, K. and Hasegawa, A. (1992). Generation of asymptotically stable optical solitons and suppression of Gordon-Haus effect. *Electron. Lett.*, 17(1):31–33.

Kogelnik, H. and Shank, C. V. (1971). Stimulated emission in a periodic structure. *Appl. Phys. Lett.*, 18(4):152–154.

Kolimbiris, H. (2004). *Fiber Optics Communications*. Pearson Education, Upper Saddle River, NJ.

Kondo, J., Kondo, A., Aoki, K., Takatsuji, S., Mitomi, O., Imaeda, M., Kozuka, Y., and Minakata, M. (2002). 40-Gb/s x-cut $LiNbO_3$ optical modulator with two-step back-slot structure. *J. Lightwave Technol.*, 20(12):2110–2114.

Korotky, S. K. and Alferness, R. C. (1988). Waveguide electrooptic devices for optical communication. In Miller, S. E. and Kaminow, I. P., editors, *Optic Fiber Telecommunications II*, chapter 11, pages 421–465. Academic Press, San Diego.

Korotky, S. K. and Veselka, J. J. (1996). An RC network analysis of long term Ti:$LiNbO_3$ bias stability. *J. Lightwave Technol.*, 14(12):2687–2697.

Koshiba, M., Tsuji, Y., and Nishio, M. (1999). Finite-element modeling of broad-band traveling-wave optical modulators. *IEEE Trans. Microwave Theory & Tech.*, 47(9):1627–1633.

Koyama, F. and Oga, K. (1988). Frequency chirping in external modulators. *J. Lightwave Technol.*, 6(1):87–93.

Kubota, K., Noda, J., and Mikami, O. (1980). Traveling wave optical modulator using a directional coupler $LiNbO_3$ waveguide. *IEEE J. Quantum Electron.*, QE-16(7):754–760.

Kumar, S., Mauro, J. C., Ranghavan, S., and Chowdhury, D. Q. (2002). Intrachannel nonlinear penalties in dispersion-managed transmission systems. *IEEE J. Sel. Top. Quantum Electron.*, 8(3):626–631.

Kuri, T. and Kitayama, K. (2002). Optical heterodyne detection of millimeter-wave-band radio-on-fiber signals with a remote dual-mode local light source. *IEEE Trans. Microwave Theory & Tech.*, 49(10):2025–2029.

Kuri, T. and Kitayama, K. (2003). Optical heterodyne detection technique for densely multiplexed millimeter-wave-band radio-on-fiber systems. *J. Lightwave Technol.*, 21(12):3167–3179.

Lachs, G., Henning, R. E., Choi, Y., and Zaidi, S. M. (1990). Multichannel image-rejection coherent detection system. *J. Lightwave Technol.*, 8(12):1874–1881.

Langenhorst, R., Piper, W., Eiselt, M., Rohde, D., and Weber, H. G. (1991). Balanced phase and polarization diversity coherent optical receiver. *IEEE Photon. Technol. Lett.*, 3(1):80–82.

Lanne, S. and Corbel, E. (2004). Practical considerations for optical polarization-mode dispersion compensators. *J. Lightwave Technol.*, 22(4):1033–1040.

Laroia, R., Farvardin, N., and Tretter, S. A. (1994). On optimal shaping of multidimensional constellations. *IEEE Trans. Info. Theory*, 40(4):1044–1056.

Lau, K. Y. and Yariv, A. (1985). Ultra-high speed semiconductor lasers. *IEEE J. Quantum Electron.*, QE-21(2):121–138.

Lax, M. (1967a). Classical noise. V. Noise in self-sustained oscillators. *Phys. Rev.*, 160(2):290–307.

Lax, M. (1967b). Quantum noise. X. Density-matrix treatment of field and population-difference fluctuations. *Phys. Rev.*, 157(2):213–231.

Leclerc, O. and Desurvire, E. (1998). Effect of synchronous modulation on the soliton optical phase. *Opt. Lett.*, 23(18):1453–1455.

Lee, C. L. and Lai, Y. (2002). Evolutionary programming synthesis of optimal long-period fiber grating filters for EDFA gain flattening. *IEEE Photon. Technol. Lett.*, 14(11):1557–1559.

Lee, J.-S. and Shim, C.-S. (1994). Bit-error-rate analysis of optically preamplified receivers using an eigenfunction expansion method in optical frequency domain. *J. Lightwave Technol.*, 12(7):1224–1229.

Lee, S. S., Garner, S. M., Chuyanov, V., Zhang, H., Steier, W H., Wang, F., Dalton, L. R., Udupa, A. H., and Fetterman, H. R. (2000). Optical intensity modulator based on a novel electrooptic polymer incorporating a high $\mu\beta$ chromophore. *IEEE J. Quantum Electron.*,

36(5):527–532.

Lee, C.-C., Chen, Y.-K., Chang, C.-H., Feng, K.-M., Tzeng, S.-L., and Chi, S. (2002a). Hybrid 10-Gb/s, 2.5-Gb/s, 64-QAM, and AM-VSB high-capacity wavelength-division-multiplexing transport systems using SMF and LEAF fibers. *IEEE Photon. Technol. Lett.*, 14(2):230–232.

Lee, J., Nam, S., Lee, S. H., and Jeong, J. (2002b). A complete small-signal equivalent circuit model of cooled butterfly-type 2.5 Gbps DFB laser modules and its application to improve high frequency characteristics. *IEEE Trans. Adv. Packaging*, 25(4):543–548.

Leibrich, J., Wree, C., and Rosenkranz, W. (2002). CF-RZ-DPSK for suppression of XPM on dispersion-managed long-haul optical WDM transmission on standard single-mode fiber. *IEEE Photon. Technol. Lett.*, 14(2):215–217.

Lenz, G., Eggleton, B. J., Giles, C. R., Madsen, C. K., and Slusher, R. E. (1998). Dispersive properties of optical filters for wdm systems. *IEEE J. Quantum Electron.*, 34(12?):1390–1402.

Leonberger, F. J. and Donnelly, J. P. (1990). Semiconductor integrated optic devices. In Tamir, T., editor, *Guided-Wave Optoelectronics*, pages 317–396. Springer-Verlag, Berlin, 2 edition.

Li, C.-S. and Tong, F. (1996). Crosstalk and interference penalty in all-optical networks using static wavelength routers. *J. Lightwave Technol.*, 14(6):1120–1126.

Lichtman, E. (1995). Limitations imposed by polarization-dependent gain and loss on all-optical ultralong communication systems. *J. Lightwave Technol.*, 13(5):906–913.

Lin, C., Kogelnik, H., and Cohen, L. G. (1980). Optical-pulse equalization of low-dispersion transmission in single-mode fibers in the 1.3-1.7-μm spectral region. *Opt. Lett.*, 5(11):476–478.

Lin, L. Y., Goldstein, E. L., and Tkach, R. W. (2000). On the expandability of free-space micromachined optical cross connects. *J. Lightwave Technol.*, 18(4):482–489.

Lindsey, W. C. and Simon, M. K. (1973). *Telecommunication Systems Engineering.* Prentice-Hall, Englewood Cliffs, N.J.

Linke, R. A. and Gnauck, A. H. (1988). High-capacity coherent lightwave systems. *J. Lightwave Technol.*, 6(11):1750–1769.

Liu, M. M.-K. (1996). *Principles and Applications of Optical Communications.* McGraw-Hill, New York.

Liu, X., Wei, X., Gnauck, A. H., Xu, C., and Wickham, L. K. (2002a). Suppression of intrachannel four-wave-mixing induced ghost pulses in high-speed transmissions by phase inversion between adjacent marker blocks. *Opt. Lett.*, 27(13):1177–1179.

Liu, X., Wei, X., Slusher, R. E., and McKinstrie, C. J. (2002b). Improving transmission performance in differential phase-shift-keyed systems by use of lumped nonlinear phase-shift compensation. *Opt. Lett.*, 27(18):1616–1618.

Liu, X., Wei, X., Su, Y., Leuthold, J., Kao, Y.-H., Kang, I., and Giles, R.C. (2004a). Transmission of an ASK-labeled RZ-DPSK signal and label erasure using a saturated SOA. *IEEE Photon. Technol. Lett.*, 16(6):1594–1596.

Liu, X., Xie, C., and van Wijngaarden, A. J. (2004b). Multichannel PMD mitigation and outage reduction through FEC with sub-burst-error-correction period PMD scrambling. *IEEE Photon. Technol. Lett.*, 16(9):2183–2185.

Liu, X., Xu, C., and Wei, X. (2004c). Performance analysis of time-polarization multiplexed 40-Gb/s RZ-DPSK DWDM transmission. *IEEE Photon. Technol. Lett.*, 16(1):302–304.

Liu, X., Kao, Y.-H., Movassaghi, M., and Giles, R. C. (2004d). Tolerance to in-band coherent crosstalk of differential phase-shift-keyed signal with balanced detection and FEC. *IEEE Photon. Technol. Lett.*, 16(4):1209–1211.

Liu, A., Jones, R., Liao, L., Samara-Rubio, D., Rubin, D., Cohen, O., Nicolaescu, R., and Paniccia, M. (2004e). A high-speed silicon optical modulator based on a metal semiconductor capacitor. *Nature*, 427:615–618.

Lu, P., Chen, L., and Bao, X. (2001). Statistical distribution of polarization-dependent loss in the presence of polarization-mode dispersion in single-mode fibers. *IEEE Photon. Technol.*

Lett., 13(5):451–453.

Lu, G. W., Chan, K., Chen, L.-K., and Chan, C.-K. (2004). Robustness of DPSK-WDM systems against nonlinear polarization fluctuation. *IEEE Photon. Technol. Lett.*, 16(3):927–929.

Lunardi, L. M., Moss, D. J., Chandrasekhar, S., Buhl, L. L., Lamont, M., McLaughlin, S., Randall, G., Colbourne, P., Kiran, S., and Hulse, C. A. (2002). Tunable dispersion compensation at 40-Gb/s using a multicavity etalon all-pass filter with NRZ, RZ, and CS-RZ modulation. *J. Lightwave Technol.*, 20(12):2136–2144.

Lyubomirsky, I. and Chien, C.-C. (2005). DPSK demodulator based on optical discriminator filter. *IEEE Photon. Technol. Lett.*, 17(2):492–494.

MacLeod, H. A. (2001). *Thin Film Optical Filters.* Institute of Physics, London, 3 edition.

Malyon, D. J. and Stallard, W. A. (1989). 565 Mbit/s FSK direct detection system operating with four cascaded photoionic amplifiers. *Electron. Lett.*, 25(8):495–497.

Malyon, D. J. and Stallard, W. A. (1990). Multichannel FSK experiment using an optical amplifier repeater and wavelength tunable direct detection receiver. *Electron. Lett.*, 26(25):1863–1865.

Mamyshev, P. V. and Mamysheva, N. A. (1999). Pulse-overlapped dispersion-managed data transmission and intrachannel four-wave mixing. *Opt. Lett.*, 24(21):1454–1456.

Marcum, J. I. (1960). A statistical theory of target detection by pulsed radar. *IRE Trans. Info. Theory*, IT-6(1):56–267. Earlier RAND report no. RM-753, July 1948.

Marcuse, D. (1980). Pulse distortion in single-mode fibers. *Appl. Opt.*, 19(10):1953–1660.

Marcuse, D. (1981). Pulse distortion in single-mode fibers. part 3: Chirped pulses. *Appl. Opt.*, 20(20):3573–3579.

Marcuse, D. (1990). Derivation of analytical expressions for the bit-error probability in lightwave systems with optical amplifiers. *J. Lightwave Technol.*, 8(12):1816–1823.

Marcuse, D. (1991). Calculation of bit-error probability for a lightwave system with optical amplifiers and post-detection Gaussian noise. *J. Lightwave Technol.*, 9(4):505–513.

Marcuse, D. and Lin, C. (1981). Low dispersion single-mode fiber transmission–the question of practical versus theoretical maximum transmission bandwidth. *IEEE J. Quantum Electron.*, QE-17(6):869–878.

Marembert, V., Schubert, C., Ferber, S., Schulze, K., Schmidt-Langhorst, C., Boerner, C., Kroh, M., Ludwig, R., Watanabe, S., Futami, F., Okabe, R., and Weber, H.G. (2004). Single-channel 640 Gbit/s DPSK transmission over a 160 km fibre link. In *European Conf. on Optical Commun.* postdeadline Th4.4.2.

Mårtensson, J., Berntson, A., Westlund, M., Danielsson, A., Johannisson, P., Anderson, D., and Lisak, M. (2001). Timing jitter owing to intrachannel pulse interactions in dispersion-managed transmission systems. *Opt. Lett.*, 26(2):55–57.

Marti, J. and Capmany, J. (1996). BER impairment due to laser linewidth in OFDM-OOK systems using double-cavity Fabry-Perot demultiplexers. *J. Lightwave Technol.*, 14(5):641–648.

Martinelli, M. and Chipman, R. A. (2003). Endless polarization control algorithm using adjustable linear retarders with fixed axes. *J. Lightwave Technol.*, 21(9):2089–2096.

Mason, B., Ougazzaden, A., Lentz, C. W., Glogovsky, K. G., Reynolds, C. L., Przybylek, G. J., Leibenguth, R. E., Kercher, T. L., Boardman, J. W., Rader, M. T., Geary, J. M., Walters, F. S., Peticolas, L. J., Freund, J. M., Chu, S. N. G., Sirenko, A., Jurchenko, R. J., Hybertsen, M. S., Ketelsen, L. J. P., and Raybon, G. (2002). 40-Gb/s tandem electroabsorption modulator. *IEEE Photon. Technol. Lett.*, 14(1):27–29.

Massey, J. L. (1981). Capacity, cutoff-rate and coding for a direct-detection optical channel. *IEEE Trans. Commun.*, COM-29(11):1616–1621.

Massicott, J. F., Armitage, J. R., Wyatt, R., Ainslie, B. J., and Craig-Ryan, S. P. (1990). High gain, broadband, 1.6 μm Er^{3+} doped silica fibre amplifier. *Electron. Lett.*, 26(20):1645–1646.

Massicott, J. F., Wyatt, R., and Ainslie, B. J. (1992). Low noise operation of Er^{3+} doped silica fibre amplifier around 1.6 μm. *Electron. Lett.*, 28(20):1924–1925.

Matsui, Y., Murai, H., Arahira, S., Kutsuzawa, S., and Ogawa, Y. (1997). 30-GHz bandwidth 1.55-μm strain-compensated InGaAlAs-InGaAsP MQW laser. *IEEE Photon. Technol. Lett.*, 9(1):25–27.

Mazurczyk, V. J. and Zyskind, J. L. (1994). Polarization dependent gain in Erbium doped-fiber amplifiers. *IEEE Photon. Technol. Lett.*, 6(5):616–618.

McDonough, R. N. and Whalen, A. D. (1995). *Detection of Signals in Noise.* Academic Press, San Diego, second edition.

McKinstrie, C. J. and Xie, C. (2002). Phase jitter in single-channel soliton systems with constant dispersion. *IEEE J. Sel. Top. Quantum Electron.*, 8(3):616–625. erratum vol. 8, p. 956, 2002.

McKinstrie, C. J., Xie, C., and Lakoba, T. I. (2002). Efficient modeling of phase jitter in dispersion-managed soliton systems. *Opt. Lett.*, 27(21):1887–1889.

McKinstrie, C. J., Radic, S., and Xie, C. (2003). Reduction of soliton phase jitter by in-line phase conjugation. *Opt. Lett.*, 28(17):1519–1521.

Meada, M. W. and Smith, D. A. (1990). New polarization-insensitive detection scheme based on fibre polarisation scrambling. *Electron. Lett.*, 27(1):10–12.

Mears, R. J., Reekie, L., Jauncey, I. M., and Payne, D. N. (1987). Low-noise Erbium-doped fibre amplifier operating at 1.54 μm. *Electron. Lett.*, 23(19):1026–1028.

Mecozzi, A. (1994a). Limits to long-haul coherent transmission set by the Kerr nonlinearity and noise of the in-line amplifiers. *J. Lightwave Technol.*, 12(11):1993–2000.

Mecozzi, A. (1994b). Long-distance transmission at zero dispersion: Combined effect of Kerr nonlinearity and the noise of the in-line amplifiers. *J. Opt. Soc. Amer. B*, 11(3):462–469.

Mecozzi, A. and Shtaif, M. (2001). On the capacity of intensity modulated systems using optical amplifiers. *IEEE Photon. Technol. Lett.*, 13(9):1029–1031.

Mecozzi, A. and Shtaif, M. (2004). Signal-to-noise-ratio degradation caused by polarization-dependent loss and the effect of dynamic gain equalization. *J. Lightwave Technol.*, 22(8):1856–1871.

Mecozzi, A., Moores, J. D., Huas, H. A., and Lai, Y. (1991). Soliton transmission control. *Opt. Lett.*, 16(23):1841–1843.

Mecozzi, A., Clausen, C. B., and Shtaif, M. (2000a). Analysis of intrachannel nonlinear effects in highly dispersed optical pulse transmission. *IEEE Photon. Technol. Lett.*, 12(4):292–294.

Mecozzi, A., Clausen, C. B., and Shtaif, M. (2000b). System impact of intra-channel nonlinear effects in highly dispersed optical pulse transmission. *IEEE Photon. Technol. Lett.*, 12(12):1633–1635.

Mecozzi, A., Clausen, C. B., Shtaif, M., Park, S.-G., and Gnauck, A. H. (2001). Cancellation of timing and amplitude jitter in symmetric links using highly dispersed pulses. *IEEE Photon. Technol. Lett.*, 13(5):445–447.

Meissner, E. (1989). 116 photons/bit in a 565 Mbit/s optical DPSK heterodyne transmission experiment. *Electron. Lett.*, 25(4):281–282.

Middleton, D. (1960). *An Introduction to Statistical Communication Theory.* McGraw-Hill, New York.

Milivojevic, B., Abas, A. F., Hidayat, A., Bhandare, S., Sandel, D., Noe, R., Guy, M., and Lapointe, M. (2005). 1.6-b/s/Hz 160-Gb/s 230-km RZ-DQPSK polarization multiplex transmission with tunable dispersion compensation. *IEEE Photon. Technol. Lett.*, 17(2):495–497.

Miller, D. A. B., Weiner, J. S., and Chemla, D. S. (1986). Electric-field dependence of linear optical properties in quantum well structures: Waveguide electroabsorption and sum rules. *IEEE J. Quantum Electron.*, QE-22(9):1816–1830.

Minardi, M. J. and Ingram, M. A. (1995). Adaptive crosstalk cancellation and laser frequency drift compensation in dense WDM networks. *J. Lightwave Technol.*, 13(8):1624–1635.

Minowa, J. and Fujii, Y. (1983). Dielectric mulltilayer thin-film filters for WDM transmission systems. *J. Lightwave Technol.*, 1(1):116–121.

Mitra, P. P. and Stark, J. B. (2001). Nonlinear limits to the information capacity of optical fibre communications. *Nature*, 411(6841):1027–1030.

Miyagi, M. and Nishida, S. (1979). Pulse spreading in a single-mode fiber due to third-order dispersion. *Appl. Opt.*, 18(5):678–682.

Miyamoto, Y., Masuda, H., Hirano, A., Kuwahara, S., Kisaka, Y., Kawakami, H., Tomizawa, M., Tada, Y., and Aozasa, S. (2002). S-band WDM coherent transmission of 40 × 43-Gbit/s CS-RZ DPSK signals over 400 km DSF using hybrid GS-TDFAs/Raman amplifiers. *Electron. Lett.*, 38(24):1569–1570.

Miyazaki, T. and Kubota, F. (2004). Superposition of DQPSK over inverse-RZ for 3-bit/symbol modulation-demodulation. *IEEE Photon. Technol. Lett.*, 16(12):2643–2645.

Miyazaki, Y., Tada, H., Tokizaki, S., Takagi, K., Aoyagi, T., and Mitsui, Y. (2003). Small-chirp 40-Gbps electroabsorption modulator with novel tensile-strained asymmetric quantum-well absorption layer. *IEEE J. Quantum Electron.*, 39(6):813–819.

Mizuochi, T., Ishida, K., Kobayashi, T., Abe, J., Kinjo, K., Motoshima, K., and Kasahara, K. (2003). A comparative study of DPSK and OOK WDM transmission over transoceanic distances and their performance degradations due to nonlinear phase noise. *J. Lightwave Technol.*, 21(9):1933–1943.

Mizuochi, T., Miyata, Y., Kobayashi, T., Ouchi, K., Kuno, K., Kubo, K., Shimizu, K., Tagami, H., Yoshida, H., Fujita, H., Akita, M., and Motoshima, K. (2004). Forward error correction based on block Turbo code with 3-bit soft decision for 10-Gb/s optical communication systems. *IEEE J. Sel. Top. Quantum Electron.*, 10(2):376–386.

Mollenauer, L. F., Stolen, R. H., and Gordon, J. P. (1980). Experimental observation of picosecond pulse narrowing and solitons in optical fibers. *Phys. Rev. Lett.*, 45(13):1095–1098.

Möller, L., Su, Y., Xie, C., Ryf, R., Doerr, C. R., Liu, X., and Buhl, L. L. (2004). Generation of a 160-Gb/s RZ-DPSK signal and its detection with a one-bit Mach-Zehnder interferometer. In *European Conf. on Optical Commun.* postdeadline paper Th4.4.6.

Monroy, I. T. and Tangdiongga, E. (1998). Performance evaluation of optical cross-connects by saddlepoint approximation. *J. Lightwave Technol.*, 16(3):317–323.

Moon, N. S. and Kikuchi, K. (2003). $N \times N$ multiwavelength optical cross-connect based on tunable fiber bragg gratings. *J. Lightwave Technol.*, 21(3):703–718.

Moore, R. O., Biondini, G., and Kath, W. L. (2003). Importance sampling for noise-induced amplitude and timing jitter in soliton transmission systems. *Opt. Lett.*, 28(2):105–107.

Morita, I. and Edagawa, N. (2003). 50 GHz-spaced 64 × 42.7 Gbit/s transmission over 8200km using pre-filtered CS-RZ DPSK signal and EDFA repeaters. In *European Conf. on Optical Commun.* postdeadline paper Th4.3.1.

Morton, P. A., Logan, R. A., Tanbun-Ek, T., Jr., P. F. Sciortino, Sergent, A. M., Montgomery, R. K., and Lee, B. T. (1992). 25 GHz bandwidth 1.55 μm GaInAsP p-doped strained multiquantum-well lasers. *Electron. Lett.*, 28(23):2156–2157.

Muoi, T. and Hullett, J. (1975). Receiver design for multilevel digital optical fiber systems. *IEEE Trans. Commun.*, COM-23(9):987–994.

Murakami, M., Matsuda, T., Maeda, H., and Imai, T. (2000). Long-haul WDM transmission using higher order fiber dispersion management. *J. Lightwave Technol.*, 18(9):1197–1204.

Murata, H. and Inagaki, N. (1981). Low-loss single-mode fiber development and splicing research in Japan. *IEEE J. Quantum Electron.*, QE-17(6):835–849.

Mynbaev, D. K. and Scheiner, L. L. (2001). *Fiber-Optic Communications Technology*. Prentice-Hall, Upper Saddle River, NJ.

Nagata, H. (2000). Activation energy of DC-drift of x-cut $LiNbO_3$ optical intensity modulators. *IEEE Photon. Technol. Lett.*, 12(4):386–388.

Nagata, H., Feke, G. D., Li, Y., and Bosenberg, W.R. (2004). DC drift of z-cut $LiNbO_3$ modulators. *IEEE Photon. Technol. Lett.*, 26(7):1655–1657.

Nagayama, K., Kakui, M., Matsui, M., Saitoh, I., and Chigusa, Y. (2002). Ultra-low-loss (0.1484 dB/km) pure silica core fibre and extension of transmission distance. *Electron. Lett.*, 38(20):1168–1169.

Naito, T., Chikama, T., Ishikawa, G., and Kuwahara, H. (1990). 4 Gbit/s, 223 km optical fiber transmission experiment using newly proposed direct-modulation PSK. *Electron. Lett.*, 26(20):1734–1736.

Nakazawa, M., Kimura, Y., and Suzuki, K. (1989). Efficient Er^{3+}-doped optical fiber amplifier pumped by a 1.48 μm InGaAsP laser diode. *Appl. Phys. Lett.*, 54(4):295–297.

Nakazawa, M., Kubota, H., Suzuki, K., Yamada, E., and Sahara, A. (2000). Ultrahigh-speed long-distance TDM and WDM soliton transmission technologies. *IEEE J. Sel. Top. Quantum Electron.*, 6(2):363–396.

Narimanov, E. E. and Mitra, P. (2002). The channel capacity of a fiber optics communication system: perturbation theory. *J. Lightwave Technol.*, 20(3):530–537.

Neilson, D. T., Frahm, R., Kolodner, P., Bolle, C. A., Ryf, R., Kim, J., Papazian, A. R., Nuzman, C. J., Gasparyan, A., Basavanhally, N. R., Aksyuk, V. A., and Gates, J. V. (2004). 256 × 256 port optical cross-connect subsystem. *J. Lightwave Technol.*, 22(6):1499–1509.

Nelson, L. E., Jopson, R. M., Gnauck, A. H., and Chraplyvy, A. R. (1999). Resonances in cross-phase modulation impairment in wavelength-division-multiplexed lightwave transmission. *IEEE Photon. Technol. Lett.*, 11(7):907–909.

Nicholson, G. (1984). Probability of error for optical heterodyne DPSK system with quantum phase noise. *Electron. Lett.*, 20(24):1005–1007.

Nicholson, G. and Stephens, T. D. (1989). Performance analysis of coherent optical phase-diversity receivers with DPSK modulation. *J. Lightwave Technol.*, 7(2):393–399.

Nizhizawa, H., Yamada, Y., Shibata, Y., and Habara, K. (1999). 10-Gb/s optical DPSK packet receiver proof against large power fluctuations. *IEEE Photon. Technol. Lett.*, 11(6):733–735.

Noé, R., Heidrich, H., and Hoffmann, D. (1988a). Endless polarization control systems for coherent optics. *J. Lightwave Technol.*, 6(7):1199–1208.

Noé, R., Sessa, W. B., Welter, R., and Kazovsky, L. G. (1988b). New FSK phase-diversity receiver in a 150 Mbit/s coherent optical transmission system. *Electron. Lett.*, 24(9):567–568.

Noé, R., Rodler, H., Ebberg, A., Gaukel, G., and Auracher, F. (1989). Optical FSK transmission with pattern-independent 119 photoelectrons/bit receiver sensitivity and endless polarisation control. *Electron. Lett.*, 25(12):757–758.

Noé, R., Rodler, H., Ebberg, A., Gaukel, G., Noll, B., Wittmann, J., and Auracher, F. (1991). Comparison of polarization handling methods in coherent optical systems. *J. Lightwave Technol.*, 9(10):1353–1366.

Noé, R., Hinz, S., Sandel, D., and W ust, F. (2001). Crosstalk detection schemes for polarization division multiplex transmission. *J. Lightwave Technol.*, 19(10):1469–1475.

Noé, R., Sandel, D., and Mirvoda, V. (2004). PMD in high-bit-rate transmission and means for its mitigation. *IEEE J. Sel. Top. Quantum Electron.*, 10(2):341–355.

Noguchi, K., Mitomi, O., and Miyazawa, H. (1998). Millimeter-wave Ti:$LiNbO_3$ optical modulators. *J. Lightwave Technol.*, 16(4):615–619.

Noguchi, K., Mitomi, O., Miyazawa, H., and Seki, S. (1995). A broadband Ti:$LiNbO_3$ optical modulator with a ridge structure. *J. Lightwave Technol.*, 13(6):1164–1168.

Norimatsu, S. and Ishida, O. (1994). Impact of flicker noise and random-walk noise on a phase-locked loop with finite propagation delay. *J. Lightwave Technol.*, 12(1):86–95.

Norimatsu, S. and Iwashita, K. (1991). PLL propagation delay-time influence on linewidth requirements of optical PSK homodyne detection. *J. Lightwave Technol.*, 9(10):1367–1375.

Norimatsu, S. and Iwashita, K. (1992). Linewidth requirements for optical synchronous detection systems with nonnegligible loop delay time. *J. Lightwave Technol.*, 10(3):341–349.

Norimatsu, S. and Iwashita, K. (1993). The influence of cross-phase modulation on optical FDM PSK homodyne transmission systems. *J. Lightwave Technol.*, 11(5):795–804.

Norimatsu, S., Iwashita, K., and Noguchi, K. (1990). 10 Gbit/s optical PSK homodyne transmission experiments using external cavity DFB LDs. *Electron. Lett.*, 26(10):648–649.

Norimatsu, S., Iwashita, K., and Noguchi, K. (1992). An 8 Gb/s QPSK optical homodyne detection experiment using external-cavity laser diodes. *IEEE Photon. Technol. Lett.*, 4(7):765–767.

Nussmeier, T. A., Goodwin, F. E., and Zavin, J. E. (1974). A 10.6-μm terrestrial communication link. *IEEE J. Quantum Electron.*, QE-10(2):230–235.

Nyoklak, G., Eggleton, B. J., Lenz, G., and Strasser, T. A. (1998). Dispersion penalty measurements of narrow fiber Bragg gratings at 10 Gb/s. *IEEE Photon. Technol. Lett.*, 10(11):1319–1321.

Oda, K., Takato, N., and Toba, H. (1991). A wide-FSK waveguide double-ring resonator for optical FDM transmission systems. *J. Lightwave Technol.*, 9(6):728–736.

Oda, K., Suzuki, S., Takahashi, H., and Toba, H. (1994). An optical FDM distribution experiment using a high finesse waveguide-type double ring resonator. *IEEE Photon. Technol. Lett.*, 6(8):1031–1034.

Oh, M.-C., Zhang, H., Zhang, C., Erlig, H., Chang, Y., Tsap, B., Chang, D., Szep, A., Steier, W. H., Fetterman, H. R., and Dalton, L. R. (2001). Recent advances in electrooptic polymer modulators incorporating highly nonlinear chromophore. *IEEE J. Sel. Top. Quantum Electron.*, 7(5):826–835.

Ohm, M. (2004). Optical 8-DPSK and receiver with direct detection and multilevel electrical signal. In *IEEE/LEOS Workshop on Advanced Modulation Formats.* paper FC4.

Ohm, M. and Speidel, J. (2003). Quaternary optical ASK-DPSK and receivers with direct detection. *IEEE Photon. Technol. Lett.*, 15(1):159–161.

Okoshi, T. (1982). Heterodyne and coherent optical fiber communications: Recent progress. *IEEE Trans. Microwave Theory & Tech.*, MTT-30(8):1138–1149.

Okoshi, T. (1984). Recent progress in heterodyne/coherent optical-fiber communications. *J. Lightwave Technol.*, 2(4):341–346.

Okoshi, T. (1985). Polarization-state control schemes for heterodyne or homodyne optical fiber communications. *J. Lightwave Technol.*, LT-3(6):1232–1237.

Okoshi, T. (1986). Ultimate performance of heterodyne/coherent optical fiber communications. *J. Lightwave Technol.*, LT-4(10):1556–1562.

Okoshi, T. (1987). Recent advances in coherent optical fiber communication systems. *J. Lightwave Technol.*, 5(1):44–52.

Okoshi, T. and Cheng, T. H. (1987). Four-port homodyne receiver for optical fiber communications comprising phase and polarization diversity. *Electron. Lett.*, 23(8):377–378.

Okoshi, T. and Kikchi, K. (1980). Frequency stabilization of semiconductor lasers for heterodyne-type optical communication schemes. *Electron. Lett.*, 16:179–181.

Okoshi, T. and Kikuchi, K. (1988). *Coherent Optical Fiber Communications.* KTK Scientific, Tokyo.

Okoshi, T., Ishida, O., and Kikuchi, K. (1988). Simple formula for bit-error rate in optical heterodyne DPSK systems employing polarisation diversity. *Electron. Lett.*, 24(2):120–122.

Oliver, B. M. (1961). Signal-to-noise ratio in photoelectric mixing. *Proc. IEEE*, 49(12):1960–1961.

Olshansky, R., Hill, P., Lanzisera, V., and Powazinik, W. (1987). Frequency response of 1.3 μm InGaAsP high speed semiconductor lasers. *IEEE J. Quantum Electron.*, QE-23(9):1410–1418.

Olsson, A. and Tang, C. L. (1979). Electrooptically tuned external-cavity CW semiconductor laser and FM optical communication. *IEEE J. Quantum Electron.*, QE-15(9):1085–1088.

Ono, H., Yamada, M., and Ohishi, M. (1997). Gain-flattened Er^{3+}-doped fiber amplifier for a WDM signal in the 1.57-1.60-μm wavelength region. *IEEE Photon. Technol. Lett.*, 9(5):596–598.

Ono, H., Yamada, M., and Shimizu, M. (2003). S-band Erbium-doped fiber amplifiers with a multistage configuration - design, characterization, and gain tilt compensation. *J. Lightwave Technol.*, 21(10):2240–2246.

Pan, Z., Song, Y. W., Yu, C., Wang, Y., Yu, Q., Popelek, J., Li, H., Li, Y., and Willner, A. E. (2002). Tunable chromatic dispersion compensation in 40-Gb/s systems using nonlinearly chirped fiber Bragg gratings. *J. Lightwave Technol.*, 20(12):2239–2246.

Pan, Z., Wang, Y., Song, Y., Motaghian, R., Havstad, S., and Willner, A. E. (2003). Monitoring chromatic dispersion and PMD impairments in optical differential phase shift-keyed (DPSK) systems. In *Optical Fiber Commun. Conf.*, paper WP1.

Papoulis, A. (1984). *Probability, Random Variables, and Stochastic Processes.* McGraw Hill, New York, second edition.

Paré, C., Villeneuve, A., Bélanger, P.-A., and Doran, N. J. (1996). Compensating for dispersion and the nonlinear Kerr effect without phase conjugation. *Opt. Lett.*, 21(7):459–461.

Park, C. A., Borley, S. J., Hankey, J., Debney, B. T., Ascham, S. R., Marsh, S. J., and Thorley, A. M. (1991). Field demonstration of coherent FSK system with 60 dB optical power budget. *Electron. Lett.*, 27(9):689–690.

Park, Y. K., Delavaux, J. M. P., Tench, R. E., and Cline, T. W. (1990). 1.7 Gb/s, 419 km transmission experiment using a shelf mounted FSK coherent system and packaged fiber amplifier modules. *IEEE Photon. Technol. Lett.*, 2(12):917–919.

Peddanarappagari, K. V. and Brandt-Pearce, M. (1997). Volterra series transfer function of single mode fibers. *J. Lightwave Technol.*, 15(12):2232–2241.

Peddanarappagari, K. V. and Brandt-Pearce, M. (1998). Volterra series approach for optimizing fiber-optic communications system designs. *J. Lightwave Technol.*, 16(11):2046–2051.

Pepper, D. M. and Yariv, A. (1980). Compensation for phase distortions in nonlinear media by phase conjugation. *Opt. Lett.*, 5(1):59–60.

Petermann, K. (1991). *Laser Diode Modulation and Noise.* Kluwer Academic, Dordrecht, The Netherlands.

Peyton, B. J., DiNardo, A. J., Kanischak, G. M., Arams, F. R., Lange, R. A., and Sard, E. W. (1972). High-sensitivity receiver for infrared laser communications. *IEEE J. Quantum Electron.*, QE-8(2):252–263.

Pezeshki, B., Vail, E., Kubicky, J., Yoffe, G., Zou, S., Heanue, J., Epp, P., Rishton, S., Ton, D., Faraji, B., Emanuel, M., Hong, X., Sherback, M., Agrawal, V., Chipman, C., and Razazan, T. (2002). 20-mW widely tunable laser module using DFB array and MEMS selection. *IEEE Photon. Technol. Lett.*, 14(10):1457–1459.

Phillips, M. R. and Darcie, T. E. (1997). Lightwave video transmission. In Kaminov, I. P. and Koch, T. L., editors, *Optical Fiber Telecommunications*, volume III. Academic Press, San Diego.

Pires, J. J. O. and de Rocha, J. R. F. (1992). Performance analysis of DPSK direct detection optical systems in the presence of interferometric intensity. *J. Lightwave Technol.*, 10(11):1722–1730.

Poole, C. D. and Nagel, J. A. (1997). Polarization effects in lightwave systems. In Kaminow, I. P. and Koch, T. L., editors, *Optical Fiber Telecommunications*, volume IIIA, pages 114–161. Academic Press, San Diego.

Poole, C. D. and Wagner, R. E. (1986). Phenomenological approach to polarization dispersion in long single-mode fibers. *Electron. Lett.*, 22(19):1029–1030.

Poole, C. D., Winters, J. H., and Nagel, J. A. (1991). Dynamical equation for polarization dispersion. *Opt. Lett.*, 16(6):372–374.

Prabhu, V. K. (1969). Error-rate considerations for digital phase-modulation systems. *IEEE Trans. Commun. Technol.*, COM-17(1):33–42.

Prabhu, V. K. (1976). PSK performance with imperfect carrier recovery. *IEEE Trans. Aerosp. Electron. Syst.*, AES-12(2):107–125.

Pratt, W. (1966). Binary detection in an optical polarization modulation communication channel. *IEEE Trans. Commun. Technol.*, COM-14(5):664–665.

Proakis, J. G. (2000). *Digital Communications.* McGraw Hill, New York, fourth edition.

Ralston, J. D., Weisser, S., Esquivias, I., Larkins, E. C., Rosenzweig, J., Tasker, P. J., and Fleissner, J. (1993). Control of differential gain, nonlinear gain and damping factor for high-speed application of GaAs-based MQW lasers. *IEEE J. Quantum Electron.*,

29(6):1648–1659.

Rashleigh, S. C. and Ulrich, R. (1978). Polarization mode dispersion in single-mode fibers. *Opt. Lett.*, 3(2):60–62.

Rasmussen, C., Fjelde, T., Bennike, J., Liu, F., Dey, S., Mikkelsen, B, Mamyshev, P., Serbe, P., van de Wagt, P., Akasaka, Y., Harris, D., Gapontsev, D., Ivshin, V., and Reeves-Hall, P. (2003). DWDM 40G transmission over trans-Pacific distance (10,000 km) using CSRZ-DPSK, enhanced FEC and all-Raman amplified 100 km Ultra-Wave^{TM} fiber spans. In *Optical Fiber Commun. Conf.*, postdeadline paper PD18.

Rasmussen, C., Fjelde, T., Bennike, J., Liu, F., Dey, S., Mikkelsen, B, Mamyshev, P., Serbe, P., van de Wagt, P., Akasaka, Y., Harris, D., Gapontsev, D., Ivshin, V., and Reeves-Hall, P. (2004). DWDM 40G transmission over trans-Pacific distance (10,000 km) using CSRZ-DPSK, enhanced FEC and all-Raman amplified 100 km Ultra-Wave^{TM} fiber spans. *J. Lightwave Technol.*, 22(1):203–207.

Razavi, B. (1997). Design considerations for direct-conversion receivers. *IEEE Trans. Circuits Syst. II*, 44(6):428–438.

Rebola, J. L. and Cartaxo, A. V. T. (2000). Optimization of level spacing in quaternary optical communication systems. In *Proc. SPIE: Applications of the Photonic Technology*, volume 4087, pages 49–59, Quebec City, Que., Canada.

Rice, S. O. (1944). Mathematical analysis of random noise. *Bell Syst. Tech. J.*, 23(3):283–332. vol. 24, no. 1, pp. 46–157, 1945.

Rice, S. O. (1948). Statistical properties of a sine wave plus random noise. *Bell Syst. Tech. J.*, 27(1):109–157.

Rochat, E., Dandliker, R., Haroud, K., Czichy, R. H., Roth, U., Costantini, D., and Holzner, R. (2001). Fiber amplifiers for coherent space communication. *IEEE J. Sel. Top. Quantum Electron.*, 7(1):64–80.

Rohde, M., Caspar, C., Heimes, N., Konitzer, M., Bachus, E.-J., and Hanik, N. (2000). Robustness of DPSK direct detection transmission format in standard fibre WDM systems. *Electron. Lett.*, 36(17):1483–1484.

Roudas, I., Antoniades, N., Otani, T., Stern, T. E., Wagner, R. E., and Chowdhury, D. Q. (2002). Accurate modeling of optical multiplexer/demultiplexer concatenation in transparent multiwavelength optical networks. *J. Lightwave Technol.*, 20(6):921–936.

Ryu, S. (1991). Signal linewidth broadening due to fibre nonlinearities in long-haul coherent optical fibre communication systems. *Electron. Lett.*, 27(17):1527–1529.

Ryu, S. (1992). Signal linewidth broadening due to nonlinear Kerr effect in long-haul coherent systems using cascaded optical amplifiers. *J. Lightwave Technol.*, 10(10):1450–1457.

Ryu, S. (1995). *Coherent Lightwave Communication Systems*. Artech House, Boston.

Ryu, S., Yamamoto, S., and Mochizuki, K. (1987). Polarization-insensitive operation of coherent FSK transmission system using polarization diversity. *Electron. Lett.*, 23(25):1382–1384.

Ryu, S., Yamamoto, S., Namihira, Y., Mochizuki, K., and Wakabayashi, H. (1988). First sea trail of FSK heterodyne optical transmission system using polarization diversity. *Electron. Lett.*, 24(7):399–400.

Ryu, S., Yamamoto, S., Namihira, Y., Mochizuki, K., and Wakabayashi, H. (1991a). Polarization diversity techniques for the use of coherent optical fiber submarine cable systems. *J. Lightwave Technol.*, 9(5):675–682.

Ryu, S., Yamamoto, S., Taga, H., Edagawa, N., Yoshida, Y., and Wakabayashi, H. (1991b). Long haul coherent optical fiber communication systems using optical amplifiers. *J. Lightwave Technol.*, 9(2):251–260.

Ryu, S., Miyazaki, T., Kawazawa, T., Namihira, Y., and Wakabayashi, H. (1992). Field demonstration of 195 km-long coherent unrepeated submarine cable system using optical booster amplifier. *Electron. Lett.*, 28(21):1965–1967.

Saito, S. and Kimura, T. (1964). Demodulation of phase-modulated optical maser beam by auto-correlation technique. *Proc. IEEE*, 52(9):1048.

Saito, S., Yamamoto, Y., and Kimura, T. (1980). Optical heterodyne detection of directly modulated semiconductor laser signals. *Electron. Lett.*, 16:826–827.

Saito, S., Yamamoto, Y., and Kimura, T. (1981). Optical FSK heterodyne detection experiments using semiconductor laser transmitter and local oscillator. *IEEE J. Quantum Electron.*, QE-17(6):935–951.

Saito, S., Yamamoto, Y., and Kimura, T. (1983). S/N and error rate evaluation for an optical FSK-heterodyne detection system using semiconductor lasers. *IEEE J. Quantum Electron.*, QE-19(2):180–193.

Saito, S., Yamamoto, Y., and Kimura, T. (1991). Coherent optical fiber transmission systems. In Li, T., editor, *Topics in Lightwave Transmission Systems*, pages 203–266. Academic Press, San Diego.

Saito, S., Murakami, M., Naka, A., Fukada, Y., Imai, T., Aiki, M., and Ito, T. (1992). In-line amplifier transmission experiment over 4500 km at 2.5 Gb/s. *J. Lightwave Technol.*, 10(8):1117–1124.

Saito, S., Aiki, M., and Ito, T. (1993). System performance of coherent transmission over cascaded in-line fiber amplifiers. *J. Lightwave Technol.*, 11(2):331–342.

Saleh, B. E. A. and Teich, M. C. (1991). *Fundamentals of Photonics.* Wiley-Interscience, New York.

Salgado, H. M., Ferreira, J. M., and OReilly, J. J. (1997). Extraction of semiconductor intrinsic laser parameters by intermodulation distortion analysis. *IEEE Photon. Technol. Lett.*, 9(10):1331–1333.

Sandel, D., Bhandare, S., Abas, A. F., Milivojevic, B., Noé, R., Guy, M., and Lapointe, M. (2004). Automatic tunable chromatic dispersion compensation at 40 Gb/s in ASK and DPSK, NRZ, and CSRZ 263-km transmission experiments. *IEEE Photon. Technol. Lett.*, 16(11):2568–2570.

Sargent, M., Scully, M. O., and Lamb, W. E. (1974). *Laser Physics.* Addison-Wesley, London.

Sato, K. (2002). Semiconductor light sources for 40-Gb/s transmission systems. *J. Lightwave Technol.*, 20(12):2035–2045.

Sato, K., Kuwahara, S., Miyamoto, Y., and Shimizu, N. (2002). Direct modulation of a distributed feedback laser for 40-Gb/s very-short-reach optical links. In *Optical Fiber Commun. Conf.*, Paper ThF2.

Savory, S. and Hadjifotiou, A. (2004). Laser linewidth requirements for optical DQPSK systems. *IEEE Photon. Technol. Lett.*, 16(3):930–932.

Schiess, M. and Carlden, H. (1994). Evaluation of the chirp parameter of a Mach-Zehnder intensity modulator. *Electron. Lett.*, 30(18):1524–1525.

Schöpflin, A., Kugelmeier, S., Gottwald, E., Felicio, D., and Fischer, G. (1990). PSK optical homodyne system with nonlinear phase-locked loop. *Electron. Lett.*, 26(6):395–396.

Schwartz, M., Bennett, W. E., and Stein, S. (1966). *Communication Systems and Techniques.* McGraw-Hill, New York.

Senior, J. (1992). *Optical Fiber Communications.* Prentice Hall, Upper Saddle River, NJ, 2 edition.

Shake, I., Takara, H., Mori, K., Kawanishi, S., and Yamabayashi, Y. (1998). Influence of inter-bit four-wave mixing in optical TDM transmission. *Electron. Lett.*, 34(16):1600–1601.

Shamai (Shitz), S. and Bar-David, I. (1995). The capacity of average and peak-power-limited quadrature Gaussian channels. *IEEE Trans. Info. Theory*, 41(4):1060–1071.

Shannon, C. E. (1948). A mathematical theory of communication. *Bell Syst. Tech. J.*, 27(7, 10):379–423, 623–646.

Shi, B. and Sundstrom, L. (2000). A 200-MHz IF BiCMOS signal component separator for linear LINC transmitters. *IEEE J. Solid-State Circuits*, 35(7):987–993.

Shi, Y., Wang, W., Bechtel, J. H., Chen, A., Garner, S., Kalluri, S., Steier, W. H., Chen, D., Fetterman, H. R., Dalton, L. R., and Yu, L. (1996). Fabrication and characterization of high-speed polyurethane-disperse red 19 integrated electrooptic modulators for analog system applications. *IEEE J. Sel. Top. Quantum Electron.*, 2(2):289–299.

Shi, Y., Zhang, C., Zhang, H., Bechtel, J. H., Dalton, L. R., Robinson, B. H., and Steier, W. H. (2000). Low (sub-1-Volt) halfwave voltage polymeric electro-optic modulator achieved controlling chromophore shape. *Science*, 288:119–122.

Shibata, N., Nosu, K., Iwashita, K., and Azuma, Y. (1990). Transmission limitations due to fiber nonlinearities in optical FDM systems. *IEEE J. Sel. Areas Commun.*, 8(6):1068–1077.

Shibutani, M. and Yamazaki, S. (1989). A study on an active square-law combining method for a polarization-diversity coherent optical receiver. *IEEE Photon. Technol. Lett.*, 1(7):182–183.

Shikada, M., Takano, S., Fujita, S., Mito, I., and Minemura, K. (1988). Evaluation of power penalties caused by feedback noise of distributed feedback laser diodes. *J. Lightwave Technol.*, 6(3):655–659.

Shimada, S., editor (1995). *Coherent Lightwave Communications Technology*. Kluwer Academic, Dordrecht, The Netherlands.

Shirasaki, M., Nishimoto, H., Okiyama, Y., and Touge, T. (1988). Fibre transmission properties of optical pulses produced through direct phase modulation of DFB laser diode. *Electron. Lett.*, 24(8):486–488.

Shiu, D.-S. and Kahn, J. M. (1999). Shaping and nonequiprobable signaling for intensity-modulated signals. *IEEE Trans. Info. Theory*, 45(5):2661–2668.

Shum, P. and Ghafouri-Shiraz, H. (1996). Analysis of bit error rate in an optical soliton communication system. *Opt. Laser Technol.*, 28(7):535–547.

Shum, P., Ghafouri-Shiraz, H., and Yu, S. F (1997). Analysis of a DPSK soliton transmission system. *Opt. Laser Technol.*, 29(7):411–414.

Sinsky, J. H., Adamiecki, A., Burrus, C. A., Chandrasekhar, S., Leuthold, J., and Wohlgemuth, O. (2003). A 40-Gb/s integrated balanced optical front end and RZ-DPSK performance. *IEEE Photon. Technol. Lett.*, 15(8):1135–1137.

Siuzdak, J. and van Etten, W. (1989). BER evaluation for phase and polarization diversity optical homodyne receivers using noncoherent ASK and DPSK demodulation. *J. Lightwave Technol.*, 7(4):584–599.

Siuzdak, J. and van Etten, W. (1991). BER performance evaluation for CPFSK phase and polarization diversity coherent optical receivers. *J. Lightwave Technol.*, 9(11):1583–1592.

Smit, M. K. and van Dam, C. (1996). PHASAR-based WDM-devices: Principles, design and applications. *IEEE J. Sel. Top. Quantum Electron.*, 2(2):236–250.

Smith, D. (1987). Techniques for multigigabit coherent optical transmission. *J. Lightwave Technol.*, LT-5(10):1466–1478.

Smith, J. G. (1971). The information capacity of amplitude- and variance-constrained scalar Gaussian channels. *Info. & Control*, 18(3):203–219.

Smith, P. J., Shafi, M., and Kaiser, C. P. (1995). Optical heterodyne binary-DPSK systems: A review of analysis and performance. *IEEE J. Sel. Areas Commun.*, 13(3):557–568.

Smith, N. J., Doran, N. J., Forysiak, W., and Knox, F. M. (1997). Soliton transmission using periodic dispersion compensation. *J. Lightwave Technol.*, 15(10):1808–1822.

Spano, P., Piazzolla, S., and Tamburrini, M. (1983). Phase noise in semiconductor lasers: A theoretical approach. *IEEE J. Quantum Electron.*, QE-19(7):1195–1199.

Spilker, Jr., J. J. (1977). *Digital Communications by Satellite*. Prentice Hall, Englewood Cliff, NJ.

Stark, J. B., Mitra, P., and Sengupta, A. (2001). Information capacity of nonlinear wavelength division multiplexing fiber optic transmission line. *Opt. Fiber Technol.*, 7(4):275–288.

Stein, E. M. and Stein, J. C. (1991). Stock price distribution with stochastic volatility: An analytical approach. *Rev. Financial Studies*, 4(4):727–752.

Stein, S. (1964). Unified analysis of certain coherent and noncoherent binary communications systems. *IEEE Trans. Info. Theory*, IT-10(1):43–51.

Stern, T. (1960). Some quantum effects in information channels. *IRE Trans. Info. Theory*, IT-6(4):435–440.

Stolen, R. (1979). Polarization effects in fiber Raman and Brillouin lasers. *IEEE J. Quantum Electron.*, QE-15(10):1157–1160.

Stolen, R. H. (1980). Nonlinearity in fiber transmission. *Proc. IEEE*, 68(10):1232–1236.

Striegler, A. G. and Schmauss, B. (2004). Compensation of intrachannel effects in symmetric dispersion-managed transmission systems. *J. Lightwave Technol.*, 22(8):1877–1882.

Su, Y., Liu, X., and Leuthold, J. (2004). Wide dynamic range 10-Gb/s DPSK packet receiver using optical-limiting amplifiers. *IEEE Photon. Technol. Lett.*, 16(1):296–298.

Suematsu, Y., Arai, S., and Kishino, K. (1983). Dynamic single-mode semiconductor lasers with a distributed reflector. *J. Lightwave Technol.*, LT-1(1):161–176.

Suematsu, Y., Iga, K., and Arai, S. (1992). Advanced semiconductor lasers. *Proc. IEEE*, 80(3):383–397.

Sugiyama, M. (2003). Hi perf $LiNbO_3$ modulators. In *Optical Fiber Commun. Conf.*, paper FP1.

Sugiyama, M., Doi, M., Taniguchi, S., Nakazawa, T., and Onaka, H. (2002). Driver-less 40 Gb/s $LiNbO_3$ modulator with sub-1 V drive voltage. In *Optical Fiber Commun. Conf.*, postdeadline paper FB6.

Sun, L. and Ye, P. (1990). Optical homodyne receiver based on an improved balance phase-locked loop with the data-to-phaselock crosstalk suppression. *IEEE Photon. Technol. Lett.*, 2(9):678–679.

Sun, C. K., Orazi, R. J., Pappert, S. A., and Burns, W. K. (1996). A photonic-link millimeter-wave mixer using cascaded optical modulators and harmonic carrier generations. *IEEE Photon. Technol. Lett.*, 8(9):1166–1168.

Sun, Y., Sulhoff, J. W., Srivastava, A. K., Zyskind, J. L., Strasser, T. A., Pedrazzani, J. R., Wolf, C., Zhou, J., Judkins, J. B., Espindola, R. P., and Vengsarkar, A. M. (1997). 80 nm ultra-wideband Erbium-doped silica fibre amplifier. *Electron. Lett.*, 33(23):1965–1967.

Sunnerud, H., Andrekson, P. A., and Karlsson, M. (2003). Optimum receiver decision point in presence of PMD in fiber-optic communication systems. *IEEE Photon. Technol. Lett.*, 15(11):1651–1653.

Sunnerud, H., Xie, C., Karlsson, M., Samuelsson, R., and Andrekson, P. A. (2002). A comparison between different PMD compensation techniques. *J. Lightwave Technol.*, 20(3):368–378.

Suzuki, M. and Edagawa, N. (2003). Dispersion-managed high-capacity ultra-long-haul transmission. *J. Lightwave Technol.*, 21(4):916–929.

Suzuki, M., Morita, I., Edagawa, N., Yamamoto, S., Taga, H., and Akiba, S. (1995). Reduction of Gordon-Haus timing jitter by periodic dispersion compensation. *Electron. Lett.*, 31(23):2027–2029.

Takachio, N., Iwashita, K., Nakanishi, K., and Koike, S. (1989). Chromatic dispersion equalization in an 8 Gbit/s 202 km CPFSK transmission experiment. In *Int'l Opt. Commun. Conf.* paper PDA-13.

Takahashi, H., Suzuki, S., Kato, K., and Nishi, I. (1990). Arrayed-waveguide grating for wavelength division multi/demultiplexer with nanometre resolution. *Electron. Lett.*, 26(2):87–88.

Takahashi, T., Imai, T., and Aiki, M. (1994). Automatic compensation technique for timewise fluctuating polarisation mode dispersion in in-line amplifier systems. *Electron. Lett.*, 30(4):348–349.

Takahashi, H., Oda, K., and Toba, H. (1996). Impact of crosstalk in an arrayed-waveguide multiplexer on $N \times N$ optical interconnection. *J. Lightwave Technol.*, 14(6):1097–1105.

Takushima, T., Douke, T., Wang, X., and Kikuchi, K. (2002). Dispersion tolerance and transmission distance of a 40-Gb/s dispersion management soliton transmission system. *J. Lightwave Technol.*, 20(3):360–367.

Tamburrini, M., Spano, S., and Piazzolla, S. (1983). Influence of semiconductor-laser phase noise on coherent optical communication systems. *Opt. Lett.*, 8(3):174–176.

Tang, J. (2001a). The multispan effects of Kerr nonlinearity and amplifier noises on Shannon channel capacity for a dispersion-free nonlinear optical fiber. *J. Lightwave Technol.*,

19(8):1110–1115.

Tang, J. (2001b). The Shannon channel capacity of dispersion-free nonlinear optical fiber transmission. *J. Lightwave Technol.*, 19(8):1104–1109.

Tang, J. (2002). The channel capacity of a multispan DWDM system employing dispersive nonlinear optical fibers and an ideal coherent optical receiver. *J. Lightwave Technol.*, 21(7):1095–1101.

Taylor, M. G. (1993). Observation of new polarization dependence effect in long haul optically amplified system. *IEEE Photon. Technol. Lett.*, 5(10):1244–1246.

Taylor, M. G. (2004). Coherent detection method using DSP for demodulation of signal and subsequent equalization of propagation impairments. *IEEE Photon. Technol. Lett.*, 16(2):674–676.

Tench, R. E., Delavaux, J.-M. P., Tzeng, L. D., Smith, R. W., Buhl, L. L., and Alferness, R. C. (1987). Performance evaluation of waveguide phase modulators for coherent systems at 1.3 and 1.5 μm. *J. Lightwave Technol.*, LT-5(4):492–501.

Testa, F., Merli, S., and Pagnan, P. (1994). Fabry-Perot filter bandwidth effects on end to end transmission performances in a multiwavelength transport network. *IEEE Photon. Technol. Lett.*, 6(8):1027–1030.

Tkach, R. W. and Chraplyvy, A. R. (1986). Regimes of feedback effects in 1.5-μm distributed feedback lasers. *J. Lightwave Technol.*, LT-4(11):1655–1661.

Tkach, R. W., Chraplyvy, A. R., Forghieri, F., Gnauck, A. H., and Derosier, R. M. (1995). Four-photon mixing and high-speed WDM systems. *J. Lightwave Technol.*, 13(5):841–849.

Toba, H., Oda, K., Nakanishi, K., Shibata, N., Nosu, K., Takato, N., and Fukuda, M. (1990). A 100-channel optical FDM transmission/distribution at 622 Mb/s over 50 km. *J. Lightwave Technol.*, 8(9):1396–1401.

Toba, H., Oda, K., and Nosu, K. (1991). Design and performance of FSK-direct detection scheme for optical FDM systems. *J. Lightwave Technol.*, 9(10):1335–1343.

Tokle, T., Davidson, C. R., Nissov, M., Cai, J.-X., Foursa, D., and Pilipetskii, A. (2004). 6500 km transmission of RZ-DQPSK WDM signals. *Electron. Lett.*, 40(7):444–445.

Tomizawa, M., Kisaka, Y., Ono, T., Miyamoto, Y., and Tada, Y. (2002). Statistical design of polarization mode dispersion on high-speed transmission systems with forward error correction. *IEICE Trans. Commun.*, E85-B(2):454–462.

Tonguz, O. K. and Wagner, R. E. (1991). Equivalence between preamplified direct detection and heterodyne receivers. *IEEE Photon. Technol. Lett.*, 3(9):835–837.

Tran, A. V., Chae, C.-J., and Tucker, R. S. (2003). A bidirectional optical add-drop multiplexer with gain using multiport circulators, fiber bragg gratings, and a single unidirectional optical amplifier. *IEEE Photon. Technol. Lett.*, 15(7):975–977.

Tsao, H.-W., Wu, J., Yang, S.-C., and Lee, Y.-H. (1990). Performance analysis of polarization-insensitive phase diversity optical FSK receivers. *J. Lightwave Technol.*, 8(3):385–395.

Tsao, H.-W., Kuo, C.-C., Yang, S.-C., Wu, J., and Lee, Y.-H. (1992). Theoretical analysis of phase diversity optical homodyne receiver with FSK demodulation. *IEEE Trans. Commun.*, 40(4):795–809.

Tsuritani, T., Ishida, K., Agata, A., Shimomura, K., Morita, I., Tokura, T., Taga, H., Mizuochi, T., and Edagawa, N. (2003). 70 GHz-spaced 40 × 42.7 Gbit/s transmission over 8700 km using CS-RZ DPSK signal, all-Raman repeaters and symmetrically dispersion-managed fiber span. In *Optical Fiber Commun. Conf.*, postdeadline paper PD23.

Tsuritani, T., Ishida, K., Agata, A., Shimomura, K., Morita, I., Tokura, T., Taga, H., Mizuochi, T., Edagawa, N., and Akiba, S. (2004). 70 GHz-spaced 40 × 42.7 Gbit/s transmission over 9400 km using prefilterred CS-RZ DPSK signal, all-Raman repeaters, and symmetrically dispersion-managed fiber span. *J. Lightwave Technol.*, 22(1):215–224.

Tsuzuki, K., Yasaka, H., Ishibashi, T., Ito, T., Oku, S., Iga, R., Kondo, Y., and Tohmori, Y. (2004). 10-Gbit/s, 100-km SMF transmission using an InP-based *n-i-n* Mach-Zehnder modulator with a driving voltage of 1.0 V_{pp}. In *Optical Fiber Commun. Conf.*, osa. postdeadline paper PDP14.

Tucker, R. S. (1985). High-speed modulation of semiconductor lasers. *J. Lightwave Technol.*, LT-3(6):1180–1192.

Tucker, R. S. and Kaminow, I. (1984). High-frequency characteristics of directly modulated InGaAsP ridge waveguide and buried heterostructure lasers. *J. Lightwave Technol.*, LT-2(4):385–393.

Tur, M. and Goldstein, E. L. (1991). Probability distribution of phase-induced intensity noise generated by distributed-feedback lasers. *Opt. Lett.*, 15(1):1–3.

Turin, G. L. (1960). The characteristic function of Hermitian quadratic forms in complex normal variables. *Biometrika*, 47:199–201.

Turitsyn, K. S., Derevyanko, S. A., Yurkevich, I. V., and Turitsyn, S. K. (2003). Information capacity of optical fiber channels with zero average dispersion. *Phys. Rev. Lett.*, 91(203901).

Turner, E. H. (1966). High-frequency electro-optic coefficients of Lithium Niobate and Lithium Tantalate. *IEEE J. Quantum Electron.*, 2(4):128.

Uetsuka, H. (2004). AWG technologies for dense WDM applications. *IEEE J. Sel. Top. Quantum Electron.*, 10(2):393–402.

Urick, V. J., Qiu, J. X., Dexter, J. L., Williams, K J., Bucholtz, F., and McDermitt, C. (2004). Giga-symbol-per-second 16-QAM transmission over a single-channel Raman-amplified 100 km link. In *Optical Fiber Commun. Conf.*, paper ThM2.

Vaa, M., Golovchenko, E. A., Rahman, L., Mohs, G., Patterson, W., and Pilipetskii, A. (2004). Dense WDM RZ-DPSK transmission over transoceanic distances without use of periodic dispersion management. In *European Conf. on Optical Commun.* postdeadline paper Th4.4.4.

Vahala, K., Harder, C., and Yariv, A. (1983). Observation of relaxation resonance effects in the field spectrum of semiconductor lasers. *Appl. Phys. Lett.*, 42(3):211–213.

Vahala, K. and Yariv, A. (1983a). Semiclassical theory of noise in semiconductor lasers-Part I. *IEEE J. Quantum Electron.*, QE-19(6):1096–1101.

Vahala, K. and Yariv, A. (1983b). Semiclassical theory of noise in semiconductor lasers-Part II. *IEEE J. Quantum Electron.*, QE-19(6):1102–1109.

Vareille, G., Becouarn, L., Pecci, P., Tran, P., and Marcerou, J. F. (2003). 8370 km with 22 dB spans ULH transmission of 185*10.709 Gbit/s RZ-DPSK channels. In *Optical Fiber Commun. Conf.*, postdeadline paper PD20.

Vasic, B. and Djordjevic, I. B. (2002). Low-density parity check codes for long-haul optical communication systems. *IEEE Photon. Technol. Lett.*, 14(8):1208–1210.

Vengsarkar, A. M., Pedrazzani, J. R., Judkins, J. B., Lemaire, P. J., Bergano, N. S., and Davidson, C. R. (1996). Long-period fiber-grating-based gain equalizers. *Electron. Lett.*, 21(5):336–338.

Vodhanel, R. S. (1989). 5 Gbit/s direct optical DPSK modulation of a 1530-nm DFB laser. *IEEE Photon. Technol. Lett.*, 1(8):218–220.

Vodhanel, R. S. and Enning, B. (1988). 1 Gbit/s bipolar optical FSK transmission experiment over 121 km of fibre. *Electron. Lett.*, 24(3):163–165.

Vodhanel, R. S., Enning, B., and Elrefaie, A. F. (1988). Bipolar optical FSK transmission experiments at 150 Mbit/s and 1 Gbit/s. *J. Lightwave Technol.*, 6(10):1549=1553.

Vodhanel, R. S., Elrefaie, A. E., Iqbal, M. Z., Wagner, R. E., Gimlett, J. L., and Tsuji, S. (1990). Performance of directly modulated DFB lasers in 10-Gb/s ASK, FSK, and DPSK lightwave systems. *J. Lightwave Technol.*, 8(9):1379–1386.

Waarts, R. G., Friesem, A. A., Lichtman, E., Yaffe, H. H., and Baurn, R.-P. (1990). Nonlinear effects in coherent multichannel transmission through optical fibers. *IEEE Proc.*, 78(8):1344–1368.

Walker, G. R., Steele, R. C., and Walker, N. G. (1990). Optical amplifier noise figure in a coherent optical transmission system. *J. Lightwave Technol.*, 8(9):1409–1413.

Walker, N. G. and Carroll, J. E. (1984). Simultaneous phase and amplitude measurements on optical signals using a multiport junction. *Electron. Lett.*, 20(23):981–983.

Walker, N. G. and Walker, G. R. (1990). Polarization control for coherent communications. *J. Lightwave Technol.*, 8(3):438–458.

Walker, R. G. (1987). High-speed electrooptic modulation in GaAs/GaAlAs waveguide devices. *J. Lightwave Technol.*, LT-5(10):1444–1453.

Walker, R. G. (1991). High-speed III-V semiconductor intensity modulators. *IEEE J. Quantum Electron.*, 27(3):654–667.

Walklin, S. and Conradi, J. (1997). Effect of Mach-Zehnder modulator DC extinction ratio on residual chirp-induced dispersion in 10-Gb/s binary and AM-PSK duobinary lightwave systems. *IEEE Photon. Technol. Lett.*, 9(10):1400–1402.

Walklin, S. and Conradi, J. (1999). Multilevel signaling for increasing the reach of 10 Gb/s lightwave systems. *J. Lightwave Technol.*, 17(11):2235–2248.

Wandernoth, B. (1992). 20 photon/bit 565 Mbit/s PSK homodyne receiver using synchronisation bits. *Electron. Lett.*, 28(4):387–388.

Wang, J. and Kahn, J. M. (2004a). Conventional DPSK versus symmetrical DPSK: Comparison of dispersion tolerances. *IEEE Photon. Technol. Lett.*, 16(6):1585–1587.

Wang, J. and Kahn, J. M. (2004b). Impact of chromatic and polarization-mode dispersions on DPSK systems using interferometric demodulation and direct detection. *J. Lightwave Technol.*, 22(2):362–371.

Wang, J. and Petermann, K. (1992). Small signal analysis for dispersive optical fiber communication systems. *J. Lightwave Technol.*, 10(1):96–100.

Watanabe, S., Chikama, T., Naito, T., and Kuwahara, H. (1989). 560 Mbit/s optical PSK heterodyne detection using carrier recovery. *Electron. Lett.*, 25(9):588–590.

Watanabe, S., Naito, T., Chikama, T., and Kuwahara, H. (1992). Phase noise cancellation in a carrier recovered optical heterodyne receiver. *J. Lightwave Technol.*, 10(12):1963–1973.

Way, W. I. (1998). *Broadband Hybrid Fiber/Coax Access System Technologies.* Academic Press, San Diego.

Way, W. I., Wagner, S. S., Choy, M. M., Lin, C., Menendez, R. C., Tohme, H., Yi-Yan, A., von Lehman, A. C., Spicer, R. E., Andrejco, M., Saifi, M. A., and Lemberg, H. L. (1990). Simultaneous distribution of multichannel analog and digital video channels to multiple terminals using high-density WDM and a broad-band in-line Erbium-doped fiber amplifier. *IEEE Photon. Technol. Lett.*, 2(9):665–668.

Wedding, B. and Haslach, C. N. (2001). Enhanced PMD mitigation by polarization scrambling and forward error correction. In *Optical Fiber Commun. Conf.*, paper WAA1.

Wegener, L. G. L., Povinelli, M. L., Green, A. G., Mitra, P. P., Stark, J. B., and Littlewood, P. B. (2004). The effect of propagation nonlinearities on the information capacity of WDM optical fibre systems: Cross-phase modulation and four-wave mixing. *Physica D*, 189:81–99.

Wei, H. and Plant, D. V. (2004). Intra-channel nonlinearity compensation with scaled translational symmetry. *Opt. Express*, 12(18):4282–4296.

Wei, X. and Liu, X. (2003). Analysis of intrachannel four-wave mixing in differential phase-shift keying transmission with large dispersion. *Opt. Lett.*, 28(23):2300–2302.

Wei, X., Liu, X., and Xu, C. (2003a). Numerical simulation of the SPM penalty in a 10-Gb/s RZ-DPSK system. *IEEE Photon. Technol. Lett.*, 15(11):1636–1638.

Wei, X., Liu, X., and Xu, C. (2003b). Q factor in numerical simulations of DPSK with optical delay demodulation. E-print: `http://arXiv.org/physics/0304002`.

Weisser, S., Larkins, E. C., Czotscher, K., Benz, W., Daleiden, J., Esquivias, I., Fleissner, J., Ralston, J. D., Romero, B., Sah, R. E., Schönfelder, A., and Rosenzweig, J. (1996). Damping-limited modulation bandwidths up to 40 GHz in undoped short-cavity $In_{0.35}Ga_{0.65}$As-GaAs multiple-quantum-well lasers. *IEEE Photon. Technol. Lett.*, 8(5):608–610.

Welford, D. and Mooradian, A. (1982). Output power and temperature dependence of the linewidth of CW (GaAl)As diode lasers. *Appl. Phys. Lett.*, 40(10):865–867.

Wen, Y. J., Nirmalathas, A., and Lee, D.-S. (2004). RZ/CSRZ-DPSK and chirped NRZ signal generation using a single-stage dual-electrode Mach-Zehnder modulator. *IEEE Photon.*

Technol. Lett., 16(11):2466–2468.

Westphah, F.-J. and Strebel, B. (1988). Optical heterodyne image rejection receiver. *Electron. Lett.*, 24(7):440–442.

Willner, A. E. (1990). Simplified model of an FSK-to-ASK direct-detection system using a Fabry-Perot demodulator. *IEEE Photon. Technol. Lett.*, 2(5):363–366.

Willner, A. E. and Hoanca, B. (2002). Fixed and tunable management of fiber chromatic dispersion. In Kaminow, I. P. and Li, T., editors, *Optical Fiber Telecommunications*, volume IV. Academic Press, San Diego.

Willner, A. E., Desurvire, E., Presby, H. M., Edwards, C. A., and Simpson, J. R. (1990). Use of LD-pumped Erbium-doped fiber preamplifiers with optimal noise filtering in a FDMA-FSK 1 Gb/s star network. *IEEE Photon. Technol. Lett.*, 2(9):669–672.

Willner, A. E., Nezam, S. M. R. M., Yan, L., Pan, Z., and Hauer, M. C. (2004). Monitoring and control of polarization-related impairments in optical fiber systems. *J. Lightwave Technol.*, 22(1):106–125.

Winters, J. H. (1990). Equalization in coherent lightwave systems using a fractionally spaced equalizer. *J. Lightwave Technol.*, 8(10):1487–1491.

Winters, J. H. and Gitlin, R. D. (1990). Electrical signal processing techniques in long-haul fiber-optic systems. *IEEE Trans. Commun.*, 38(9):1439–1453.

Winzer, P. J. and Kim, H. (2003). Degradations in balanced DPSK receivers. *IEEE Photon. Technol. Lett.*, 15(9):1282–1284.

Winzer, P. J., Kogelnik, H., Kim, C. H., Kim, H., Jopson, R. M., Nelson, L. E., and Ramanan, K. (2003). Receiver impact on first-order PMD outage. *IEEE Photon. Technol. Lett.*, 15(10):1482–1484.

Wood, T. H. (1988). Multiple quantum well (MQW) waveguide modulators. *J. Lightwave Technol.*, 6(6):743–757.

Wooten, E. L., Kissa, K. M., Yi-Yan, A., Murphy, E. J., Lafaw, D. A., Hallemeier, P. F., Maack, D., Attanasio, D. V., Fritz, D. J., McBrien, G. J., and Bossi, D. E. (2000). A review of Lithium Niobate modulators for fiber-optic communications systems. *IEEE J. Sel. Top. Quantum Electron.*, 6(1):69–82.

Wozencraft, J. M. and Jacobs, I. M. (1965). *Principles of Communication Engineering*. John Wiley & Sons, New York.

Wree, C., Hecker-Denschlag, N., Gottwald, E., Krummrich, P., Leibrich, J., Schmidt, E.-D., Lankl, B., and Rosenkranz, W. (2003a). High spectral efficiency 1.6-b/s/Hz transmission (8×40 Gb/s with a 25-GHz grid) over 200-km SSMF using RZ-DQPSK and polarization multiplexing. *IEEE Photon. Technol. Lett.*, 15(9):1303–1305.

Wree, C., Leibrich, J., Eick, J., Rosenkranz, W., and Mohr, D. (2003b). Experimental investigation of receiver sensitivity of RZ-DQPSK modulation format using balanced receiver. In *Optical Fiber Commun. Conf.*, paper ThE5.

Wu, M. and Way, W. I. (2004). Fiber nonlinearity limitations in ultra-dense WDM systems. *J. Lightwave Technol.*, 22(6):1483–1498.

Wyner, A. D. (1988). Capacity and error exponent for the direct detection photon channel - Part I and II. *IEEE Trans. Info. Theory*, 34(6):1449–1471.

Xie, C. and Mollenauer, L. F. (2003). Performance degradation induced by polarization-dependent loss in optical fiber transmission systems with and without polarization-mode dispersion. *J. Lightwave Technol.*, 21(9):1953–1957.

Xie, C., Moller, L., Haunstein, H., and Hunsche, S. (2003). Comparison of system tolerance to polarization-mode dispersion between different modulation formats. *IEEE Photon. Technol. Lett.*, 15(8):1168–1170.

Xie, C., Kang, I., Gnauck, A. H., Moller, L., Mollenauer, L. F., and Grant, A. R. (2004). Suppression of intrachannel nonlinear effects with alternate-polarization formats. *J. Lightwave Technol.*, 22(3):806–812.

Xu, C. and Liu, X. (2002). Postnonlinearity compensation with data-driven phase modulators in phase-shift keying transmission. *Opt. Lett.*, 27(18):1619–1621.

Xu, C., Mollenauer, L. F., and Liu, X. (2002). Compensation of nonlinear self-phase modulation with phase modulators. *Electron. Lett.*, 38(24):1578–1579.

Xu, C., Liu, X., Mollenauer, L. F., and Wei, X. (2003). Comparison of return-to-zero differential phase-shift keying and on-off keying long-haul dispersion managed transmission. *IEEE Photon. Technol. Lett.*, 15(4):617–619.

Xu, C., Liu, X., and Wei, X (2004). Differential phase-shift keying for high spectral efficiency optical transmissions. *IEEE J. Sel. Top. Quantum Electron.*, 10(2):281–293.

Yagi, M., Satomi, S., Tanaka, S., Ryu, S., and Asano, S. (2005). Field trial of automatic chromatic dispersion compensation for 40-Gb/s-based wavelength path protection. *IEEE Photon. Technol. Lett.*, 17(1):229–231.

Yamamoto, Y. (1980). Receiver performance evaluation of various digital optical modulation-demodulation systems in 0.5-10μm-wavelength region. *IEEE J. Quantum Electron.*, QE-16(11):1251–1259.

Yamamoto, Y. (1984a). AM and FM quantum noise in semiconductor lasers -Part I: Theoretical analysis. *IEEE J. Quantum Electron.*, QE-19(1):34–46.

Yamamoto, Y. (1984b). AM and FM quantum noise in semiconductor lasers -Part II: Comparison of theoretical and experimental results for AlGaAs lasers. *IEEE J. Quantum Electron.*, QE-19(1):47–58.

Yamamoto, Y. and Haus, H. A. (1986). Preparation, measurement and information capacity of optical quantum states. *Rev. Mod. Phys.*, 58(4):1001–1020.

Yamamoto, Y. and Kimura, T. (1981). Coherent optical fiber transmission systems. *IEEE J. Quantum Electron.*, QE-17(6):919–935.

Yamamoto, S., Edagawa, N., Taga, H., Yoshida, Y., and Wakabayashi, H. (1993). Observation of BER degradation due to fading in long-distance optical amplifier system. *Electron. Lett.*, 29(2):209–210.

Yamazaki, S. and Emura, K. (1990). Feasibility study on QPSK optical-heterodyne detection system. *J. Lightwave Technol.*, 8(11):1646–1653.

Yamazaki, S., Ono, T., and Ogata, T. (1993). Compensation for chromatic dispersion and nonlinear effect in high dispersive coherent optical repeater transmission system. *J. Lightwave Technol.*, 11(4):603–611.

Yariv, A. (1997). *Optical Electronics in Modern Communications.* Oxford Univ. Press, Oxford, 5 edition.

Yeh, C.-H., Lee, C.-C., and Chi, S. (2004). 120-nm bandwidth Erbium-doped fiber amplifier in parallel configuration. *IEEE Photon. Technol. Lett.*, 16(7):1637–1639.

Yeung, R. W. (2002). *A First Course in Information Theory.* Kluwer Academic, Dordrecht, The Netherlands.

Yoshikane, N. and Morita, I. (2004a). 1.14 b/s/Hz spectrally-efficient 50 $\times$ 85.4 Gb/s transmission over 300 km using copolarized CS-RZ DQPSK signals. In *Optical Fiber Commun. Conf.*, postdeadline paper PDP38.

Yoshikane, N. and Morita, I. (2004b). 160% spectrally-efficient 5.12 Tb/s (64 $\times$ 85.4 Gb/s RZ DQPSK) transmission without polarisation demultiplexing. In *European Conf. on Optical Commun.* postdeadline paper Th4.4.3.

Yoshikane, N. and Morita, I. (2005). 1.14 b/s/Hz spectrally-efficient 50 $\times$ 85.4 Gb/s transmission over 300 km using copolarized CS-RZ DQPSK signals. *J. Lightwave Technol.*, 23(1):108–114.

Yu, J., Rolland, C., Somani, A., Bradshaw, S., and Yevick, D. (1996). Phase-engineered III-V MQW Mach-Zehnder modulator. *IEEE Photon. Technol. Lett.*, 8(8):1018–1020.

Yu, M., Kan, C., Lewis, M., and Sizmann, A. (2002). Statistics of signal-to-noise ratio and path-accumulated power due to concatenation of polarization-dependent loss. *IEEE Photon. Technol. Lett.*, 14(10):1418–1420.

Yuen, H. P. and Chan, V. W. S. (1983). Noise in homodyne and heterodyne detection. *Opt. Lett.*, 8(3):177–179. Erratum: 8(6):345.

Zhang, X., Larson, L. E., Asbeck, P. M., and Nanawa, P. (2001). Gain/phase imbalance-minimization techniques for LINC transmitters. *IEEE Trans. Microwave Theory & Tech.*,

49(12):2507–2516.

Zhu, B., Leng, L., Gnauck, A. H., Pedersen, M. O., Peckham, D., Nelson, L. E., Stulz, S., Kado, S., Gruner-Nielsen, L., Lingle, R. L., Knudsen, S., Leuthold, J., Doerr, C., Chandrasekhar, S., Baynham, G., Gaarde, P., Emori, Y., and Namiki, S. (2002). Transmission of 3.2 Tb/s (80 × 42.7 Gb/s) over 5200 km of UltraWave^{TM} fiber with 100-km dispersion-managed spans using RZ-DPSK format. In *European Conf. on Optical Commun.*, Copenhagen, Denmark. postdeadline paper PD4.2.

Zhu, B., Nelson, L. E., Stulz, S., Gnauck, A. H., Doerr, C., Leuthold, J., Grüner-Nielsen, L., Pederson, M. O., Kim, J., Lingle, R., Emori, Y., Ohki, Y., Tsukiji, N., Oguri, A., and Namiki, S. (2003). 6.4-Tb/s (160 × 42.7 Gb/s) transmission with 0.8 bit/s/Hz spectral efficiency over 32 × 100 km of fiber using CSRZ-DPSK format. In *Optical Fiber Commun. Conf.*, postdeadline paper PD19.

Zhu, B., Nelson, L. E., Stulz, S., Gnauck, A. H., Doerr, C., Leuthold, J., Grüner-Nielsen, L., Pederson, M. O., Kim, J., and Lingle, R. (2004a). High spectral density long-haul 40-Gb/s transmission using CSRZ-DPSK format. *J. Lightwave Technol.*, 22(1):209–214.

Zhu, Y., Cordina, K., Jolley, N., Feced, R., Kee, H., Rickard, R., and Hadjifotiou, A. (2004b). 1.6 bit/s/Hz orthogonally polarized CSRZ - DQPSK transmission of 8 × 40 Gbit/s over 320 km NDSF. In *Optical Fiber Commun. Conf.*, paper TuF1.

Index